THE BOOK OF CATERPILLARS

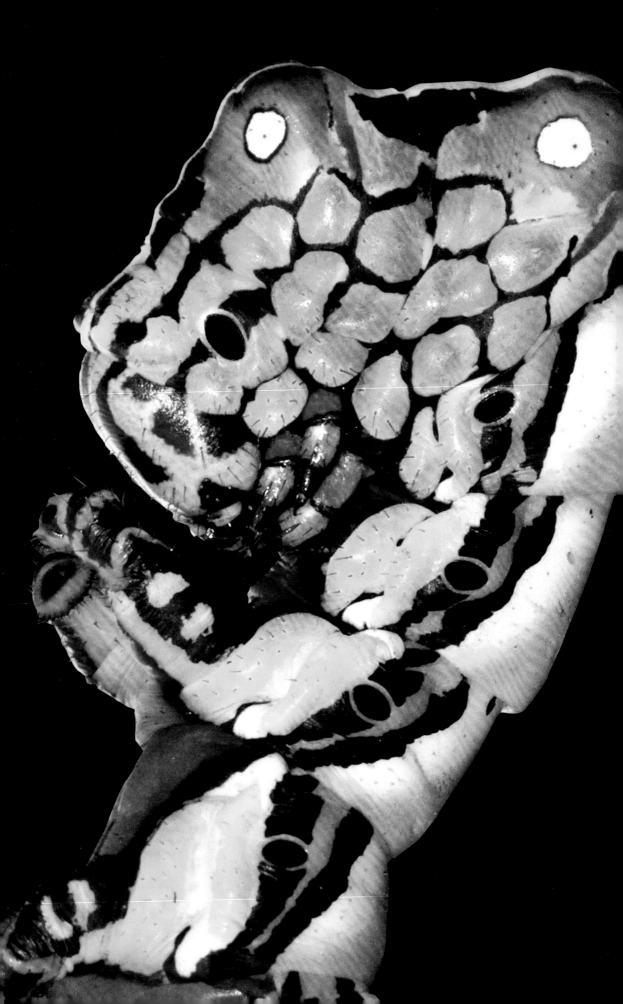

THE BOOK OF
CATERPILLARS

A LIFE-SIZE GUIDE TO SIX HUNDRED
SPECIES FROM AROUND THE WORLD

EDITED BY
DAVID G. JAMES

CONTRIBUTORS
DAVID ALBAUGH, BOB CAMMARATA, ROSS FIELD, HAROLD GREENEY,
JOHN HORSTMAN, DAVID JAMES, SALLY MORGAN, TONY PITTAWAY,
JAMES A. SCOTT, ANDREI SOURAKOV, MARTIN TOWNSEND, KIRBY WOLFE

THE UNIVERSITY OF CHICAGO PRESS

Chicago

DAVID G. JAMES is an associate professor in the Department of Entomology, Washington State University, based at the Irrigated Agriculture Research and Extension Center, Prosser.

The University of Chicago Press, Chicago 60637

© 2017 Quarto Publishing plc

Published 2017

MIX
Paper from
responsible sources
FSC® C008047

Printed in China

26 25 24 23 22 21 20 19 18 17   1 2 3 4 5

ISBN-13: 978-0-226-28736-2 (cloth)
ISBN-13: 978-0-226-28753-9 (e-book)
DOI: 10.7208/chicago/9780226287539.001.0001

Library of Congress Cataloging-in-Publication Data

Names: James, David G., editor.
Title: The book of caterpillars : a life-size guide to six hundred species from around the world / edited by David G. James.
Description: Chicago : The University of Chicago Press, 2017. | Includes bibliographical references and index.
Identifiers: LCCN 2017013376 | ISBN 9780226287362 (cloth : alk. paper) | ISBN 9780226287539 (e-book)
Subjects: LCSH: Caterpillars. | Caterpillars—Classification. | Caterpillars—Conservation.
Classification: LCC QL542 .B66 2017 | DDC 595.7813/92—dc23
LC record available at https://lccn.loc.gov/2017013376

This book was conceived and designed by
Ivy Press
An imprint of The Quarto Group
The Old Brewery, 6 Blundell Street
London N7 9BH, United Kingdom
T (0)20 7700 6700 F (0)20 7700 8066
www.QuartoKnows.com

Publisher SUSAN KELLY
Creative Director MICHAEL WHITEHEAD
Editorial Director TOM KITCH
Commissioning Editor KATE SHANAHAN

Produced by 3REDCARS
Editors RACHEL WARREN CHADD, JOHN ANDREWS
Designer JANE McKENNA
Illustrator BILL DONOHOE

JACKET IMAGES
Species from the following pages: Bob Cammarata 60, 527, 613; Ross Field 47, 85, 152, 165, 220, 228, 256; Harold Greeney 67, 107, 213, 226, 237; John Horstman/itchydogimages 74, 195, 209, 214, 243, 253, 266, 278, 300, 309, 317, 319, 585; David James 57, 105, 125, 135, 138, 170, 179, 191, 241, 260, 282; David Liebman 41, 132; Tony Pittaway 441; James A Scott 286; Shutterstock/ Eric Isselee 180; Shutterstock/Sari O'Neal 533; Shutterstock/ Kamieniak Sebastian 591; Shutterstock/xpixel 568; Leroy Simon 375, 391, 410, 417, 504; Andrei Sourakov 154, 286; © J Voogd 182, 540; Wolfgang Wagner 38, 53, 65, 118, 177; Roger Wasley 269, 338; Wikimedia Commons/Bernard Dupont 245; Kirby Wolfe 431.

LITHOCASE IMAGES
David Liebman 39; Shutterstock/Roger Meerts 509; Wolfgang Wagner 201.

CONTENTS

INTRODUCTION

Caterpillars—the immature stage of moths and butterflies—are diverse and remarkable, with an extraordinary range of survival techniques that have helped make the Lepidoptera one of the most successful insect groups. After beetles, it is the second largest order on the planet; at least 160,000 species have been identified and described, with thousands more undescribed. Lepidoptera are also very widespread, occupying every continent except Antarctica, in habitats ranging from rocky mountain slopes to tropical rain forests, and from waste ground to woollen clothes. Their ecological significance, too, is immense. As larvae, they are mostly prodigious herbivores, hosts for parasitic flies and wasps, and potential food for birds, reptiles, and mammals. As adults, they are vital pollinators.

The myriad colors, forms, patterns, and sizes of different caterpillars are all part of their arsenal against predation as they grow, pupate, and perform the magic trick of metamorphosis—transformation into a butterfly or moth. Some caterpillars are cleverly disguised in the colors of their habitat, and others are strikingly colored and patterned, announcing to predators that they are unpalatable or even toxic. Certain species have stinging spines, others can pull mammal-like faces, while many Papilionidae butterflies can puff up their front end to look like a snake's head, complete with eyespots and an everted organ that mimics a forked tongue.

All caterpillars, however, share the same basic body plan of a large head, small thorax with six true legs, a comparatively huge ten-segment abdomen, and a large gut where all the material they consume is processed. In most species, a pair of thick, fleshy prolegs is present on half of the abdominal segments, enabling the caterpillar to move around, while

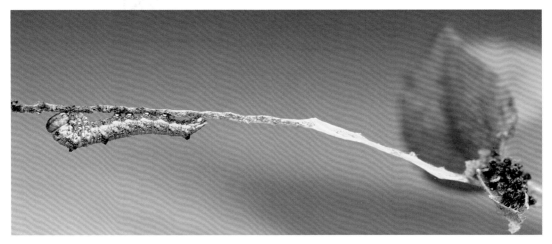

breathing is conducted through tiny pores, called spiracles, on the sides of the body. A caterpillar feeds for much of its life, using scissor-like jaws, or mandibles, to snip off and grind up tiny piece after tiny piece of foodstuff. As its body expands, it molts, often changing appearance. Most caterpillars develop through five instars (stages), shedding their skin at each stage. From egg hatch to maturity, they will increase in mass by up to 1,000 times.

In Lepidoptera, all development occurs at the caterpillar stage, which can take as little as ten days or, if suspended to escape extreme heat or cold, may last a few years, and up to seven years in the Arctic Woolly Bear (*Gynaephora groenlandica*). When the caterpillar pupates, all the necessary cells are present to be reorganized during metamorphosis into a moth or butterfly, whose life is usually much shorter.

TOP **A moth caterpillar** of the Lasiocampidae family is convincingly camouflaged as a patch of moss on tree bark in Pu'er, Yunnan, China. Many Lepidopteran larvae have an extraordinary ability to blend into their surroundings.

ABOVE **The California Sister** (*Adelpha californica*) caterpillar, when young, rests immune from predators on piers it creates from its frass (excreta).

SELECTION CRITERIA

Despite being so numerous, many Lepidoptera are relatively unknown and undescribed, especially at the larval stage. More than 70 families and 55,000 species comprise the "microlepidoptera" group of very small moths, with minute caterpillars that have been rarely, if ever, studied or photographed. This book, therefore, focuses on the caterpillars of larger moths and butterflies, which have received most attention from scientists and photographers. The 600 species that are described here reveal the enormous diversity of form, coloration, and adaptation that exists among these creatures. They range in size from large (6 in / 150 mm) hawkmoths and Saturniidae larvae, such as the Hickory Horned Devil (*Citheronia regalis*), to tiny (⅜ in / 10 mm) moth caterpillars like the Case-bearing Clothes Moth (*Tinea pellionella*), with a full panoply of spiny, hairy, striped, and variously patterned and ornamented larvae in between, from every continent where Lepidoptera live. Some of the caterpillars feature unusual adaptations or live in extreme habitats; others are the subject of scientific research, or are culturally significant, or economically important.

HOW THE BOOK WORKS

The larval life and ecology of 600 species are described in text and images in two sections—Butterfly Caterpillars and Moth Caterpillars. While not strictly a taxonomic division, this reflects common practice, as all butterfly species are generally considered members of the superfamily Papilionoidea, while the more numerous moth species account for all other Lepidoptera.

Each caterpillar is shown life size at maturity, together with a line drawing of the adult butterfly or moth. Some have also been magnified to highlight their detail. All images are of live caterpillars, as, unlike adult butterflies and moths, caterpillars cannot be pinned and photographed because they rapidly lose their coloration after death. A distribution map indicates each species' range. The entry heading may be the species' common name, accompanied by its Latin name (the genus + species name), or, where there is no accepted common name, only the Latin name. Below the heading, the "authority" is given, that is, who first described the species and the date when it was described. Parentheses are used to show that a genus name has changed since it was first described, while square brackets indicate discrepancies and uncertainties about the author or date.

An information box above each entry briefly summarizes key details about the species—its family, range, habitat, host plants or material, a notable fact, and its conservation status. Each species has been checked against the IUCN (International Union for the Conservation of Nature) Red List of Threatened Species, but as relatively few Lepidoptera have been assessed, many species are listed as "Not evaluated," although this

is often modified by local expert information and regional or national assessments. A few vulnerable species may also be described as being on an appendix of CITES (Convention on International Trade in Endangered Species of Wild Fauna and Flora); this means they are subject to an international agreement restricting trade in specimens.

SPREADING THE WORD

Anybody can study Lepidoptera, and finding and keeping caterpillars should be as much a feature of a young child's life as rearing tadpoles. Watching these insects develop and metamorphose can be an inspirational experience. Yet, in many places, species numbers are dwindling as a result of habitat destruction, agricultural development, pesticide use, and climate change. School classroom programs for rearing caterpillars, popular in the United States, Europe, and Australia, do much to stimulate interest in Lepidoptera and create awareness of the threat to their survival. Further research is also required to help better manage their conservation.

Very few species cause significant damage, despite their reputation as "pests" for feeding on cultivated plants, and, arguably, any damage is vastly outweighed by the value of butterflies and moths as pollinators. Both adults and caterpillars, in all their wondrous forms, play a further vital role. They live in such a variety of habitats and are so sensitive to change within those habitats, that scientists increasingly view the insects as an important bellwether of environmental health. For without the caterpillar as Lepidoptera progeny, plant regulator, and food for many creatures, ecosystems would collapse.

ABOVE **The caterpillar of the Pale Tussock** (*Calliteara pudibunda*) greatly outshines its dull-colored adult, with its conspicuous, flower-like tufts of yellow hair. These are part of its defense mechanism and a warning to potential predators. The hairs are both urticating and detach easily, making the caterpillar distinctly unpalatable.

WHAT IS A CATERPILLAR?

Whether hairy, spiny, ridged, or smooth, the world's caterpillars all share one common trait, reflected in Eric Carle's children's classic, *The Very Hungry Caterpillar*. Typically described as "eating machines," they may increase their body mass by up to 1,000 times as they mature. They are the developmental stage of butterflies and moths and have a simple goal—to eat, grow, and become an adult. While a butterfly or moth sometimes survives only long enough to reproduce, the larval period may last days, weeks, months, two to three years, or occasionally even longer in species that are dormant during winter or hot summers.

STRUCTURE

Butterflies and moths have the caterpillar in common. While the adults can often be distinguished from each other by the structure of the antennae and the way the wings are held at rest, there is no simple physical characteristic that distinguishes a butterfly caterpillar from a moth caterpillar. Despite the extreme diversity of color and form in the hundreds of thousands of species, all caterpillars share the same basic features, built on the standard insect plan of head, thorax, and abdomen. The head is large, the thorax (the middle section between the head and the abdomen) is small, and the whole body is long and tubular.

The head

The epicranium, a hard head capsule with a triangular front plate or "frons," has a characteristic inverted, Y-shaped line extending down from the top of the head; this line distinguishes the caterpillar from any other grub. In most caterpillars, the head is conspicuous, although in families such as Lycaenidae it may be retracted into the thorax. There are six simple, lateral eyes (stemmata) to help the caterpillar distinguish between dark and light and give it some spatial awareness. There is a short antenna on each side of the mouth, and the mouthparts consist of a pair of jaws, or mandibles, bounded by an upper flap (labrum) and lower structure (labium). The mandibles swing from side to side, "shearing" through vegetation, and often bear small, sharp, toothlike projections. Located centrally on the lower side of the head is the labial spinneret, the secretory structure through which modified salivary glands discharge silk that is used by the larvae in various ways—sometimes to bind foliage or create a silk web, or during pupation to suspend a chrysalis or construct a cocoon.

ABOVE **The head of a caterpillar** is its control center, containing its key sensory features and the organs it needs to feed. The head above is that of the spiny Atlas Moth caterpillar (*Attacus atlas*).

Thorax, abdomen, and legs

The thorax is small, muscular, and made up of three segments, each bearing a pair of true, jointed legs. The abdomen, consisting of ten segments, is the largest part of a caterpillar and where food is digested and processed. There are pairs of spiracles (respiratory pores) on all of the abdominal segments except for the last two. The abdominal legs or prolegs are quite

BELOW **All caterpillars**, like this Mulberry Silkworm (*Bombyx mori*), have a three-part body—head, thorax, and abdomen. The true legs are jointed legs, while the fleshy prolegs, present in most species, lack musculature.

11

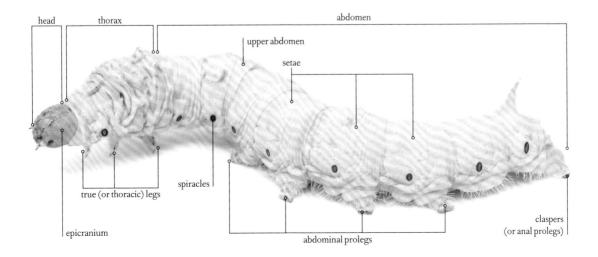

head | thorax | abdomen

upper abdomen

setae

true (or thoracic) legs

spiracles

epicranium

abdominal prolegs

claspers (or anal prolegs)

different from the true legs, being fleshy and barrel-shaped, and bearing hooks or crochets at the base. Most caterpillars have four pairs of prolegs on the third to sixth abdominal segments and another pair on combined segments nine and ten. Geometridae caterpillars, however, have only two pairs of prolegs, one on the sixth abdominal segment and the other on the tenth, producing a characteristic walking pattern that has given them the nickname of "inchworms" or "loopers." Limacodidae larvae, the so-called "slug" caterpillars, have suckers instead of prolegs and secrete a liquefied silk lubricant to help them glide along.

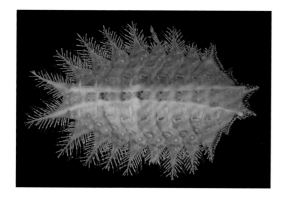

12

TOP **The Crowned Slug** (*Isa textula*), seen from above, is, like other Limacodidae caterpillars, named for its sluglike gait. Its form is also characteristically flat, and it has stinging spines and hairs.

ABOVE **The underside** of the Crowned Slug reveals vertical "muscles" that undulate to create motion, either forward or backward, helping the caterpillar move along using its suckers, lubricated by the liquid silk it secretes from salivary glands.

Setae, spines, and shields

Caterpillars are clothed with hairlike structures called setae, which serve to protect, act as sensors, or secrete substances; for instance, the setae of some species of Pieridae butterfly caterpillars in their early stages produce droplets of fluid, which appear to help deter predators and parasitoids. Further types of ornamentation include fleshy filaments, hardened cones, branching spines, and thoracic shields, all with primarily defensive functions.

DISTINCTIVE LARVAE

Other insects have a similar larval stage, but caterpillars can usually be distinguished from other larvae by their characteristic Y-shaped head marking, more diverse patterning (grubs are frequently quite dull), and by their abdominal prolegs, as most other larvae have stocky true legs but no abdominal legs, or no legs at all. The larvae of sawflies (insects of the order Hymenoptera, which also includes bees and wasps) are very caterpillar-like but have a single lateral eye (not six) and have six to eight (rather than five or fewer) pairs of prolegs.

RANGE AND DIVERSITY

Caterpillars occupy a vast range of habitats, from seed pods to kitchen pantries, and from hot deserts to mountains and even into the Arctic Circle.

Adaptation to such different environments has led to extraordinary diversity in appearance and survival strategies. More than half of all species are relatively unstudied "microlepidoptera," the often pale-colored, featureless, and very wormlike larvae of tiny moths, many of which feed concealed within stems, fruits, seeds, and other foodstuffs and materials.

By contrast, the caterpillars of macro-moths and butterflies are often colorful, with showy features such as bristles, spines, and filaments, and make no attempt to conceal themselves. Bright, so-called aposematic coloring is often a "warning" to potential predators that the caterpillars are or might be bad tasting. Very hairy or spiny caterpillars are equally unpalatable to predators such as birds; arming the spines with toxic chemical secretions adds a further layer of defense.

The heads of caterpillars also show incredible diversity in coloration, patterning, and shape, again as a defense, some resembling "faces," with horns, false eyes, nose, and mouth. Others protect their head by having "head-like" posteriors, presumably a bid to fool predators (at least half of the time) into attacking the less vulnerable end. Many species, however, are cryptically colored to blend with their environment. Some even change hue according to the part of the host plant they are feeding on, such as certain lycaenid butterfly caterpillars, which are green when consuming leaves but become red, yellow, or orange if they eat flower buds and petals.

13

BELOW **The caterpillar** of the silkmoth *Automeris larra*, like many Saturniidae silkmoth larvae, is large and intimidating when full grown. Its flamboyant spines can also deliver a painful sting.

FROM EGGS
TO PUPATION

Like all eggs, butterfly and moth eggs are fragile and attractive to predators. The relatively slow-moving larvae that hatch from them are also vulnerable and, in order to survive their complex phases of development and reach pupation, the final stage before adulthood, they must deploy a remarkable range of strategies that have evolved to meet the challenges of their habitats.

EGGS—LAYING AND HATCHING

Using visual and olfactory stimuli, female butterflies and moths often carefully select a spot on or close to a specific host plant, where their miniscule eggs can hatch in safety, although some moths distribute eggs randomly, conferring the benefits of a broad host plant range. Differing in size and shape according to species, eggs may be laid singly or in glued-together masses of up to 1,000, on upper or lower surfaces of leaves, on buds or flowers, encircled around twigs, on the ground, on rocks, or on other non-plant substrates. Being so small and often cryptically colored, perhaps

BELOW LEFT **The developing larva** is apparent through the transparent shell of this mature egg of the Western Tiger Swallowtail (*Papilio rutulus*) and is only hours from hatching.

BELOW RIGHT **Eggs of some species** such as the California Tortoiseshell (*Nymphalis californica*) are laid in large masses. This mass of about 250 eggs will produce caterpillars that are gregarious for most of their lives.

resembling plant parts, fungi, detritus, or even bird droppings, caterpillar eggs are rarely found by casual observers.

The eggs usually develop rapidly, hatching within two to ten days, depending on temperature. Sometimes, though, they are programmed to delay hatching, spending adverse weather conditions—extreme cold or heat—in a state of developmental arrest, known as diapause. They then hatch only when the host plants they feed on reappear.

LARVAL STAGES

The caterpillar hatches by cutting a hole in the shell with its mandibles and, according to species, may consume the entire eggshell on the way out or leave the empty eggshell with a telltale exit hole. The new larva immediately sets about feeding and protecting itself. It may move to a safer location on the plant, cover itself with a silk-tied leaf shelter, or, in the case of gregarious larvae, join with its siblings in creating an extensive silk-web nest. Caterpillars at the newly hatched stage, known as the first instar, usually feed rapidly, often doubling in size within a few days. Once the larval "skin," or integument, tightens and appears stretched, with a swelling at the head caused by the larger, inelastic head capsule of the next instar, the larva is nearing its first molt, or ecdysis. Before molting, larvae find a site hidden from predators, spin a small pad of silk to which they attach their claspers, and remain motionless for 12 to 48 hours.

Molting, which takes only minutes, begins at the head end, with the integument splitting and slipping backward along the body as the larva moves slightly forward. In most species, the next instar consumes the old integument and soon resumes feeding. Newly molted larvae often show temporary paler coloration, which disappears within 2 to 12 hours. Some species have four or six instars, but most have five, molting four times. However, where larval development is interrupted multiple times by diapause, seven to nine instars can occur.

15

ABOVE **Successive instars** can differ considerably in appearance as well as size. This saturniid moth caterpillar, *Arsenura batesii,* has molted and left its spectacular, tentacled fourth instar skin behind, becoming cryptic and sticklike in its fifth, final instar.

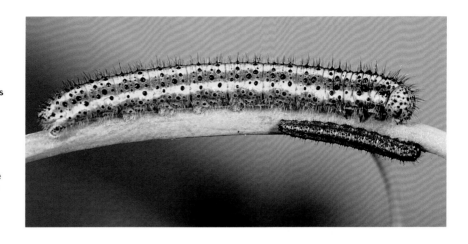

RIGHT **Two caterpillars of the Checkered White** (*Pontia protodice*), a tiny second instar and a much larger fifth instar, demonstrate how fast these larvae grow. The fifth instar is just nine days older than the second instar.

FEEDING AND GROWING

Caterpillars are programmed to eat as much as possible in order to grow and mature. The period from egg hatch to pupation may be as little as ten days, although in species with multiple dormancies caterpillars can live for two to three years, with one Arctic species taking up to seven years to complete development. An approximate doubling of length occurs in each successive instar. Between 60 and 80 percent of the total plant mass eaten by a developing caterpillar is consumed in the final instar. Size is relative, however, as the largest saturniid silkmoth and hawkmoth caterpillars grow up to 6 in (150 mm) in length, while the final instars of "micromoths" may reach only 3/16 in (5 mm).

Species also grow at different rates and in different seasons, depending on their preferred food. Some feed only on leaf buds, others on young leaves, mature leaves, flower buds, flowers, seeds, or even stems. Buds, flowers, and seeds are more nutritious (generally with more nitrogen) than leaves or stems, promoting faster growth but within a shorter growth period. Food sources such as grasses and evergreen needles are low in nutrition but hugely abundant over vast areas, so caterpillars exploiting these resources grow slowly but with little competition.

DEALING WITH ENVIRONMENTAL EXTREMES

Caterpillars, like all insects, are cold blooded and depend on environmental conditions to achieve the optimum body temperatures for development. For the majority of species, the range of body temperatures favoring development is 59–86°F (15–30°C). When temperatures remain below 41°F (5°C), with periods below 32°F (0°C) and limited, low-angle sunshine, many caterpillars are unable to develop. To survive long, hard winters, they have to change their physiology and enter a dormant state of suspended animation, or diapause. In late summer or fall, some caterpillars prepare for overwintering by seeking refuges, such as curled leaves, seed pods, under rocks, or other sheltered locations, where they will be buffered against the elements. Here, a lowered metabolic rate and radical biochemical changes, including synthesis of a kind of "antifreeze," protect them against extreme cold. Species living in hot, dry Mediterranean or desert climates, where temperatures frequently reach 100–115°F (38–45°C) and plant life is often sparse, face a similar challenge, entering summer dormancy, or estivation, and delaying pupation and adult emergence until fall, when conditions are better for survival and reproduction.

Caterpillars may reenter diapause multiple times if environmental stimuli signal the onset of unfavorable conditions. Post-diapause, checkerspot (*Euphydryas*) butterfly caterpillars recommence feeding in late winter or early spring on fresh host plant growth, but if a lack of moisture affects that growth, the larvae become dormant, potentially living for two to three years with only short periods of development annually. Caterpillars living at high elevations, such as those of the Arctic Fritillary (*Boloria chariclea*), depend on timely snowmelt to enable them to feed and complete development in time for the normal midsummer flight period. In late spring, after diapause, these caterpillars appear to measure day length to determine if they can complete development in time. If not, they overwinter twice, as an early instar then a late instar. Climate, elevation, and food plant also affect the number of broods developed during a year.

17

BELOW **Preparing to pupate**, the Common Wood Nymph (*Cercyonis pegala*) hangs in an inverted J shape. This well-camouflaged but vulnerable prepupal stage may last from 12 to 48 hours depending on the temperature.

PREPARING TO PUPATE

When nearing maturity and pupation, full-fed larvae often change color, most "shrink" to a certain degree, and some enter a wandering phase, the "wanderers" seeking sites away from the host plant. Some go underground, some hide, and others build a protective cocoon or blend in with the background, either through coloration or by creating a broken outline. This high degree of crypsis, and the talent of wandering prepupal caterpillars for finding secluded pupation sites, means the particularly vulnerable pupal stage of butterfly and moth metamorphosis is the least likely to be seen.

RIGHT **Caterpillars of the Arctiinae** subfamily spin their own hairs into a cocoon held together by silk. This helps to protect the pupa within by making access more difficult for parasitoids. Fluff around the exit hole on the left shows that here the moth has already eclosed.

THE MIRACLE OF METAMORPHOSIS

Perhaps the most celebrated trait of Lepidoptera is their capacity to metamorphose—changing their body structure and appearance so completely that larvae and adults look as if they are two quite separate species. While most insects metamorphose, some practice "incomplete metamorphosis," with no pupal stage; larvae hatch from eggs and are usually a miniature version of the adult. Insects undergoing complete metamorphosis, which also include beetles, flies, and wasps, are considered more highly evolved. Fossil records suggest that metamorphosis began to occur up to 300 million years ago and conferred an evolutionary advantage on metamorphosing species, because their different forms and habitats ensured that adults and larvae did not compete for the same resources.

MAKING THE CHANGE

Pupation describes the transition of a species from active eating machine (caterpillar) to the immobile, non-feeding preparatory stage (pupa), which will ultimately yield the adult butterfly or moth. The term "chrysalis" is

BELOW **At pupation**, lichen moth caterpillars of the *Cyana* genus weave meshwork baskets around themselves. The basket is made of the caterpillar's own body hairs and is constructed in two stages, with a base and an upper half, which is loosely hinged to the lower, long side of the basket. Ultimately, the two parts are pulled together to completely enclose the developing pupa inside.

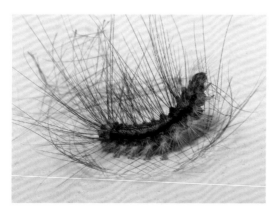

generally used for hard-cased butterfly pupae or the casing itself, while many moths spin a protective outer silk "cocoon" around themselves. Pupae are formed in one of five basic modes: loose on the ground, within a silken cocoon or leaf shelter, underground in an earthen cell, hanging by the terminal end (cremaster) attached to a silk pad, or attached upright by the cremaster with a supporting silk girdle. Loose pupae are common in moths but rare in butterflies. Skippers—butterflies from the Hesperiidae family—commonly form pupae within tied leaf or grass shelters, while hanging pupae are characteristic of Nymphalidae butterflies, and girdled pupae are found in species from the butterfly families Papilionidae, Pieridae, and Lycaenidae. While some moth larvae spin cocoons on leaves, twigs, or branches, many burrow in leaf litter or to varying depths in the ground. Several species incorporate protective materials with their silk into the cocoon to strengthen it, such as chewed bark and their own stinging setae. Others add twigs or bits of vegetation to help disguise the cocoon.

When a pupation site is selected, and silk pads, shelters, or cocoons are complete, the prepupal larva shrinks a little and waits motionless for the final molt to occur. The outer skin then softens, splits, and falls away,

ABOVE **The Lime Butterfly** (*Papilio demoleus*) makes its remarkable transformation from the final "pharate" pupal stage to flying adult. The cells within the pupa have regrouped into adult form but initially remain enclosed within the chrysalis. As this shell becomes more transparent, the maturing adult uses its feet to break free and extricate itself. Next, it hangs from the pupal shell or nearby substrate as the new wings dry and stiffen, and prepares its proboscis for sucking nectar from flowers. It may take up to an hour before it is ready for its first flight.

20

ABOVE LEFT TO RIGHT
The final caterpillar molt into a pupa or chrysalis is complete in just a few minutes. Here, the larval skin of the Two-tailed Swallowtail (*Papilio multicaudata*) is intact in the left image, but in the center image it has peeled back to the lower part of the body. In the right image, the skin has dropped away leaving the new soft pupa to harden.

leaving the pupal case that has formed beneath. In most species, final coloration of the pupa matches its immediate environment. Even the Monarch butterfly (*Danaus plexippus*), which is brightly colored at all larval stages to alert predators to the toxins it contains, does not advertise this fact at pupation; its green pupa blends with the foliage around it. While pupae may occasionally wriggle if disturbed, they are generally unobtrusive, remaining hidden or camouflaged to avoid predation at this crucial stage.

Larvae that form hanging pupae adopt a characteristic J shape. After 12 to 48 hours, the larval skin splits behind the head, revealing not another caterpillar integument, or skin, but a fleshy, soft integument, usually green, yellow, or orange. With much wriggling, the larval skin moves down the body, revealing increasingly more of the soft, new pupa. Once the shed skin reaches the terminal segment, the pupal cremaster probes and seeks the silk pad spun earlier by the prepupal larva. With hanging pupae this is a critical phase; if the cremaster fails to make contact with the silk pad to which it attaches with tiny hooks, the soft pupa will fall, and likely perish. After attachment, more wriggling usually results in the shed skin dropping away, and eventually the pupa stops moving, hardens, and assumes the coloration that allows it to blend in with its environment.

THE TRANSFORMATION

Within the pupa a remarkable process takes place. Hormones trigger the release of enzymes that break down the larval structure into a sort of "soup," containing tiny, disc-shaped groups of cells, present but suppressed at the larval stage, which will now develop into adult body parts. In some species, this metamorphosis is rapid, taking as little as five to seven days. In others, pupae oversummer and overwinter in a dormant state, or diapause, sometimes for two or three years. A few days before a butterfly or moth emerges, or ecloses, the pupa darkens, indicating the advanced stage of development. On the final day, as the shell becomes increasingly transparent, first the patterns and color of the wings, then the rest of the body, can be seen. At this "pharate" pupa stage, eclosion is just hours away, and pupae that will become females may well already have males, attracted by pheromones, in attendance. Indeed, in a number of *Heliconius* butterfly species, males are known to compete for a chance to mate not only with newly eclosed females but also with pupae at the pharate stage.

In many butterfly species, adult eclosion is synchronized to occur early in the morning, often soon after dawn, presumably to optimize a successful emergence, post-eclosion drying of wings, and inaugural flight. Night-flying moths often eclose around dusk. Eclosion begins with the butterfly or moth pushing with its feet against the shell covering the legs, antennae, and proboscis; to soften the toughest cocoons, such as those made with silk and chewed bark by "kitten" moths (*Furcula* species), the adult first ejects an acid solution. Once the legs are free, the adult grabs hold of the shell, pulling out the rest of the body until the entire butterfly or moth is fully extricated. Hanging and girdled pupae species, which emerge substantially aided by gravity, then usually hang from the pupal shell or a nearby support, while butterflies and moths eclosing from pupae on or near the ground wander for a short while to find an appropriate support site. Once a site has been chosen, the adult begins pumping its wings to full size and sets about zipping together the two parts of the coiled proboscis to form a tube for sipping nectar—a vulnerable period lasting from 5 to 15 minutes. The wings may remain limp and flaccid for another hour or so, depending on the ambient temperature—then, its amazing transformation complete, the adult flies.

BELOW **A newly eclosed butterfly**, Lorquin's Admiral (*Limenitis lorquini*), hangs beneath its discarded chrysalis, with beautifully marked wings still folded as it waits to make its first flight. The North American species, a member of the Nymphalidae family, is usually on the wing between April and October, depending on its region.

21

RIGHT **Feeding damage** by early instar caterpillars of the Pink-edged Sulphur (*Colias interior*) on *Vaccinium* (blueberry) host plants is distinctive. This type of feeding, leaving stems and veins untouched, is known as skeletonization.

VORACIOUS EATERS

A caterpillar is designed primarily to consume. How much it eats and the quality of its food determine its growth rate and health, as well as adult size and reproductive success. In a caterpillar's final instar, the amount of food consumed increases fourfold, providing the stored water, fat, and protein to carry it through the pupal stage to adulthood. For species that do not feed as moths or butterflies, it is the last chance to build the reserves needed to survive the rest of their short life.

EATING AND DIGESTION

Caterpillars chew leaves, or other food, using serrated mandibles, or jaws, that move from side to side, while a pair of sensory organs below the mandibles taste the food and push it back into the mouth, where it mixes with saliva. From there, it enters the digestive system, basically a long tube—the caterpillar version of an alimentary canal. The chewed food is stored in the crop, a pouch-like organ, before entering the largest section of the tube, the midgut, to be digested and absorbed. Indigestible food accumulates in the hindgut and rectum, and is expelled through the anus in small, hard pellets called frass. Most caterpillars do not drink water, extracting it instead from their food; one notable exception is the Drinker moth caterpillar (*Euthrix potatoria*), which imbibes water droplets.

SOURCES AND FEEDING TACTICS

Caterpillars usually feed on plants, consuming all parts or specializing on leaves, buds, stems, flowers, or seeds. Some develop on a single host plant species, while others are highly polyphagous, feeding on a wide variety of plants. Where hosts contain toxic chemicals, caterpillars have evolved to neutralize them and even use them as a defense against predators. Caterpillars can overcome a plant's physical barriers, too, such as hairs or sticky, toxic latex. Monarch larvae (*Danaus plexippus*), for instance, cut milkweed leaf veins or the petiole to prevent the flow of latex into leaf parts.

Ants, aphids, or scale insects provide a diet for carnivorous caterpillars, such as Moth Butterfly larvae (*Liphyra brassolis*), which enter the nests of the Green Tree Ant (*Oecophylla smaragdina*) and devour the ant larvae. Some caterpillars feed on dung and others feed on fungi, shells, feathers, or fibers—the preferred host material of the insatiable Case-bearing Clothes Moth larva (*Tinea pellionella*), a pest of worldwide renown.

Unpalatable or well-defended caterpillars feeding in groups do not hide and are usually "messy" feeders, leaving large areas of damaged leaves. Palatable caterpillars usually conceal evidence of feeding, consuming entire leaves or severing half-eaten leaves and letting them drop from the plant. Others feed within plants, such as the Subflexa Straw Moth caterpillar (*Heliothis subflexus*), which shelters inside the lantern-shaped husks that enclose the physalis fruits it consumes.

HUNGRY PESTS

Despite the large amount of plant material devoured by a typical caterpillar, its impact on a well-developed host plant is usually minimal and mostly escapes notice. However, a large hawkmoth (Sphingidae) caterpillar feeding on a small, herbaceous species can destroy many plants. An apple orchard is easily damaged by opportunistic species such as the Codling moth (*Cydia pomonella*), as is a field of cabbages by the Cabbage Looper (*Trichoplusia ni*). Population explosions of the Forest Tent Caterpillar (*Malacosoma disstria*) can defoliate thousands of acres of trees. A number of moth larvae but relatively few butterfly caterpillars are economic pests; of butterfly larvae, the Cabbage White (*Pieris rapae*) is perhaps the most widely distributed culprit.

23

BELOW **Gregarious larvae** like these caterpillars of the Fire-rim Tortoiseshell *(Aglais milberti)* on Stinging Nettle *(Urtica dioica)* consume entire leaves and small plants. The caterpillars produce copious amounts of webbing that provides support for the larvae and protection from predators.

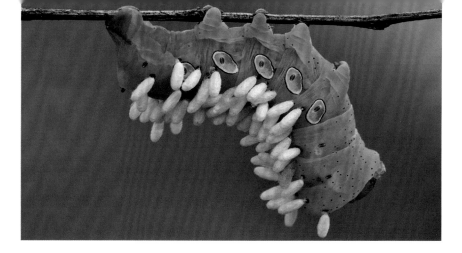

RIGHT **Parasitic wasps** in the Braconidae family are major parasitoids of caterpillars. Braconid maggots spend their 10 to 14-day lives eating the insides of a caterpillar, such as this Pandora Sphinx (*Eumorpha pandorus*), and break out *Alien*-like to pupate in cocoons, from which tiny wasps will emerge.

CATERPILLAR DEFENSES

From every 100 eggs laid by a female moth or butterfly, very few—perhaps just one to five—will survive to become an adult. Biotic (natural, living) enemies and abiotic (mostly climatic) factors combine to ruthlessly decimate populations of eggs, caterpillars, and pupae. Consequently, every species is engaged in an ever-evolving "arms race," which pits each new larval defense strategy against improved counterstrategies by its predators. Multiple means of defense are employed by virtually all caterpillar species, with individual tactics often changing in importance during development.

The caterpillar's natural enemies range from birds and mammals to other insects, such as praying mantids, beetles, lacewings, and spiders. Parasitic flies and wasps, which lay eggs that develop inside the caterpillars, literally consuming them alive, pose possibly the greatest threat and sometimes completely wipe out caterpillar populations.

CONCEALMENT AND EVASION

Various tactics help species escape non-parasitoid predation. Relatively small caterpillars, such as some lycaenids, hide by burrowing into the host plant itself. Others use their host plant to create a refuge; some skipper (Hesperiidae) caterpillars bind leaves together with their silk to create an individual "nest" or fashion bivouacs from sections of leaf flipped over and tied down with silk.

Larvae that are active at night can avoid diurnal enemies such as birds and larger predatory insects. The many caterpillars that rest by day concealed at the base of their host plants are among the hardest to detect. Camouflage, or crypsis, is equally effective for diurnal feeders. Many leaf-eating caterpillars are green, blending into the foliage, while those that feed on flowers may match the red, yellow, or white markings of their flower food, or the colors of other plant parts. Grass-feeding caterpillars are often green with paler stripes, while a number of Geometridae moth caterpillars are convincingly twiglike in appearance. The Camouflaged Looper moth caterpillar (*Synchlora aerata*) takes disguise a step further, adorning its body with petals and other plant fragments.

Some species have bolder markings, often white or yellow, that break up the background color, blurring their outline. Posture can also change a caterpillar's appearance. For instance, mid-instar caterpillars of the California Sister (*Adelpha californica*) rest in a Loch Ness Monster-type posture, making them disappear against the lobed oak leaves of their host tree. Others mimic bird droppings, such as the early instars of many swallowtail species of the Papilionidae family, which are black or dark brown with a white "saddle."

FRASS DEFENSES

Reducing telltale odors is a further protective measure. Most invertebrate enemies of caterpillars find their prey by scent. One likely significant source is caterpillar feces, called frass. Some skipper and pierid caterpillars use their anal comb to fling frass for distances of up to 40 times their own body length.

ABOVE **The Pagoda Bagworm Moth Caterpillar** (*Pagodiella hekmeyri*), like other bagworms, spends its larval life inside a mobile home of silk, leaf, and other fragments, adding extensions as it grows. The shelter, built from available construction materials within the species' habitat, and extended as the caterpillar develops, effectively conceals it from predators.

Rather than disposing of frass, however, some caterpillars, including web-building species, substantially contaminate their nests with it. As a result, the frass odor of these species may somehow be neutralized or disguised. Early instars of the California Sister eat around the midrib of a leaf, then use frass pellets silked together to extend this midrib "pier." Species such as the Zebra Mosaic (*Colobura dirce*) and Staff Sergeant (*Athyma selenophora*) create frass chains and barriers that appear to deter intruders such as ants.

THREATS, SCARE TACTICS, AND CHEMICAL DETERRENTS

When concealment fails, some caterpillars display sudden movements to try to scare an attacker, such as head-jerking and thrashing the anterior part of the body from side to side. This tactic is most effective when performed in unison by a large group of spiny caterpillars, such as

mid-instar Mourning Cloak (*Nymphalis antiopa*) larvae. Late instar California Sister caterpillars thrash, and display and move their mandibles as if to bite. Mature caterpillars of some swallowtail butterfly and hawkmoth species have eyespots on the thorax, which are enlarged when the larva is threatened, often giving it the appearance of a small snake. Similarly, the head capsules of some larvae and the pupal heads of many Hesperiidae species have markings or modifications that resemble a vertebrate face, which again could deter a potential attacker. Other nymphalid caterpillars have elongated horns at one or both ends of the body, which are waved around in a threatening manner if a predator approaches.

The spiny armature of many caterpillars turns soft, palatable larvae into prickly, tongue-stabbing mouthfuls that only a few predators can tolerate. Spines, in combination with other tactics such as thrashing, mandible-baring, curling, and dropping, are all likely deterrents, and make it more difficult for parasitic wasps to alight on the caterpillar and insert their eggs.

Caterpillars of many species across many families engage in chemical defense, by using toxic chemicals sequestered from host plants or by

BELOW **When under threat**, mature caterpillars of certain species, like this Common Rippled Hawkmoth (*Eupanacra mydon*), puff up their anterior segments, conceal their true legs, and enlarge their eyespots, giving them the appearance of a small snake.

BOTTOM **Paired horns**, spines, false eyespots, a "nose," and black mandibles combine to create an intimidating "face" on the head of this fifth instar California Sister (*Adelpha californica*) butterfly caterpillar.

producing noxious compounds from benign chemicals. The best-known example is probably the Monarch (*Danaus plexippus*), whose caterpillars sequester cardenolides or cardiac glycosides from milkweed host plants. These plant poisons make Monarch larvae, pupae, and adults unpalatable to vertebrate predators. The striking, yellow, black, and white banding of Monarch larvae is quickly recognized by birds as indicating distastefulness; as a result, similarly marked caterpillars may also be avoided.

First instars of most Pieridae butterflies carry oily droplets on the tips of their dorsal setae; these droplets contain chemicals that repel ants and other predators. Swallowtail butterfly caterpillars possess a unique chemical defense in the form of an eversible forked, fleshy gland, called an osmeterium, located in a slit behind the head and colored yellow, orange, or red. When threatened, the caterpillar shoots out the gland, which resembles a snake tongue and glistens with an odiferous secretion that repels predators. Many caterpillars from other butterfly and moth families possess a similar eversible fleshy "neck" gland, located ventrally beneath the head on the anterior margin of the first segment. These organs (called adenosma) also contain chemicals that appear to repel ants and other predators.

SAFETY IN NUMBERS

Aggregation, or gregariousness—group feeding and resting—is a behavioral tactic to reduce the odds of any single individual being attacked; this is often practiced in early instars before other defense methods develop. Communal larvae may also build silken webs, supports, and platforms to help keep the community together.

In some caterpillars, including the California Tortoiseshell (*Nymphalis californica*), the less important rear end strongly resembles the crucial head end in a bid to confuse predators. Early instars of this species feed and rest communally, and the striking appearance of twice the number of "heads" in a community may well reduce the risk of real heads being attacked by a predator.

ABOVE **Gregarious behavior** by final instar caterpillars of the Banded Swallowtail (*Papilio demolion*) may enhance camouflage on a host plant leaf, but if this fails, simultaneous eversion of odiferous and snake tongue-like osmeteria may deter predator attacks.

ANT BODYGUARDS

Some caterpillars, particularly those of the Lycaenidae butterfly family, have developed a defense strategy based on recruiting ant bodyguards to repel threats from parasitoids and other predators. Although ants are a significant natural enemy of most other butterfly larvae, lycaenid caterpillars have evolved to produce something that many ant species love—sugar-rich honeydew—and most of the larvae have a functional "honeydew" gland, producing sugars and amino acids, which the ants consume. The physical presence and activity of ants swarming over the caterpillars and substrate effectively prevent parasitoids and predators from attacking the caterpillars. A persistent predator, such as a spider or parasitoid wasp, will eventually be overpowered by ants and discarded.

CATERPILLARS AND PEOPLE

Butterflies are celebrated in literature and art, and moths make an occasional sinister appearance, but their larvae feature more rarely and play quite singular and disparate roles in popular culture. Some species are still best known as destructive pests, but the caterpillar has important uses, too, as a centuries-old producer of fine silk and, increasingly, as a nutritious food.

CATERPILLARS IN POPULAR CULTURE

Many of today's children and their parents are familiar with *The Very Hungry Caterpillar*, created by Eric Carle, the eponymous hero of which consumes ever-increasing amounts of unlikely food, pupates, and becomes a glorious butterfly. An earlier celebrity is the hookah-smoking caterpillar of Lewis Carroll's *Alice's Adventures in Wonderland*, first drawn by the Victorian artist John Tenniel, the insect later appearing as a surreal blue animation in Disney's 1951 *Alice in Wonderland* movie, and then in CGI form, voiced by Alan Rickman, in Tim Burton's 2010 movie of the same name. In a song from the 1952 film musical *Hans Christian Andersen*, Danny Kaye's "Inchworm," with its curious looping gait, is "measuring the marigolds," while the Scottish singer Donovan sang— somewhat inaccurately—"Caterpillar sheds its skin to find a butterfly within" in his 1967 hit "There is a Mountain."

BELOW **The larger than life** caterpillar of *Alice's Adventures in Wonderland*, drawn in black and white by the illustrator John Tenniel, is a beautifully surreal image of an insect otherwise largely unrepresented in literature, and one that, once seen, is almost impossible to forget.

LEFT **A "good" pest**, the caterpillar of the Cactus Moth (*Cactoblastis cactorum*) was first introduced into Australia to control prickly pear cacti (*Opuntia* spp.) and later used similarly in other places, including South Africa and the Caribbean. Now, the rapid spread of the moth species in the United States is said to threaten cactus industries and the survival of animals that feed on cactus.

29

In North America, the Banded Woolly Bear caterpillar (*Pyrrharctia isabella*) is sometimes credited with the ability to forecast weather. While the larva is naturally black at each end and copper gold in between, the extent of black banding seems to vary annually. More extensive black banding is said to predict a harsher winter. The population size of another woolly bear, the caterpillar of the Ranchman's Tiger Moth (*Platyprepia virginalis*), is held by some to indicate the outcome of US presidential elections. When the caterpillars are common in California, it is said, a Democrat is voted president, when uncommon, a Republican wins. Despite an avalanche of opinion polls to the contrary, the Ranchman's Tiger Moth caterpillar accurately predicted Donald Trump's win in 2016.

CATERPILLARS AS PESTS

Mention "caterpillar" to a gardener, forester, or farmer and the response is likely to be negative. Hungry caterpillars of a small number of widespread pest species can have a huge impact on humans by feeding on agricultural crops, stored products, forest trees, and garden plants. Caterpillars of the Case-bearing Clothes Moth (*Tinea pellionella*) are notorious for munching holes in household materials. Millions of dollars continue to be spent on pesticides annually to control pest caterpillar species throughout the world.

Yet most species do no damage to the things we grow and cherish, and some have even been employed to kill unwanted plants. The Cactus Moth (*Cactoblastis cactorum*), for instance, was introduced into Australia from South America in 1925, so that its caterpillars could control invasive prickly pear cacti (*Opuntia* spp.), which they did with spectacular success.

 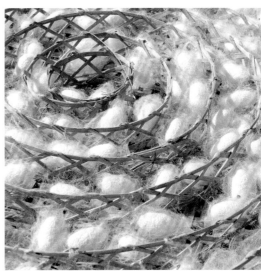

ABOVE LEFT **Thousands of white cocoons** formed by Mulberry Silkworms (*Bombyx mori*) are collected and treated so that the precious silk produced by their salivary glands—a long, continous filament enclosing each pupa—can be unwound.

ABOVE RIGHT **Before the silk** can be spun, the cocoons are briefly steamed or boiled to soften the natural gum also secreted by the silkworm to bind its silk together. In the traditional process, the pupae perish. A new, though more expensive, technique allows the silkmoth to eclose before its cocoon is used.

CATERPILLARS AS SILK-PRODUCERS

Sericulture, the farming of the Mulberry Silkworm (*Bombyx mori*) species for its silk, has been practiced in China for more than 5,000 years and in Europe from around 550 CE, when legend has it the first silkworm eggs were smuggled by monks into Constantinople. Originally transported along the Silk Roads, which connected East and West, silk was for centuries the most luxurious fabric available—beautiful but practical, lightweight yet strong, cool in hot weather, and with excellent dyeing properties.

At the start of the silk-making process, the larvae are fed mulberry leaves and develop through each instar until they spin their silken cocoons. Each cocoon is made up of a continuous filament up to 4,000 ft (1,200 m) long, composed of fibroin protein, held together with a gummy fluid called sericin. To soften the gum, the cocoons are treated with hot air, steam, or boiling water, then several cocoons are carefully unwound simultaneously to create a single strand of raw silk. It takes as many as 2,500 silkworm cocoons to produce just 1 lb (around 450 g) of silk. Significant "wild" silk production is also obtained from the Chinese Tussah Silkmoth (*Antheraea pernyi*) and the Suraka Silkmoth (*Antherina suraka*).

A more esoteric use of silk is the lost art of cobweb painting that originated in the sixteenth century. The intricate process involved collecting caterpillar or spider silk, layering it over a frame, then painting it with a fine-tipped, woodcock feather brush. The transparent effect of the webbing gave the images an ethereal glow. For its elasticity and tensile strength, artists in the Tyrolean Alps favored webbing produced by *Yponomeuta evonymellus* larvae. A fine, 200-year-old Tyrolean cobweb painting of the Virgin Mary can be seen at Chester Cathedral in England.

CATERPILLARS AS FOOD

Across the world humans have consumed caterpillars for thousands of years. Today an estimated two billion people eat insects, including caterpillars, as part of their daily diet. These range from the witchetty grubs, usually the larvae of the cossid moth (*Endoxyla leucomochla*), eaten by indigenous Australian people, to the crispy, dried cuchamás (green caterpillars) of Mexico, where at least 67 Lepidoptera species are consumed. In Asia, the Bamboo Borer (*Omphisa fuscidentalis*) is such a popular deep-fried dish that the larvae are now bred commercially by caterpillar farmers, which helps protect the population in the wild. In southern Africa, close to 40 species of caterpillars are harvested for food. Those regularly consumed include the Mopane Worm (*Gonimbrasia belina*), an important source of protein for many people. The protein content of Lepidoptera larvae varies between 14 and 68 percent, which is comparable and often exceeds that of raw beef (19 to 26 percent) or raw fish (16 to 28 percent); the Mopane Worm is particularly protein rich.

Because caterpillars are so nutritious, supplying healthy fats, protein, vitamins, minerals, and fiber, the United Nations is actively promoting edible insects as a way of combating world hunger. Cultivating caterpillars for consumption is also more environmentally friendly than raising animals for food, because the larvae are about three times more efficient at converting feed into edible product. Caterpillars emit fewer greenhouse gases and less ammonia than cattle or pigs, and farming them requires significantly less land and water. Caterpillar gathering and rearing, whether at household level or on an industrial scale, also offers important livelihood opportunities for people in both developing and developed countries.

31

LEFT **Deep-fried bamboo larvae** are a popular and nutritious snack in Thailand and other parts of eastern Asia. The larvae of the Bamboo Borer (*Omphisa fuscidentalis*) are collected en masse as they diapause on bamboo, and are also increasingly farmed.

RIGHT **As part of caterpillar outreach** initiatives, project leader Lee Dyer shows a large, stinging, flannel moth caterpillar, (*Megalopyge* sp.) to an international group of scientists gathered in the Atlantic Forest, a rain forest in Bahia State, Brazil, rich in biological diversity.

RESEARCH AND CONSERVATION

Caterpillars, reared quite easily and quickly, are rewarding study subjects, helping scientists make important discoveries about biology, genetics, plant chemistry, and even the effects of climate change. These insects are models of adaptability, and their interactions with their habitat provide fascinating insights into how organisms adjust to a changing environment.

CATERPILLARS AND CLIMATE CHANGE

A long-term research and outreach project launched in Costa Rica in 1995 by scientists funded by the Earthwatch Institute and extended to centers in Ecuador, Brazil, Arizona, Louisiana, and Nevada monitors caterpillars to investigate how climate change affects plant chemistry and interactions between plants, herbivores, and parasitoids. Data from the project suggests that a warming climate can put the life cycles of caterpillars and parasitoids out of sync; for example, caterpillars pupate earlier, before the parasitoids that attack them are fully developed. As a result, more caterpillars survive, contributing to outbreaks and consuming more plant matter, and parasitoid populations fall. Whether this is a permanent situation, or if parasitoids will eventually "catch up" remains unclear.

Some United States butterfly species such as the Sachem skipper (*Atalopedes campestris*) have increased their geographical range in response to a warming climate. Once restricted to California and the southern states, the Sachem is now common in Oregon and expanding northward through Washington State. In contrast, alpine specialists such as the Astarte Fritillary (*Boloria astarte*), with a range from northwestern North America to northeastern Siberia, could run out of "cool space" and become extinct.

LEPIDOPTERA CONSERVATION

Worldwide, habitat loss is the principal threat to Lepidoptera, with some species in steep decline. At the root of the problem are urban development, agricultural expansion, and forest clearance, erasing the natural, wild terrain where eggs are laid, caterpillars feed, develop, and pupate, and adult moths and butterflies eclose and breed. Around 10 percent of butterfly species in Europe face extinction, according to the United Kingdom charity Butterfly Conservation. In the United Kingdom, especially the south, moth numbers have declined by up to 40 percent over the past 50 years. Populations of the most familiar North American butterfly, the Monarch (*Danaus plexippus*), have contracted by 80 to 90 percent in two decades.

In response, conservation groups have launched community efforts to save endangered butterflies, such as the Richmond Birdwing (*Ornithoptera richmondia*)—the focus of recovery projects in Australia. In the United States, more than 15,000 waystations containing nectar and caterpillar host plants for Monarch butterflies have been established. A reduction in the use of pesticides is helping, and some farmers in Europe, the United States, and New Zealand are incorporating a nature reserve element within their landscape strategy, which could further boost Lepidoptera numbers.

Rearing caterpillars is at the center of United States penitentiary-based conservation efforts aimed at restoring populations of Taylor's Checkerspot butterfly (*Euphydryas editha taylori*) in Oregon and Washington State. Thousands of Monarch caterpillars are reared annually by inmates at Washington State Penitentiary, who tag the butterflies, before release, to help provide data on migration routes and destinations. One prison inmate reportedly said, "Watching a caterpillar transform itself into a butterfly proves to me that I can change too," showing that, even in the more unlikely places, the miracle of Lepidoptera metamorphosis remains a source of inspiration.

33

LEFT **Butterfly gardening**—growing the flowering plants that different species favor, as well as the host plants they need to breed—is a conservation trend that is gaining momentum and has great potential to stabilize or even reverse current population declines.

THE CATERPILLARS

BUTTERFLY CATERPILLARS

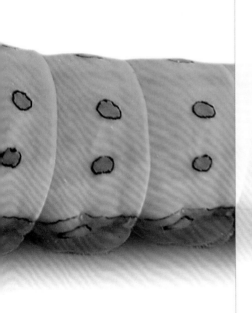

The order Lepidoptera includes around 160,000 species, of which fewer than 12 percent—just under 19,000 species in the superfamily Papilionoidea—are classified as butterfly species. This chapter describes 246 butterfly caterpillars from six of the seven Papilionoidea families: Papilionidae, Hesperiidae, Pieridae, Riodinidae, Lycaenidae, and Nymphalidae. Butterfly species are largely distinguished from moth species by adult features, such as the structure of the antennae and the way the wings are held at rest. The larvae can look much like those of moths but rarely spin cocoons, as many moth caterpillars do.

All Papilionidae caterpillars have forked organs (osmeteria) on the prothoracic segment, which they evert to produce an unpleasant odor if the larvae are threatened. Most Hesperiidae larvae have large heads, and many build leaf shelters. Pieridae species, which include the notorious "whites" that feed on cruciferous vegetables, have distinctively angled pupae with a silk girdle at the first abdominal segment. Riodinidae species, native to South America and southern areas of North America, are similar to those of Lycaenidae but lack a "honey gland," which Lycaenidae caterpillars use to attract and appease ants. The larvae of Nymphalidae, the largest butterfly family with more than 6,000 species and a dozen subfamilies, are, however, highly variable.

FAMILY	Papilionidae
DISTRIBUTION	Eastern Europe to the Middle East, Armenia to Turkmenistan
HABITAT	Rocky slopes, olive groves, and vineyards
HOST PLANTS	Pipevine (*Aristolochia* spp.)
NOTE	Brightly colored caterpillar that hides in leaf bags
CONSERVATION STATUS	Near threatened

ADULT WINGSPAN
2⅛–2⅜ in (54–60 mm)

CATERPILLAR LENGTH
Up to 1⅞ in (48 mm)

38

ARCHON APOLLINUS
FALSE APOLLO
(HERBST, 1789)

Female False Apollo butterflies lay round, green eggs on the underside of leaves, and the young caterpillars appear in April and May. At first they are gregarious, living together in leaf webs, but as they get older they move apart and shelter in individual leaf bags, a feature not seen in other European swallowtail species. The mature caterpillars crawl to the ground, where they pupate just below the surface in a loose cocoon. They overwinter and emerge in spring. The adults are on the wing from March to April; a single generation is produced annually.

Archon apollinus is a variable species with up to five subspecies. Characteristically, the adults lose their wing scales as they age, leaving transparent areas of wing, especially on the forewings; in older specimens the forewings may be completely transparent. The species is under threat from herbicide sprays that kill its host plants, surviving better in the less intensively farmed regions of Turkmenistan.

Actual size

The False Apollo caterpillar has a cigar shape with a brown-black body covered in short, black hairs. Dorsally, there are two rows of red spots and a central double row of white spots, although the white may be absent in some individuals. Laterally, there is also a row of red spots.

FAMILY	Papilionidae
DISTRIBUTION	United States, south to Central America
HABITAT	Forests and meadows
HOST PLANTS	Pipevine (*Aristolochia* spp.)
NOTE	Unpalatable caterpillar whose color can vary according to temperature
CONSERVATION STATUS	Not evaluated, but common

ADULT WINGSPAN
2¾–5 in (70–130 mm)

CATERPILLAR LENGTH
2–2¾ in (50–70 mm)

BATTUS PHILENOR
PIPEVINE SWALLOWTAIL
(LINNAEUS, 1771)

39

Females of the Pipevine Swallowtail lay clusters of up to 20 eggs, so the young caterpillars feed in groups. This enables them to resist plant defenses, such as hairs on the plant's surface called trichomes. Older caterpillars are solitary and eat leaves, stems, and seeds. The larvae can consume up to half of their host plant's foliage, and significantly increase plant mortality and reduce seed production. The aristolochic acid found in host plant foliage is converted by the caterpillars into chemical defenses and passed by them to pupae, adult butterflies, and eggs. Hence all stages are unpalatable to predators, and even parasitoids are deterred from attacking them.

The caterpillars can be polymorphic. Red caterpillars have been found to occur at temperatures greater than 86°F (30°C), and in Texas their numbers increased with rising daily temperatures. The larvae thermoregulate by climbing on to non-host vegetation to avoid excessive heat. Very similar caterpillars are found in very different-looking, but related, "birdwing" butterflies from the Old World, which also feed on pipevines.

Actual size

The Pipevine Swallowtail caterpillar has fleshy filament projections on each segment—the longest on the thoracic segments, especially the first. Dorsally, there are paired orange-red spots. The larvae are either black or smoky red; black is the norm, but the red phenotype occurs in western Texas and Arizona, where temperatures are highest. Larvae in Florida have longer projections than those in California; the two populations are considered different subspecies.

FAMILY	Papilionidae
DISTRIBUTION	Timor, southeastern New Guinea, and northern and northeastern Australia
HABITAT	Open forests and savannah woodlands
HOST PLANTS	Pipevine (*Aristolochia* spp. and *Pararistolochia* spp.)
NOTE	Unpalatable caterpillar that emits a sweet odor if disturbed
CONSERVATION STATUS	Not evaluated, but common in coastal tropical and subtropical areas

ADULT WINGSPAN
2¾ in (70 mm)

CATERPILLAR LENGTH
1⁹⁄₁₆–1¹¹⁄₁₆ in (40–43 mm)

CRESSIDA CRESSIDA
CLEARWING SWALLOWTAIL
(FABRICIUS, 1775)

40

The Clearwing Swallowtail caterpillar hatches from yellow eggs with vertical rows of raised orange dots. When feeding on small-leaved species of pipevine, the caterpillar will often consume the entire plant and then wander on the ground in search of more vines. The food plants are poisonous, and the caterpillar becomes unpalatable to predators, passing on this trait to the adult butterfly. When disturbed, the caterpillar also emits a sweet smell from its extended osmeterium as a warning to predators. Generally there is only one caterpillar per plant, as the female will avoid laying on plants that already have an egg or a caterpillar.

Pupation will often occur off the vine, on grass or a nearby tree trunk, the pupa upright and supported by a central silken girdle and an attachment by the cremaster. The adults fly low and slowly but are capable of rapid flight if disturbed. There are several broods each year, and the caterpillars are more numerous in the wet season because of better food quality and quantity and warmer temperatures. *Cressida cressida* is the only species in the *Cressida* genus.

The Clearwing Swallowtail caterpillar has variable reddish-brown coloring, mottled with white or creamy-white markings. The abdominal segments three, four, and seven often have a broken transverse white band, and both the thoracic and abdominal segments have short, rounded dorsolateral and lateral tubercles. There is a yellow osmeterium on the prothorax, and the head is brownish black.

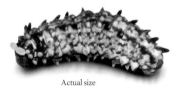

Actual size

FAMILY	Papilionidae
DISTRIBUTION	Throughout eastern United States
HABITAT	Normally in moist woodlands, but will visit flowers in open areas
HOST PLANTS	Pawpaw (*Asimina* spp.)
NOTE	Caterpillar protected by the host plant chemicals it ingests
CONSERVATION STATUS	Not evaluated, but common

ADULT WINGSPAN
2½–4¹/₁₆ in (64–104 mm)

CATERPILLAR LENGTH
1¾–2⅛ in (45–55 mm)

EURYTIDES MARCELLUS
ZEBRA SWALLOWTAIL
(CRAMER, 1777)

41

Zebra Swallowtail caterpillars hatch in the spring and the fall, and feed on the underside of the host plant leaves. Like most other swallowtail larvae, they have osmeteria, which are everted when the caterpillars are disturbed. These organs look like horns and have a strong smell, produced by the isobutyric and 2-methyl butyric acids that the larvae absorb from the host plant. The smell effectively repels ants and spiders. Younger larvae may also fall from the plant if disturbed.

The caterpillars usually pupate on the underside of leaves of the host plant. When daylight hours are shorter, some pupae go into diapause. In Florida, there are two generations per year. The spring generation is smaller and lighter colored than the fall generation, which also has proportionately longer tails and wider black stripes. The Zebra Swallowtail is the only temperate species of an otherwise tropical group of kite swallowtails. Close relatives of this butterfly species occur in the Caribbean and South America. The latest classification suggests it should be placed in the genus *Protographium*.

The Zebra Swallowtail caterpillar is green, often with yellow, black, and white narrow bands between segments, though larvae can be quite variably colored. It has a single wide, multicolored band separating the thorax from the abdomen, the head is beige, and the forked osmeterium behind it (everted only when the larva is disturbed or threatened) is yellow.

Actual size

FAMILY	Papilionidae
DISTRIBUTION	Northeastern India, southern China, Japan, Korea, Chinese Taipei, and most of Southeast Asia
HABITAT	Low-level tropical and subtropical rain forests, and urban areas where larval food plants occur
HOST PLANTS	Camphor Laurel (*Cinnamomum camphora*), other *Cinnamomum* spp., *Lindera* spp., and *Neolitsea* spp.
NOTE	Humped caterpillar of moderate-sized blue swallowtail butterfly
CONSERVATION STATUS	Not evaluated, but common and widespread

ADULT WINGSPAN
3⅛ in (80 mm)

CATERPILLAR LENGTH
1¹¹⁄₁₆ in (43 mm)

42

GRAPHIUM SARPEDON
COMMON BLUEBOTTLE
(LINNAEUS, 1758)

Eggs and early stage caterpillars of the Common Bluebottle are typically found on saplings and new growth, often at low heights. When not feeding, the caterpillar in all instars rests along the midrib on the upper leaf surface. The caterpillar stage lasts from two to five weeks depending on the temperature. The larvae are usually very sluggish and will pupate near where they feed. The pupa, which is wedge-shaped and pale green, is suspended upright and supported with a silken girdle and the cremaster.

The striking adults—black with a semitranslucent medial blue-green band extending to their wingtips—often congregate at muddy puddles or urine-soaked soil. They are swift-flying butterflies and common in reserves, urban parks, and where Camphor Laurel trees are planted as ornamentals. There are as many as eight sibling species previously considered subspecies of the Common Bluebottle. These range from India to Australia and have a similar appearance and biology.

Actual size

The Common Bluebottle caterpillar tapers to the posterior and is green, with a prominent yellow, transverse band connecting the third pair of thoracic spines. These spines are white with black and yellow rings at their base. There are also two pairs of shorter thoracic spines and a pair on the anal segment. There is a pale, translucent yellow osmeterium in the prothoracic segment.

FAMILY	Papilionidae
DISTRIBUTION	Europe, across central Asia to China, Korea, and Japan
HABITAT	Woodlands, orchards, field margins, dry grassland, parks, and gardens
HOST PLANTS	Shrubs, including Blackthorn (*Prunus spinosa*) and other *Prunus* spp., hawthorn (*Crataegus* spp.), apple (*Malus* spp.), and *Sorbus* spp.
NOTE	Well-camouflaged caterpillar that lays a silken trail
CONSERVATION STATUS	Not evaluated, but considered endangered in some parts of its range

ADULT WINGSPAN
2¹¹⁄₁₆–3½ in (69–90 mm)

CATERPILLAR LENGTH
2 in (50 mm)

IPHICLIDES PODALIRIUS
SCARCE SWALLOWTAIL
(LINNAEUS, 1758)

43

Scarce Swallowtail caterpillars hatch from pale, round eggs laid singly by the female butterfly on the underside of the host plant leaves. The larvae live sedentary and solitary lives, spinning a silk cushion on leaves on which to rest. When they move across the host plant to feed, they leave a trail of silk threads that allow them to retrace the route to their resting place. The pupa is either yellow green or brown. Caterpillars produced early in the season are yellow green and found on the host plant, while those pupating late in the season are brown and found in the leaf litter, where they overwinter. This difference is known as a seasonal polymorphism.

The adults are on the wing from March to October, and there are three generations a year. Despite its common name, the species is widespread and generally plentiful, but in some areas it is declining due to loss of habitat, especially its preferred Blackthorn hedges.

The Scarce Swallowtail caterpillar is apple green with a speckled appearance. There is a single yellow dorsal line and also a yellow lateral line below the spiracles. A series of short, oblique, backward-pointing yellow lines run along the side. The body is covered in many small, raised dark spots, each bearing a short seta.

Actual size

FAMILY	Papilionidae
DISTRIBUTION	New Guinea and surrounding islands, far northeastern Australia
HABITAT	Rain-forest clearings where host plants are in abundance
HOST PLANTS	Pipevine (*Aristolochia* spp.)
NOTE	Voracious caterpillar that is toxic to predators
CONSERVATION STATUS	Not evaluated, but widespread; international export is restricted, as for all *Ornithoptera* (birdwing) species

ADULT WINGSPAN
4⅞–6 in (125–150 mm)

CATERPILLAR LENGTH
2⁹⁄₁₆–2¾ in (65–70 mm)

ORNITHOPTERA PRIAMUS
NEW GUINEA BIRDWING
(LINNAEUS, 1758)

44

The New Guinea Birdwing caterpillar is dark brown and glossy, giving its smooth, dry skin a wet appearance. It is also covered with apparently sharp (but actually soft), dark brown to black spines, which suggest that it would make a painful meal, helping to protect against predation. There is a white saddle on the fourth abdominal segment. The spines on the fifth segment are orange tan tipped in black and are the largest spikes on the body.

Although *Ornithoptera priamus* is one of the largest butterfly species in the world, its caterpillar is smaller than you might expect. This is because the adult's body is small in relation to the wings, which are huge, especially in the case of females. The early instars can be found feeding in small groups but as they get larger, requiring more food, they venture off on their own. Since pipevine host plants are toxic, the caterpillar, chrysalis, and adult butterfly are toxic to predators.

Later instar caterpillars can defoliate an entire pipevine and in some cases will resort to cannibalism if no other food is available. When disturbed, the caterpillar will extend an organ, called an osmeterium, from behind its head that resembles the forked tongue of a snake. This releases a foul smell into the air in a bid to discourage predators. The chrysalis mimics not only a dead leaf but in some cases also the flower of the host plant.

Actual size

FAMILY	Papilionidae
DISTRIBUTION	Eastern Australia
HABITAT	Lowland and upland subtropical rain forests
HOST PLANTS	Birdwing vines (*Pararistolochia praevenosa* and *P. laheyana*)
NOTE	Caterpillar no longer in decline thanks to host plant renewal
CONSERVATION STATUS	Not evaluated; rated in a report by Environment Australia in 2002 as not at risk in the south of its range, but of some concern in northern areas

ADULT WINGSPAN
4⅛–4½ in (105–115 mm)

CATERPILLAR LENGTH
2⁵⁄₁₆–2¾ in (58–70 mm)

ORNITHOPTERA RICHMONDIA

RICHMOND BIRDWING

(GRAY, [1853])

45

After hatching, Richmond Birdwing caterpillars consume their eggshell and then feed on soft leaves, although they will cannibalize unhatched eggs. There are usually five instars, or six if nutrient concentrations in the leaves are low. The osmeterium, found on all instars, produces a volatile odor when the caterpillar is alarmed, and this is thought to repel predators. One to three generations are produced annually depending on the altitude and seasonal rainfall, and the caterpillar stage lasts from 22 to 46 days.

The caterpillar leaves its food plant to pupate on nearby foliage and overwinters as a pupa. The presence and spread of the South American weed *Aristolochia elegans*, upon which the butterflies lay their eggs, are serious threats to the species, as the leaves of the plant are toxic to the caterpillar. At one point, the Richmond Birdwing had disappeared from about 65 percent of its range and, until about 1997, continued to decline rapidly in areas where it was considered to be stable. A program of reintroducing host plants, with significant community involvement, is now aiding recovery of the species.

The Richmond Birdwing caterpillar is dark blackish brown or pale brown gray, and each segment has a dorsolateral row of long spines and a ventrolateral row of shorter spines. The spines are orange brown in the middle and tipped black, except the long spines on segment four, which have a large, central, pale yellow-orange area that often extends to the base. The head is dark brown with a yellow "collar" and a yellow osmeterium.

Actual size

FAMILY	Papilionidae
DISTRIBUTION	Northern, eastern, and southern Australia, and New Guinea
HABITAT	*Citrus* orchards, river valleys, open moist lowland forests, and eucalypt woodlands
HOST PLANTS	Cultivated and native *Citrus* spp.
NOTE	Day-feeding caterpillar that is a minor pest of *Citrus*
CONSERVATION STATUS	Not evaluated, but common

ADULT WINGSPAN
4–4¼ in (100–110 mm)

CATERPILLAR LENGTH
2⅛–2⁹⁄₁₆ in (55–65 mm)

PAPILIO AEGEUS
ORCHARD SWALLOWTAIL
DONOVAN, 1805

46

Orchard Swallowtail caterpillars hatch from eggs laid singly on young *Citrus* shoots a few days earlier. Although the caterpillars are camouflaged, blending in well with *Citrus* foliage, they are frequently parasitized by wasps and flies. To deter predators, the caterpillars evert their forked, bright red osmeterium, which emits a *Citrus* smell. The caterpillars feed diurnally, resting by night on the upper side of leaves. Development from egg hatch to pupation takes 24 to 27 days under warm temperatures, with short day lengths hastening development and inducing pupal dormancy.

In southern areas of Australia the pupae overwinter, but in northern areas breeding continues year-round. The Orchard Swallowtail is a minor pest of young *Citrus* plants in inland irrigated areas of New South Wales, Victoria, and South Australia, and can also defoliate young *Citrus* plants in home gardens. Damage is rarely noticed on mature *Citrus* trees. A number of subspecies occur in tropical areas north of Australia.

The Orchard Swallowtail caterpillar is green with paired spines along the body dorsally. There are variable lateral and dorsal oblique bands and spots. The thorax has a brown area ventrolaterally, extending obliquely backward and upward to form a band on the first abdominal segment. Ventrally, the prolegs and body are white, and the head is black.

Actual size

FAMILY	Papilionidae
DISTRIBUTION	Eastern Australia
HABITAT	Eucalypt open forests and suburban gardens
HOST PLANTS	Mainly *Citrus* spp., but also other genera of Rutaceae
NOTE	Caterpillars that can cause minor damage to *Citrus* trees
CONSERVATION STATUS	Not evaluated, but common

ADULT WINGSPAN
2⅝–2¹³⁄₁₆ in (67–72 mm)

CATERPILLAR LENGTH
1½ in (38 mm)

PAPILIO ANACTUS
DAINTY SWALLOWTAIL
MACLEAY, 1826

47

The Dainty Swallowtail caterpillar eggs are laid singly on the edge of young leaves and shoots. They hatch in a few days, and the caterpillars consume their eggshell. The larvae feed openly during the day and in the early instars are camouflaged in a "bird dropping" pattern. All instars have a fleshy osmeterium that is everted when the larva becomes alarmed by potential threats. The osmeterium produces a *Citrus*-like, odoriferous secretion. In the northern half of its range, the species breeds throughout the year.

Pupation takes place on twigs of the host tree but often well away from where the caterpillars have been feeding. Pupae produced in the fall in southern areas overwinter and do not emerge until at least the following spring. The species has expanded its normal range by breeding in urban gardens that have planted *Citrus*. Male butterflies patrol in open areas up to 6 ft (1.8 m) above the ground and can also be found on the top of hills.

The Dainty Swallowtail caterpillar is blue black with numerous small, bluish-white spots, a middorsal row and lateral rows of large, yellow patches, and a white, broken ventrolateral line. Each segment has a subdorsal short, black spine. The head is black, and the prothorax has a slit containing an orange bifid osmeterium.

Actual size

FAMILY	Papilionidae
DISTRIBUTION	From Mexico south to Argentina, including the Caribbean, with a small population in southern Florida
HABITAT	Open areas and secondary growth
HOST PLANTS	*Zanthoxylum* spp. and *Citrus* spp.
NOTE	Caterpillars that resemble bird droppings in early instars
CONSERVATION STATUS	Not evaluated, but locally common

ADULT WINGSPAN
5¼–5½ in (134–140 mm)

CATERPILLAR LENGTH
3–4 in (76–100 mm)

48

PAPILIO ANDROGEUS
ANDROGEUS SWALLOWTAIL
CRAMER, [1775]

Androgeus Swallowtail caterpillars hatch from green eggs laid singly on leaf tips of the host plant and are initially orange brown with a shiny skin. Typical of swallowtails, they resemble bird droppings, with characteristic streaks of blue throughout the brown areas, and feed openly on leaf upper surfaces. Within a month, the larvae reach full size and spend most of their non-feeding time on twigs, where they are well camouflaged. When fully developed, the caterpillars pupate on twigs. In parts of the range, pupae overwinter, and the adults fly in the spring.

The butterflies were unknown in Florida before 1976, but they are strong fliers and may have strayed into the state from the Caribbean. However, the Florida population is similar morphologically to the population in South America. The larvae were first discovered feeding on *Citrus*, so they could have arrived as caterpillars or eggs on imported *Citrus* varieties. The larvae were once considered a pest on *Citrus*, but they inflict far less damage than related species of the *Heraclides* subgenus, such as the Giant Swallowtail (*Papilio cresphontes*) or the Lime Swallowtail (*P. demoleus*).

The Androgeus Swallowtail caterpillar is mainly brown with characteristic streaks of blue gray. It has elongated, cream-colored patches on the thorax that extend to the front of the brown head and a creamy dorsal saddle from the first to fourth abdominal segments. The prolegs and final three abdominal segments also bear white markings. When threatened, the larva can swell the thoracic region to expose long, orange, forked osmeteria, giving it a snakelike appearance.

Actual size

FAMILY	Papilionidae
DISTRIBUTION	Florida, the Bahamas, Hispaniola, and Cuba
HABITAT	Tropical hardwood hammock
HOST PLANTS	Torchwood (*Amyris elemifera*); in captivity has also fed on Wild Lime (*Zanthoxylum fagara*)
NOTE	Caterpillar that exposes forked white osmeteria when disturbed
CONSERVATION STATUS	Not evaluated, but listed as federally endangered in the United States (subspecies *ponceanus*) and uncommon elsewhere in its range

ADULT WINGSPAN
3⅜–5 in (86–130 mm)

CATERPILLAR LENGTH
2⅜–3⅛ in (60–80 mm)

PAPILIO ARISTODEMUS

SCHAUS' SWALLOWTAIL

ESPER, 1794

49

Early instar Schaus' Swallowtail larvae, like those of other members of their genus, resemble bird or lizard droppings—dark, with white uric acid splashes toward one end. Thus camouflaged, young caterpillars rest openly on the upper surface of leaves, where bird droppings land. In later instars, when too large to resemble droppings, they tend to feed at night and rest on twigs during the day. Approaching maturity, larvae develop white bands toward the middle of the body, which probably serve to break up their outline and reduce the chances of detection by predators.

Like other swallowtail larvae, the caterpillar has an osmeterium, which it uses as a form of defense. The chemicals secreted from the osmeterium (aliphatic acids, esters, monoterpene hydrocarbons, and sesquiterpenes) have been shown to be highly repellent to worker ants and probably offer good protection from other predators. A captive propagation program, ongoing for many years at the University of Florida, is helping to ensure the long-term survival of Schaus' Swallowtail, which, due to mosquito spraying, habitat loss, and imported fire ants, has seen its range in the United States reduced to a number of islands in the Florida Keys.

The Schaus' Swallowtail caterpillar, unlike related swallowtail larvae in the subgenus *Heraclides*, lacks a complete saddle—the white patch around its center—and is also a little more colorful, with a number of small, round blue spots. Otherwise, it is typical of the genus: brown black with extensive white patches on the first thoracic segment and the final two abdominal segments. A yellow lateral stripe of uneven width extends throughout the thoracic area and the first six abdominal segments. The forked osmeterium is white.

Actual size

FAMILY	Papilionidae
DISTRIBUTION	Arabian peninsula, across southern Asia to China, Southeast Asia, Papua New Guinea, and Australia; recent introductions into Turkey, Dominican Republic, Puerto Rico, Jamaica, and Cuba
HABITAT	Diverse, including savannahs, forests, riverbeds, and gardens, up to 6,900 ft (2,100 m) elevation
HOST PLANTS	*Citrus* spp. and scurf pea (*Cullen* spp.)
NOTE	Caterpillar that invades *Citrus* and is an occasional pest
CONSERVATION STATUS	Not evaluated, but common over a wide geographic range

ADULT WINGSPAN
2¹³⁄₁₆ in (72 mm)

CATERPILLAR LENGTH
1⅝ in (42 mm)

PAPILIO DEMOLEUS
LIME SWALLOWTAIL
LINNAEUS, 1758

50

Young caterpillars of the Lime Swallowtail resemble bird droppings and are usually ignored by predators. Late instars are mostly green and camouflaged on their food plants until their last molt, when they seek shelter as they feed. Development from egg to adult takes 26 to 59 days. In colder climates the pupae may overwinter, extending the generation time to several months. Known as the Chequered Swallowtail in Australia, where it feeds on scurf peas, the species will develop on *Citrus* species if caterpillars are placed on the plant.

The Lime Swallowtail has increased its range in the past decade or so, due to its strong flight and the expansion of agriculture and urbanization. It has also been recorded from Portugal (a single adult, in 2012) and continues to spread, becoming the most widespread swallowtail in the world. The caterpillars are a pest in *Citrus* stock nurseries from the Middle East to India, and they are a potential threat to *Citrus* growing elsewhere.

Actual size

The Lime Swallowtail caterpillar is cylindrical, tapering toward the rear, and pale green with lateral and two subdorsal rows of pink spots edged with brown, a white ventrolateral line, and a pair of short, fleshy spines behind the head. A black band occurs on the fourth and fifth segments. An orange-red osmeterium can be everted from the prothoracic segment.

FAMILY	Papilionidae
DISTRIBUTION	Southern India and Sri Lanka, northeast India to southern China, southern Japan, South Korea, and Southeast Asia (Malay Peninsula, Indonesia, and the Philippines)
HABITAT	Open forests
HOST PLANTS	Members of Rutaceae, including *Citrus* spp., *Euodia* spp., *Fortunella* spp., *Toddalia* spp., and *Zanthoxylum* spp.
NOTE	Caterpillar that develops into a snake's head look-alike
CONSERVATION STATUS	Not evaluated, but very common

ADULT WINGSPAN
4–4⅝ in (100–120 mm)

CATERPILLAR LENGTH
2–2⅛ in (50–55 mm)

PAPILIO HELENUS
RED HELEN
LINNAEUS, 1758

51

From the tiny second instar to the fourth, Red Helen caterpillars resemble bird droppings as a camouflage defense against predation. The final (fifth) instar sees a dramatic transformation both in color and pattern as the larvae take on the typical snake's head mimicry of many *Papilio* species. Should it be threatened, though, the caterpillar inflates a red, fleshy, forked appendage—the osmeterium—from behind its head. This can be quite startling and also smells unpleasant. When not feeding, the larvae rest exposed on branches and twigs on the host plant, relying on their color to remain undetected.

The larval period lasts from 26 to 30 days. The chrysalis is green (when among fresh twigs and leaves) or brown (when on woody or dry branches) and deeply curved outward, supported by a single-strand silk harness. The pupal period lasts 14 to 22 days. The Red Helen butterfly is a large, black swallowtail with white spots on its hindwings, often seen cruising pathways or clearings in the forest, nectaring at flowers, or mud-puddling. Within its tropical range, the species occurs year round.

The Red Helen caterpillar is a deep green color, with a mottled brown band separating the thorax and abdomen and two oblique brown and white bands mid-abdomen that meet dorsally. The underbelly is white, while the head and ventral thorax are dark brown. The eyespots, complete with slit pupils, are connected by a green "stained glass" pattern mid-thorax.

Actual size

FAMILY	Papilionidae
DISTRIBUTION	South America, from eastern Brazil through Paraguay and Uruguay to Argentina
HABITAT	Dry thorn scrub
HOST PLANTS	*Berberis ruscifolia*
NOTE	Eye-catching caterpillar that has distinctive green and brown markings
CONSERVATION STATUS	Not evaluated, but locally rare

ADULT WINGSPAN
2⅜–3½ in (60–90 mm)

CATERPILLAR LENGTH
1¾–2⅛ in
(45–55 mm)

52

PAPILIO HELLANICHUS

PAPILIO HELLANICHUS

HEWITSON, 1868

The large *Papilio hellanichus* caterpillars live solitary lives on *Berberis ruscifolia* bushes. Their effective cryptic coloring helps them to blend with their surroundings, making it very difficult for predators to spot them. It is the only species of the Papilionidae that feeds on this plant species and, as a result, its distribution is limited to dry scrub where the host plant is found. There are two generations a year, although in wet years, many adults fail to emerge from their pupae, and sightings are rare.

The adult butterfly, one of a number of Papilionidae species known as "swallowtails" for the forked appearance of their hind wings, is also susceptible to disturbance by humans, so it is only found in remote areas. Captive breeding could help boost its numbers, but this species has proved to be incredibly difficult to breed—largely because its food plant does not grow well away from its natural habitat, and the caterpillars are reluctant to feed on other plants.

The *Papilio hellanichus* caterpillar has distinctive markings. It is mostly green with a brown ventral surface, legs, and prolegs. There is a large eyespot on the thorax, and an oblique band of brown and cream runs across the thorax to the abdomen, with a lateral brown loop on the abdomen.

Actual size

FAMILY	Papilionidae
DISTRIBUTION	Corsica, Sardinia
HABITAT	Grassy slopes with low-growing trees and shrubs up to 4,920 ft (1,500 m) elevation
HOST PLANTS	Umbelliferae, including Corsican Rue (*Ruta corsica*), Giant Fennel (*Ferula communis*), and *Peucedanum paniculatum*
NOTE	Striking green and black caterpillar that has an orange osmeterium
CONSERVATION STATUS	Least concern, with an increasing population

ADULT WINGSPAN
2¹³⁄₁₆–3 in (72–76 mm)

CATERPILLAR LENGTH
1⁹⁄₁₆ in (40 mm)

PAPILIO HOSPITON
CORSICAN SWALLOWTAIL
GENÉ, 1839

53

Corsican Swallowtail caterpillars hatch from eggs laid singly on leaves of the host. They feed on the leaves and then overwinter as pupae on the ground on rocks, wood, or stems. There is one generation a year, with a possible partial second generation. Corsican adults are on the wing in July, while the Sardinian population flies earlier in May. The males show hilltopping behavior while they await the arrival of the females.

The Corsican Swallowtail is endemic to just two islands, Corsica and Sardinia, where it is protected and subject to various conservation measures. For example, it is listed in Appendix 1 of CITES, which makes the collection and trade in the species illegal. The species has been threatened by local farmers burning the host plants, which are poisonous to livestock, but in recent years habitat protection and management measures, such as controlled fires to remove overgrown scrub and open up the habitat, have resulted in a population increase. The caterpillar is readily confused with that of *Papilio machaon*.

The Corsican Swallowtail caterpillar is largely green. There are bands of black with blue, orange, and yellow marks along the dorsal surface. Laterally, there is a line of orange and white marks. A large, orange osmeterium lies behind the head. The legs are blue.

Actual size

FAMILY	Papilionidae
DISTRIBUTION	Western North America, from British Columbia to Arizona
HABITAT	Hilltops, rocky slopes, canyons, riverbanks, and roadsides
HOST PLANTS	Desert parsley (*Lomatium* spp.)
NOTE	Caterpillar that grows fast on short-lived host plants
CONSERVATION STATUS	Not evaluated, but common

ADULT WINGSPAN
2⅜–2⁹⁄₁₆ in (60–65 mm)

CATERPILLAR LENGTH
1⁹⁄₁₆–2 in (40–50 mm)

54

PAPILIO INDRA
INDRA SWALLOWTAIL
REAKIRT, 1866

Indra Swallowtail caterpillars hatch in late spring from eggs laid singly on desert parsleys up to a week earlier. The first instar consumes its eggshell and develops rapidly, pupating within 20 to 30 days, according to temperature. Most growth occurs in the final instar, when the caterpillar doubles in size. The larvae come in two color forms—banded or black. Banded caterpillars tend to rest openly, whereas black caterpillars rest concealed within the foliage. Bold-banding in some individuals may serve as a real or false indication of distastefulness.

Rapid development as caterpillars is critical to this species as its host plants senesce quickly in late spring and early summer. There are usually five instars but sometimes only four under hot, dry conditions. The pupae oversummer and overwinter, spending 11 to 12 months in this stage. Male butterflies are more often seen than females, which tend to stay on the upper reaches of rocky slopes where their host plants grow.

The Indra Swallowtail caterpillar is black with or without bold, white to pink bands, one on each segment. The body is smooth, with large white, yellow, or pink spots on each segment. The head is yellow with two wide, frontal, black bands in an inverse V-shape.

Actual size

FAMILY	Papilionidae
DISTRIBUTION	Borneo, the Philippines
HABITAT	Clearings in tropical rain forests
HOST PLANTS	*Citrus* spp.
NOTE	Large caterpillar that in early stages resembles a bird dropping
CONSERVATION STATUS	Not evaluated

PAPILIO LOWII

GREAT YELLOW MORMON

DRUCE, 1873

ADULT WINGSPAN
4½–4⅞ in (115–125 mm)

CATERPILLAR LENGTH
Up to 2⅜ in (60 mm)

55

Great Yellow Mormon caterpillars hatch from creamy, yellow, round eggs with a roughened surface, which are laid singly by the female on the underside of host plant leaves. The newly hatched caterpillars feed first on the eggshell then on the leaves. When young, the larvae resemble bird droppings, which provides perfect camouflage. Mature caterpillars, with their two distinctive eyespots, look very different. When fully developed, the caterpillar pupates as a green chrysalis suspended from a stem by a silk girdle. There are several generations annually.

The adults are on the wing in the canopy, only venturing lower in the forest to feed and to lay eggs. They are variable in appearance and very similar to the Great Mormon (*Papilio memnon*), which also has variable markings; in some classifications, *P. lowii* is considered a subspecies of *P. memnon*. Confusingly, the species is also sometimes known as the Asian Swallowtail, a name shared with *P. xuthus*.

The Great Yellow Mormon caterpillar is mostly green with white-edged, brown, oblique bands on the abdomen and two eyespots on the thorax. The eyespots are joined by a brown transverse band. There is a second transverse band on the first abdominal segment, which, when viewed from above, creates the illusion of a shield. The head and underside are brown.

Actual size

FAMILY	Papilionidae
DISTRIBUTION	North America, Eurasia, and North Africa
HABITAT	Hillsides, open areas, canyons, fens, and shrub-steppe
HOST PLANTS	Variety of Umbelliferae species, including Fennel (*Foeniculum vulgare*); also Tarragon (*Artemisia dracunculus*)
NOTE	Charismatic species with many subspecies and forms
CONSERVATION STATUS	Not evaluated, but some subspecies may be threatened

ADULT WINGSPAN
4–4¼ in (100–110 mm)

CATERPILLAR LENGTH
1⁹⁄₁₆–2 in (40–50 mm)

56

PAPILIO MACHAON
OLD WORLD SWALLOWTAIL
LINNAEUS, 1758

Old World Swallowtail caterpillars feed primarily on the foliage of their host plants. Defense is based on crypsis, with young caterpillars sporting a "bird dropping" saddle and older caterpillars blending in well with their host plants. Birds reportedly reject the caterpillars of some Old World Swallowtail subspecies, so distastefulness may be another defense. Braconid wasps often parasitize the larvae. Their development is rapid, taking only three weeks from egg hatch to pupation. The pupae from late summer caterpillars overwinter.

Up to three broods are produced annually, depending on latitude and elevation, and the pupae may remain dormant for two or more years. Adults sip nectar from thistles, zinnia, phlox, and rudbeckia. The English Swallowtail (*Papilio machaon britannicus*) is a renowned subspecies confined to some Norfolk fens, and the Oregon Swallowtail (*P. machaon oregonius*) is that state's official insect. The Swedish botantist Carl Linnaeus (1707–78) named *P. machaon* after Machaon, the physician to the Greeks in the Trojan War.

The Old World Swallowtail caterpillar is smooth and whitish green, becoming green, with transverse, intersegmental, black bands and alternating, black and yellow spots arranged transversely on each segment to form bands. The green head is prominently marked with four black stripes and—sometimes—large spots. The prolegs are green, each with a black spot.

Actual size

FAMILY	Papilionidae
DISTRIBUTION	Western North America, from British Columbia to Guatemala
HABITAT	Canyon bottoms, roadsides, riparian areas, shrub-steppe, parks, and gardens
HOST PLANTS	Cherry and plum (*Prunus* spp.), ash (*Fraxinus* spp.), and serviceberry (*Amelanchier* spp.)
NOTE	Rarely seen caterpillar that is among the largest in North America
CONSERVATION STATUS	Not evaluated, but common within its range

PAPILIO MULTICAUDATA

TWO-TAILED TIGER SWALLOWTAIL

W. F. KIRBY, 1884

ADULT WINGSPAN
4⅝–5 in (120–130 mm)

CATERPILLAR LENGTH
2–2⅜ in (50–60 mm)

57

The Two-tailed Tiger Swallowtail caterpillar hatches from large eggs laid singly on the upper side of host plant leaves. The first instar consumes the eggshell before feeding on terminal leaves. Second instars spin a thin silk mat on the upper side of a leaf to which they return when not feeding. Older caterpillars produce similar mats, curving them into an upwardly convex shape; such "nest" leaves are not eaten. Later instars eat large gaps in leaves, or entire leaves, and sometimes rest on the underside of a leaf.

Camouflage protects the caterpillars, as the defensive chemicals they emit—mainly from the orange, forked osmeterium—also protect them. Development is rapid, taking about 33 days from egg hatch to pupation. There are four or five instars; environmental cues may dictate the number. Pupation occurs at the base of the host plant, on the trunk, or on a stem, and the pupa overwinters. There are one to three broods annually according to elevation and latitude.

The Two-tailed Tiger Swallowtail caterpillar is bright green with two to six small, blue spots on segments three, four, and seven to ten. Between the third and fourth segments there is a black transverse band anteriorly edged in white. There are complex false eyespots dorsally on segment three, mostly yellow and narrowly outlined in black with a blue dot in the center.

Actual size

FAMILY	Papilionidae
DISTRIBUTION	India, Southeast Asia (southern China, Chinese Taipei, and Indonesia)
HABITAT	Open forest
HOST PLANTS	Members of Rutaceae, including *Toddalia asiatica*, *Euodia meliifolia*, *Zanthoxylum rhetsa*, and *Citrus* spp.
NOTE	Caterpillar with menacing eyespots, exhibiting classic mimicry of swallowtail larvae
CONSERVATION STATUS	Not evaluated, but common within its range

ADULT WINGSPAN
4–4⅞ in (100–125 mm)

CATERPILLAR LENGTH
2 in (50 mm)

58

PAPILIO NEPHELUS
YELLOW HELEN
BOISDUVAL, 1836

Early instar caterpillars are bird-dropping mimics with earthy colors and a glossy, even slimy finish. The fifth and final instar caterpillar is a complete departure in appearance—eyespots and a widened thorax successfully emulate a menacing "snake's head" appearance, while the caterpillar's real head is tucked safely beneath. All stages possess an eversible, malodorous osmeterium to startle and repel potential threats. The chrysalis is angular and green and can be mistaken easily for foliage.

The Yellow Helen (or Black and White Helen, according to subspecies) belongs to a clade of related swallowtail butterflies, the *helenus* group, or Helens, which are typically large, predominantly black butterflies with long tails and a large, white area on the hindwings. Like other *Papilio* species, *P. nephelus* has evolved specific kinds of protective adaptation, such as crypsis (larval and pupal coloration), chemical (osmeterium secretions that change with age and size), and mimicry (bird-dropping and eyespot markings and posture). These progressively protect the species against different types and sizes of predator, from ants to birds, throughout its life cycle.

Actual size

The Yellow Helen caterpillar is mottled green. On the prominent thoracic shield, eyespots with pupils are connected by a green band, and a similar band with lighter, sinuous markings traverses the thoracic boundary. There are two sets of marbled, oblique bars on each side of the abdomen, but only the cranial pair meets dorsally in a distinct V-shape. The caudal set taper toward the dorsum but do not connect.

FAMILY	Papilionidae
DISTRIBUTION	Eastern and central North America, south to Peru
HABITAT	Open woods, grassland, and desert
HOST PLANTS	Several, including carrot (*Daucus* spp.), parsley (*Petroselinum* spp.), Dill (*Anethum graveolens*), and Parsnip (*Pastinaca sativa*); Turpentine Broom (*Thamnosma montana*) in Arizona
NOTE	Caterpillar with a snake-tongue secret weapon for repelling predators
CONSERVATION STATUS	Not evaluated, but common

ADULT WINGSPAN
3⅛–4¼ in (80–110 mm)

CATERPILLAR LENGTH
2 in (50 mm)

PAPILIO POLYXENES
BLACK SWALLOWTAIL
FABRICIUS, 1775

59

The Black Swallowtail caterpillar, a familiar sight in American home gardens, is black when young with a cream saddle resembling a bird dropping. As it grows, the caterpillar develops an orange Y-shaped osmeterium—a feature of swallowtail larvae—which pops out behind the head when it is disturbed, emitting chemicals and a pungent odor that repel predators such as ants. Its appearance, resembling a snake's forked tongue, frightens off birds, which also drop the caterpillars after sampling their bad-tasting skin.

The pupae hibernate and are green if attached to smooth, leafy places, or brown if attached to rough twigs or bark. When adult Black Swallowtails emerge, they prefer to fly to hilltops to mate. There are several generations a year. In the Sonoran and Mojave deserts of southwestern North America, males and females are mostly yellow; elsewhere females usually lose their initial yellow band and become mimics of the Pipevine Swallowtail (*Battus philenor*), which is poisonous to birds as an adult.

The Black Swallowtail caterpillar is pale to dark green with black transverse bands enclosing yellow (sometimes orange) spots—markings that mimic the poisonous caterpillars of the Monarch (*Danaus plexippus*) and Queen (*Danaus gilippus*) butterflies.

Actual size

FAMILY	Papilionidae
DISTRIBUTION	United States, from New England to Florida, west to Texas and Colorado
HABITAT	Marshlands, parks, gardens, meadows, and woodland edges
HOST PLANTS	Spicebush (*Lindera benzoin*) and Sassafras (*Sassafras albidum*)
NOTE	Caterpillar that is a structure-building master of mimicry
CONSERVATION STATUS	Not evaluated, but may be rare in periphery of its range

ADULT WINGSPAN
3–4 in (76–100 mm)

CATERPILLAR LENGTH
2⅛ in (55 mm)

PAPILIO TROILUS
SPICEBUSH SWALLOWTAIL
LINNAEUS, 1758

60

Spicebush Swallowtail caterpillars spend their complete larval cycle within the leaves of the tree upon which the eggs were originally deposited. Young larvae are bird-dropping mimics that vary in coloration from green to brown, occasionally with white saddles across abdominal segments three and eight. When the caterpillar is inactive, a leaf enclosure protects it from predation. By the fourth instar, the body color changes to lime green, transforming the caterpillar into nature's most convincing snake look-alike. Large eyespots and a retractable "forked tongue" (osmeterium) complete the illusion.

The larva pupates within a silken harness, which hangs from the underside of a low-lying leaf of the host tree. The pupa can be green or brown, depending on the seasonal coloration of the surrounding foliage. The adult Spicebush Swallowtail butterfly also employs mimicry to avoid predation by closely imitating the Pipevine Swallowtail (*Battus philenor*), a similarly patterned but foul-tasting butterfly, which shares its range.

Actual size

The Spicebush Swallowtail caterpillar's most recognizable late-instar form is lime green above, tan below, with a yellow stripe separating the colors. Four orange eyespots are present on the upper thorax. The front two thoracic spots are large and contain black "pupils" with white highlights. A series of smaller blue spots extend vertically along the abdomen.

FAMILY	Papilionidae
DISTRIBUTION	Northeastern Australia (Queensland and Northern Territory), New Guinea, and Solomon Islands
HABITAT	Rain forests and wet upland areas
HOST PLANTS	Doughwood (*Melicope* spp.) and *Citrus* spp.
NOTE	Well-camouflaged caterpillar that becomes a spectacular blue adult
CONSERVATION STATUS	Not evaluated, but quite common in its range

ADULT WINGSPAN
4–4¼ in (100–110 mm)

CATERPILLAR LENGTH
2⅜–2⁹⁄₁₆ in (60–65 mm)

PAPILIO ULYSSES
ULYSSES SWALLOWTAIL
LINNAEUS, 1758

61

Ulysses Swallowtail caterpillars hatch from eggs laid singly, or occasionally in groups of two or three, on the young growth of host plants or on the underside of mature leaves. The larvae prefer to feed on young foliage and are often found on the regrowth produced after a tree has been felled or the branches cut. They feed singly, and there is usually only one per plant. When not feeding, the caterpillar rests on a silk pad attached to the upper side of a leaf. It feeds only on adjacent foliage.

Natural enemies, including predators, parasitoids, and diseases, strongly regulate this species. Just prior to pupation, the caterpillars are particularly vulnerable to parasitic wasps. Adults in flight are an impressive sight, the brilliant blue upper side visible as a series of bright blue flashes in sunlight. Breeding continues throughout the year but is most abundant during the wet season.

The Ulysses Swallowtail caterpillar is dark green, becoming paler green laterally with a series of small, blue subdorsal and dorsolateral spots. The first abdominal segment has a broad, white, transverse dorsal band speckled with minute green spots. The remaining abdominal segments each have two variably sized, white dorsal spots. The head is pale green.

Actual size

FAMILY	Papilionidae
DISTRIBUTION	Northern Myanmar, northern Mongolia, Russian Far East, China, Chinese Taipei, Korea, Japan, Ogasawara Islands, Guam, and Hawaii
HABITAT	Woods, urban and suburban areas, and orange orchards
HOST PLANTS	Rutaceae, such as *Phellodendron amurense*, *Poncirus trifoliata*, *Zanthoxylum* spp., and cultivated *Citrus* spp.
NOTE	Caterpillar that was once worshipped as a god in Japan
CONSERVATION STATUS	Not evaluated, but common

ADULT WINGSPAN
1¾–4 in (45–100 mm)

CATERPILLAR LENGTH
2–2⅜ in (50–60 mm)

PAPILIO XUTHUS

ASIAN SWALLOWTAIL

LINNAEUS, 1767

62

Like many *Papilio* species, the Asian Swallowtail caterpillar starts out looking like a bird dropping and ends up resembling a snake. The larvae feed initially on the upper surface of the leaf (where bird droppings land), but in a change regulated by juvenile hormones (JHs) they become green and snakelike, sometimes rearing up with a swelling thorax. The larvae can defoliate *Citrus* trees. Those developing later in the year may overwinter as pupae, but consequent adults eclosing in the spring tend to be much smaller than the summer generations.

The Asian Swallowtail is among the best-studied butterflies, especially in Japan, where researchers have conducted many experiments on learning and color preferences in *Papilio xuthus* adults and, more recently, larvae. The latest research on the genetics of color patterning found that seven genes are involved in regulating color changes between different larval stages. The species was an ancient celebrity, too; according to the eighth-century *Nihon Shoki* ("Chronicles of Japan"), a green, *Citrus*-eating caterpillar, now identified as *P. xuthus*, was once worshipped as a God of the Everlasting World.

The Asian Swallowtail caterpillar is green with eyespots on the second thoracic segment, a broad, dark dorsal band separating thorax and abdomen, and several transverse bands that are darker green, and sometimes green and white. There is also a longitudinal white stripe with a thick, black border running along the prolegs. The head is green and, when disturbed, long, yellow, paired osmeteria are everted behind it.

Actual size

FAMILY	Papilionidae
DISTRIBUTION	Europe, across Asia Minor east to Mongolia, Russian Far East, and northwest China
HABITAT	Subalpine meadows and dry scrubby hillsides, mostly at 2,460–6,600 ft (750–2,000 m) elevation
HOST PLANTS	Stonecrop (*Sedum* spp.)
NOTE	Black-and-orange caterpillar that mimics a large millipede
CONSERVATION STATUS	Vulnerable

ADULT WINGSPAN
2¾–3⅜ in (70–85 mm)

CATERPILLAR LENGTH
Up to 2 in (50 mm)

PARNASSIUS APOLLO

APOLLO
(LINNAEUS, 1758)

63

The female Apollo lays up to 150 eggs on the host plant, either singly or in small groups, which then overwinter. When hatched the following spring, the young caterpillars are entirely black, making them difficult to spot. As they develop, they gain orange spots and become similar in appearance to a millipede found in the same habitat. Both the millipede and the caterpillars produce a foul-smelling liquid to deter predators—an example of Müllerian mimicry, a phenomenon first described in the nineteenth century by German naturalist Fritz Müller (1821–97).

The caterpillar pupates on the ground in a loose cocoon. There is one generation annually and the adults are on the wing from late April to September. Long prized by collectors for its beauty, the butterfly is now classified as vulnerable due to overcollection, disease, and loss of habitat. Conservationists hope that Apollo numbers will be boosted by the introduction of measures such as captive breeding and habitat protection, plus the listing of the species in Appendix II of CITES to strictly control trade in its specimens.

The Apollo caterpillar is velvety black with a row of lateral, orange-red spots arranged in pairs, with one spot larger than the other. The head and body are covered in short, black hairs.

Actual size

FAMILY	Papilionidae
DISTRIBUTION	Western North America, from Alaska to New Mexico
HABITAT	Mountains, upper elevations to highest Arctic-alpine
HOST PLANTS	Stonecrop (*Sedum* spp.)
NOTE	Aposematic caterpillar that emits a foul odor when disturbed
CONSERVATION STATUS	Not evaluated, but common within its range

ADULT WINGSPAN
2⅜–2⁹⁄₁₆ in (60–65 mm)

CATERPILLAR LENGTH
1³⁄₁₆ in (30 mm)

64

PARNASSIUS SMINTHEUS
MOUNTAIN PARNASSIAN
DOUBLEDAY, [1847]

Mountain Parnassian females lay their eggs singly on and around stonecrop plants. The eggs overwinter and the caterpillars hatch in late spring but do not consume their eggshells. The caterpillars, which are solitary, bask openly in sunshine but feed nocturnally, primarily on leaf tips. No nests are made. The larvae are easily found by searching patches of *Sedum* in mountainous areas. When disturbed, the caterpillars twitch violently, drop to the ground seeking cover, and may emit a bad-smelling chemical from their yellow osmeterium. Their aposematic coloration also suggests they may be toxic.

Caterpillar development is slow, taking 10 to 12 weeks. There are five instars, and pupation occurs in a loose cocoon constructed among ground debris. The flight period of the adult—a mostly white butterfly with some red spots and dark markings—is late May to early September depending on latitude and elevation. Other *Parnassius* species occur in North America and Eurasia, all confined to mountainous habitats.

The Mountain Parnassian caterpillar is jet black with bristly setae and four rows of contrasting yellow-gold spots that develop in intensity as the caterpillar matures. The lateral row of spots has two per segment (except segments one to three, with a single spot per segment), a larger spot anteriorly, and a smaller one posteriorly.

Actual size

FAMILY	Papilionidae
DISTRIBUTION	The Balkans, the Middle East
HABITAT	Dry grassland and meadows up to 3,300 ft (1,000 m) elevation in disturbed areas
HOST PLANTS	Birthwort (*Aristolochia* spp.)
NOTE	Distinctive spiky caterpillar that has black, white, and red markings
CONSERVATION STATUS	Not evaluated

ADULT WINGSPAN
2¹⁄₁₆– 2⁷⁄₁₆ in (52–62 mm)

CATERPILLAR LENGTH
1³⁄₁₆–1³⁄₈ in (30–35 mm)

ZERYNTHIA CERISY
EASTERN FESTOON
(GODART, [1824])

Eastern Festoon caterpillars hatch from creamy yellow eggs laid singly or in small groups by the female butterfly on the underside of leaves of the host plant, generally in shade. The larvae feed on the leaves and pupate at the base of the plant or in stones and crevices, protected within a silken web. The species overwinters as a pupa.

There is a single generation each year, and the butterflies are seen on the wing from March to July—those at higher elevations are the later flyers. In some parts of its range, the species is declining due to loss of habitat, often as a result of agricultural intensification and the use of herbicides. There are a number of subspecies found on Mediterranean islands, including Crete. The adult Cretan Eastern Festoon is less red in color and lacks a tail on its wings—different enough for some to class it as a subspecies or even a separate species (*Zerynthia cretica*).

Actual size

The Eastern Festoon caterpillar is colorful and spiky. The body is black with stripes of white running lengthwise, from which orange-red tubercles arise, each bearing many short spines. The head is brown.

FAMILY	Hesperiidae
DISTRIBUTION	From Mexico south through Central America and most of the Caribbean; South America, south to northern Argentina
HABITAT	Lowland humid forests, forest edges, and adjacent second growth, and gardens around human dwellings
HOST PLANTS	Various species of cultivated and wild *Citrus* spp., and species of the closely related genus *Zanthoxylum*
NOTE	Striped caterpillar often seen feeding on *Citrus* trees in gardens
CONSERVATION STATUS	Not evaluated, but unlikely to become endangered

ADULT WINGSPAN
2–2⅜ in (50–60 mm)

CATERPILLAR LENGTH
1⁵⁄₁₆–1½ in (33–38 mm)

ACHLYODES BUSIRUS
GIANT SICKLEWING
(CRAMER, 1779)

66

Like most skippers, the caterpillars of Giant Sicklewings build leaf shelters and use an anal comb to forcibly expel frass from their anus and away from their shelters. Early shelters are made with a triangular flap excised from the edge of a leaf, while the larger caterpillars pull the surfaces of two adjacent leaves together to form a pocket. Newly hatched caterpillars do not leave their shelters, but instead feed on the surface of the leaf inside their shelter, leaving the dorsal leaf cuticle intact to create small "windows." The caterpillars pupate inside their last larval shelter, and the dark brown chrysalis is covered with a fine, white, waxy powder.

The scientific species name is a reference to King Busiris, a figure from Greek and Egyptian mythology, who was, by some accounts, slain by Heracles shortly after he escaped bondage in the king's dungeon. Adult butterflies are fast fliers and very wary as they land, with wings spread on the ground, probing the soil for minerals.

Actual size

The Giant Sicklewing caterpillar is deep maroon purple in ground color, with white and yellow stripes transversing the body along most of its length. The head is roughly heart-shaped and strongly rugose. It is dark brown near the apex, merging into black around the clypeus and below.

FAMILY	Hesperiidae
DISTRIBUTION	From western Mexico and extreme southern Texas to Ecuador, perhaps straying as far south as Bolivia
HABITAT	Forest edges, treefall gaps, and second growth in lowland and foothill, humid and semi-humid forests
HOST PLANTS	*Citrus* spp. and *Zanthoxylum* spp.
NOTE	Shelter-dwelling caterpillar that becomes an extremely agile adult flier
CONSERVATION STATUS	Not evaluated, but not considered threatened

ADULT WINGSPAN
2⅛–2⁹⁄₁₆ in (55–65 mm)

CATERPILLAR LENGTH
1⅞–2⅛ in (48–55 mm)

ACHLYODES PALLIDA

PALE SICKLEWING
(R. FELDER, 1869)

67

Pale Sicklewing caterpillars hatch from round, creamy-white eggs deposited singly on the underside of freshly developing leaves. All instars are frequently found on cultivated *Citrus* in, and around, gardens and readily noticed by their roughly square leaf shelters. These are made by simply excising a small portion of leaf from the margin and flipping it onto the upper surface of the leaf. Larger instars may sew two adjacent leaves together to form a shallow pocket, and this is usually where pupation occurs.

When threatened, or when their shelter is pried open, larger larvae often rear back their head and attempt to bite the intruder with their formidable jaws. Adults of this widespread and common skipper species are extremely fast and wary fliers. They frequently visit wet sand, carrion, dung, flowers, and even laundry hung out to dry. *Achlyodes pallida* and *A. busirus* are the only two species of their genus; other former members have recently been assigned to *Eantis*.

Actual size

The Pale Sicklewing caterpillar has a reddish-brown, heavily granulated, and heart-shaped head, blackish basally, giving a masked appearance. The dark bluish-gray to greenish-gray body is simply patterned with a series of short, yellow dashes and hatch marks, forming a line above the spiracles that runs the length of the body. The skin at the front and rear is lightly tinged yellow.

FAMILY	Hesperiidae
DISTRIBUTION	Southwestern United States, mainland Mexico
HABITAT	Arid mountains
HOST PLANTS	Century plant (*Agave* spp.)
NOTE	One of the largest skipper caterpillars in the world
CONSERVATION STATUS	Not evaluated, but common in most of its range

ADULT WINGSPAN
2–2⅜ in (50–61 mm)

CATERPILLAR LENGTH
1¾ in (45 mm)

AGATHYMUS ARYXNA
ARIZONA GIANT-SKIPPER
(DYAR, 1905)

The Arizona Giant-skipper caterpillar lives only inside the thick leaves of plants such as *Agave utahensis* in southeastern California, Arizona, and Mexico. The mother drops each egg from the top of the plant from where it usually falls into the leaf base. The young caterpillar then crawls up to the leaf tip, bores into it, and hibernates inside its burrow. After hibernation it moves back to the leaf base and and chews another burrow to live in. It then makes a trapdoor on the bottom of the leaf and pupates inside the burrow. The adult crawls out of its trapdoor, expands its wings, and flies in August and September to mate and lay eggs.

An even larger Mexican species (the Gusano del Maguey, *Aegiale hesperiaris*) lives in giant *Agave* plants, which are used to make tequila and mescal. Introduced as a marketing ploy, the caterpillars can sometimes be seen floating just under the cork in bottles of those spirits. The larvae are also roasted, then packed in barrels to be sold as a snack food.

The Arizona Giant-skipper caterpillar is a cream brownish green or bluish white, with a black collar and rear. The body is smooth with very fine, short setae, and some individuals are lightly spotted in green. The smooth head is brown or orangish and unmarked.

Actual size

FAMILY	Hesperiidae
DISTRIBUTION	Occasionally Texas; Mexico, south to Costa Rica and Panama, and east to the Greater Antilles; South America, south to Bolivia and southern Brazil
HABITAT	Most tropical areas below 3,300 ft (1,000 m) elevation
HOST PLANTS	Pea family (Fabaceae), particularly coral trees (*Erythrina* spp.)
NOTE	Caterpillar that is an accomplished excrement thrower and tent builder
CONSERVATION STATUS	Not evaluated, but common and unlikely to become endangered

ADULT WINGSPAN
2⅛–2⁹⁄₁₆ in (55–65 mm)

CATERPILLAR LENGTH
1½–1¾ in (38–44 mm)

ASTRAPTES ALARDUS
FROSTED FLASHER
(STOLL, 1790)

69

The caterpillars of the widespread Frosted Flasher are predictably found anywhere within their range that *Erythrina* plants occur. Their distinctive, peaked tents, pyramidal in shape and folded to the upper surface of the leaf along a narrow strip of leaf, are easily spotted, even from a distance. When the caterpillars have developed to the fourth or fifth instar, they build a pocket-like shelter between two leaves, where they later pupate. The caterpillars are skillful frass-throwers, with mature larvae able to fling frass more than 3 ft (1 m).

Males perch in patches of sunlight, lying in wait for females. When not searching for oviposition sites or guarding perches, adults tend to rest upside down under large leaves, with their wings closed. Frosted Flashers are probably among those species slowly increasing their range, due largely to their use of coral trees as host plants. The trees are commonly planted in large numbers in both agricultural and populated areas, facilitating invasion by this species.

The Frosted Flasher caterpillar is rather sluglike in shape, with a bulbous, brown head decorated with two prominent, bright orange "eyespots." The cervical area and prothoracic shield are well sclerotized and bright crimson, contrasting with a green body covered in fine yellow speckling. The claspers are similar in coloration to the first thoracic section.

Actual size

FAMILY	Hesperiidae
DISTRIBUTION	Southern United States to Brazil, extending north seasonally
HABITAT	Pastures, prairies, gardens, lawns, parks, and roadsides
HOST PLANTS	Grasses, including meadow grass (*Poa* spp.), fescue (*Festuca* spp.), and Bermuda Grass (*Cynodon dactylon*)
NOTE	Common lawn-feeding caterpillar now heading north with climate warming
CONSERVATION STATUS	Not evaluated, but common

ADULT WINGSPAN
1³⁄₁₆–1⅜ in (30–35 mm)

CATERPILLAR LENGTH
1⅛–1³⁄₁₆ in (28–30 mm)

ATALOPEDES CAMPESTRIS
SACHEM
(BOISDUVAL, 1852)

70

The Sachem caterpillar silks grass blades together to make shelters that become more complex with each successive instar. First instars simply weave a few silken strands into a vague "nest," whereas prepupal fifth instars construct untidy but tightly woven, silk-lined shelter tubes. Pupation occurs on the ground within a silk cocoon lined with woollike material produced by ventral abdominal glands. Defense is based on concealment, although small predators, such as pirate bugs (Anthocoridae), can find their way into shelters and kill larvae. There are two to three broods a year with successive generations increasingly abundant.

Originally a subtropical species, *Atalopedes campestris* is extending its range in North America, apparently heading farther north as the climate warms. Despite the lack of a defined winter diapause stage, the caterpillars appear able to survive very low temperatures. Adults—among the skipper butterflies, named for their darting flight—are commonly seen in backyards visiting flowers, including marigolds, asters, and buddleia.

The Sachem caterpillar is olive brown or gray with coalescing tiny, black spots and setae. Each segment has five transverse folds posteriorly, and there is a distinct, dark brown, middorsal stripe. The head and collar are shiny black, the collar with an anterior white margin.

Actual size

FAMILY	Hesperiidae
DISTRIBUTION	Southwestern United States, northern Mexico
HABITAT	Arid canyons and washes
HOST PLANTS	*Bouteloua curtipendula* var. *caespitosa*
NOTE	Caterpillar that builds aerial nests in one native grass species
CONSERVATION STATUS	Not evaluated, and apparently secure, but possibly quite rare at the periphery of its range

ADULT WINGSPAN
1¼–1⅝ in (32–42 mm)

CATERPILLAR LENGTH
1³⁄₁₆ in (30 mm)

ATRYTONOPSIS VIERECKI
VIERECK'S SKIPPER
(SKINNER, 1902)

71

The Viereck's Skipper caterpillar lives in a tube formed of leaves it joins together with silk produced by a spinneret beneath its head. The larvae eat only the *caespitosa* variety of *Bouteloua curtipendula* as it is leafier than the main species and its large clumps provide more cover for their nests. The caterpillars grow to mature size by late summer, then hibernate in silked-leaf nests at the base of the grass clump. In spring they pupate, and adults emerge a week or two later, in May.

Tiny Trichogrammatidae wasps, which have a wingspan of only ¹⁄₃₂ in (1 mm), lay an egg in many of the *Atrytonopsis vierecki* eggs. The wasp larva eats the contents of the egg and then produces a wasp instead of a caterpillar. Parasitic wasps attack the eggs of most butterfly and moth species. The desert valleys of southwestern United States and Mexico have a dozen skippers of the *Atrytonopsis* genus; their caterpillars all live on large grasses.

The Viereck's Skipper caterpillar is tan with a darker heart-band on top, but becomes bluish green after feeding on green grass. It has a narrow, black collar, and the rear is tan. The head is tan with a distinctive, black Christmas-tree patch on the front that differs from the head pattern of numerous other similar grass-feeding skippers.

Actual size

FAMILY	Hesperiidae
DISTRIBUTION	India, China, Southeast Asia, the Philippines, Indonesia, Papua New Guinea, Australia, and Fiji
HABITAT	Openings and edges of deciduous and evergreen tropical forests
HOST PLANTS	*Terminalia* spp., *Pongamia* spp., *Rhyssopterys* spp., and *Combretum* spp.
NOTE	Caterpillar that may defoliate food plant, compelling adults to migrate
CONSERVATION STATUS	Not evaluated, but widespread and common

ADULT WINGSPAN
1⅞–2⅟₁₆ in (48–52 mm)

CATERPILLAR LENGTH
1¾ in (45 mm)

BADAMIA EXCLAMATIONIS
BROWN AWL
(FABRICIUS, 1775)

72

The Brown Awl caterpillar, also known as the Narrow-winged Awl, forms an open shelter on young foliage by joining the edges of leaves together with silk. Feeding outside the shelter occurs during the day and night. The caterpillar grows fast and has moist, sticky frass. When about to pupate, the caterpillar may descend close to the ground, particularly if the plant has been defoliated, and construct a tubular cell from a leaf. The cell is heavily lined with silk, and the pupa is attached to the silken pad by the cremaster and a silk girdle. There may be several generations a year.

Populations of this insect can be very large, and the caterpillars can strip their food supply, forcing migration of the adults. Migrations may extend beyond the species' normal breeding areas, but, where fresh food is located, a return migration may also occur. The adults have a rapid audible flight.

Actual size

The Brown Awl caterpillar is cylindrical, variable in color from yellow to purplish black suffused with yellow, and has a broad, black middorsal line and segments with black, transverse lines. The head is yellow with two black, transverse bands and many tiny, black spots. The prolegs are whitish, encircled with black.

FAMILY	Hesperiidae
DISTRIBUTION	Tropical Mexico to French Guiana and south along the Andes to Colombia, and probably at least northern Ecuador
HABITAT	Humid montane forest borders and second growth
HOST PLANTS	*Achyranthes* spp.
NOTE	Caterpillar whose early instars construct distinctive, manhole-cover-like shelters
CONSERVATION STATUS	Not evaluated, but not considered threatened

ADULT WINGSPAN
1⅛–1⁵⁄₁₆ in (28–34 mm)

CATERPILLAR LENGTH
1⅛–1¼ in (28–32 mm)

BOLLA GISELUS
BOLLA GISELUS
(MABILLE, 1883)

73

Bolla giselus caterpillars hatch from tiny, brown eggs laid singly, usually on the underside of young host plant leaves. After consuming most of their eggshell, they excise a roughly oval-shaped flap from the center of the leaf, flipping this lid over onto the upper or the lower surface of the leaf. The young larvae rest within this shelter and throw their frass considerable distances with the aid of an anal comb. Feeding damage around these shelters is distinctive, as young larvae eat only one side of the leaf, leaving clear, window-like perforations.

Older larvae tend to sew two entire leaves together, resting within this pocket when not feeding. Copious silk laid within the shelter can cause it to bow strongly outward, giving the caterpillar more room to maneuver within. Pupation occurs within the last larval shelter. Adults are fast fliers, feeding at many of the common roadside flowers, pausing frequently to bask in the sun with their wings spread.

Actual size

The *Bolla giselus* caterpillar is simply patterned, yellowish green with minute, yellowish speckling and short, pale setae scattered sparsely over most of its body. Laterally it has a thin, yellow, spiracular line. The head is dark brown and bulbous, roughly heart-shaped, and covered in small, pale setae similar to, but shorter than, those on the body.

FAMILY	Hesperiidae
DISTRIBUTION	Northeast India, into Southeast Asia (Malay Peninsula, the Philippines, Indonesia, and Hong Kong)
HABITAT	Forests, parks, and gardens
HOST PLANTS	*Schefflera lurida*, *Schefflera octophylla*, *Trevesia sundaica*, *Embelia garciniaefolia*, and *Horsfieldia* spp.
NOTE	Caterpillar that unusually resembles the adult in color and patterning
CONSERVATION STATUS	Not evaluated, but considered uncommon regionally except where host plants abound

ADULT WINGSPAN
1¾–2 in (45–50 mm)

CATERPILLAR LENGTH
2–2⅛ in (50–55mm)

BURARA GOMATA
PALE GREEN AWLET
(MOORE, 1865)

74

The Pale Green Awlet caterpillar makes a leaf shelter from the very early stages by folding a chewed leaf on itself and stitching it with silk. The larva comes out to feed on adjoining leaves. By the later stages, a whole leaf is folded in half, and eventually the mature caterpillar will pupate within this shelter. The pupa is white with black and yellow markings and is anchored to the leaf by its tail end and a silken girdle. Eggs are laid in batches, so collectively the larvae will often defoliate entire branches of their host plant, leaving just the folded pupal shelters.

This skipper species is one rare example where the caterpillar resembles the adult in its color scheme and patterning. The striking adults (bright orange with black-and-white striped wings and abdomen) usually fly very early or late in the day, nectaring on flowers, mud-puddling, or resting on the underside of leaves. As some of the host plants make attractive hedges, the complete life cycle of *Burara gomata* can be observed in urban parks and gardens.

Actual size

The Pale Green Awlet caterpillar has a distinctive black, yellow, and white dots-and-dashes pattern. The head capsule is orange and bears two rows of four black spots——"spider eyes." Three black, longitudinal lines run dorsally along the thoracic segments. Late instar larvae are rotund, with the abdominal girth nearly double that of the head.

FAMILY	Hesperiidae
DISTRIBUTION	Northern South America (Suriname and the Guianas), south through Amazonia to Peru and Bolivia
HABITAT	Primary lowland and foothill forests up to about 2,625 ft (800 m) elevation
HOST PLANTS	Unknown species of understory sapling
NOTE	Caterpillar that expertly propels frass away from its leaf shelters
CONSERVATION STATUS	Not evaluated

ADULT WINGSPAN
1⅜–1⁹⁄₁₆ in (35–40 mm)

CATERPILLAR LENGTH
1¹¹⁄₁₆–1⅞ in (43–48 mm)

CABIRUS PROCAS

CABIRUS PROCAS
(CRAMER, 1777)

75

The sluglike caterpillar of *Cabirus procas* builds a wonderfully intricate shelter on the leaves of its host plant, resting inside while not feeding. A roughly square flap is excised from the leaf margin and folded over onto the upper surface of the leaf. A heavy layer of silk is laid down in concentric circles, on both the upper and lower inner surfaces of the shelter, causing it to bow outward. When threatened, the caterpillar will often rapidly rattle its head against the shelter lid.

While resting within their leaf shelter, larvae periodically extend their abdomen out from under the lid and expertly fire a fecal pellet away from their home. Larger caterpillars can propel frass more than 3 ft (1 m). For a skipper, the adult *Cabirus procas* is a fairly weak flier, usually perching on the underside of leaves, wings open. Though strikingly different in coloration, both sexes appear to be mimics of day-flying moths.

The *Cabirus procas* caterpillar has an almost perfectly transparent skin, with a very faint wash of yellow on the posterior segments. This reveals the delicate white tracheoles radiating outward from the spiracles and makes its overall color vary with the contents of its gut. The head is dull orange and distinctly heart-shaped.

Actual size

FAMILY	Hesperiidae
DISTRIBUTION	Southern United States, through most of the Caribbean and Central America, south to Argentina
HABITAT	Widespread, often in disturbed and cultivated areas, in humid, semi-humid, and strongly seasonal habitats
HOST PLANTS	Marantaceae, mostly ornamental *Canna* spp.; also *Calathea* spp., *Maranta* spp., and *Thallia* spp.
NOTE	Common caterpillar that in large numbers can severely damage ornamentals
CONSERVATION STATUS	Not evaluated, but not likely to become endangered

ADULT WINGSPAN
2–2⅛ in (50–55 mm)

CATERPILLAR LENGTH
1½–1¾ in (38–45 mm)

CALPODES ETHLIUS
BRAZILIAN SKIPPER
(STOLL, 1782)

76

Actual size

Brazilian Skipper caterpillars hatch from eggs laid either singly or in small rows of two to six eggs; they are, however, heavily parasitized by tiny wasps. The young caterpillars make simple shelters by rolling a small portion of the leaf margin, often with a notch cut at each end of the roll to facilitate bending of the leaf. Feeding from one end of the shelter usually results in a triangular flap, folded either to the top or bottom of the leaf. Pupation occurs in the final larval shelter.

The eggs, larvae, and pupae are easy to find on ornamental *Canna* species, and outbreaks can cause considerable damage before they eventually collapse in a generation or two. The Brazilian Skipper is probably more common than records suggest, as it appears to be somewhat crepuscular, often flying in the evening. It is apparently still expanding its range and was recently found to have colonized the Galapagos Islands.

The Brazilian Skipper caterpillar has a light, caramel-brown or orange-brown head, roughly triangular in shape. Variably sized, oval, black markings over the stemmata look like eyes. The body is dull, translucent green with a transparent cuticle that leaves the weblike network of trachea clearly visible as they expand below the cuticle from each spiracle.

FAMILY	Hesperiidae
DISTRIBUTION	North Africa; southern, central, and eastern Europe to the Caucasus and Armenia
HABITAT	Juniper-oak woodlands, steppes, and dry grassy slopes
HOST PLANTS	Perennial Yellow Woundwort (*Stachys recta*) and *Sideritis scordioides*
NOTE	Hairy caterpillar that feeds on developing seeds
CONSERVATION STATUS	Not evaluated, but classed as near threatened in Europe

ADULT WINGSPAN
1⅛–1⁵⁄₁₆ in (28–34 mm)

CATERPILLAR LENGTH
Up to ¾ in (20 mm)

CARCHARODUS LAVATHERAE
MARBLED SKIPPER
(ESPER, 1783)

77

The female Marbled Skipper lays her eggs, singly, on fading flower spikes of the host plant. The young caterpillars feed on the developing seeds found with the withered flowers, spinning a loose web around the sepals. At about the third instar, they overwinter as larvae and complete their growth in late spring. The caterpillars then move to the ground, where they pupate, commonly in the leaf litter.

There is usually one generation a year, although occasionally two have been reported in areas of Italy and northern Greece. The adults, the most brightly colored of their genus, are on the wing from May to August, depending on the region, and are found at altitudes of 650–5,250 ft (200–1,600 m). The declining population reported in some areas is due mainly to the loss of habitat. The species has a narrow range of host plants, and their disappearance results in fragmentation of the butterfly population.

Actual size

The Marbled Skipper caterpillar has a dark brown body, with a mottled appearance due to the number of small, white spots. It is covered in long, fine, white hairs, and a yellow-brown stripe runs along both sides. The head is brown.

FAMILY	Hesperiidae
DISTRIBUTION	Across much of northern North America and northern Eurasia, including Japan
HABITAT	Forest openings, mountain meadows, stream banks, and moist lowlands
HOST PLANTS	Grasses, including Purple Reedgrass (*Calamagrostis purpurascens*), Brome (*Bromus* spp.), and Reed Canary Grass (*Phalaris arundinaceae*)
NOTE	Night-feeding grass caterpillar that builds a tubular nest
CONSERVATION STATUS	Not evaluated, but vulnerable in parts of its range

ADULT WINGSPAN
¾–1 in (20–25 mm)

CATERPILLAR LENGTH
1–1⅛ in (25–28 mm)

78

CARTEROCEPHALUS PALAEMON
ARCTIC SKIPPER
(PALLAS, 1771)

In North America, the eggs of the Arctic Skipper are laid singly on grasses in early summer and hatch within seven days. Development from first to fifth instar takes about six weeks, with fifth instars feeding slowly and entering dormancy after three weeks within a rolled grass-blade nest. After overwintering, the caterpillars leave the nest and pupate on a nearby grass blade. Protection is based on concealment, nocturnal feeding, and flinging frass to deter predators. Adults readily nectar on flowers and live for up to three weeks.

Early caterpillar instars pull together the edges of grass blades, forming hollow, open-ended, tubular nests and tying them together with silk crossties. Feeding caterpillars often clip grass blades, leaving the outer 2–3 in (51–76 mm) tips to fall. The life history of *Carterocephalus palaemon* varies in Europe and Asia. Populations have declined in recent decades, and the species appears vulnerable to habitat disturbance and climate change. It disappeared from England—where it was known as the Chequered Skipper—in 1976.

Actual size

The Arctic Skipper caterpillar is green with lighter green and indistinct white stripes. There is a bold, white stripe laterally bordered on both sides by green. The head is light cinnamon brown and strongly divided vertically.

FAMILY	Hesperiidae
DISTRIBUTION	Indonesia, Papua New Guinea, and scattered areas of Australia
HABITAT	Rain forests and urban gardens
HOST PLANTS	Many species of palm (Arecaceae)
NOTE	Caterpillars that can be minor pests of ornamental palms
CONSERVATION STATUS	Not evaluated, but common

CEPHRENES AUGIADES
ORANGE PALM-DART
(C. FELDER, 1860)

ADULT WINGSPAN
1⁷⁄₁₆–1⅝ in (37–41 mm)

CATERPILLAR LENGTH
1⁹⁄₁₆–2 in (40–50 mm)

79

The Orange Palm-dart caterpillar forms two types of shelter—tubular (by rolling the leaf pinnae) or flat, created between overlapping leaves—using silk to bind them. The caterpillar emerges at night to feed at the edges of the pinnae. As the caterpillar grows, it constructs new, larger shelters. Considerable damage to fronds can occur when larval numbers are high. Several generations can be completed each year, with the development of the caterpillar slowing in the cooler months, and an extra sixth instar may be completed.

Pupation generally occurs in the final shelter, but sometimes the caterpillar will leave the palm and pupate in leaf litter at the base of the tree. At pupation, the pale brown, cylindrical pupa becomes covered with a white, waxy powder. In recent years the range of the Orange Palm-dart has expanded as a result of widespread planting of ornamental palms outside the natural habitat of the butterfly.

Actual size

The Orange Palm-dart caterpillar is cylindrical, translucent pale bluish green with a darker middorsal line. The head is cream with variable brown lateral and central bands. The posterior segments have prominent setae.

FAMILY	Hesperiidae
DISTRIBUTION	The Andes of Colombia, Ecuador, and Peru, probably also northern Bolivia
HABITAT	Montane landslides, treefall gaps, and other areas dominated by *Chusquea* bamboo
HOST PLANTS	Bamboo (*Chusquea* spp.)
NOTE	Nondescript caterpillar that is frequently parasitized by tachinid flies
CONSERVATION STATUS	Not evaluated, but not considered threatened

ADULT WINGSPAN
2⅛–2⁹⁄₁₆ in (55–65 mm)

CATERPILLAR LENGTH
1⅛–1⁵⁄₁₆ in (28–34 mm)

80

DION CARMENTA
DION CARMENTA
(HEWITSON, 1870)

The *Dion carmenta* caterpillar hatches from a round, whitish egg laid singly on the underside of a leaf by the female during only a three- or four-second pause. At first, the larvae have black heads with white, unadorned bodies, slowly developing a reddish tinge to thoracic segments as they grow but otherwise changing little during ontogeny. Larvae of all instars construct and dwell in shelters built on the food plant and forcibly eject frass with the aid of an anal comb.

Pupation occurs on the food plant in a shelter formed by silking multiple leaves into a tube. Development, from oviposition to eclosion, lasts 128 to 147 days. Adults, which are encountered infrequently, feed on a variety of flowers, generally at forest edges and clearings. Females oviposit while flying rapidly over large patches of their host, touching down briefly on the upper side of leaves but otherwise maintaining rapid and erratic flight.

Actual size

The *Dion carmenta* caterpillar is whitish, with the dark green gut contents sometimes showing through the middle abdominal segments. Older larvae become chalky white. The prothoracic shield is prominent, shiny black, and extends to the spiracular area. There is often a reddish cast to anterior thoracic segments, and the head is bulbous and shiny black.

FAMILY	Hesperiidae
DISTRIBUTION	Southern Mexico and the Greater Antilles, south to Paraguay and northern Argentina
HABITAT	Humid and semi-humid clearings, forest borders, and river margins
HOST PLANTS	*Citrus* spp. and *Zanthoxylum* spp.
NOTE	Widespread and common caterpillar on cultivated *Citrus* in gardens
CONSERVATION STATUS	Not evaluated

ADULT WINGSPAN
1½–1⅞ in (38–48 mm)

CATERPILLAR LENGTH
1½–1¾ in (38–44 mm)

EANTIS THRASO
SOUTHERN SICKLEWING
(HÜBNER, [1807])

81

Southern Sicklewing caterpillars hatch from eggs laid singly, usually on the underside of very young leaves, but sometimes wedged between two leaf buds; the females pause only momentarily, perching above and curling their abdomen down and under to find a suitable site. Like most species feeding on plants of family Rutaceae, the larvae consume mainly the young, new shoots of their host plant, presumably to avoid noxious chemicals that develop as the leaves mature. When disturbed, the caterpillars often rear back, occasionally vomiting and trying to bite the intruder.

All instars build and rest inside leaf shelters constructed out of host leaves, expelling their frass a distance many times their own body length with the help of an anal comb. Young larvae excise a small portion of leaf from the margin to create a shelter "lid," folded to the dorsal surface of the leaf. Later instars usually silk together two overlapping leaf blades. Pupae are attached by heavy silking to the ventral surface of the shelter. Adults are fast fliers and frequently take nectar from the flowers of their larval host plants.

Actual size

The Southern Sicklewing caterpillar is lime green with a large, heart-shaped, yellowish-green or ivory-colored head. Its only readily apparent markings are irregular, orange-yellow stripes running the length of the body just above the spiracular line. The black mandibles contrast with the pale head and are partially covered by a bright white clypeus.

FAMILY	Hesperiidae
DISTRIBUTION	El Salvador, Nicaragua, south to northern Peru and Amazonian Brazil
HABITAT	Roadsides, clearings, and forest gaps in humid and semi-humid Andean foothill forests
HOST PLANTS	Various species of grasses (Poaceae)
NOTE	Caterpillar that builds well-concealed, rudimentary, grass-blade shelters
CONSERVATION STATUS	Not evaluated, but not considered threatened

ADULT WINGSPAN
1⅜–1¾ in (35–45 mm)

CATERPILLAR LENGTH
1⅛–1⁷⁄₁₆ in (28–36 mm)

ENOSIS UZA
BLUE-SPOTTED SKIPPER
(HEWITSON, 1877)

82

Actual size

The Blue-spotted Skipper caterpillar has a skin that is almost perfectly transparent, appearing white toward the front and rear. The weblike patterns of the white tracheoles radiating outward from the spiracles contrast with the dark green contents of its gut. The head is black and roundly triangular.

Blue-spotted Skipper caterpillars hatch from round, whitish eggs laid singly, usually on the underside of leaves. The larvae build simple shelters on their host leaves, creating ever larger structures as they develop and grow at each instar. Young larvae roll the leaf edge into a tube, and larger instars may roll the entire leaf into a tube, making it barely distinguishable from surrounding grass blades. Frass is expelled from the shelter with the aid of an anal comb. Pupation likely takes place in the final larval shelter.

In the right habitat, the caterpillars can be abundant, but are so well hidden in their camouflaged shelters that a practiced eye is required to find them. The adults are rapid fliers, pausing frequently to bask on the upper surfaces of leaves or on rocks in the sun, and feeding at moist sand and a variety of flowers in open, sunny habitats. *Enosis uza* and other members of its genus are known as grass skippers, reflecting their habitat and family.

FAMILY	Hesperiidae
DISTRIBUTION	Southern Canada, most of the United States
HABITAT	Open, flowery places at lower to middle elevations, including parks, gardens, and watercourses
HOST PLANTS	Deervetch (*Lotus* spp.), locust (*Robinia* spp.), and Wild Licorice (*Glycyrrhiza lepidota*)
NOTE	Head-twitching, frass-ejecting skipper caterpillar
CONSERVATION STATUS	Not evaluated, but common

ADULT WINGSPAN
2–2⅛ in (50–55 mm)

CATERPILLAR LENGTH
1⅜–1⁹⁄₁₆ in (35–40 mm)

EPARGYREUS CLARUS
SILVER-SPOTTED SKIPPER
(CRAMER, 1775)

83

Silver-spotted Skipper caterpillars hatch from eggs laid singly on the terminal leaves of host plants, typically only one egg per plant. Initially, the first instar larva cuts a small terminal leaf twice, both cuts on the same side and perpendicular to the midrib. It then folds the flap over the upper surface to make a shelter by silking it in place. In later instars the larger caterpillar silks together two or three leaves to create its home. Caterpillars spend most of their time in shelters, only leaving them to feed.

The pupa is suspended horizontally from the ceiling of a shelter with three silk threads. There are five instars, and the pupa overwinters. Protection is based on concealment, nocturnal feeding, frass ejection, a large red ventral gland (likely emitting repellent chemicals) on segment one, and head twitching when disturbed. Ejecting frass up to 40 body lengths away from the nest is thought to put predators "off the scent."

Actual size

The Silver-spotted Skipper caterpillar is green gold, banded transversely with olive-green spots and streaks. There is a transverse, black bar and broken black lines on each segment. There are bright orange, false eyespots on the black head. The first segment is red orange with a black collar dorsally.

FAMILY	Hesperiidae
DISTRIBUTION	Western North America, from British Columbia to Baja California
HABITAT	Open oak woodlands
HOST PLANTS	Oak (*Quercus* spp.)
NOTE	Oak leaf-folding caterpillar of a spring-flying butterfly
CONSERVATION STATUS	Not evaluated, but common

ADULT WINGSPAN
1¹⁵⁄₁₆–1¾ in (40–44 mm)

CATERPILLAR LENGTH
1–1⅛ in (25–28 mm)

ERYNNIS PROPERTIUS

PROPERTIUS DUSKYWING

(SCUDDER & BURGESS, 1870)

84

Actual size

Propertius Duskywing caterpillars hatch through the top of eggs laid singly on both surfaces of young oak leaves and buds some seven to eight days earlier. Newly hatched caterpillars construct shelters, creating leaf flaps by cutting two parallel channels in a leaf, then folding over and silking down the flap. The larvae emerge to feed primarily at night, skeletonizing nearby leaves, and rest by day in the shelter. Prepupal caterpillars form a silked cocoon in a folded leaf and overwinter within it. Pupation occurs in the spring, with adults emerging within a couple of weeks.

Survival of the caterpillars is based on concealment, although some predators, such as minute pirate bugs (Anthocoridae), are able to enter the shelters. There is usually a single annual generation of Propertius Duskywing butterflies appearing in early spring, but sometimes a partial second generation occurs. Males often congregate on moist mud, and both sexes nectar on many kinds of spring flowers.

The Propertius Duskywing caterpillar is whitish green with well-defined, dorsolateral, white stripes and white spotting. The head is orange, either with indistinct brown and reddish markings or, in some individuals, bolder and distinctive markings. Such variation in head coloring is characteristic of this species.

FAMILY	Hesperiidae
DISTRIBUTION	Areas of northeastern and eastern Australia
HABITAT	Tropical and subtropical rain forests
HOST PLANTS	*Wilkiea* spp. and *Steganthera laxiflora*
NOTE	Caterpillar that feeds on tough leaves
CONSERVATION STATUS	Not evaluated, but rated not at risk in the north of its range, although of some concern (lower risk) in southern areas, according to a 2002 report by Environment Australia

ADULT WINGSPAN
2⅛–2⅜ in (54–61 mm)

CATERPILLAR LENGTH
1¾ in (45 mm)

EUSCHEMON RAFFLESIA
REGENT SKIPPER
(W. S. MACLEAY, 1826)

85

The Regent Skipper caterpillar forms a shelter by joining two leaves by silk. The leaves remain flat, one over the top of the other. The larvae feed outside the shelter, usually for only a short period after dusk, although small caterpillars may also feed in the early morning. The caterpillar takes several months to complete development, and there is only one generation a year, possibly two in northern areas, the adults being more numerous during and just after the wet season.

Pupation takes place in the final larval shelter, with the pupa horizontal but upside down, supported by a silken girdle under the mesothorax and an attachment by the cremaster. The adults are active from late afternoon to dusk. The species is the only member of its genus, and the butterfly is unique in having the same wing-linking mechanism as that of many moths— a bristle (frenulum) at the base of the hind wings that links to a hook (retinaculum) on the underside of the forewing, enabling stable flight.

Actual size

The Regent Skipper caterpillar is greenish gray with white ventrolateral and lateral stripes and two white dorsolateral stripes, the latter close together and enclosing back areas broken by white lines. It also has yellow dorsolateral patches on abdominal segments seven to ten. The prothorax and mesothorax are yellow, the latter with a pair of short, fleshy dorsolateral tubercles. The head is black with prominent white spots.

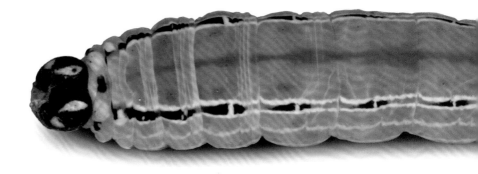

FAMILY	Hesperiidae
DISTRIBUTION	The Andes of Venezuela, Colombia, Ecuador, Peru, and Bolivia
HABITAT	Humid Andean cloud forests, especially where bamboo abounds
HOST PLANTS	Bamboo (*Chusquea* spp.)
NOTE	Skipper caterpillar that has unusually weak frass-throwing skills
CONSERVATION STATUS	Not evaluated, but unlikely to become endangered

ADULT WINGSPAN
1⅞₁₆–1½ in (36–38 mm)

CATERPILLAR LENGTH
1¹⁄₁₆–1³⁄₁₆ in (27–30 mm)

FALGA JECONIA
FALGA JECONIA
(BUTLER, 1870)

Actual size

Like the larvae of other skippers, the caterpillars of *Falga jeconia* build and inhabit tentlike shelters made from their host plant leaves. The larvae use their anal comb to fling frass away from their leafy abodes—as a form of shelter sanitation and possibly to put predators off the scent. The only detailed natural history study of *F. jeconia* found that, although the caterpillars have weak frass-throwing abilities, they are found almost exclusively over mountain streams. The fascinating possibility that this species uses mountain streams as their natural sewer system has yet to be explored.

The natural histories of the three additional species included within the genus *Falga* are completely unknown, and the yellow and black adults are generally poorly represented in collections. Indeed, even at locations where their caterpillars may be found by the thousands, adults are rarely seen, although the reasons for this are poorly understood.

The *Falga jeconia* caterpillar has a cream or ivory-colored head with contrasting red-brown mouthparts and blackish stemmata. The sides of the head are sometimes washed with reddish brown. The body often appears greenish due to ingested plant material being visible through the transparent skin; otherwise, it is whitish or yellow white. The entire body is sparsely covered with short, pale golden setae.

FAMILY	Hesperiidae
DISTRIBUTION	Southern Europe (Portugal, Spain, Italy, Greece, and Bulgaria), North Africa, Turkey, the Middle East, and southern Asia east to India
HABITAT	Hot, arid areas, including deserts, but also seasonal riverbeds, coastal areas, and dunes
HOST PLANTS	Grasses, including *Aeluropus* spp. and *Panicum* spp.
NOTE	Arid-zone caterpillar that lives on dry grasses
CONSERVATION STATUS	Not evaluated, but usually uncommon

ADULT WINGSPAN
1³⁄₁₆–1¼ in (30–32 mm)

CATERPILLAR LENGTH
1¹⁄₁₆–⅞ in (18–22 mm)

GEGENES NOSTRODAMUS
DINGY SWIFT
(FABRICIUS, 1793)

87

Dingy Swift caterpillars hatch from eggs laid singly on the grass host plants seven to ten days earlier by the female butterfly. The larvae silk together grass blades to form a tube within which they feed and rest, enabling them to avoid detection by most predators. Their development takes three to six weeks, according to temperature, and pupation occurs within a silked-together grass tube. There are two to three generations from May to October, and the species overwinters as a caterpillar. In the most southerly parts of the species' range, generations occur throughout the year.

Adult Dingy Swifts are fast-flying skipper butterflies and easily missed but fortunately have the habit of returning to the same spot for perching, so patient watching is usually rewarded. Because of the inconspicuous nature of the species, individuals may have been overlooked, and their geographic range may be greater than currently known. There are four species in the genus *Gegenes*; *G. nostrodamus* was the first to be described.

Actual size

The Dingy Swift caterpillar is light green with indistinct darker striping. Laterally, there is a low pale stripe, and the head is marked with pinkish orange and white stripes. Short, white filaments protrude from the posterior.

FAMILY	Hesperiidae
DISTRIBUTION	Western United States, west of the Rockies, from Washington State to northern Mexico
HABITAT	Riverbanks, roadsides, and canyons in arid country
HOST PLANTS	Mallows, including Common Mallow (*Malva sylvestris*), Cheeseweed (*Malva parviflora*), and Hollyhock (*Althaea rosea*)
NOTE	Nocturnal-feeding, nest-building, green caterpillar
CONSERVATION STATUS	Not evaluated, but usually common

ADULT WINGSPAN
1¾–2 in (45–50 mm)

CATERPILLAR LENGTH
1–1⅛ in (25–28 mm)

HELIOPETES ERICETORUM
NORTHERN WHITE SKIPPER
(BOISDUVAL, 1852)

88

Northern White Skipper caterpillars hatch from eggs laid singly during summer on the underside of terminal leaves of the host plant. Development from egg-laying to pupation takes about seven weeks, with the pupal period lasting up to four weeks, even under warm conditions. Early instars hide in the furls of young terminal leaves of mallows and construct small, folded leaf nests silked together at the leaf edges. These instars feed halfway through leaves, creating "window panes," with round holes produced later as the larvae mature. Nests become larger as the caterpillars grow, with pupation taking place in the final nest.

Feeding is mostly nocturnal, and larvae shoot frass away from themselves to confuse enemies. Predation by pirate bugs (Anthocoridae) is common, however. The fast-flying, relatively large skipper adults may undertake seasonal migrations in spring and fall. The males are often seen sipping moisture from muddy patches, while females are frequently spotted on flowers.

The Northern White Skipper caterpillar is variegated pale bluish green and pale yellow. There are numerous white speckles on the body, and the many pale setae are long enough to appear shaggy. The black head is densely clothed in setae, creating a hoary appearance. The collar is white with small, brown speckles.

Actual size

FAMILY	Hesperiidae
DISTRIBUTION	Western North America, from British Columbia and Alberta south to New Mexico
HABITAT	Grassland prairies to mountain meadows
HOST PLANTS	Grasses, including brome (*Bromus* spp.), ryegrass (*Lolium* spp.), and bluestem grass (*Andropogon* spp.)
NOTE	Caterpillar whose prepupal dormancy protects the adult from extreme heat
CONSERVATION STATUS	Not evaluated, but common in most areas

ADULT WINGSPAN
1–1³⁄₁₆ in (25–30 mm)

CATERPILLAR LENGTH
1³⁄₁₆–1¼ in (30–32 mm)

HESPERIA COLORADO
WESTERN BRANDED SKIPPER
(SCUDDER, 1874)

89

Western Branded Skipper eggs are laid singly at the base of grasses and on nearby surfaces. The eggs generally remain dormant and overwinter, although some hatch into first instars, which then overwinter, protected by a few strands of silk. The caterpillars develop in spring, and in early summer reach the final instar, when they stop feeding and remain dormant for between four and six weeks before pupating. The late instar dormancy appears to be a mechanism for avoiding adult emergence in hot conditions. The adults eclose in late summer to early fall.

The *Hesperia colorado* caterpillar may have five or six instars and lives in untidy tubular grass nests comprised of blades silked together. Pupation takes place in a silken cocoon within the final nest. The cocoon is liberally decorated with flocculent material produced by ventral abdominal glands, which appears to be a way of repelling moisture. Caterpillar defense is based on concealment, although pirate bugs (Anthocoridae) often invade the nests to attack the larvae.

The Western Branded Skipper caterpillar is olive brown to gray orange with pale patches and stripes on the light brown head. Five transverse ridges occur on the posterior half of each segment. There is a black dorsal collar on the first segment with anterior white edging.

Actual size

FAMILY	Hesperiidae
DISTRIBUTION	Southern Canada, northern United States, Eurasia, and North Africa
HABITAT	Alpine meadows and grasslands
HOST PLANTS	Sheeps Fescue (*Festuca ovina*)
NOTE	Slow-developing caterpillar that builds grass nests
CONSERVATION STATUS	Not evaluated, but vulnerable and threatened in some locations

ADULT WINGSPAN
1–1³⁄₁₆ in (25–30 mm)

CATERPILLAR LENGTH
1³⁄₁₆–1¼ in (30–32 mm)

HESPERIA COMMA

COMMON BRANDED SKIPPER

(LINNAEUS, 1758)

90

Common Branded Skipper caterpillars hatch in spring from overwintering eggs laid singly on grasses or on nearby surfaces. The larvae eat the tops of the eggs, leaving the remainder intact, and then feed on the edges of grass blades. Development from first to sixth instar in North American populations takes only four weeks, with the sixth instar entering dormancy for about two and a half months during summer. This dormancy is a mechanism to avoid adult emergence in hot conditions and may not occur in cooler northern habitats, for example in England, where the species is known as the Silver-spotted Skipper.

Grass blades silked together are used as nests throughout development, with increasing complexity in each successive instar. Frass is generally stored within shelters, not flung out as is the case with some other skipper caterpillars. Pupation takes place in a silken cocoon within the final larval nest lined with a flocculent material produced by ventral abdominal glands.

Actual size

The Common Branded Skipper caterpillar is dark brown with a reddish-purple cast and an indistinct, middorsal, dark stripe. The head is dark brown black with two vertical, pale orange stripes and an inverted V at its base. A black collar anteriorly edged in white is present on the first segment.

FAMILY	Hesperiidae
DISTRIBUTION	Western North America, from British Columbia to New Mexico
HABITAT	Sageland, canyons, and xeric grasslands
HOST PLANTS	Grasses, including brome (*Bromus* spp.), meadow grass (*Poa* spp.), and needle grass (*Stipa* spp.)
NOTE	Caterpillar that becomes dormant in summer heat
CONSERVATION STATUS	Not evaluated, but usually common

ADULT WINGSPAN
1⅜–1⁷⁄₁₆ in (35–37 mm)

CATERPILLAR LENGTH
1³⁄₁₆–1¼ in (30–32 mm)

HESPERIA JUBA
JUBA SKIPPER
(SCUDDER, 1874)

91

The adult female Juba Skipper lays her eggs singly on inflorescences of dead grasses or at the base of grasses, and sometimes on soil and other substrates. Caterpillars generally hatch from the eggs after ten days, although eggs laid in late fall will overwinter. Some first and second instar larvae also overwinter. Development is completed in early spring, and the first generation of adults fly in April to May. Caterpillars of the second generation develop quickly until the fifth instar, which then enters summer dormancy, resting in silken grass shelters and not feeding. Second-generation adults emerge in late August to September.

Larval defense is based on concealment, although pirate bugs (Anthocoridae) often enter nests and kill larvae. Pupation occurs in the final larval nest in a silken, flocculent-lined, moisture-repelling cocoon. The adult Juba Skipper is commonly seen feeding on various flower blooms in spring and fall, with males sometimes imbibing moisture from damp soil.

Actual size

The Juba Skipper caterpillar is dark, orangish brown, peppered with tiny, black setae, and has six, well-defined, transverse folds on the posterior half of each segment. The textured black head has pale, vertical, parallel stripes with a pale inverted V at their base. There is a dorsal, anteriorly white-margined, black collar on the first segment.

FAMILY	Hesperiidae
DISTRIBUTION	Eastern Australia
HABITAT	Gullies and swamps, where the host plant occurs as a dense understory
HOST PLANTS	Tall Saw-sedge (*Gahnia clarkei*)
NOTE	Caterpillar that occurs in colonies, often several per sedge clump
CONSERVATION STATUS	Not evaluated, but locally common in suitable habitat

ADULT WINGSPAN
1¼–1⁷⁄₁₆ in (32–36 mm)

CATERPILLAR LENGTH
1⁹⁄₁₆ in (39 mm)

HESPERILLA PICTA
PAINTED SEDGE-SKIPPER
(LEACH, 1814)

The Painted Sedge-skipper caterpillar constructs a loose, cylindrical, vertical shelter in young foliage near the top of sedge. The larva rests in the shelter during the day and emerges at night to feed at the end of the leaves. The larvae overwinter when small. Pupation occurs in the final larval shelter, usually a loose structure high in the plant, close to a stem, and with an opening at the top. The pupa, like the caterpillar, is vertical, with the head up. There are no more than two generations a year.

The genus contains 13 species, all endemic to the Australian fauna region, with caterpillars of all species feeding on sedges. The colorful, fast-flying adults feed at flowers but bask on the sedge in sunlit areas with their wings partially open. They generally stay close to their breeding areas, where their caterpillar food plants form dense stands in tall, open forests.

Actual size

The Painted Sedge-skipper caterpillar is yellowish green with a dark gray middorsal line, a white subdorsal line, a white dorsolateral line, and faint white, transverse lines. The posterior segment is tinged pink. The head is pale brown with a narrow, pale reddish-brown, median longitudinal, V-shaped band.

FAMILY	Hesperiidae
DISTRIBUTION	Southeastern United States, west to Arizona, south through Mexico and Central America to Colombia and Venezuela
HABITAT	Forest edges and clearings
HOST PLANTS	Various grasses, including St. Augustine Grass (*Stenotaphrum secundatum*), Silver Plumegrass (*Erianthus alopecuroides*), Corn (*Zea mays*), *Echinochloa povietianum*, and *Sorghum* spp.
NOTE	Cryptic larva that hides inside a rolled grass leaf
CONSERVATION STATUS	Not evaluated, but common

ADULT WINGSPAN
1¼–1¾ in (32–45 mm)

CATERPILLAR LENGTH
1⁹⁄₁₆–2 in (40–50 mm)

LEREMA ACCIUS

CLOUDED SKIPPER

(J. E. SMITH, 1797)

93

The Clouded Skipper caterpillar, a minor pest of corn and other economically important grasses, hatches from eggs laid singly on the host plant and consumes the empty eggshell. To construct a shelter, it makes a groove in the side of the leaf, causing it to curl, then connects its edges using silk threads to create a tube. The mature caterpillar uses the entire length of the leaf, fastening it into a tube with six or seven silk bands. By day, the caterpillar stays inside, sometimes feeding on the ends of the tube. At night, it ventures out and may eat its own tube and other leaves as well. If its feeding causes too much damage to its shelter, the caterpillar makes a new one.

When crawling, the caterpillar moves its head from side to side, laying down silk fibers, as it is unable to cling to the naked leaf surface. It ejects frass, expelling it some distance—a strategy designed to mislead parasitic wasps and predators, which use the scent of frass to locate their victims. While this species is probably tropical in origin, its northern limit changes with the climate; adults fly north during the summer.

Actual size

The Clouded Skipper caterpillar is light green and covered with minute, shiny hairs and a glaucous, frost-like overcolor. It has a darker dorsal line and subdorsal line and lighter-colored spiracles. The surface of the head is granular, and the background color is white, with one median and four paired (two frontal and two lateral), reddish-brown, vertical stripes.

FAMILY	Hesperiidae
DISTRIBUTION	Eastern Australia
HABITAT	Moist coastal heathlands with sandy soils
HOST PLANTS	Native iris (*Patersonia* spp.)
NOTE	Caterpillar that is covered in white, waxy powder
CONSERVATION STATUS	Not evaluated, but can be locally common

ADULT WINGSPAN
1¼ in (32 mm)

CATERPILLAR LENGTH
1 in (25 mm)

MESODINA HALYZIA
EASTERN IRIS-SKIPPER
(HEWITSON, 1868)

94

Actual size

The Eastern Iris-skipper caterpillar hatches from an egg laid singly on the host plant. It builds an elaborate, cylindrical shelter using silk and three or more leaves from the plant. The leaves are drawn together into a vertical position and bound and heavily lined with silk—a process that can take several days. Larger shelters are constructed as the caterpillar grows. It rests in the shelter, head down during the day, then emerges from the opening at the bottom of the shelter at dusk to feed on the nearby foliage. The characteristic V-shaped notch in the leaves indicates the presence of the caterpillar. There is probably only one generation a year in the south of the range but two in northern areas.

Generally, there is only one caterpillar per plant. Pupation occurs in the final shelter, the pupa suspended head down and attached by the cremaster to the silk-lined shelter. Adults feed at flowers but generally stay close to their breeding areas. The genus contains five species, all endemic to Australia and all using *Patersonia* species as food plants.

The Eastern Iris-skipper caterpillar is pale green with a dense covering of white, waxy powder. The head is black, covered with white, waxy powder, and has numerous long, white hairs.

FAMILY	Hesperiidae
DISTRIBUTION	Scattered areas in southern Australia
HABITAT	Low rainfall, open eucalyptus woodlands with a heath understory
HOST PLANTS	Mainly sword-sedges (*Lepidosperma* spp.)
NOTE	Caterpillar that constructs tubular shelters from many leaves
CONSERVATION STATUS	Not evaluated, but localized and generally uncommon

ADULT WINGSPAN
1⅜ in (35 mm)

CATERPILLAR LENGTH
1¾ in (44 mm)

MOTASINGHA TRIMACULATA
LARGE BROWN SKIPPER
(TEPPER, 1882)

95

The Large Brown Skipper caterpillar hatches from an egg laid singly on its food plant in late spring and early summer. It constructs a vertical shelter by joining and twisting 20 or more of the needlelike *Lepidosperma* leaves with silk. The shelter is open at the top, and the caterpillar rests in the shelter during the day, emerging at night to feed on tips of the foliage. New shelters are built as the caterpillar grows. On young plants, the shelter is close to the ground, but on larger clumps of the sedge it is midway up the plant. There is only one generation a year.

The caterpillar overwinters and completes growth by late winter to mid-spring, then pupates in the final shelter, the head orientated upward. Adults fly in the warmer months of spring and summer, and the males exhibit strong hilltopping behavior where they set up mating territories. The entire subfamily to which this species belongs is restricted to the Australian faunal region. The genus *Motasingha* contains two species, both confined to southern or western areas of Australia.

Actual size

The Large Brown Skipper caterpillar is semitranslucent, yellowish green or olive green, with a dark green middorsal line down the abdomen. The prothoracic plate and the anal plate are reddish brown, the latter with brown spots and several white posterior setae. The rugose head and mouthparts are black.

FAMILY	Hesperiidae
DISTRIBUTION	North Africa, southern Europe, and eastern Europe
HABITAT	Hot, dry meadows and stony grasslands, and Mediterranean maquis and garrigue scrubland up to 5,250 ft (1,600 m) elevation
HOST PLANTS	*Phlomis* spp.
NOTE	Caterpillar that rests in leaf shelters to avoid extreme heat
CONSERVATION STATUS	Not evaluated, but in decline due to loss of habitat

ADULT WINGSPAN
⅞–1¹/₃₂ in (22–26 mm)

CATERPILLAR LENGTH
⁹/₁₆–¾ in (15–20 mm)

96

MUSCHAMPIA PROTO
LARGE GRIZZLED SKIPPER
(OCHSENHEIMER, 1816)

Actual size

Large Grizzled Skipper females lay their eggs near the base of the host plant and sometimes on stones close by. Unusually for skippers the egg overwinters, and the caterpillars hatch and feed the following spring, building leaf-and-silk shelters near the tips of shoots of the host plant. Although some pupate in early summer, many mature larvae, to avoid the hottest months of the year, spin a protective cocoon between host plant leaves a short distance above the ground and rest within it for an extended period before pupating. This strategy is known as prepupal dormancy.

As a result, the butterflies may be on the wing at various times from April to October. Early fliers are often the only species active during the summer heat. There is one generation annually. The Large Grizzled Skipper, also known as the Sage Skipper, is declining due to loss of its dry habitat through development and loss of traditional patterns of farming and land management.

The Large Grizzled Skipper caterpillar has a brown head and white, mottled body. Both the head and body are covered in short, white setae. There is a single dark dorsal stripe.

FAMILY	Hesperiidae
DISTRIBUTION	Eastern Australia
HABITAT	Rain forests and open eucalypt forests, both on the coast and in nearby mountains
HOST PLANTS	More than 20 species from eight families, commonly Kurrajong (*Brachychiton populneus*) and Butterwood (*Callicoma serratifolia*)
NOTE	Caterpillar that, when young, can remain dormant for many months
CONSERVATION STATUS	Not evaluated, but common throughout much of its range

ADULT WINGSPAN
1⁹⁄₁₆–1⁵⁄₈ in (39–41 mm)

CATERPILLAR LENGTH
1¼ in (32 mm)

NETROCORYNE REPANDA
BRONZE FLAT
FELDER & FELDER, [1867]

97

Bronze Flat caterpillars hatch from eggs that are laid singly, generally on the upper surface of the new season's larger leaves. These larvae construct a shelter on the upper surface of the leaf by upending and hinging a circular disk cut from the edge of the leaf. The caterpillar secures the disk with silk and may remain inactive for months. Feeding on leaves begins at night during the spring. Larger caterpillars form twisted cylindrical shelters out of single leaves, the shelter having a distinctive exit hole at the top.

The larva pupates within this shelter. The leaf forming the shelter dies, but the base of the petiole is fastened to the twig with silk, preventing the shelter containing the pupa from falling. In the cooler southern parts of the range there is only one generation a year, but the generation time is variable, even in warmer regions. The adult butterflies fly fast and rest on sunlit foliage with their wings spread fully out.

Actual size

The Bronze Flat caterpillar is bluish gray with broad, lateral bands of yellow, black, and gray, and a middorsal black line edged in white. The prothorax and seventh to ninth abdominal segments are yellow with dorsal and lateral black spots, and the head is black with a rugose surface.

FAMILY	Hesperiidae
DISTRIBUTION	The Andes of Venezuela, Colombia, Ecuador, Peru, and Bolivia
HABITAT	Andean cloud forests and adjacent second growth
HOST PLANTS	Brambles (*Rubus* spp.)
NOTE	Dull-colored caterpillar that builds a fascinating leaf shelter
CONSERVATION STATUS	Not evaluated, but unlikely to be endangered

ADULT WINGSPAN
1½–1⅝ in (38–41 mm)

CATERPILLAR LENGTH
1–1³⁄₁₆ in (25–30 mm)

NOCTUANA HAEMATOSPILA
RED-STUDDED SKIPPER
(FELDER & FELDER, 1867)

98

At all larval stages, caterpillars of the Red-studded Skipper build shelters out of their host plant leaves, leaving them to feed and hiding within them during much of the day. First instars make a roughly round or oval-shaped cut that begins away from the leaf margin; this section of leaf is then flipped onto the dorsal surface of the leaf, rather like a manhole cover, and tightly silked to the leaf surface. Later instars excise a roughly trapezoidal-shaped section of leaf from the margin and silk this piece to the upper surface of the leaf.

The adult Red-studded Skipper is a flashy and often-illustrated species, fairly common across its broad range. The head capsules of newly molted larvae are pale cream or ivory colored prior to hardening, and larvae remain safely within their leaf shelters during this period. Larvae of all ages forcibly expel frass from the anus, with later instars documented to expel over a distance of more than 3 ft (1 m). The life cycle, from oviposition to eclosion, lasts around 135 days.

The Red-studded Skipper caterpillar
has a shiny, black to deep brown, roughly heart-shaped head with short, pale setae visible only under a dissecting microscope. The body is dull orange brown or orange green, and is sparsely covered with tiny, pale spots. The pronotal shield is shiny black, roughly rectangular, and well developed. Laterally, there is a thin, orange spiracular line running along most of the abdomen.

Actual size

FAMILY	Hesperiidae
DISTRIBUTION	Western North America, from British Columbia to Baja California, and east to Colorado
HABITAT	Most grassy areas, including forest roadsides, meadows, yards, and shrub-steppes, from sea level to 8,200 ft (2,500 m) elevation
HOST PLANTS	Grasses, including Bermuda Grass (*Cynodon dactylon*), Common Wild Oat (*Avena fatua*), and Bluebunch Wheatgrass (*Pseudoroegneria spicta*)
NOTE	Night-feeding caterpillar that lives in concealed grass-blade nests
CONSERVATION STATUS	Not evaluated, but usually common throughout its range

ADULT WINGSPAN
1–1³⁄₁₆ in (25–30 mm)

CATERPILLAR LENGTH
¾–1 in (20–25 mm)

OCHLODES SYLVANOIDES
WOODLAND SKIPPER
(BOISDUVAL, 1852)

99

Woodland Skipper caterpillars hatch after seven to ten days from eggs laid singly on the underside of dead grass blades. They consume their eggshells and, without further feeding, construct overwintering shelters by tying the edges of a grass blade together with silk strands. In the spring, the larvae take five weeks to develop to the final instar, which then lasts about another month. Feeding caterpillars build nests by pulling together the edges of a grass blade and tying it into a tube, forcibly expelling frass to confuse predators. Larvae leave their nest at night to feed on its tip and sides.

Pupation occurs in a newly tied, grass-blade nest containing much flocculent material. Here, the pupa hangs with its ventral side upward, attached to the nest top by a silk thread and cremaster. Adult Woodland Skippers can be abundant, sometimes seen in their hundreds visiting late season nectar sources such as rabbitbrush (*Ericameria* spp., *Chrysothamnus* spp.) and knapweed (*Centaurea* spp.).

The Woodland Skipper caterpillar is olive green to cinnamon brown with a dark dorsal stripe. Laterally, there are one or two olive-brown stripes bordered with white. Numerous tiny, black setae cover the body. The head is bifurcated dorsally and is orange brown (sometimes white) with black markings.

Actual size

FAMILY	Hesperiidae
DISTRIBUTION	Southeastern Australia, including northwest Tasmania
HABITAT	Generally open woodland in alpine environments at 3,300–5,250 ft (1,000–1,600 m) elevation, but near sea level in Tasmania
HOST PLANTS	Tall Sedge (*Carex appressa*) and occasionally other *Carex* spp.
NOTE	Caterpillar that overwinters in a leaf shelter, sometimes under snow
CONSERVATION STATUS	Not evaluated, but vulnerable in Tasmania, although locally common on the mainland

ADULT WINGSPAN
1–1³⁄₁₆ in (25–30 mm)

CATERPILLAR LENGTH
1³⁄₁₆ in (30 mm)

OREISPLANUS MUNIONGA
ALPINE SEDGE-SKIPPER
(OLLIFF, 1890)

Actual size

The Alpine Sedge-skipper caterpillar hatches from an egg laid singly on the underside of a host plant leaf. It constructs a cylindrical shelter by joining several leaf tips together with silk, leaving the shelter open at the top. When it outgrows the first shelter, it rebuilds. The caterpillar develops slowly during the fall to early summer, when conditions are cool and the plants may be under snow. Feeding occurs at night on leaves above the shelter, and the larva rests in the shelter during the day. Pupation occurs in early summer in the final shelter, which is usually lower in the sedge clump. There is only one generation a year.

Oreisplanus munionga caterpillars suffer heavy parasitism, with a large, orange ichneumonid wasp emerging from many pupae. Adults fly close to their food plants and breed in areas where the plant density is high. The species belongs to a subfamily of skipper butterflies (Trapezitinae) that is restricted to the Australian region. There are two species in the genus, both of which occur in temperate areas.

The Alpine Sedge-skipper caterpillar is greenish brown with a dark middorsal line and white subdorsal and lateral lines. There are a few pale hairs at the end of the abdomen. The head is brown with a median, longitudinal, V-shaped band, bordered by a narrow brown band and a black dorsolateral band.

FAMILY	Hesperiidae
DISTRIBUTION	Across Africa, the Middle East, Greece (Rhodes and Kos), southwest Turkey, and into northwest India
HABITAT	Damp forests and forest edges, wetlands, parks, and occasionally grasslands
HOST PLANTS	*Imperata arundinacea*, *Ehrharta erecta*, and *Panicum miliaceum*
NOTE	Green caterpillar that lives its life in a leaf tube
CONSERVATION STATUS	Not evaluated

ADULT WINGSPAN
1⅞ in (40 mm)

CATERPILLAR LENGTH
1⅜–1⁹⁄₁₆ in (35–40 mm)

PELOPIDAS THRAX
MILLET SKIPPER
(HÜBNER, [1821])

101

Millet Skipper caterpillars hatch from round eggs laid singly on leaves by the female butterfly. The young caterpillars each construct a shelter from a single leaf, rolling the edges into a tube held in place by silk. They feed on one edge of the leaf and make a new roll as they run out of food. As they develop, the larvae use several leaves for their tube shelters. The caterpillars develop through six instars and then pupate within the leaf tube.

In parts of the species' range there are two generations. The first is on the wing from June to July and the second from September to October, although in tropical Africa the Millet Skipper, also know as the White-branded Swift, is seen for much of the year. The adult is a fast-flying butterfly seen around flowers and puddles, with aggressive males that hilltop and may defend their territories. The species is known to migrate. The genus *Pelopidas* comprises ten species, mostly found in Africa and South Asia.

Actual size

The Millet Skipper caterpillar is mostly green with a banded brown head. The thorax is yellow green, and there are several yellow bands on the abdomen, a dark dorsal stripe, and several pale lateral stripes. Many tiny spots create a mottled appearance.

FAMILY	Hesperiidae
DISTRIBUTION	Southern Canada, United States, and Mexico
HABITAT	Watercourses, parks, shrub-steppe, waste ground, and field edges
HOST PLANTS	Lambsquarters (*Chenopodium album*), Russian Thistle (*Salsola kali*), and pigweed (*Amaranthus* spp.)
NOTE	Caterpillar that feeds and builds nests on common garden weeds
CONSERVATION STATUS	Not evaluated, but usually common

ADULT WINGSPAN
1–1³⁄₁₆ in (25–30 mm)

CATERPILLAR LENGTH
¾–1 in (20–25 mm)

102

PHOLISORA CATULLUS
COMMON SOOTYWING
(FABRICIUS, 1793)

Common Sootywing adult females lay well-camouflaged eggs singly, usually on the upperside of an older, mid-sized leaf. Caterpillars hatch five to six days later, and development to pupation takes only 22 days with about four days in each instar. The larvae rest in the upper new growth of host plants, in nests built by cutting a leaf inward in two places, then folding the loose flap over and silking it in place along the margin. Older instars bend entire leaves inward, toward the midvein, fastening them in place with silk.

The larvae feed nocturnally on leaf edges away from the shelter, and plant leaves near nests show considerable feeding damage. Older instars shoot frass to deter predators. The final instar overwinters, with pupation taking place in spring. In the north of the species range, there are two generations a year with adults flying from April to September. In the south, breeding and flight continue year-round.

The Common Sootywing caterpillar is medium to dark green, yellower anteriorly with numerous white spots. The setae on the head and body are short and plentiful, imparting a fuzzy appearance. There is an indistinct, dark, middorsal stripe. The head on segment one is black, and the collar is black, edged anteriorly in white.

Actual size

FAMILY	Hesperiidae
DISTRIBUTION	Southern Rocky Mountains, south to western Mexico
HABITAT	Mountain valley bottoms
HOST PLANTS	Hay grasses (tall, wide-leafed grasses, such as *Dactylis* spp. and *Poa* spp.)
NOTE	Caterpillars that build silked tube nests in grasses
CONSERVATION STATUS	Not evaluated, but common in most of its range

ADULT WINGSPAN
1¼–1¹¹⁄₁₆ in (32–43 mm)

CATERPILLAR LENGTH
1³⁄₁₆ in (30 mm)

POANES TAXILES
TAXILES SKIPPER
(W. H. EDWARDS, 1881)

103

The Taxiles Skipper caterpillar eats any of several dozen species of tall, skinny, wide-leaf grasses mostly growing at the bottom of valleys—the kinds of grasses farmers grow for hay. The larva attaches silk threads on each side of the top of a wide leaf, and each thread shrinks and makes the leaf curl. When the leaf is curled into a tube, the edges are silked shut to create a leaf-tube nest where the caterpillar rests when not feeding. Half-grown caterpillars hibernate in a silk nest, develop further in the spring, and then pupate in another silked-leaf nest.

Like other grass-feeding skippers, and also satyrids (Satyrinae), the caterpillar has mandibles without teeth (like scissor blades) to cut through tough grass leaves; other species have toothed mandibles to saw through leaves. A single generation of adults emerges in early summer. The adults, named for their darting flight, are avid flower visitors commonly found near moist oases at the bottom of desert canyons. There are more than 3,500 skipper species worldwide.

Actual size

The Taxiles Skipper caterpillar is brownish green to reddish tan with hundreds of tiny, reddish dots, some weak longitudinal stripes, a narrow, black collar, and orange-brown head. The dorsal stripe is usually the most prominent, and the head is distinctly pubescent. The terminal segment is usually browner than the rest.

FAMILY	Hesperiidae
DISTRIBUTION	Western United States (Washington State, Oregon, and California)
HABITAT	Coastal prairies, pine savannahs, woodland openings, and heath meadows
HOST PLANTS	Grasses, including Idaho Fescue (*Festuca idahoensis*), Red Fescue (*Festuca rubra*), and California Oatgrass (*Danthonia californica*)
NOTE	Nest-building, night-feeding caterpillar
CONSERVATION STATUS	Not evaluated, but when assessed for the US Endangered Species Act was judged to be potentially threatened, though secure within occupied habitats

ADULT WINGSPAN
¾–1 in (20–25 mm)

CATERPILLAR LENGTH
1³⁄₁₆–1 in (21–25 mm)

104

POLITES MARDON
MARDON SKIPPER
(W. H. EDWARDS, 1881)

Mardon Skipper caterpillars hatch from eggs laid singly at the base of grasses. They develop quickly, pupating five to six weeks after egg hatch. Early instars eat through grass blades partially from the side, while later instars eat through the tender grass tips, causing much of the blade to drop to the ground. Nests are untidy, vertical shelters created by silking blades together, and caterpillars leave them at night to feed. Frass is stored within shelters—possibly to conceal it from predators. Prepupal larvae construct a stronger, final shelter, typically horizontal and near the ground, for pupation. Overwintering occurs either as caterpillar or pupa.

The Mardon Skipper was recently considered for endangered species listing but subsequently found to be secure in its occupied habitats. Adults fly low to the ground, feeding on flowers such as dandelions (*Taraxacum* spp.), vetches (*Vicia* spp.), and asters (*Aster* spp.). There is only one generation annually, and the flight period is two to four weeks.

Actual size

The Mardon Skipper caterpillar is dark brown to gray black with limited white speckling and a middorsal, black stripe. Each segment has three transverse folds posteriorly. The head is black with prominent white-tan, parallel stripes and smaller stripes laterally. The collar on segment one is black, edged in white anteriorly.

FAMILY	Hesperiidae
DISTRIBUTION	Areas of North America, northern Europe, and northern Asia
HABITAT	Alpine, mountain tops above the tree line, talus slopes, and also moist meadows
HOST PLANTS	Cinquefoil (*Potentilla* spp.) and wild strawberry (*Fragaria* spp.)
NOTE	Mountaintop caterpillar taking up to two years to complete development
CONSERVATION STATUS	Not evaluated, but never occurs in large numbers

ADULT WINGSPAN
1–1³⁄₁₆ in (25–30 mm)

CATERPILLAR LENGTH
⅞–1 in (23–25 mm)

PYRGUS CENTAUREAE
ALPINE GRIZZLED SKIPPER
(RAMBUR, [1842])

105

Adult female Alpine Grizzled Skippers lay eggs singly on the underside of host plant leaves, where they hatch after nine to ten days. The first caterpillar instar chews a hole at the top of the egg, through which it emerges leaving the rest of the shell uneaten. Caterpillars build folded-leaf nests throughout development, in which they remain except when feeding on adjacent leaves, mostly by night, always returning to shelters. They eject frass from nests, presumably to help divert predators, although pirate bugs (Anthocoridae) seem not to be duped by this ploy and likely take a significant number of larvae.

In some locations, complete development of the species may last two years. In captivity, development from the second to the fifth instar takes about six weeks, with growth in the fifth instar slowing and pupation occurring about a month later. A mature caterpillar or pupa may overwinter. The single generation of Alpine Grizzled Skippers flies during June to July in areas of high elevation.

Actual size

The Alpine Grizzled Skipper caterpillar is orange brown, and there are two indistinct, dark dorsal lines. The first segment collar is brown, and the posterior segment is orange, as are the spiracles. The head is dark brown to black and densely clothed in pale setae.

FAMILY	Hesperiidae
DISTRIBUTION	Europe, from southern and central England, east into Turkey and Russia; also Mongolia, northeast China, and Japan
HABITAT	Chalk grassland, open woodlands with glades and tracks, cuttings and embankments, and derelict quarries
HOST PLANTS	Rosaceae, including *Agrimonia* spp., *Alchemilla* spp., and *Potentilla* spp.
NOTE	Greenish caterpillar that feeds and rests within its leaf shelter
CONSERVATION STATUS	Not evaluated, but experiencing population decline in some parts of its range

ADULT WINGSPAN
1–1⅛ in (25–28 mm)

CATERPILLAR LENGTH
1¹¹⁄₁₆ in (18 mm)

106

PYRGUS MALVAE
GRIZZLED SKIPPER
(LINNAEUS, 1758)

Grizzled Skipper caterpillars hatch from dome-shaped eggs laid singly on the underside of leaves of the host plant. The young larva spins a thin silken web over the upper surface of leaves in which it shelters and feeds. The older caterpillar, which rests much of the time, lives within a folded leaf that is secured in place by silken threads. It grows relatively slowly, reaching its final instar after about two months. The larvae pupate near the base of the plant within a loose cocoon. The pupa overwinters, and the adult emerges the following spring.

There is usually a single generation on the wing between May and June, but in good years there may be a second generation in late summer. Overall, the species is in decline, mostly as a result of the intensification of farming and the lack of traditional management techniques such as coppicing and livestock grazing, which have led to a loss of host plants.

The Grizzled Skipper caterpillar is pale green with bands of straw-yellow and several yellow-green stripes on the dorsal and lateral surfaces. Numerous tiny, yellow dots create a mottled appearance. The head is black. Both the head and body are covered in short, white hairs.

Actual size

FAMILY	Hesperiidae
DISTRIBUTION	The Andes of Colombia, Ecuador, Peru, and Bolivia
HABITAT	Humid, mid-elevation cloud forests, forest edges, and regenerating forests
HOST PLANTS	*Vismia* spp.
NOTE	Caterpillar with white hairs often stained orange by *Vismia* sap
CONSERVATION STATUS	Not evaluated, but not likely to be endangered

PYRRHOPYGE PAPIUS
SHOULDER-STREAKED FIRETIP
HOPFFER, 1874

ADULT WINGSPAN
2–2³⁄₁₆ in (50–56 mm)

CATERPILLAR LENGTH
1⅞–2⅛ in (48–54 mm)

107

The strikingly banded caterpillar of the Shoulder-streaked Firetip hides away inside a cleverly crafted leaf shelter while not feeding. Small caterpillars carefully excise a manhole-shaped circle from the center of a leaf, then feed on the isolated tissue of this flap, which enables them to avoid gumming up their mandibles with the thick, bright orange latex of their host plant. After hatching, the rather slow-growing caterpillars may take as long as 110 days before they pupate, their rate of growth likely curbed by the compounds they ingest from their chemically defended host plant.

There are around 40 species of fat-bodied, small-winged adult firetips—*Pyrrhopyge* species—all extremely fast fliers and most frequently seen dashing about with an audible buzzing of their wings. They descend to the ground to feed at feces, rotting fruit, and urine-enriched soils, only rarely visiting flowers. For those species with described caterpillars, all are known to build larval shelters and have the ability to forcibly expel their frass away from their homes.

Actual size

The Shoulder-streaked Firetip caterpillar is dark maroon to deep red in ground color with thin, but bright orange, yellow-orange, or yellow, intersegmental stripes on the abdomen. It is covered in long, silky hairs, especially on the thorax and head, these hairs being bright white apically and crimson basally. The caterpillar's heavily armored head bears strong vertical ridges.

FAMILY	Hesperiidae
DISTRIBUTION	Southwestern United States, south to northern Mexico
HABITAT	Open pinewoods in lower mountains
HOST PLANTS	Grasses growing in large clumps, such as needlegrass (*Stipa* spp.) and Side Oats Grama (*Bouteloua curtipendula*)
NOTE	Caterpillar that is semitranslucent
CONSERVATION STATUS	Not evaluated, but apparently secure within its range

ADULT WINGSPAN
1–1³⁄₁₆ in (25–30 mm)

CATERPILLAR LENGTH
1³⁄₁₆ in (30 mm)

STINGA MORRISONI
MORRISON'S SKIPPER
(W. H. EDWARDS, 1878)

108

Actual size

The Morrison's Skipper caterpillar is dull greenish tan with a very narrow, black collar, and the body is semitranslucent, revealing some internal organs. Some *Stinga morrisoni* caterpillars also have a vague, dark dorsal stripe on the body. The bifurcated head is solid black in some individuals, varying to orange brown with several paler, vertical stripes.

Morrison's Skipper caterpillars live in large clumps of many kinds of grass, which afford them protection from predators. Each caterpillar silks several grass leaves together and remains inside that nest, where its strange, semitranslucent body attracts no attention. The caterpillar makes larger and larger vertical nests as it becomes full grown, and then hibernates in a leaf nest over the winter, pupating in the spring. There is a single annual generation, except in west Texas, where there are two generations.

Adults emerge in May, and males fly to nearby hilltops waiting for females and the opportunity to mate. Both sexes visit a variety of flowers for nectar. The butterflies have a silver arrowhead mark on the underside of each hindwing, giving rise to the alternative common name of Arrowhead Skipper. There is only one species of *Stinga*, but hundreds of other species of skipper caterpillars also eat grasses or sedges.

FAMILY	Hesperiidae
DISTRIBUTION	Timor, New Guinea, and northern and eastern Australia
HABITAT	Lowland open forests and paperbark woodlands
HOST PLANTS	Blady Grass (*Imperata cylindrica*) and Guinea Grass (*Panicum maximum*)
NOTE	Caterpillar that constructs a shelter from one rolled leaf blade
CONSERVATION STATUS	Not evaluated, but locally common in northern areas

ADULT WINGSPAN
⅞ in (22 mm)

CATERPILLAR LENGTH
1 in (25 mm)

SUNIANA LASCIVIA
DINGY GRASS-DART
(ROSENSTOCK, 1885)

109

The Dingy Grass-dart caterpillar constructs a shelter from a single leaf blade by rolling the edges and joining them with silk. This refuge is in the upper part of the leaf, and the larva emerges from the bottom of the shelter at night to feed on the leaf below. Eventually, the shelter, and the leaf above, are left drooping down, with only the uneaten midrib preventing the structure from falling. After consuming the shelter, the caterpillar will move to another leaf and construct a new one.

Pupation occurs in the final larval shelter, or a new construction at the base of the plant, but only after the larva has plugged both ends of the shelter with silk. There is only one generation a year in the south of the range, but the species breeds throughout the year in tropical areas, with several generations completed annually. There are three species in the *Suniana* genus, all of them from the same Timor, New Guinea, and Australia region.

Actual size

The Dingy Grass-dart caterpillar is pale green with a darker middorsal line. Its anal segment is rounded and has short hairs. The head is pale brown with a narrow reddish lateral band surrounding more central, elongated longitudinal patches.

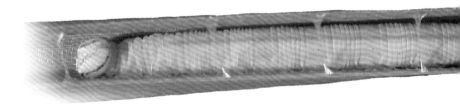

FAMILY	Hesperiidae
DISTRIBUTION	From Nicaragua south through Colombia and Amazonia to southern Brazil
HABITAT	Humid and semi-humid forests and forest borders, generally below 2,625 ft (800 m) elevation
HOST PLANTS	*Inga* spp.
NOTE	Caterpillar that builds distinctive, triangular, perforated shelters
CONSERVATION STATUS	Not evaluated, but not considered threatened

ADULT WINGSPAN
1³⁄₁₆–1⁹⁄₁₆ (30–40 mm)

CATERPILLAR LENGTH
1⁹⁄₁₆–2 in (40–50 mm)

110

TELEMIADES ANTIOPE
PLÖTZ'S TELEMIADES
(PLÖTZ, 1882)

Plötz Telemiades caterpillars hatch from eggs laid singly, usually on fresh leaves of their host and almost always on small seedlings in the understory. Their shelters are unique, beautiful, and easily recognized from a distance. Young larvae excise an elongated, delicate triangle from the leaf margin, flipping it to the top of the leaf and resting below it. The final shelter is almost invariably formed at the tip of a leaf and is also roughly triangular, but the caterpillar creates intricate channels and perforations across its surface as it develops within.

The mature caterpillar, colored in subtle complementary shades of blue, orange, and gray, pupates within its final larval shelter, forming a glossy, red-brown chrysalis. Adults are very fast fliers, almost invisible to the naked eye when at full speed and only visible when they land, wings spread wide, on the undersurface of a leaf. The genus *Telemiades* contains some 17 species with a number of subspecies, all in Central and South America.

The Plötz's Telemiades caterpillar is elongated, tapering to the front and rear, and roughly trapezoidal in cross section. It is split-toned, grayish green to dark gray above, with turquoise highlights and whitish hatch marks. Laterally it is bright orange with white and pale yellow highlights. The head is pale gray with two large, dark gray spots near the top, resembling eyes.

Actual size

FAMILY	Hesperiidae
DISTRIBUTION	The Andes of Venezuela, south to Peru
HABITAT	Edges of upper subtropical and lower temperate cloud forests
HOST PLANTS	Amazonvine (*Stigmaphyllon* spp.)
NOTE	Caterpillar that builds precise, triangular shelters from host plant leaves
CONSERVATION STATUS	Not evaluated, but not considered threatened

ADULT WINGSPAN
1⁷⁄₁₆–1¹¹⁄₁₆ in (37–43 mm)

CATERPILLAR LENGTH
1⁹⁄₁₆–1¾ in (40–45 mm)

THEAGENES ALBIPLAGA

MERCURIAL SKIPPER

(FELDER & FELDER, [1867])

111

Mercurial Skipper caterpillars, like many other larvae of the family Hesperiidae, rest within shelters they build with silk and portions of their host plant. During early instars, caterpillars of this species excise a thin strip of leaf from the margin, curling it over onto the upper surface to create a shelter. As the larvae grow larger, a new shelter is built as the smaller one is outgrown, each successive shelter becoming less elongate and more triangular. The final instars generally sew together the overlapping margins of adjacent leaves, chewing numerous holes in the walls of the otherwise completely sealed refuge.

The adults—among the most distinctive and widespread skippers in South America—are rapid fliers, zipping about erratically low to the ground and periodically settling to feed at puddles or moist sand. To date, details of the life cycle and fascinating shelter-building behavior of the larvae of *Theagenes albiplaga* have not been published.

Actual size

The Mercurial Skipper caterpillar is simply patterned, green to yellow green with small, yellowish speckling and tiny, pale setae scattered sparsely over most of the body. The head is more boldly patterned, orange and dark brown, roughly heart-shaped, and heavily reticulated with irregular bumps and grooves.

FAMILY	Hesperiidae
DISTRIBUTION	North America
HABITAT	Open woods and brushy areas
HOST PLANTS	Legumes such as wild pea (*Lathyrus* spp.) and vetch (*Vicia* spp.); also tick-trefoil (*Desmodium* spp.), bush clover (*Lespedeza* spp.), clover (*Trifolium* spp.), and Birdsfoot Trefoil (*Lotus corniculatus*)
NOTE	Caterpillar resembling many other species in the large skipper family
CONSERVATION STATUS	Not evaluated, but common in North America

ADULT WINGSPAN
1¼–1⅞ in (32–47 mm)

CATERPILLAR LENGTH
1⅜ in (35 mm)

THORYBES PYLADES
NORTHERN CLOUDYWING
(SCUDDER, 1870)

112

Northern Cloudywing caterpillars, found on numerous legumes, especially those with tendrils such as the wild pea and vetch, construct a nest by silking one or more of the host plant leaves together. The narrow neck of the caterpillar is characteristic of most Hesperiidae species, allowing them to maneuver the head to silk the inside of the leaf nest. The caterpillars venture out to feed on leaves but rest inside most of the time. They have a comb on the rear that they use to snap their dung far away, to avoid fouling the nest.

Mature, fully fed caterpillars hibernate in a silk nest in the litter until pupation. The adults fly from March to September in the south of their range, where there may be two generations, and between May and July farther north, where there is one generation. Male adults wait near the ground in tiny clearings between bushes for females to arrive for mating. The Northern Cloudywing is one of North America's most common skippers.

The Northern Cloudywing caterpillar varies from sandy tan, peppered with tiny black spots and white spots, to rich reddish brown. Two light dorsolateral lines are present, and there may be a dark dorsal stripe. The head and collar are black with a dense covering of very short setae.

Actual size

FAMILY	Hesperiidae
DISTRIBUTION	Southern Canada, northern United States, Europe (including southern United Kingdom and southern Scandinavia), North Africa, and central Asia
HABITAT	Hayfields, meadows, pastures, and grassy waste ground
HOST PLANTS	Grasses, including Reed Canary Grass (*Phalaris arundinaceae*), Timothy Hay (*Phleum pretense*), and Orchard Grass (*Dactylis glomerata*)
NOTE	Night-feeding caterpillar that rests by day in grass nests
CONSERVATION STATUS	Not evaluated, but common or abundant in most locations

ADULT WINGSPAN
¾–1 in (20–25 mm)

CATERPILLAR LENGTH
¹¹⁄₁₆–⅞ in (18–22 mm)

THYMELICUS LINEOLA
EUROPEAN SKIPPERLING
(OCHSENHEIMER, 1808)

113

Female European Skipperlings (known as Essex Skippers in the United Kingdom) lay eggs in strings of up to 20, end to end, but not quite touching, delicately glued to the concave side of a grass blade. Eggs overwinter and hatch over an extended period in spring. Larvae leave eggshells uneaten, except for the escape hole. Caterpillars construct nests in the upper third of grass hosts, where a single grass blade is pulled into a tube and stitched together with tidy silk crossties. The larvae are nocturnal feeders and spend the days inactive within nests.

Development of the caterpillars from hatching takes about seven weeks, and the adults emerge after about a week in the pupa. European Skipperlings have reached pest status on hay and pasture grasses in some parts of Canada and the eastern United States. There is also some concern that they are displacing some native skippers in parts of the western United States.

The European Skipperling caterpillar is blue green with numerous dark speckles covering the body, each with a tiny, dark seta. Body segments are strongly creased with lateral folds, and small spots are restricted anteriorly and posteriorly. The head capsule is bifurcated and whitish with dark, vertical stripes.

Actual size

FAMILY	Hesperiidae
DISTRIBUTION	Eastern Australia
HABITAT	Moist coastal and subcoastal woodlands up to 4,600 ft (1,400 m) elevation
HOST PLANTS	Mat rush (*Lomandra* spp.)
NOTE	Caterpillar that is common in urban landscapes
CONSERVATION STATUS	Not evaluated, but locally common

ADULT WINGSPAN
1⅝–1¹³⁄₁₆ in (42–46 mm)

CATERPILLAR LENGTH
1⁷⁄₁₆–1⁹⁄₁₆ in (36–40 mm)

114

TRAPEZITES SYMMOMUS
SPLENDID OCHRE
HÜBNER, 1823

Splendid Ochre caterpillars hatch from eggs that are seldom deposited on the green leaves of the food plant but more often on dry leaves, flowers, or seed heads. The larva forms an initial shelter by joining new leaves with silk. In later instars, it rolls one or more dead or green leaves into a cylindrical tube, emerging to feed on the end of the sedge blades, usually during the night. The characteristic V-notches and chewed tip of the blade often indicate the presence of a caterpillar in a shelter lower down in the sedge clump.

The caterpillars usually pupate in the final shelter or in leaf litter under the sedge. In the cooler areas of southeastern Australia, there is only one generation a year, but there are two generations in warmer regions. The genus contains 18 species and is endemic to the Australia faunal region, as is the entire subfamily (Trapezitinae). The caterpillars of *Trapezites symmomus* are the largest in the genus.

The Splendid Ochre caterpillar is cylindrical, hairless, and pinkish brown with a darker middorsal line and fainter, longitudinal subdorsal lines. The head has a rough surface, is notched at the top, and colored reddish brown, with a yellow inverted Y-shaped band.

Actual size

FAMILY	Hesperiidae
DISTRIBUTION	From eastern United States south to southern and central Argentina
HABITAT	Forest edges, meadows, and urban areas, where hosts often grow as weeds
HOST PLANTS	Pea family (Fabaceae), including Beaked Butterfly Pea (*Centrosema virginianum*), Kudzu (*Pueraria montana*), and beggar's ticks (*Desmodium* spp.); also vine legumes such as beans (*Phaseolus* spp.) and hog peanuts (*Amphicarpa bracteata*)
NOTE	Caterpillar that rests and pupates within its leaf shelter
CONSERVATION STATUS	Not evaluated, but common

ADULT WINGSPAN
1¾–2⅜ in (45–60 mm)

CATERPILLAR LENGTH
1⁹⁄₁₆–2 in (40–50 mm)

URBANUS PROTEUS
LONG-TAILED SKIPPER
(LINNAEUS, 1758)

115

The small caterpillar of the Long-tailed Skipper is a common pest of crops such as beans and ornamentals, including *Wisteria*. It can eat a substantial amount of foliage—more than 21 sq ft (2 m²)—mostly (at least 90 percent) during its last two instars. The larvae also roll leaves into shelters, in which they rest; when fully grown they line the shelter with silk and pupate in it. Such shelters protect them from visually hunting predators but do not fool parasitoids, such as *Bassus* braconid wasps, *Palmisticus* eulophid wasps, and shiny green *Chrysotachina* tachinid flies, which hunt mostly by smell, locating caterpillar droppings.

The Long-tailed Skipper frequently co-occurs with other *Urbanus* species. In the southeastern United States it flies together with *U. dorantes*, and their caterpillars feed on the same plants; in Costa Rica or northern Argentina, more than ten *Urbanus* species may be found in a small area. Adults can be difficult to tell apart, but the caterpillars have distinguishing traits, such as spots on their heads or stripes on their bodies.

The Long-tailed Skipper caterpillar is yellow green, with black speckling, a dark, middorsal line on its back, and paired, dorsal, yellow-orange lateral bands. Heads can be black, or black with red patches, but are mostly dark red with a large, central, black patch on the front and around the eyes (stemmata). The prolegs are orange red, while ventrally the caterpillar is translucent green.

Actual size

FAMILY	Hesperiidae
DISTRIBUTION	The Andes of Venezuela, Colombia, Ecuador, Peru, and Bolivia
HABITAT	Andean temperate and upper tropical forest edges and roadsides
HOST PLANTS	Various genera and species of grasses, including *Cenchrus tristachyus* and *Paspalum* spp.
NOTE	Caterpillar that builds a well-camouflaged, tubelike shelter
CONSERVATION STATUS	Not evaluated, but not likely to be endangered

ADULT WINGSPAN
1½–1¾ in (38–45 mm)

CATERPILLAR LENGTH
1–1³⁄₁₆ in (25–30 mm)

VETTIUS CORYNA
SILVER-PLATED SKIPPER
(HEWITSON, 1866)

Actual size

The Silver-plated Skipper caterpillar has a translucent, greenish-white head, with black bands running laterally from the ocular area and meeting dorsally. Two white bands run anterior to the black bands. The clypeus is bright white or yellow white, contrasting with the black mandibles. The simply patterned body is elongate, emerald green to pale yellow, and bears four narrow, pale, powdery white, longitudinal stripes.

Silver-plated Skipper caterpillars build and rest inside shelters created by making two cuts on opposite sides of a grass blade that almost meet at the midvein. The opposing leaf margins are drawn together with silk to form a shallow pocket or narrow tube. Lastly, a tiny cut is made near the base of the leaf midvein, causing the portion of the blade forming the shelter to sag downward into a vertical position. Older caterpillars have an eversible, pale, purple-red, ventral prothoracic "neck" gland. When disturbed, the caterpillar rears back onto its claspers and, with its head tipped back, everts the gland.

There are some 22 species in the genus *Vettius*, all with showy colorful butterflies. *Vettius coryna* males guard low perches throughout the day, dashing out with a flash of silver to challenge anything flying past, even species many times their size. Females search for oviposition sites during periods of full sun, touching down briefly on the upper surfaces of narrow-bladed grasses, curling their abdomen underneath the blade to lay single eggs, pausing occasionally to bask.

FAMILY	Pieridae
DISTRIBUTION	Western North America, from northwest Canada to Mexico
HABITAT	Riparian and open habitats, meadows, montane summits, slopes, canyons, and shrub-steppe
HOST PLANTS	Rockcress (*Arabis* spp.), wintercress (*Barbera* spp.), and Tumble Mustard (*Sisymbrium altissimum*)
NOTE	Cryptic caterpillar that becomes a dazzling spring butterfly
CONSERVATION STATUS	Not evaluated, but considered secure within its range

ADULT WINGSPAN
1³⁄₁₆–1⅜ in (30–35 mm)

CATERPILLAR LENGTH
¼–1 in (20–25 mm)

ANTHOCHARIS SARA

SARA ORANGETIP

LUCAS, 1852

117

Sara Orangetip females lay their eggs singly, usually one per host plant; the eggs hatch in about four days. The caterpillars develop rapidly, taking only 16 to 20 days from egg hatch to pupation. They feed preferentially on flowers, buds, and seedpods, and then move on to leaves and stems, systematically consuming the host plant as they move downward, finally eating the large basal leaves. The caterpillars often rest on seedpods or stems where their slender green bodies blend in well with the narrow, green plant parts.

The caterpillars go through five instars, and the pupa oversummers and overwinters, spending 10 to 11 months in this stage. About 10 percent of pupae take two to three years before producing adults. Orangetip butterflies are avid flower visitors, seeking nectar from spring flowering plants such as *Phlox*, mustards, and fiddlenecks. Closely related *Anthocharis* species occur elsewhere in North America and Europe, and all emerge early in spring.

Actual size

The Sara Orangetip caterpillar is light green dorsally and darker green ventrally. There is a prominent, lateral, white stripe that extends all along the body and onto the head. The body is densely clothed with short setae and tiny, black spots, and the head is green.

FAMILY	Pieridae
DISTRIBUTION	Europe, North Africa, temperate Asia, Korea, and Japan
HABITAT	Scrubby grasslands, roadsides, meadows, and woodland edges
HOST PLANTS	Blackthorn (*Prunus spinosus*), Bird Cherry (*Prunus padus*), hawthorn (*Crataegus* spp.), and Apple (*Malus pumila*)
NOTE	Communal caterpillar that falls to the ground if disturbed
CONSERVATION STATUS	Not evaluated, but usually common

ADULT WINGSPAN
2–2¾ in (50–70 mm)

CATERPILLAR LENGTH
1⁹⁄₁₆–2 in (40–50 mm)

118

APORIA CRATAEGI
BLACK-VEINED WHITE
(LINNAEUS, 1758)

Black-veined White caterpillars hatch in July from yellow, spindle-shaped eggs laid by the female butterfly in batches of 50 to 200 on the upper surfaces of host plant leaves. The young larvae feed for a while, then enter hibernation as second or third instars in September. During the early instars, caterpillars live and feed communally but gradually become more independent with age and are solitary in the final instar. Pupation occurs on a twig or branch, with the chrysalis attached vertically by the cremaster and a silken girdle.

The adult butterflies nectar at a variety of flowers, and males often congregate on urine-tainted soil or animal dung. There is a single generation annually, flying from May to August, and drier habitats are preferred. There are about 30 related species in the genus *Aporia*, most of which are limited to Southeast Asia. *Aporia crataegi* used to occur in southern England but became extinct in the 1920s.

Actual size

The Black-veined White caterpillar is black dorsally with orange-brown markings. Ventrally, the coloration is whitish, and there is a single black stripe on each side. Long, white or tan-colored setae cover the body. The head and terminal segment are black.

FAMILY	Pieridae
DISTRIBUTION	Much of Australia, Papua New Guinea, Solomon Islands, Fiji, and Indonesia (Java, West Papua)
HABITAT	Wherever food plants grow in tropical, subtropical, and temperate areas, including the arid zone of central Australia
HOST PLANTS	Caper bushes (*Capparis* spp.) and Currant Bush (*Apophyllum anomalum*)
NOTE	Caterpillar that can defoliate plants, including commercial caper crops
CONSERVATION STATUS	Not evaluated, but common

ADULT WINGSPAN
2⅛ in (55 mm)

CATERPILLAR LENGTH
1⁵⁄₁₆ in (34 mm)

BELENOIS JAVA

CAPER WHITE
(LINNAEUS, 1768)

119

Caper White caterpillars hatch in clusters of up to 100 individuals and feed gregariously on the leaves of caper bushes, leaving only the midrib. The caterpillars complete their development in about three weeks. Although many larvae die, particularly as a result of disease and parasitism from tachinid flies, over a season complete defoliation of a large tree can occur, with hundreds of butterflies produced from a single tree. Other nearby trees may remain unaffected. The caterpillars can be minor pests in commercial caper crops.

Caterpillars pupate on the leaves and stems of the food plants but might leave a defoliated plant to pupate. The pupa is attached to the plant via a silken central girdle and anal hooks into a silken pad. Adults migrate particularly during late spring, often flying hundreds of miles over a few days. The genus is large and mainly found in tropical Africa and Southwest Asia, with only this species occurring in Southeast Asia and Australia.

The Caper White caterpillar is cylindrical, brown or olive green with numerous small, yellow, raised dots on the head and body from which arise a fringe of white hairs. The head is black with a white, inverted V-shaped mark.

Actual size

FAMILY	Pieridae
DISTRIBUTION	Africa, India, Sri Lanka, Myanmar, and parts of China
HABITAT	Mountain meadows, grassland, gardens, and parks
HOST PLANTS	*Cassia* spp. and *Senna* spp.
NOTE	Well-camouflaged green caterpillar that produces a strong migrant
CONSERVATION STATUS	Not evaluated

ADULT WINGSPAN
2⅛–2⅝ in (54–66 mm)

CATERPILLAR LENGTH
Up to 1¾ in (45 mm)

CATOPSILIA FLORELLA
AFRICAN MIGRANT
(FABRICIUS, 1775)

120

The female African Migrant lays her eggs singly on flower buds and young shoots of the host plants. The pale, elongated eggs are laid vertically on the surface so they look as if they are standing up. The young, green caterpillars emerge and feed on the buds, their color providing excellent camouflage. The older caterpillars then feed on the leaves, often defoliating the plant. Pupation occurs on stems or leaves. Like the larvae, the chrysalis is also green with a yellow line along the side, giving it the appearance of a leaf with a midrib.

The adults, which are on the wing all year round, are powerful fliers, and they may migrate over long distances, hence the common name of the species. Those that breed in South Africa migrate in a northwest direction from summer to fall. Mass migrations have been reported in Tanzania, with butterflies flying north in November, then east in March, and finally returning south in May.

Actual size

The African Migrant caterpillar is predominantly green in color with bands of tiny, black, hair-bearing tubercles. There is a distinctive black and pale yellow stripe along both sides. The head is green with tiny, black spots.

FAMILY	Pieridae
DISTRIBUTION	India, Southeast Asia, southern China, Chinese Taipei, and northern and eastern Australia
HABITAT	Mainly subtropical and tropical in a variety of habitats; adults often migrate to cooler temperate areas
HOST PLANTS	Senna (*Senna* spp.) and cassia (*Cassia* spp.)
NOTE	Colorful but well-camouflaged caterpillar of a widespread butterfly
CONSERVATION STATUS	Not evaluated, but common in some locations

ADULT WINGSPAN
2¹⁄₁₆ in (53 mm)

CATERPILLAR LENGTH
1⁷⁄₁₆ in (37 mm)

CATOPSILIA PYRANTHE
MOTTLED EMIGRANT
(LINNAEUS, 1758)

121

Mottled Emigrant caterpillars feed openly on the leaves of their host plants, usually on the upper surface of new or recent growth, completing development in as little as four weeks under warm conditions. Breeding may be continual throughout the year but is often seasonal in many locations. The caterpillars can be found in urban areas where the food plants are cultivated as street trees or in gardens. Pupation occurs on a leaf or stem of the food plant.

In Australia, *Catopsilia pyranthe* is also known as the White Migrant, and, as suggested by the species' common name, the butterflies can be migratory. Pale and dark forms of the butterfly can occur, depending on seasonal conditions. The genus *Catopsilia* contains six species, ranging from Africa through Southeast Asia to Australia. All are known to be migratory, although migration may occur into areas where food plants are absent and, as a result, no caterpillars will be found. Migratory flights usually last only a few weeks.

Actual size

The Mottled Emigrant caterpillar is cylindrical and green with a yellow lateral line to the body, edged above with small, raised, black spots. Smaller black spots cover the entire dorsal and lateral surfaces of the caterpillar, including the head.

FAMILY	Pieridae
DISTRIBUTION	North America, including Mexico
HABITAT	Montane meadows, alfalfa fields, gardens, parks, and pastures
HOST PLANTS	Legumes, including Alfalfa (*Medicago sativa*), clover (*Trifolium* spp.), vetch (*Vicia* spp.), Birdsfoot Trefoil (*Lotus corniculatus*), and lupine (*Lupinus* spp.)
NOTE	Caterpillar that can occur in millions around Alfalfa fields
CONSERVATION STATUS	Not evaluated, but common

ADULT WINGSPAN
1¾–2 in (45–50 mm)

CATERPILLAR LENGTH
1⅜–1⁹⁄₁₆ in (35–40 mm)

122

COLIAS EURYTHEME
ORANGE SULPHUR
BOISDUVAL, 1852

Before the Orange Sulphur caterpillar hatches, its white, spindle-shaped egg turns yellow, then orange red. First instars consume most of the eggshell after hatching. During the early instars, the caterpillars skeletonize leaves by feeding between veins on both sides of the midrib. Later instars consume entire leaves from the edge or tip. Development from egg hatch to pupation takes about two to four weeks, but sometimes longer, depending on the temperature and host plant. The caterpillar may overwinter with reduced growth, although the species is unable to survive the winter in northerly areas.

Colias eurytheme caterpillars are highly camouflaged on their host plants, but natural enemies, including predatory bugs, parasitic wasps, and birds, take a great toll on populations. Disease is also an important regulator of numbers. The caterpillars of all *Colias* species are very similar and easy to confuse. Increased acreages of Alfalfa in the western United States have contributed to the great abundance of Orange Sulphurs in this region.

Actual size

The Orange Sulphur caterpillar is dark green with a thick, spiracular, white stripe that has intermittent red spots or lines. There are large numbers of tiny, black dots and short, pale setae, and some individuals have a vague dark stripe middorsally and a yellow dorsolateral stripe. The head is light green, peppered with tiny, black dots.

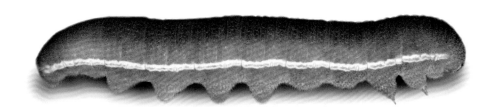

FAMILY	Pieridae
DISTRIBUTION	Northern and eastern Australia, southeastern New Guinea
HABITAT	Savannahs and paperbark woodlands
HOST PLANTS	Mistletoe (*Amyema* spp., *Decaisnina* spp., *Dendrophthoe* spp., and *Muellerina* spp.)
NOTE	Gregarious caterpillar that sometimes defoliates the food plant
CONSERVATION STATUS	Not evaluated, but common in tropical and subtropical regions

ADULT WINGSPAN
2⅞₆ in (62 mm)

CATERPILLAR LENGTH
1⁹⁄₁₆ in (40 mm)

DELIAS ARGENTHONA
SCARLET JEZEBEL
(FABRICIUS, 1793)

123

Scarlet Jezebel caterpillars hatch from eggs laid in closely packed, irregular groups of up to 50 on young leaves and stems of the mistletoe host plant. The larvae first devour their eggshells, then consume the young soft growth. Later instars feed on older leaves. They develop synchronously and can cause defoliation of small mistletoe plants. Growth is completed in three to four weeks, and there are several generations annually.

The caterpillars often pupate together on the mistletoe but also singly on the mistletoe or nearby on the host tree. The pupa is attached to a silken pad by the cremaster and a silk girdle. In the northern areas of the species' range, breeding occurs in the cool months, but adults are found all year. The adults feed at flowers, especially on the mistletoe host plants, and males will often fly on hilltops.

Actual size

The Scarlet Jezebel caterpillar is cylindrical, variable in color from yellowish brown to greenish pink with a black anal segment. Sparse, white dorsolateral spots give rise to long, white hairs. There are smaller ventrolateral hairs, and the entire body is covered with many tiny, white hairs. The head is black with short, white hairs.

FAMILY	Pieridae
DISTRIBUTION	Eastern Australia
HABITAT	Montane rain forests in the north and coastal rain forests in the south
HOST PLANTS	Mistletoe (*Amyema* spp., *Muellerina* spp.)
NOTE	Gregarious caterpillar from a species-rich butterfly genus
CONSERVATION STATUS	Not evaluated, but common

ADULT WINGSPAN
2³⁄₁₆ in (56 mm)

CATERPILLAR LENGTH
1½ in (38 mm)

124

DELIAS NIGRINA
BLACK JEZEBEL
(FABRICIUS, 1775)

Black Jezebel caterpillars hatch in batches of up to 90 and feed gregariously. Their development time is synchronous within the batch and takes three to five weeks depending on the temperature. High temperatures can be detrimental to their survival. Caterpillars descend on silken threads if disturbed, ascending the thread later to resume feeding. The larvae pupate singly, usually on the mistletoe food plant but occasionally on the host tree. The pupae are attached in an upright position via a silken girdle and the hooklike cremaster.

Several generations are completed annually, with the adult butterflies being more common in the winter months, particularly in tropical and subtropical locations. The adults are occasionally migratory in the fall and have been recorded well outside their normal breeding areas. The large, colorful *Delias* genus contains at least 165 species, which originated in Australia and spread to Southeast Asia and India.

The Black Jezebel caterpillar is cylindrical and colored olive green or dark brown, with a row of dorsolateral yellow spots from which arise long, white hairs. The head is black with white hairs.

Actual size

FAMILY	Pieridae
DISTRIBUTION	Western North America, from Alaska to New Mexico
HABITAT	Montane slopes, meadows, canyons, shrub-steppe, desert washes, and beaches
HOST PLANTS	Rockcress (*Arabis* spp.), tansymustards (*Descurainia* spp.), and Tumble Mustard (*Sisymbrium altissimum*)
NOTE	Caterpillar with a pupal period lasting 10 to 11 months
CONSERVATION STATUS	Not evaluated, although the US Fish and Wildlife Service lists subspecies *insulana* as a species of concern

ADULT WINGSPAN
1⁹⁄₁₆–1¾ in (40–45 mm)

CATERPILLAR LENGTH
1³⁄₁₆–1⅜ in (30–35 mm)

EUCHLOE AUSONIDES
LARGE MARBLE
(LUCAS, 1852)

125

Large Marble caterpillars emerge from eggs that are laid singly on unopened buds of cruciferous host plants. Early instar larvae station themselves vertically among flower clusters to feed on buds and flowers, and cover this part of the plant with loosely spun silk. The caterpillars do not make nests. Later instars move on to leaves and seedpods. The larvae are usually solitary with only one per host plant. Prior to pupation, they become purplish and wander in search of a suitable pupation site.

Development from egg hatch through five instars to pupation takes roughly three weeks. The pupa oversummers and overwinters, spending 10 to 11 months in this stage. Survival appears to depend on camouflage. However, minute pirate bugs (Anthocoridae) common on crucifers probably kill many of the larvae, and parasitic wasps parasitize late instars and also pupae. Other closely related *Euchloe* species occur in North America and Europe.

The Large Marble caterpillar is yellow with broad, purple-gray, longitudinal stripes. The body is covered with large, black spots, each bearing a single seta. There is a bold, white ventrolateral stripe bordered below with a broken, yellow line. The head is gray with black spots, and the legs and claspers are also gray.

Actual size

FAMILY	Pieridae
DISTRIBUTION	Africa, south and Southeast Asia, and Australia
HABITAT	Gardens, parks, grassland, scrub, and open forest
HOST PLANTS	Silk plant (*Albizia* spp.), leadtree (*Leucaena* spp.), *Senna* spp., and other members of Fabaceae
NOTE	Green caterpillar that rests well camouflaged on many plants
CONSERVATION STATUS	Not evaluated

ADULT WINGSPAN
1⁹⁄₁₆ in (40 mm)

CATERPILLAR LENGTH
1–1³⁄₁₆ in (25–30 mm)

126

EUREMA HECABE
COMMON GRASS YELLOW
(LINNAEUS, 1758)

Actual size

Common Grass Yellow caterpillars hatch from spindle-shaped eggs laid singly several days earlier on the upper side of leaves of the host plant. The newly emerged caterpillars consume the eggshell first before moving on to the leaves. The young larvae are green and covered in tiny tubercles. As they develop, the larvae rest alongside the midrib of leaves, which action provides excellent camouflage. The pupa is found hanging from the stem of the host plant.

Although it prefers grassy habitats, this species does not feed on grass, and its common name is more likely to have been derived from the way the butterfly adult flies slowly, staying close to the ground. The Common Grass Yellow is also widespread, as its name suggests, and a migrant species that occurs in many different habitats due to the distribution of its host plants. There are about 18 subspecies.

The Common Grass Yellow caterpillar is green with a thin, dark dorsal stripe and a lateral, creamy yellow stripe. Each segment has vertical ribs and is covered in tiny tubercles, giving a textured appearance. The body is covered in short hairs.

FAMILY	Pieridae
DISTRIBUTION	Europe, North Africa, across the Middle East and northern Asia
HABITAT	Various, including grassland, woodlands, parks, and gardens
HOST PLANTS	Alder Buckthorn (*Frangula alnus*) and Buckthorn (*Rhamnus cathartica*)
NOTE	Pale green caterpillar that is well camouflaged among leaves
CONSERVATION STATUS	Not evaluated, but widespread and common

ADULT WINGSPAN
2⅜–2¹⁵⁄₁₆ in (60–74 mm)

CATERPILLAR LENGTH
1⁵⁄₁₆ in (33 mm)

GONEPTERYX RHAMNI
BRIMSTONE
(LINNAEUS, 1758)

127

Brimstone caterpillars hatch from pale, yellow-green eggs laid singly some 10 to 14 days earlier on the youngest leaves of the host plant. The young larvae move to the upperside of the leaf to feed. At rest, they lie along the midrib of the leaf, and if disturbed they raise the front half of their body. Their main natural enemies are birds and parasitic wasps. The larval stage takes about a month. When the caterpillars are ready to pupate, they attach themselves to the underside of a leaf or stem by a silk pad and girdle.

Like the caterpillar, the green pupa is well camouflaged, as it resembles a curled leaf. The emerging, long-lived butterfly feeds on nectar in the summer and then hibernates until early spring; it is often the first butterfly to be seen each year. Many claim that the conspicuous yellow wings of the adult Brimstone gave rise to the word butterfly from "butter-colored fly."

Actual size

The Brimstone caterpillar is green with tiny, black spots, the shade of green matching that of the host plant on which it lives. Its body is covered with fine, short hairs. The caterpillar is very similar to that of the Cabbage White (*Pieris rapae*).

FAMILY	Pieridae
DISTRIBUTION	East Andean slopes, from Venezuela to Bolivia
HABITAT	Second growth and forest edges of humid Andean cloud forests
HOST PLANTS	*Cardamine* spp.
NOTE	Well-camouflaged caterpillar that it is nearly impossible to find
CONSERVATION STATUS	Not evaluated, but not considered threatened

ADULT WINGSPAN
1⁹⁄₁₆–1¾ in (40–44 mm)

CATERPILLAR LENGTH
1⁵⁄₁₆–1³⁄₁₆ in (24–30 mm)

128

LEPTOPHOBIA ELEONE
SILKY WANDERER
(DOUBLEDAY, 1847)

The slow-moving caterpillars of the Silky Wanderer hatch from bright orange, cylindrical eggs laid singly on the upper side of mature host plant leaves. The well-disguised eggs are similar in size and color to the host plant seeds, which are also scattered across the upper surfaces of leaves after being thrown from their explosively opening seedpods. The larvae, too, are perfectly colored to disappear from sight while resting on the host leaves. Young larvae are found near the apex of their small, herbaceous hosts; as they grow, they feed on more mature leaves, closer to the ground.

When ready to pupate, the final instars wander from their host plant, sometimes for considerable distances, and build their pupae in sheltered and well-hidden locations such as under dead leaves or branches, usually close to the ground. Silky Wanderer adults are fairly weak fliers and feed from a variety of common roadside flowers.

Actual size

The Silky Wanderer caterpillar is almost uniformly light green, perfectly matching the coloration of its host plant. It is sparsely covered in short, soft, pale setae, giving it a soft, fuzzy appearance. The only hint of color is a yellowish wash across the dorsum, extending laterally to the spiracular area, broken supra-spiracularly by a narrow, darker green line.

FAMILY	Pieridae
DISTRIBUTION	Southern and central United States, northern Mexico
HABITAT	Weedy and disturbed locations, arid lands, parks, and gardens
HOST PLANTS	Weedy Asteraceae: beggarticks (*Bidens* spp.) and marigolds (*Dyssodia* spp.)
NOTE	Aggressive caterpillar that develops into tiny, migrant adult
CONSERVATION STATUS	Not evaluated, but common within its range

ADULT WINGSPAN
¾–1 in (20–25 mm)

CATERPILLAR LENGTH
⅝ in (16 mm)

NATHALIS IOLE

DAINTY SULPHUR

BOISDUVAL, 1836

129

Dainty Sulphur caterpillars hatch from eggs laid singly on young leaves; first instars consume the eggshells after hatching. Feeding by first instars causes "windowpaning" of leaves, producing apparent holes between veins—in reality a layer of transparent "cuticle." Later instars consume entire leaves and flower petals from the edge. Resting on stems or leaf midribs, the caterpillars feed mainly at night and do not build nests. Their defense appears to be based on camouflage and intimidatory behavior in the final instar—waving the head end around when threatened. The caterpillars also emit chemicals from a ventral gland and setae, which may have a protective function.

There are four instars, and development from egg hatch to pupation takes about three weeks, with adults emerging after another two weeks. The annual northerly spread of Dainty Sulphur butterflies in North America is facilitated by a broad host range, rapid development, and the use of watercourses and roadways as travel routes.

The Dainty Sulphur caterpillar is green with indistinct, pale, vermiform markings, and middorsal and spiracular stripes that may be pale, red, or bold magenta. A pair of enlarged pink, red, or magenta protuberances are present behind the head. The head, true legs, and claspers are green, and the spiracles are white.

Actual size

FAMILY	Pieridae
DISTRIBUTION	Western North America, from British Columbia to Arizona
HABITAT	Low to high elevation coniferous forests
HOST PLANTS	Pine (*Pinus* spp.), firs, such as *Pseudotsuga menziesii* and *Abies grandis*, and Western Hemlock (*Tsuga heterophylla*)
NOTE	Caterpillar that can defoliate pine forests
CONSERVATION STATUS	Not evaluated, but common

ADULT WINGSPAN
1¾–2 in (45–50 mm)

CATERPILLAR LENGTH
1–1³⁄₁₆ in (25–30 mm)

130

NEOPHASIA MENAPIA
PINE WHITE
(FELDER & FELDER, 1859)

Pine White caterpillars hatch in spring from eggs laid in angled rows of 3 to 25 or more along conifer needles during late summer and fall; the eggs then overwinter. The first instars do not consume their own eggshells but do feed on adjacent unhatched eggs. Early instars feed gregariously—typically four to six on a needle, most with their heads toward the tip. Later instars generally feed alone and are superbly camouflaged. Frass is flicked away by young caterpillars, but older ones simply drop it. When disturbed, young caterpillars may drop, suspended by silk. Older caterpillars regurgitate food and wave their heads around when threatened.

Development from egg hatch to pupation takes about two months. Pine Whites normally exist at low population levels but occasionally "explode" into major outbreaks, with caterpillars defoliating large areas of forest, and butterflies so numerous that they create the impression of a living snowstorm.

Actual size

The Pine White caterpillar is dark conifer green with numerous, small, white spots, and distinct lateral (thick) and dorsolateral (thin), yellowish or whitish stripes. There are two very short, taillike projections on the posterior segment. The head is green with yellowish spots, and the true legs are black.

FAMILY	Pieridae
DISTRIBUTION	The Andes of Venezuela and Colombia, south to central Peru
HABITAT	Cloud and upper subtropical forests and forest borders, often along streams
HOST PLANTS	Unknown species of mistletoe (Loranthaceae)
NOTE	Plump, dull-colored caterpillar that feeds in very large groups
CONSERVATION STATUS	Not evaluated, but not considered threatened

ADULT WINGSPAN
3¼–3⁷⁄₁₆ in (82–88 mm)

CATERPILLAR LENGTH
2⅛–2⅜ in (55–60 mm)

PEREUTE CALLINICE
PEREUTE CALLINICE
(FELDER & FELDER, 1861)

131

Pereute callinice caterpillars hatch from elongate, yellow eggs, laid in untidy clusters on the underside of leaves of their host plant. The larvae are gregarious and, sometime during the fourth or fifth instar, they begin to migrate daily, single file, down to the base of the plant to which their own host is attached. They are, then, most frequently found while resting in large groups during the day on tree trunks, usually on alder species. It is not known, however, if their host plant, mistletoe, is a specialist parasite on alders. The caterpillar likely relies on its dull coloration and sparse pale setae to camouflage it as a piece of moldy, dead leaf.

The butterfly adults are most frequently seen chasing each other above the sunny canopy or, later in the day, found feeding on the ground at puddles or water seeps. While at rest, with the bright colors of the upper wing surfaces hidden, they are, like their caterpillars, quite cryptic.

The *Pereute callinice* caterpillar is rather plain, nearly uniform dark brown across its entire body, with a dark brown or blackish head. It is sparsely covered with short, soft, yellowish setae, of varying lengths but generally longer posteriorly. Older caterpillars may take on a purplish cast to the underlying brown.

Actual size

FAMILY	Pieridae
DISTRIBUTION	Southeast Canada; northeastern, central, and southern United States; south to South America
HABITAT	Gardens, open spaces, disturbed areas, watercourses, glades, and seashores
HOST PLANTS	Senna (*Senna* spp., *Cassia* spp.), clover (*Trifolium* spp.), and *Chamaecrista* spp.
NOTE	Caterpillar camouflaged yellow or green, according to its host plant
CONSERVATION STATUS	Not evaluated, but common

ADULT WINGSPAN
2⁹⁄₁₆–3¹⁄₁₆ in (65–78 mm)

CATERPILLAR LENGTH
1⁹⁄₁₆–1¾ in (40–45 mm)

PHOEBIS SENNAE
CLOUDLESS SULPHUR
(LINNAEUS, 1758)

132

The Cloudless Sulphur caterpillar is either green or yellow. The green form has a yellow lateral stripe and characteristic tripled, blue, lateral dashes that in some individuals form blue, transverse bands. The head is green or yellow with raised black spots.

Cloudless Sulphur caterpillars hatch from eggs laid singly on young leaves or flower buds of host plants six days earlier. The larvae feed on foliage, buds, and flowers but do not build shelters. The younger instars are not capable of diapause, so those occurring in the more northerly areas are often killed by freezing temperatures in fall. The caterpillars, which rest beneath leaf petioles, rely on camouflage for protection from natural enemies. If they eat mostly leaves, they are green, but if they feed primarily on yellow *Senna* flowers (which they prefer), they become yellow.

Most larvae pupate on their host plant. Adults, which likely eclose some 10 to 14 days later, are migratory and occur in New England and the upper Midwest of the United States in many years, breeding from midsummer to fall. In the south, many generations are produced year-round, but only one or two in the north. Adults have very long proboscises and are adapted to deep-throated nectar sources.

Actual size

FAMILY	Pieridae
DISTRIBUTION	Europe, North Africa, South Africa, and Asia
HABITAT	Farms, vegetable gardens, croplands, and pastures
HOST PLANTS	Crucifers (*Brassica* spp.), including cabbage, cauliflower, and Brussels sprouts
NOTE	One of few agriculturally significant butterfly caterpillar pests
CONSERVATION STATUS	Not evaluated, but widespread and common

ADULT WINGSPAN
2½–3 in (63–76 mm)

CATERPILLAR LENGTH
1⅜–1⁹⁄₁₆ in (35–40 mm)

PIERIS BRASSICAE

LARGE WHITE

(LINNAEUS, 1758)

133

Large White caterpillars hatch from eggs laid in irregular batches of 30 to 100 on the underside of host plant leaves. They eat the empty shells, then spin a fine layer of silk over the leaf and feed gregariously, eating only the cuticle. Later they perforate the leaf. Large White caterpillars feed and rest at the same time. The caterpillars tend to congregate in a line along the edge of a leaf, devouring it as they move backward. Late instars disperse and feed singly. About 30 days after hatching, the final instar seeks a place for pupation.

The larvae usually pupate under a ledge or on a tree trunk, with the ivory white and gray pupae fixed upright. Sometimes pupae are attached to the host plant; the pupae are then green in color. Considerable numbers of Large White caterpillars fail to pupate because of parasitism by ichneumonid wasps, which lay their eggs in caterpillars; the developing wasp larvae feed on and eventually kill their host.

The Large White caterpillar is gray green, darkest dorsally, with three yellow, longitudinal stripes. The edges of the stripes are poorly defined and blend into the ground color. The ventral surface is greenish, and the prolegs are brown. Short, black spots occur all over the body, each bearing a fine seta.

Actual size

FAMILY	Pieridae
DISTRIBUTION	Europe, Asia, Africa, North America, Australia, and New Zealand
HABITAT	All open habitats, especially disturbed areas, croplands, gardens, and parks
HOST PLANTS	Native and commercial crucifers (*Brassica* spp.)
NOTE	Most economically damaging butterfly caterpillar pest in the world
CONSERVATION STATUS	Not evaluated, but widespread and common

ADULT WINGSPAN
1¾–2 in (45–50 mm)

CATERPILLAR LENGTH
1³⁄₁₆ in (30 mm)

134

PIERIS RAPAE
CABBAGE WHITE
(LINNAEUS, 1758)

Cabbage White caterpillars hatch from eggs laid singly on the underside of host plant leaves; a Cabbage White female lays up to 750 eggs in a lifetime. The larvae develop rapidly, taking 15 to 20 days from egg hatch to pupation. Most—85 percent—of their consumption of host plant leaves occurs in the final instar. Young caterpillars make holes in leaves, while older individuals eat leaves from the edge. There are five instars and no nests are made. When temperatures are cool and day lengths less than 13 hours, the developing caterpillars produce diapausing pupae that overwinter.

Parasitic wasps are important regulators of Cabbage White populations, parasitizing caterpillars and pupae. Camouflage helps protect the caterpillars. They also secrete droplets from their setae that deter ants but may attract wasps. Cabbage Whites are found mostly in disturbed urban and agricultural environments and are much less common in wilderness and other undisturbed areas.

Actual size

The Cabbage White caterpillar is green, peppered with tiny, black setae with a distinct yellow middorsal line. On each segment there are ten small, white protuberances, each bearing a short, pale seta with a droplet. A yellow dash occurs on each segment near the spiracle. The head is green with many small setae. The true legs and prolegs are also green.

FAMILY	Pieridae
DISTRIBUTION	Western North America, from British Columbia to New Mexico
HABITAT	Arid lands, shrub-steppe, deserts, canyons, and watercourses
HOST PLANTS	Rock cress (*Arabis* spp.), tansymustard (*Descurainia* spp.), and Tumble Mustard (*Sisymbrium altissimum*)
NOTE	Camouflaged caterpillar that develops rapidly
CONSERVATION STATUS	Not evaluated, but common within its range

ADULT WINGSPAN
1¾–2 in (45–50 mm)

CATERPILLAR LENGTH
1³⁄₁₆ in (30 mm)

PONTIA BECKERII
BECKER'S WHITE
(W. H. EDWARDS, 1871)

135

Becker's White eggs are laid singly on the flowers and seedpods of host plants. The caterpillars hatch after three days and partially eat their eggshell. Their development is rapid, with pupation occurring just 14 days after egg hatch. The whole cycle, from egg to emerging butterfly, takes little more than three weeks. The caterpillars feed on all parts of the host plant, although young instars prefer flowers. No nests are made, and survival is based on camouflage. Prepupal caterpillars wander, and pupation may occur on or off the host plant.

The pupa is attached by a silk girdle, usually to a twig, and resembles a bird dropping. The number of generations produced by this species appears to depend on the quality of the host plant. Caterpillars feeding on good-quality hosts produce a further generation, while poor-quality hosts result in diapausing pupae. Males are aggressive and sometimes "fight" to mate with newly emerged females.

Actual size

The Becker's White caterpillar is bright yellowish green with distinct intersegmental yellow bands. Large, black dots occur on each segment, each carrying a long, white seta, giving the caterpillar a hairy appearance. The head is white and yellow with black spots. The caterpillar turns pinkish brown prior to pupation.

FAMILY	Pieridae
DISTRIBUTION	Western North America, from Alaska and British Columbia to California and New Mexico
HABITAT	Open areas in mountains and high plains
HOST PLANTS	Mustard family, such as rockcress (*Arabis* spp.), tansymustard (*Descurainia* spp.), whitlow grass (*Draba* spp.), and peppergrass (*Lepidium* spp.)
NOTE	Species whose life cycle may be accelerated at high altitudes
CONSERVATION STATUS	Not evaluated, but usually common within its range

ADULT WINGSPAN
1½–2⅟₁₆ in (38–53 mm)

CATERPILLAR LENGTH
1⅜ in (35 mm)

136

PONTIA OCCIDENTALIS
WESTERN WHITE
(REAKIRT, 1866)

Western White caterpillars hatch from single yellowish eggs that turn orange; to ensure the offspring have enough to eat, the adult female lays only on a plant where no other butterfly eggs are present. The young caterpillars feed on the host plant leaves, also ingesting the mustard oils they contain, which make the larvae distasteful to birds. The caterpillars may have only four instars and can reach pupation within a week as an adaptation to the unpredictable and ephemeral climate of higher elevations. The pupae hibernate attached to a twig by the abdomen tip and a silken belt.

There is just one generation at high altitudes but several at lower altitudes. Western White caterpillars eat dozens of plants in the mustard family (Cruciferae) but prefer to eat flowers and young fruits, so, unlike Cabbage White (*Pieris rapae*) caterpillars, they are not pests on cabbage. Caterpillars of several closely related species—especially *P. protodice*—are similar to the Western White.

The Western White caterpillar is bluish gray with vivid yellow and white bands, and is covered with black spots and hairs, while the head is mostly bluish gray with an extension of a yellowish stripe. The setae are short and arise from raised bumps giving the caterpillars a spotted appearance.

Actual size

FAMILY	Pieridae
DISTRIBUTION	Western North America, from northwest Canada to southern California
HABITAT	Rocky desert steppe, subalpine ridges, hills, and canyons
HOST PLANTS	Rock cress (*Arabis* spp.), tansymustard (*Descurainia* spp.), and Tumble Mustard (*Sisymbrium altissimum*)
NOTE	Spectacularly colored caterpillar of a high desert butterfly
CONSERVATION STATUS	Not evaluated, but widespread in western United States; more scattered in Canada

ADULT WINGSPAN
1⅜–1⁹⁄₁₆ in (35–40 mm)

CATERPILLAR LENGTH
1³⁄₁₆ in (30 mm)

PONTIA SISYMBRYII
SPRING WHITE
(BOISDUVAL, 1852)

137

The eggs of the Spring White are laid singly on the flowers of host plants and hatch within two to three days. First instars do not eat their eggshells but will cannibalize any nearby eggs. They feed ravenously, consuming mostly leaves but also flowers and seedpods; a single individual is capable of defoliating a plant. The caterpillars feed and rest positioned parallel to stems, seedpods, or leaf veins and do not construct nests. They are well camouflaged in early instars but develop bright aposematic colors in later instars, suggesting that they are distasteful to predators.

Spring White caterpillars grow rapidly through five instars and reach pupation about 20 days after hatching. There is only one spring generation a year, and the pupa oversummers and overwinters. Pupal diapause for up to four years has been reported. The caterpillars are vulnerable to viral and bacterial disease and infections.

Actual size

The Spring White caterpillar is strikingly colored, with each segment banded in bright yellow, porcelain white, and black. The body is covered in fine setae, and the black head is speckled with white dots. The true legs are black, and the prolegs are white.

FAMILY	Riodinidae
DISTRIBUTION	Southwestern Canada, western United States, and Mexico
HABITAT	Desert canyons, arid flats, banks, and roadsides
HOST PLANTS	Buckwheat (*Eriogonum* spp.)
NOTE	Slow-growing caterpillar that becomes a fiery-colored, fall butterfly
CONSERVATION STATUS	Not evaluated, although subspecies *Apodemia mormo langei* is critically imperiled

ADULT WINGSPAN
1¼ in (32 mm)

CATERPILLAR LENGTH
1¹⁄₁₆ in (27 mm)

138

APODEMIA MORMO
MORMON METALMARK
(FELDER & FELDER, 1859)

In northern parts of its range, the Mormon Metalmark caterpillar develops very slowly during spring and summer, spending much time in apparent dormancy. Late instars may make loose silk nests to rest in. Pupation occurs in late summer, and the butterfly flies in early fall. The eggs develop embryos but then diapause and overwinter. Caterpillar survival appears to depend on concealment in refugia, but the bright purple and gold coloration may be aposematic. The caterpillars feed on many buckwheat species, chewing small holes through the leaves.

Approximately 15 closely related metalmark butterfly species occur from Canada to Brazil. The seasonality of these species (including *Apodemia mormo*) varies according to latitude, with more southerly species having earlier and longer flight periods. Young caterpillars—instead of eggs—may overwinter in lower latitude populations. The subspecies *A. mormo langei*, which occurs in the Antioch Dunes of California, is endangered due to loss of habitat and host plants.

Actual size

The Mormon Metalmark caterpillar is purple with bold, paired, black, raised spots dorsally. Yellow or gold-orange spots occur between the black spots. Two rows of gold spots are present laterally, with long, white setae arising. The intensity of colors and markings varies geographically. The head is black.

FAMILY	Riodinidae
DISTRIBUTION	Trinidad and the Amazon basin, south to central Brazil and east-central Peru
HABITAT	Margins of streams, ponds, and oxbow lakes
HOST PLANTS	Aquatic *Montrichardia* spp.
NOTE	Beautiful caterpillar rarely seen outside of its leaf shelter
CONSERVATION STATUS	Not evaluated, but not considered threatened

ADULT WINGSPAN
1⁷⁄₁₆–1⁹⁄₁₆ in (36–40 mm)

CATERPILLAR LENGTH
1–1³⁄₁₆ in (25–30 mm)

HELICOPIS CUPIDO
SPANGLED CUPID
(LINNAEUS, 1758)

139

Spangled Cupid caterpillars live and feed within the tightly rolled new leaves of their host plants. To prevent these leaves from unrolling as they develop, the larvae silk them tightly closed. Repeated oviposition by adults results in the presence of several generations of larvae living together within these shelters. As adults, both sexes are usually found resting under the leaves of tall vegetation growing around the edges of lagoons or slow-moving backwaters, flying only infrequently, usually to swirl about chasing other *Helicopis cupido* butterflies.

The most remarkable feature of Spangled Cupid caterpillars is their large cluster of balloon setae on the prothoracic shield. This and similar structures are unique to the Riodinidae, and they are known from both myrmecophilous (ant attended) and non-myrmecophilous genera, although rare overall. The exact function of these setae is unknown, but they may store and disperse noxious defensive chemicals, as well as facilitate symbiotic relationships with ants. Internally, the balloon setae are filled with a spongy material consisting of a dense latticework of tiny strands.

Actual size

The Spangled Cupid caterpillar is short, stout, and densely covered with soft, downy, white setae. Its dull yellowish head is usually completely hidden by these fur-like setae and by a large clump of pink, balloon-like setae sprouting from the dorsum of the prothorax.

FAMILY	Riodinidae
DISTRIBUTION	Western Ecuador
HABITAT	Lower temperate and foothill forests and forest borders
HOST PLANTS	Unknown genera and plants of the family Rubiaceae
NOTE	Caterpillar whose coloring provides excellent camouflage against its host plant
CONSERVATION STATUS	Not evaluated, but not considered threatened

ADULT WINGSPAN
1⅛–1⁵⁄₁₆ in (29–33 mm)

CATERPILLAR LENGTH
1¹¹⁄₁₆–⅞ in (18–22 mm)

140

LEUCOCHIMONA AEQUATORIALIS
ECUADORIAN EYEMARK
(SEITZ, 1913)

The Ecuadorian Eyemark caterpillar hatches from a minute, disk-shaped egg laid on areas of new leaf growth. Like many other members of the family Riodinidae, the larva is unremarkable in appearance, closely matching the ground color of leafy portions of its host plant so that it is nearly invisible when not in motion. Even then, the caterpillar generally moves so slowly that it remains hard to detect. So far as is known, all instars share the same appearance and habits, beginning life feeding on the newest leaf growth and moving to more mature foliage as later instars.

Adult Ecuadorian Eyemarks are most commonly seen flitting along the edges of forests in a soft, bouncy flight. They pause frequently to perch either on the upper or lower surface of leaves, usually with their wings held partially open but occasionally, when perched below, with wings fully spread. The species has not been formally described.

Actual size

The Ecuadorian Eyemark caterpillar is essentially a uniform light green, including the head. The dorsum bears thin, indistinct, brownish lines along the intersegmental sutures, providing the only color other than the short, slightly curved, pale brownish setae that sparsely cover most of the body.

FAMILY	Riodinidae
DISTRIBUTION	Northern Amazon, from Suriname south to Brazil and west to northern Peru
HABITAT	Humid lowland forest, favoring light gaps and forest edges
HOST PLANTS	*Inga* spp.
NOTE	Tiny caterpillar that forms a protective association with ants
CONSERVATION STATUS	Not evaluated, but not considered threatened

ADULT WINGSPAN
1⅛–1⁵⁄₁₆ in (29–33 mm)

CATERPILLAR LENGTH
¾–1⁵⁄₁₆ in (20–24 mm)

NYMPHIDIUM CACHRUS
NYMPHIDIUM CACHRUS
(FABRICIUS, 1787)

141

The most reliable way to find *Nymphidium cachrus* caterpillars is to look for concentrations of ants on the new growth of the host plant. Invariably, large numbers of ants signal the presence of the larvae, to which they are avidly attending. Specialized balloon-shaped setae behind the head and eversible organs on the rear of the abdomen help the caterpillars call for and appease their attending ants, who, in turn, protect the caterpillars from parasitic wasps and invertebrate predators. Indeed, the ants themselves would be predators if they were not instead offered the "nectar" produced by these eversible abdominal glands.

Both the caterpillars and the ants that protect them consume nectar produced by extrafloral nectaries on the host plant leaves. These nectaries are designed to attract ants, which, in turn, protect the young shoots of the plants from herbivores. Thus, by offering their own bribes to the ants, the caterpillars have, in fact, infiltrated the plant's own protective association and are allowed to stay and feed there. The adult, which ecloses from a green pupa, is sometimes known as the Firestreak for the touches of fiery orange on its brown-bordered, white wings.

The *Nymphidium cachrus* caterpillar is robust, tanklike, and trapezoidal in cross section. It has a caramel-brown head and predominantly light green body that is washed with pink, especially along the sides. Its most notable features are the specialized, balloon-shaped setae on the top of the prothorax that aid in its interactions with protective ants.

Actual size

FAMILY	Riodinidae
DISTRIBUTION	Western Amazon, from Colombia south to Bolivia
HABITAT	Humid Amazonian forest, especially in tree gaps and at river edges
HOST PLANTS	*Bauhinia* spp.
NOTE	Sluglike caterpillar that forms a protective relationship with ants
CONSERVATION STATUS	Not evaluated, but not considered threatened

ADULT WINGSPAN
1–1⅛ in (26–29 mm)

CATERPILLAR LENGTH
¾–⅞ in (20–23 mm)

142

PROTONYMPHIDIA SENTA
PROTONYMPHIDIA SENTA
(HEWITSON, 1853)

Actual size

Protonymphidia senta caterpillars hatch from small, flattened, white eggs laid singly on the extrafloral nectaries of their host plant. The larvae feed on the secretions of the nectaries as well as plant tissue. Even in the world of caterpillars, which are hardly renowned for their speed, these insects are decidedly slow-moving. This is not accidental, however, as they rely largely on their lack of movement and general resemblance to a bud of new plant growth to escape detection. In addition, the caterpillars are almost continually protected by a swarm of ants, which they reward with drops of nutritious liquid excreted from specialized glands on their abdomen.

The larvae form brown pupae, and the eclosing adults also feed from the extrafloral nectaries of the host plant unmolested by the resident ants. Research suggests that the female deliberately oviposits where the ants are present. This species is considered distinctive within the family Riodinidae and is the only member of its very recently erected genus.

The *Protonymphidia senta* caterpillar is short, stout, and sluglike. Its small, caramel-brown head is partially hidden by its fleshy thorax. The body is green, washed with maroon, especially laterally, and bears many minute, pale setae.

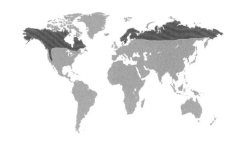

FAMILY	Lycaenidae
DISTRIBUTION	Canada, western United States (Washington State to California), and northern Eurasia
HABITAT	High, windswept rocky ridges and scree slopes
HOST PLANTS	Saxifrage (*Saxifraga* spp.), primrose (*Primula* spp.), and legumes (Fabaceae)
NOTE	Northern hemisphere caterpillar probably vulnerable to climate warming
CONSERVATION STATUS	Not evaluated, but widespread and common

ADULT WINGSPAN
1–1⅛ in (25–28 mm)

CATERPILLAR LENGTH
⅜ in (10 mm)

AGRIADES GLANDON
ARCTIC BLUE
(DE PRUNNER, 1798)

143

The eggs of the Arctic Blue are laid singly under leaves or on flowers; the larvae eat buds and flowers. The caterpillars develop to the second or third instar and then overwinter. In some areas and seasons, they may overwinter twice. The caterpillars produce considerable frass, which may betray their presence. They feed mainly at night and hide by day. Predators and parasitoids appear to take a great toll on eggs and young larvae.

Arctic Blue caterpillars are distinguished by their deep magenta-purple coloration, which camouflages them surprisingly well among the red stalks and stems of their saxifrage host plants. Other lycaenid caterpillars may have reddish or purplish instars, but few have the color intensity of Arctic Blues. The coloration may vary on other host plants, with green larvae often found on legume host plants.

Actual size

The Arctic Blue caterpillar is dark purplish red with a contrasting white ventrolateral stripe, a black dorsal stripe bordered white on each side, and black, diagonal marks dorsolaterally. Its setae are numerous, short, and dark. The head is black but usually hidden from view when the caterpillar is resting.

FAMILY	Lycaenidae
DISTRIBUTION	Across Europe, except the northernmost regions, Turkey to Turkmenistan, and western China
HABITAT	Calcareous grasslands, heathlands, and open woodlands
HOST PLANTS	Various, including *Erodium* spp., *Helianthemum* spp., and *Geranium* spp.
NOTE	Caterpillar that has a mutually beneficial relationship with ants
CONSERVATION STATUS	Not evaluated, but quite common and widespread

ADULT WINGSPAN
⅞–1⅛ in (22–28 mm)

CATERPILLAR LENGTH
⅜–⁹⁄₁₆ in (10–15 mm)

ARICIA AGESTIS
BROWN ARGUS
([DENIS & SCHIFFERMÜLLER], 1775)

144

Actual size

The female Brown Argus flies close to the ground in search of host plants on which to oviposit and lays her white eggs singly on the underside of leaves, close to the midrib. The newly hatched caterpillars remain on the underside, chewing the leaf, but leave the upper surface intact, creating a see-through window. The species has a mutually beneficial relationship with ants. During the day, the caterpillars rest and are milked by ants for a sugary exudate. In return, the ants provide protection from predators and parasitoids.

The Brown Argus caterpillar is green and plump in shape with a dorsal and lateral reddish stripe. The body is covered in short, white setae.

There are two generations across much of the range but only one brood in northerly parts. The caterpillars from the second generation overwinter and in spring pupate at the base of the food plant. Sometimes, the ants bury the pupae shallowly in the ground. The adult butterflies are on the wing in early summer and again later in summer, where they are seen in groups on sunny, south-facing slopes.

FAMILY	Lycaenidae
DISTRIBUTION	Southern and western United States, Mexico, and south to Venezuela
HABITAT	Desert flats, gullies, roadsides, weedy areas, and salt marshes
HOST PLANTS	Chenopodiaceae, including saltbush (*Atriplex* spp.), Russian thistle (*Salsola* spp.), Shoreline Purslane (*Sesuvium portulacastrum*), and Black Pigweed (*Trianthema portulacastrum*)
NOTE	One of the smallest butterfly caterpillars in the world
CONSERVATION STATUS	Not evaluated, but secure though possibly rare in parts of its range

ADULT WINGSPAN
$1\frac{1}{16}$ in (18 mm)

CATERPILLAR LENGTH
$\frac{7}{16}$ in (11 mm)

BREPHIDIUM EXILIS

WESTERN PYGMY BLUE

(BOISDUVAL, 1852)

145

The tiny Western Pygmy Blue caterpillar has numerous stubby setae, giving it a frosted appearance. The eggs hatch in four to five days, and the emerging caterpillars feed on leaves, flowers, and seeds, often hiding in bracts. Concealment and camouflage help protect them, as does attendance by ants, which deter predators and parasitoids. The caterpillars develop rapidly, with pupation occurring after about three weeks; the adults emerge eight to ten days later. There is no dormancy stage, and breeding continues year round in the southern United States and Mexico.

Because they blend in so well with their surroundings, caterpillars of Western Pygmy Blues are hard to find on host plants, but the presence of attending ants may provide clues. The tiny adults are inconspicuous in flight. Males spend much time searching for females, which often remain in hiding on host plants. *Brephidium exilis* disperses northward every summer, sometimes reaching as far as Idaho and Manitoba in Canada.

Actual size

The Western Pygmy Blue caterpillar is light green with a black, shiny head, and an indistinct, ventrolateral, white line and numerous splotchy, white markings laterally. Dorsally, yellowish-white markings form two lines of small chevrons. A red form of the caterpillar also occurs in which the dorsal stripe is dark red and intermittent.

FAMILY	Lycaenidae
DISTRIBUTION	Southern Africa; also introduced into Mediterranean Europe and Morocco
HABITAT	Dry slopes, gardens, and parks
HOST PLANTS	*Geranium* spp. and *Pelargonium* spp.
NOTE	Hairy caterpillar that is a horticultural pest in many countries
CONSERVATION STATUS	Not evaluated, but increasing its range

ADULT WINGSPAN
%₁₆–1 in (15–25 mm)

CATERPILLAR LENGTH
½ in (13 mm)

CACYREUS MARSHALLI
GERANIUM BRONZE
BUTLER, 1897

146

Actual size

Geranium Bronze caterpillars hatch from white, sea urchin-shaped eggs, that are laid by the female butterfly singly on leaves near the flower buds of the host plants. The larvae feed on the flower buds, first piercing a hole through the sepals and then burrowing in. While consuming the contents, they remain hidden within the buds. Once the buds have been devoured, the well-camouflaged caterpillars move on to the more mature leaves. They pupate either on the food plant or within fallen leaves beneath the plant. The pupa is pale yellow brown, hairy, and attached via silk threads.

The Geranium Bronze caterpillar is usually green but can be yellow. It is covered in many short, white setae and has several pink stripes running along the length of its body.

The bronze and white adults fly year-round in warmer areas but elsewhere during only the summer months. The species is native to South Africa but has been introduced to the Mediterranean region as a result of imported potted geranium plants carrying eggs or caterpillars. It is described as a pest species in these areas as it does considerable damage to ornamental plants.

FAMILY	Lycaenidae
DISTRIBUTION	Southern Canada, western United States
HABITAT	Openings in coniferous forests
HOST PLANTS	Pine (*Pinus* spp.)
NOTE	Wonderfully cryptic caterpillar that is found among pine needles
CONSERVATION STATUS	Not evaluated, but common and secure within its range

ADULT WINGSPAN
1–1³⁄₁₆ in (25–30 mm)

CATERPILLAR LENGTH
⁹⁄₁₆ in (15 mm)

CALLOPHRYS ERYPHON
WESTERN PINE ELFIN
(BOISDUVAL, 1852)

147

Western Pine Elfin caterpillars hatch from eggs laid singly on the soft, new-growth branch tips of the host trees. The eggs are typically tucked deep into a crevice close to the base of a slanted pine needle. The first instar feeds on new-growth needles at the pine branch tips, mining the surface initially, then consuming the needles. Yellow frass is produced in great quantities by the caterpillars and may betray their presence. However, they are colored and striped identically to the host needles, which creates superb camouflage. The larvae are solitary and do not make nests.

The caterpillars develop from egg hatch to pupation in 33 to 39 days, and the pupae overwinter. The adults eclose and fly in spring. Males emerge before females and are territorial, returning repeatedly to the same perching spot; they also like to sip moisture from damp soil and puddles. The Western Pine Elfin may occur occasionally on fir, spruce, and larch trees.

Actual size

The Western Pine Elfin caterpillar is forest green to dark green with four, bold, contrasting, white stripes dorsally and laterally. The body is finely covered with tan-colored setae, and the head is green. The spiracles are also tan colored. Early instars are cinnamon brown with less distinct striping.

FAMILY	Lycaenidae
DISTRIBUTION	North America
HABITAT	Bluffs, open fields, forest edges, and dry, rocky open places
HOST PLANTS	Juniper (*Juniperus* spp.) and cedar (*Thuja* spp., *Cupressus* spp.)
NOTE	Green caterpillar that turns pink just before pupation
CONSERVATION STATUS	Not evaluated, but at least one subspecies is thought vulnerable

ADULT WINGSPAN
1–1³⁄₁₆ in (25–30 mm)

CATERPILLAR LENGTH
⁹⁄₁₆ in (15 mm)

CALLOPHRYS GRYNEUS
JUNIPER HAIRSTREAK
(HÜBNER, [1819])

148

Juniper Hairstreak caterpillars are well adapted to feeding on junipers and cedars and do so throughout the United States. The superb camouflage of this caterpillar blends in particularly well with its host plants, and is clearly the basis of its survival strategy. Development from egg hatch to pupation takes 30 to 50 days depending on temperature. The caterpillars feed entirely on juniper needles, excavating holes on the upper surfaces in early instars, then consuming the needles when more mature. The larvae are solitary, and cannibalism can occur in captivity.

A great deal of comparative work has been done on the adults of *Callophrys gryneus* and on analysis of blend zones where subspecies meet. Nevertheless, dialogue continues as to whether this is truly a single species. Much less attention has been given to the comparative biology of the immature stages. The caterpillars usually change color to fuchsia pink and wander just prior to pupation. Overwintering occurs as a pupa.

The Juniper Hairstreak caterpillar is bright, shiny, dark forest green punctuated by bold, highly contrasting, white markings, thickly covered with short, blond setae. There is a ventrolateral line of bold, white bars, one per segment, each bar enlarged anteriorly. A second bold, white marking occurs dorsolaterally on each segment, and the head is green.

Actual size

FAMILY	Lycaenidae
DISTRIBUTION	Western North America, from British Columbia to New Mexico
HABITAT	Sagebrush shrub-steppe, open hillsides, canyons, and washes
HOST PLANTS	Buckwheat (*Eriogonum* spp.)
NOTE	Well-camouflaged caterpillar of Wyoming's tiny, green State butterfly
CONSERVATION STATUS	Not evaluated, but less common in Canada than the United States

ADULT WINGSPAN
⅞–1 in (23–25 mm)

CATERPILLAR LENGTH
⅝ in (16 mm)

CALLOPHRYS SHERIDANII
SHERIDAN'S HAIRSTREAK
(W. H. EDWARDS, 1877)

149

The Sheridan's Hairstreak caterpillar develops rapidly, pupating about a month after the egg is laid. Early instars feed on the upper surfaces of new buckwheat leaves, creating grooves and holes halfway through the leaf, leaving extensive areas of small, yellow spots. The caterpillars blend well with their host plants, their coloring providing protective camouflage. The larvae live solitary lives and do not build nests or shelters. The pupal stage occupies 10 to 11 months of the year; the pupa oversummers and overwinters.

In spring, males of this species from arid zones are the first to emerge and perch on rocks or bare ground in canyon bottoms, challenging passing insects in search of females. There are a number of similar, green hairstreak species in North America and Eurasia, and about half a dozen subspecies of *Callophrys sheridanii*. Higher elevation forms of *C. sheridanii* tend to have caterpillars that are brightly colored with red markings, which may be a consequence of feeding on red flowers.

The Sheridan's Hairstreak caterpillar is green with numerous, short, stubbly setae and a distinct, yellow ventrolateral stripe below the spiracles; there are also two indistinct, yellow stripes dorsally. Viewed from above, the caterpillar displays a broken line of yellowish spots. Mature caterpillars become pink before pupation.

Actual size

FAMILY	Lycaenidae
DISTRIBUTION	Western North America, from southern Canada to Mexico
HABITAT	Low to high elevation clearings in conifer forests
HOST PLANTS	Dwarf mistletoe (*Arceuthobium* spp.)
NOTE	Caterpillar that feeds on conifer-infesting mistletoes
CONSERVATION STATUS	Not evaluated, but possibly rare at the periphery of its range

ADULT WINGSPAN
1–1¼ in (25–32 mm)

CATERPILLAR LENGTH
¹¹⁄₁₆ in (17 mm)

CALLOPHRYS SPINETORUM
THICKET HAIRSTREAK
(HEWITSON, 1867)

150

Actual size

The Thicket Hairstreak caterpillar is well camouflaged and specializes on dwarf mistletoes as host plants. It develops rapidly, taking only a month from egg hatch to pupation. The larvae are slow moving, clinging tightly to buds or fruiting heads and only changing position after the plant part has been consumed. They feed on any part of the dwarf mistletoe plant but prefer terminal buds, first eating a round entry hole, then hollowing out the inside. Prepupal caterpillars become stationary and shrink. The caterpillars are solitary and do not build nests.

Adults spend much of their time in treetops, coming to the ground only to feed on flower nectar. Females lay their eggs in early summer, placing them deep into crevices on the host plants. Mature caterpillars can be found in midsummer by searching the terminal buds of dwarf mistletoes for circular feeding holes, although their coloring can still make them difficult to detect.

The Thicket Hairstreak caterpillar is dark green and bears a darker, green dorsal stripe. The dorsal tubercles are enlarged and mostly orange, each bearing a reddish-brown spot. Each is also bordered with a diagonal white line anteriorly; beyond the white line the segment is contrasting black. A ventrolateral row of spots is white, yellow, orange, and dark brown, and the head is dark brown.

FAMILY	Lycaenidae
DISTRIBUTION	Eastern Australia
HABITAT	Rain-forest margins in the north and tall eucalypt forests in the south
HOST PLANTS	Elderberry Panax (*Polyscias sambucifolia*) and Celery Wood (*Polyscias elegans*)
NOTE	Caterpillar that feeds on flower buds
CONSERVATION STATUS	Not evaluated, but locally common in northern areas, although rarer in the south

ADULT WINGSPAN
1¾₁₆ in (30 mm)

CATERPILLAR LENGTH
⅝ in (16 mm)

CANDALIDES CONSIMILIS
DARK PENCILLED-BLUE
WATERHOUSE, 1942

151

Dark Pencilled-blue caterpillars hatch from eggs laid on small flower buds. The larvae eat through the side of the flower buds, remaining on the outside, and consume the anthers. Older caterpillars eat more of the bud and will graze the surface of leaves if young flower buds are unavailable. The larvae develop quickly and pupate in leaf litter at the base of the shrub. There is only one generation a year in the south of the range, with pupae produced in February emerging the following November (late spring). However, in northern areas there are usually two generations a year.

The caterpillars are well camouflaged on their host plants, but their presence is evident from damaged flower buds. Unlike many lycaenid caterpillars, this species is not attended by ants. The adult male butterflies are often found on hilltops. The 29 species of the genus *Candalides* are largely confined to the New Guinea mainland and Australia.

Actual size

The Dark Pencilled-blue caterpillar is pale green or pinkish green, with lateral projections on all abdominal segments, prominent dorsal projections on the first to sixth abdominal segments, and a pair of large dorsolateral projections on the eighth segment. The thorax has a pair of subdorsal projections on each segment. The head is yellowish green and hidden under the prothorax.

FAMILY	Lycaenidae
DISTRIBUTION	East, southeast, and other scattered areas of Australia
HABITAT	Wide range, from coastal heathland to arid and subalpine woodlands
HOST PLANTS	Native and introduced plantain (*Plantago* spp.) and plants from Lamiaceae, Myoporaceae, Scrophulariaceae, and Thymelaeaceae
NOTE	Well-camouflaged caterpillar often located only by a feeding scar
CONSERVATION STATUS	Not evaluated, but locally common

ADULT WINGSPAN
1–1¼ in (25–32 mm)

CATERPILLAR LENGTH
¹¹⁄₁₆ in (17 mm)

152

CANDALIDES HEATHI
RAYED BLUE
(COX, 1873)

The Rayed Blue caterpillar feeds openly on the underside of leaves, producing irregular patches with the upper leaf cuticle and the veins remaining intact. Older caterpillars consume greater portions of the leaves. The larvae feed largely at night and rest during the day on the underside of lower leaves. They are sometimes attended by a few small, black ants but are usually unattended. The caterpillars complete development within three to five weeks.

Pupation occurs on the underside of leaves, on stems, or under litter at the base of the host plant. The pupa is attached by anal hooks and a central girdle. There is one generation a year in southern areas of Australia, with dormant pupae overwintering. However, in the warmer northern regions, pupal dormancy seems to be broken by rain, and breeding occurs throughout much of the year. Pupal dormancy can last more than 20 months.

Actual size

The Rayed Blue caterpillar is flat and broad with a weakly scalloped margin and square ends. Its base color is green with a darker middorsal line, pale yellow subdorsal chevron lines, and a pale yellow ventrolateral line, covered with numerous small, white hairs. It also has a prominent dorsal ridge on the thorax and abdominal segments one to six, consisting of protuberances topped with short, dark, bristly hairs.

FAMILY	Lycaenidae
DISTRIBUTION	Western North America, from British Columbia south to California and southeast to Texas
HABITAT	Shrubby riparian areas from sea level to high elevations
HOST PLANTS	Wide variety of shrubs and bushes, including Mountain Balm (*Ceanothus greggii*), many other *Ceanothus* spp., and Red Osier Dogwood (*Cornus sericea*)
NOTE	Caterpillar that is variably colored for better camouflage on different host plants
CONSERVATION STATUS	Not evaluated, but common within its range

ADULT WINGSPAN
1–1⅛ in (25–28 mm)

CATERPILLAR LENGTH
⁹⁄₁₆ in (14 mm)

CELASTRINA ECHO
ECHO BLUE
(W. H. EDWARDS, 1864)

153

The Echo Blue caterpillar hatches in two days from eggs that the adult female lays singly among terminal buds, pushing them as far out of sight as possible. The first instar chews round holes into buds, hollowing out the inside. The caterpillars do not fully enter the buds but often have their heads and necks extended deeply. They feed voraciously and forcibly eject frass, which prevents contamination of feeding sites and helps conceal their presence from predators; small predatory bugs take a large number of larvae. The larvae develop rapidly through four instars, taking as little as 12 to 14 days from egg hatch to pupation.

Mature caterpillars pupate under cover away from the plant, and the pupa overwinters. One to three broods are produced annually, depending on habitat latitude and elevation. The color variation apparent in the caterpillars may be influenced by host plant coloration. The Echo Blue is one of a number of closely related blue butterflies occurring throughout North America and Eurasia.

Actual size

The Echo Blue caterpillar is green but highly variable in the extent and coloration of additional markings. There are usually reddish dorsal plates on each segment. Some have a bold, white ventrolateral stripe, while others are pale green laterally, whitish below, and punctuated with a large, green spot on each segment.

FAMILY	Lycaenidae
DISTRIBUTION	Southern Florida, the Bahamas, Cuba, and the Cayman Islands
HABITAT	Open areas and suburbs where host plants grow
HOST PLANTS	Cycad, including *Zamia pumila* and *Cycas revoluta*
NOTE	Caterpillar protected by cyanogenic compounds derived from its host plants
CONSERVATION STATUS	Not evaluated, but a Florida subspecies was close to extinction due to overharvesting of its host plant, although it has lately shown a strong recovery

ADULT WINGSPAN
1⁹⁄₁₆–2⅛ in (40–54 mm)

CATERPILLAR LENGTH
1 in (25 mm)

EUMAEUS ATALA
ATALA
(POEY, 1832)

154

Atala caterpillars hatch from cream-colored eggs laid in groups on the leaf tips of the host plant and feed in groups, skeletonizing the tough cuticle of the leaves. They soon become brightly colored and gain a protective toxic chemical—cycasin—from the plants they feed on. They remain gregarious until almost the end of the larval stage. While most Lepidoptera have a fixed number of larval instars, the Atala caterpillar, if deprived of food, will pupate from as early as the end of the third instar, resulting in a smaller adult. The normal number of instars is five.

To pupate, larvae crawl away from their final feeding site and create mats of silk to which the pupae are attached. Pupation can occur in large clusters. As the pupae remain protected by the toxic chemicals sequestered by the caterpillar, a single pupa tasted and rejected by a predator can protect the whole group from harm. However, some predators can tolerate the toxic chemicals; these include assassin and ambush bugs (Reduviidae), curlytail lizards (Leiocephalidae), and the Cuban Tree Frog (*Osteopilus septentrionalis*).

Actual size

The Atala caterpillar is bright red, including head and prolegs, and bears seven pairs of bright yellow dorsal spots. Short hairs cover the body, and, typically of lycaenid caterpillars, the head is retracted into the thorax and can only be observed from underneath or when the caterpillar extends it to feed.

FAMILY	Lycaenidae
DISTRIBUTION	Northwestern United States (Washington State, Oregon)
HABITAT	Open areas in moderate to high elevation forests and shrub-steppe
HOST PLANTS	Sulphur Buckwheat (*Eriogonum umbellatum*)
NOTE	One of many *Euphilotes* caterpillars in western North America
CONSERVATION STATUS	Not evaluated

ADULT WINGSPAN
⅞–1 in (23–25 mm)

CATERPILLAR LENGTH
⁷⁄₁₆ in (11 mm)

EUPHILOTES GLAUCON
SUMMIT BLUE
(W. H. EDWARDS, 1871)

155

Summit Blue caterpillars hatch from their eggs in five days. The first instars do not eat their eggshells but feed primarily on the buds, flowers, and fruits of the host plant. When feeding on seeds, the caterpillars cut small, round holes and hollow out the insides with their extendable necks. Development from egg hatch to pupation takes around 30 days. The caterpillars are solitary and do not construct nests. The primary survival strategy appears to be camouflage, although ants may also attend and protect the caterpillars from the parasites that commonly attack them.

The pupae overwinter, with adults emerging in the spring. Adult Summit Blues associate closely with the host plant, the females flying from flower to flower, bending their abdomens to lay eggs on the blossoms of freshly opened blooms and avoiding the more mature ones. Females lay their eggs singly, almost always on the inside parts of an open flower. There is a single generation annually.

Actual size

The Summit Blue caterpillar is variable in color but usually pale cinnamon to red with a contrasting yellow to red, broken stripe dorsally and laterally. The pale setae are fine and longer laterally than dorsally. The caterpillar's coloration deepens to dark red as it approaches pupation.

FAMILY	Lycaenidae
DISTRIBUTION	North Africa, southern Europe
HABITAT	Hot, dry grasslands up to 3,600 ft (1,100 m) elevation
HOST PLANTS	Various, including *Dorycnium* spp., *Genista* spp., *Lotus* spp., and *Ononis* spp.
NOTE	Plump, pale green caterpillar that is attended by ants
CONSERVATION STATUS	Not evaluated, but under threat in much of its range

ADULT WINGSPAN
⅞–1¼ in (22–32 mm)

CATERPILLAR LENGTH
⁹⁄₁₆–¹¹⁄₁₆ in (15–18 mm)

156

GLAUCOPSYCHE MELANOPS
BLACK-EYED BLUE
(BOISDUVAL, [1828])

Black-eyed Blue caterpillars hatch from eggs laid singly by the female butterfly on a range of plants of the Fabaceae family found growing on dry grassland. The green larvae are well camouflaged and rarely spotted as they feed. As with other *Glaucopsyche* species, the caterpillars are tended by ants, particularly ants in the genus *Camponotus*. The ants give protection from parasitoids and predators in exchange for a sticky secretion (honeydew) that they milk from the caterpillars. The species overwinters as a pupa, which is found on the ground near the host plants.

The adults, which have a distinctive blue coloring with black eyespots, referenced in the common name, eclose and are on the wing between May and July, and there is a single generation annually. Although this is a widespread butterfly, it does not occur in any significant numbers in any one place. The species is under threat from the cultivation of its grassland habitats as well as habitat losses to tourism, industry, and even energy schemes.

Actual size

The Black-eyed Blue caterpillar is pale green and plump with a tapered body shape. There are dorsal and lateral stripes of dark green, pale green, and white running the length of the body. The body is also covered in short, white setae.

FAMILY	Lycaenidae
DISTRIBUTION	Australia, subtropical to temperate east coast
HABITAT	Coastal sandstone areas and understory of tall, open eucalyptus forests
HOST PLANTS	Hazel Pomaderris (*Pomaderris aspera*) and other *Pomaderris* spp.
NOTE	Caterpillar that is well camouflaged but heavily parasitized by flies
CONSERVATION STATUS	Not evaluated, but usually uncommon, though occasionally locally abundant

ADULT WINGSPAN
1–1⅛ in (26–28 mm)

CATERPILLAR LENGTH
¹¹⁄₁₆ in (18 mm)

HYPOCHRYSOPS BYZOS
YELLOW-SPOT JEWEL
(BOISDUVAL, 1832)

157

The Yellow-spot Jewel caterpillar skelctonizes the underside of the foliage as it feeds. It is difficult to spot as the line on its back looks like a leaf vein. The caterpillar pupates in a depression on the underside of the leaf, usually on one that has not been fed upon. There is one generation a year; the caterpillar stage lasts ten months with little feeding during the winter months.

Yellow-spot Jewel butterflies have brilliant red, orange, and black bands edged with iridescent green on the underside. They fly fast and are seldom seen but will bask in the sun and feed at flowers. The *Hypochrysops* genus has at least 57 species, mainly from tropical rain forests of New Guinea and Australia. Unlike most caterpillars of the genus, this species does not have a close association with ants, which can help protect against parasitic insects; tachinid flies often attack the larvae.

Actual size

The Yellow-spot Jewel caterpillar is bluish green or yellowish green and flat, with its head withdrawn beneath the thorax. There is a cream middorsal line, and darker green and brown dorsal and dorsolateral markings. The abdominal segments are well defined and lobed laterally, the lobes having a dense fringe of pale hairs.

FAMILY	Lycaenidae
DISTRIBUTION	New Guinea, northern Australia
HABITAT	Subcoastal and coastal habitats, including mangroves, that support the attendant ant
HOST PLANTS	Wide range (at least 12 families), including *Senna* spp., *Terminalia* spp., *Smilax* spp., *Cassia* spp., and commonly mangroves, such as *Flagellaria* spp. and *Ceriops* spp.
NOTE	Caterpillar that is always attended by protective Green Tree Ants
CONSERVATION STATUS	Not evaluated, but locally common

ADULT WINGSPAN
1³⁄₁₆ in (30 mm)

CATERPILLAR LENGTH
¹⁵⁄₁₆ in (24 mm)

158

HYPOLYCAENA PHORBAS
BLACK-SPOTTED FLASH
(FABRICIUS, 1793)

Actual size

Black-spotted Flash caterpillars hatch from white, pitted eggs laid by the female on the underside of a host plant leaf. The larvae are myrmecophilous—benefited by ants—and invariably attended and protected by Green Tree Ants (*Oecophylla smaragdina*). The caterpillars generally pupate in a shelter on the food plant but also occasionally in overlapping foliage from neighboring trees, often in groups. The pupa is attached into a silk pad by anal hooks and a central, silken girdle. The pupae are also attended by Green Tree Ants.

Hypolycaena phorbas caterpillars can be found throughout the year, during which time several generations are completed. They usually hide by day under a leaf or in a leaf shelter and feed at night on the leaves, young shoots, buds, and flowers of their food plant. Several caterpillars can be found in a shelter, usually in young terminal leaves. The butterfly adults feed at flowers and fly rapidly, with males defending territory from a vantage point at the end of a twig.

The Black-spotted Flash caterpillar is variable in color, with markings that are either bright green with longitudinal green, orange, or reddish dorsal bands edged with white, or dark reddish brown with white lines. The prothoracic plate is green and conceals the greenish-yellow head.

FAMILY	Lycaenidae
DISTRIBUTION	Southeastern Australia
HABITAT	Open forests and temperate eucalypt woodland
HOST PLANTS	Wattle (*Acacia* spp.)
NOTE	Gregarious caterpillar that can emit audible sounds
CONSERVATION STATUS	Not evaluated, but locally common

ADULT WINGSPAN
1¼–1⅜ in (32–35 mm)

CATERPILLAR LENGTH
¹¹⁄₁₆ in (18 mm)

JALMENUS EVAGORAS
IMPERIAL HAIRSTREAK
(DONOVAN, 1805)

159

Imperial Hairstreak caterpillars hatch in the spring from clusters of overwintering eggs laid in crevices on twigs and bark during the previous fall, usually on small wattles less than 6 ft (1.83 m) high. They feed gregariously during the day on the foliage and are always attended by numerous small, black ants, usually *Iridomyrmex* species. The ants, which probably protect the caterpillars from parasitism and predation, collect secretions from glands near the rear of the caterpillar's body. Two or three *Jalmenus evagoras* generations are completed each year, and a colony can remain on the same or nearby trees for many years.

Pupation of the caterpillars often occurs on a communal web, the pupae suspended head down by a central girdle and anal hooks. Both the mature larva and the pupa are capable of making audible sounds, a feature not uncommon, at least for pupae, in species associated with ants. A large colony of the larvae can defoliate small trees. The genus is endemic to Australia and contains 11 known species.

Actual size

The Imperial Hairstreak caterpillar is olive green to black with an orange-brown ventrolateral band and with an oblique white dorsolateral line to each segment. There are dorsal and dorsolateral tubercles on the thorax and the abdomen, and the caterpillar has fine, marginal hairs. A pair of eversible tentacle organs on the eighth abdominal segment secrete volatile substances that attract ants.

FAMILY	Lycaenidae
DISTRIBUTION	Across Europe and Africa, central and South Asia, southern parts of China, Southeast Asia, Australia, New Zealand, and Hawaii
HABITAT	Grasslands, downlands, waste grounds, and gardens
HOST PLANTS	*Cytisus* spp., *Lathyrus* spp., *Medicago* spp., and other members of family Fabaceae
NOTE	Sluglike caterpillar that kills fellow larvae
CONSERVATION STATUS	Not evaluated, but widespread and common in much of its range

ADULT WINGSPAN
1⁵⁄₁₆–1⁵⁄₁₆ in (24–34 mm)

CATERPILLAR LENGTH
³⁄₈–½ in (10–12 mm)

160

LAMPIDES BOETICUS
LONG-TAILED BLUE
(LINNAEUS, 1767)

Actual size

Long-tailed Blue caterpillars hatch from eggs laid singly by the female butterfly on the flowers and flower buds of the host plants. The larvae feed on flowers first and then move to the pods, where they burrow inside to eat the seeds. The caterpillars pupate within the withered pods and leaves, which fall to the ground. Unusually, the caterpillars are cannibalistic, attacking one another until only one caterpillar survives on each plant. In some areas, the caterpillars are tended by ants that milk them for a sugary secretion and probably provide some protection against parasitoid wasps and flies.

The adults are on the wing in summer, and in many areas there are several overlapping generations. The species is most common in the southern Mediterranean. Despite its small size, the Long-tailed Blue migrates over long distances, crossing mountain ranges and oceans—hence its wide distribution, which is steadily increasing. As eggs or larvae, the species has also been imported with ornamental plants.

The Long-tailed Blue caterpillar is sluglike, with a brown head, a brown dorsal stripe, and pale brown lateral stripes. There are several color forms that range from creamy white to pale green and dark green. The body is covered in short setae.

FAMILY	Lycaenidae
DISTRIBUTION	Africa, across southern and eastern Europe, and Asia Minor as far as the Himalayas
HABITAT	Grasslands, wastelands, parks, and gardens
HOST PLANTS	Wide range, including hawthorn (*Crataegus* spp.), *Melilotus* spp., and *Plumbago* spp.
NOTE	Plump caterpillar that feeds on flowers and seeds
CONSERVATION STATUS	Not evaluated

ADULT WINGSPAN
¾–1⅛ in (20–29 mm)

CATERPILLAR LENGTH
⅜ in (10 mm)

LEPTOTES PIRITHOUS
COMMON ZEBRA BLUE
(LINNAEUS, 1767)

161

Common Zebra Blue caterpillars hatch from eggs laid by the female butterfly close to flower buds and flowers. They feed on the flowers and later the seeds, and their complete life cycle through pupation to adult eclosion and flight takes between four and eight weeks, depending on the climate. The fast-flying species is on the wing from February to October, and there are several generations a year. The pupae of the final generation overwinter.

Common Zebra Blues tend to occur singly and in small groups, but large congregations have been recorded around nectar plants such as lucerne crops. Despite its small size, the butterfly is a strong flier and migrates over long distances, including oceans. It is common across Africa, where the caterpillars are seen on a wide range of host plants. The species' alternative common name, Lang's Short-tailed Blue, was given to a specimen that was spotted in the United Kingdom in 1938. However, it is not closely related to the various "tailed-blues" and "short-tailed blues" of the genus *Cupido*.

Actual size

The Common Zebra Blue caterpillar is variable in color, ranging from olive green to almost white. The plump body is tapered toward the hind end. There is a central, dark dorsal stripe and pale, oblique stripes along the sides. The body is covered in short hairs.

FAMILY	Lycaenidae
DISTRIBUTION	From northeastern India through areas of Southeast Asia to New Guinea and northern Australia
HABITAT	Range of tropical lowland environments, including mangroves, riparian rain forests, open forests, and urban areas
HOST PLANTS	Feeds on the larvae of Green Tree Ants (*Oecophylla smaragdina*)
NOTE	Predatory caterpillar that lives within arboreal Green Tree Ant nests
CONSERVATION STATUS	Not evaluated, although uncommon and rarely seen

ADULT WINGSPAN
2¹³⁄₁₆–3 in (71–76 mm)

CATERPILLAR LENGTH
1¹⁄₃₂– 1³⁄₁₆ in (26–30 mm)

162

LIPHYRA BRASSOLIS
MOTH BUTTERFLY
WESTWOOD, [1864]

The Moth Butterfly caterpillar hatches on or near the arboreal nest of the Green Tree Ant. It enters the nest to consume ant larvae; usually no more than one or two caterpillars feed in one nest. The caterpillar's head senses ant brood, and the antennae are used to pull ant larvae to the mouthparts so they can be consumed without attack from adult ants. Young caterpillars probably produce chemicals that allay ant aggression as small caterpillars would be easy prey for the ants. Larger caterpillars have a cuticle comprising overlapping, scale-like, setal sockets that provides a strong, flexible mechanical barrier. A caterpillar can devour all green ant larvae in one nest.

Pupation occurs within the ant nest in the caterpillar's final instar skin. Upon emerging, the butterfly is protected by white scales on its new wings that clog the mandibles of the attacking ants. In some regions of Southeast Asia, ant larvae and pupae from these easily accessible nests are harvested for human food, and the Moth Butterfly caterpillars are typically consumed as well.

Actual size

The Moth Butterfly caterpillar is orange brown, oval, flattened, and slightly convex with an upturned margin and dorsally three central transverse grooves. The head is white, and the antennae are long. Tiny setae cover the body and are denser on the lateral margin and ventral surface.

FAMILY	Lycaenidae
DISTRIBUTION	France, the Netherlands, and eastern Europe into Russia and Kazakhstan
HABITAT	Damp grasslands and meadows, and marshes
HOST PLANTS	Various docks and sorrels (*Rumex* spp.)
NOTE	Well-camouflaged green caterpillar that is found in wetland habitats
CONSERVATION STATUS	Near threatened

ADULT WINGSPAN
1⅛–1¼ in (28–32 mm)

CATERPILLAR LENGTH
¾ in (20 mm)

LYCAENA DISPAR

LARGE COPPER
(HAWORTH, 1803)

163

The female Large Copper lays eggs, either singly or in small groups, on the leaves of host plants found near water. Two weeks later the eggs hatch, and the caterpillars feed on the underside of leaves, leaving the upper surface intact. Their nibbling creates a small groove in which the young larvae rest. As they get older, they feed on the whole leaf. The young caterpillar overwinters at the base of the host plant and is able to survive for as long as two months underwater if its habitat floods in winter. It resumes feeding and pupates the following spring as a yellow-brown chrysalis attached by silk to the host plant stem.

The adults, which have beautiful, iridescent copper wings, eclose and fly from June to July. The species has experienced a severe decline in much of its range due to the loss of its wetland habitats. In some regions, notably the United Kingdom, the species became extinct, and there have been several failed attempts to reintroduce it.

Actual size

The Large Copper caterpillar is green in color, plump, and tapered toward the back. There are faint vertical white lines marking the segments and scattered white dots. The body is covered in tiny, white setae.

FAMILY	Lycaenidae
DISTRIBUTION	Northwestern United States, from Washington State to California and east to Colorado and Montana
HABITAT	Montane meadows and roadsides
HOST PLANTS	Knotweed and smartweed (*Polygonum* spp.), and docks and sorrels (*Rumex* spp.)
NOTE	Slow-moving, cryptic caterpillar that produces fast-moving, dazzling butterfly
CONSERVATION STATUS	Not evaluated, but common

ADULT WINGSPAN
1–1 ³⁄₁₆ in (25–30 mm)

CATERPILLAR LENGTH
¾ in (20 mm)

LYCAENA EDITHA
EDITH'S COPPER
(MEAD, 1878)

164

Actual size

Edith's Copper caterpillars hatch in late spring from overwintering eggs and feed voraciously, completing their development through four instars within three weeks. The larvae feed only on leaves, grazing mostly on the lower surfaces, eventually eating holes through them and later feeding on leaf edges. No nests are built. The green caterpillars depend on camouflage for protection, and attendance by ants is common, aiding defense from predators and parasitoids. The pupae hatch in early summer, and there is a single generation annually.

The range of hosts used by Edith's Copper is uncertain, and it is possible that many other species in the Polygonaceae family are utilized. The adults are strongly attracted to flowers such as Yarrow (*Achillea millefolium*) and aster (Asteraceae), and males aggressively guard low perches. *Lycaena editha* can be abundant locally and may be expanding its range northward into Canada. It is related to a number of similar *Lycaena* species occurring in high-elevation habitats.

The Edith's Copper caterpillar is bright green with small, white spots peppering the body. A distinct, middorsal, red line is present. There is a dense covering of short, orange-brown setae on the body and the green head. The spiracles are pink orange encircled in brown. Prior to pupation the red stripe fades.

FAMILY	Lycaenidae
DISTRIBUTION	Southern mainland Australia, Tasmania
HABITAT	Heathland, ranging from alpine to semi-arid inland areas
HOST PLANTS	Native peas of the Fabaceae family, including Gorse Bitter-pea (*Daviesia ulicifolia*), other *Daviesia* spp., *Aotus* spp., and *Bossiaea* spp.
NOTE	Caterpillar camouflaged to match the color of flowers it feeds on
CONSERVATION STATUS	Not evaluated, but locally common

ADULT WINGSPAN
¾ in (20 mm)

CATERPILLAR LENGTH
½ in (12 mm)

NEOLUCIA AGRICOLA

FRINGED HEATH-BLUE

(WESTWOOD, [1851])

165

Fringed Heath-blue caterpillars hatch in the early spring from eggs laid during the previous spring or early summer. Hatching coincides with the flowering of the food plants. The early instars burrow into the calyx, feeding within the flower, while the larger caterpillars consume entire flowers. The mature caterpillars are difficult to find as they blend in with the red and yellow flowers on which they are feeding. The caterpillars develop rapidly, and adults are flying by the end of spring. There is only one generation a year.

The caterpillars pupate on the stems of the food plant. The adults fly close to the ground, and males are known to fly on hilltops. The genus contains only three species, every one confined to Australia. The caterpillars of all species are usually not attended by ants, but occasionally, in Western Australia, ants have been found associated with *Neolucia agricola*.

Actual size

The Fringed Heath-blue caterpillar is weakly scalloped laterally and is variable in color, although usually green or reddish green with a broad, dark, reddish-green dorsal band edged with white and a reddish lateral band edged ventrally in white. The caterpillar has short, blunt, paired protuberances on the thorax and abdomen and numerous lateral white hairs.

FAMILY	Lycaenidae
DISTRIBUTION	Inland eastern Australia
HABITAT	Arid to semi-arid acacia woodlands
HOST PLANTS	Grey Mistletoe (*Amyema quandang*)
NOTE	Nocturnal-feeding caterpillar that is often attended by small ants
CONSERVATION STATUS	Not evaluated, but large areas of known habitat have been cleared for agriculture

ADULT WINGSPAN
1⁵⁄₁₆ in (34 mm)

CATERPILLAR LENGTH
⅞ in (22 mm)

OGYRIS BARNARDI

BRIGHT PURPLE AZURE

(MISKIN, 1890)

166

Caterpillars of the Bright Purple Azure hatch from eggs that are laid singly on the leaves and flower buds of the Grey Mistletoe host plant. Early instar caterpillars feed openly on the leaves during the day, but larger caterpillars shelter by day in borer holes or under loose bark close to the shrub. They are usually attended by a few small, black ants and emerge at night to feed. In southern areas of the species' range, there is one generation a year, but in northern areas there are several broods.

The caterpillars pupate where they shelter, attached by anal hooks into a silken pad and supported by a central silken girdle. The adults fly rapidly around the host tree, settling frequently on dead twigs or feeding at flowers. The largely Australian *Ogyris* genus has 15 species, many with blue, iridescent coloration, all associated with ants, although for three species the caterpillars are thought to be predatory on immature ants.

Actual size

The Bright Purple Azure caterpillar is pale grayish green to pinkish brown with flattened, dark brown, anterior and posterior extremities. There are pale brown, dorsal, chevron markings, and the anal and prothoracic plates are dark brown. The body is covered in brown and black, minute, secondary setae. The head is yellowish brown and hidden beneath the prothorax.

FAMILY	Lycaenidae
DISTRIBUTION	Eastern and southeastern Australia
HABITAT	Eucalypt woodlands supporting high densities of host plant
HOST PLANTS	Several species of mistletoe, mainly *Amyema* spp.
NOTE	Nocturnal-feeding caterpillar always attended by large, aggressive sugar ants
CONSERVATION STATUS	Not evaluated, but locally common in the north of its range and uncommon in the south

ADULT WINGSPAN
1⅞–2¹⁄₁₆ in (47–53 mm)

CATERPILLAR LENGTH
1¼ in (32 mm)

OGYRIS GENOVEVA

SOUTHERN PURPLE AZURE

(HEWITSON, [1853])

167

The caterpillars of the Southern Purple Azure hatch from eggs laid singly on the mistletoe host plant or under loose bark. The small caterpillars congregate under bark and are attended by large sugar ants, emerging at night to feed on the mistletoe leaves. Larger caterpillars rest in temporary ant nests in hollow limbs of the host tree or under rocks, or in galleries underground at the tree base. Up to 200 caterpillars can occur in these nests, and they follow silk trails each night to feed on mistletoe that may be more than 33 ft (10 m) away.

The caterpillars pupate in the ant nests and remain guarded by the ants. The pupa is attached by anal hooks into a silken pad and supported by a central girdle. The adults are local and fast flying, and males are strong hilltoppers. The genus *Ogyris* has 15 species, 14 from Australia and one from Papua New Guinea. All are associated with ants, but in three species the caterpillars are thought to prey on ant broods.

Actual size

The Southern Purple Azure caterpillar is yellowish brown and dorsally dull purplish brown, with pale yellow, dorsal, chevron markings. The lateral edges are scalloped, and there are some short, peripheral, bristly hairs. The head is brown and withdrawn below the prothorax. The eighth abdominal segment has a pair of eversible organs that are involved in chemical communication with ants.

FAMILY	Lycaenidae
DISTRIBUTION	United States, from Connecticut west to southeast Iowa and Missouri, south to east Texas, the Gulf Coast, and peninsular Florida, with rare strays to Michigan and Wisconsin
HABITAT	Forest and forest edges
HOST PLANTS	Oak (*Quercus* spp.)
NOTE	Cryptically green caterpillar found on the underside of leaves
CONSERVATION STATUS	Not evaluated, but common

ADULT WINGSPAN
1¼–1⅝ in (32–41 mm)

CATERPILLAR LENGTH
¾–1 in (20–25 mm)

PARRHASIUS M-ALBUM
WHITE M HAIRSTREAK
(BOISDUVAL & LECONTE, 1833)

168

Actual size

White M Hairstreak caterpillars hatch from eggs likely laid at the tip of branches of large trees, as do other hairstreak species that feed on oak. Although the species' complete life history has not yet been described, it is probable that the young larvae, like those of many other hairstreaks, feed on buds and fresh growth before moving to larger leaves. There are three generations a year. Two will pupate on the underside of host plant leaves, while the overwintering generation likely pupates in leaf litter. Before pupation, the caterpillar turns reddish brown, makes a silk pad on the leaf, and attaches itself with a girdle.

Like other lycaenid larvae, this caterpillar can be distinguished from those of other families by the fleshy lobe on each proleg, flanked on each side by rows of crochets—hardened hooklike structures that help larvae adhere to leaves and other surfaces. The White M Hairstreak is the most northerly representative of an otherwise tropical genus, whose other five members are found from Mexico to Bolivia.

The White M Hairstreak caterpillar is green, turning reddish brown prior to pupation, and, typical of lycaenid caterpillars, is stout and sluglike. It is covered with minute, white setae and has a retracted head inside the prothorax, which stays hidden unless the caterpillar is feeding. The ventral surface is closely pressed against the leaf surface and flat. Underneath, the caterpillar is lightly colored.

FAMILY	Lycaenidae
DISTRIBUTION	Within 12 sq mile (31 km²) area of the Antelope Desert of southern Oregon, United States
HABITAT	Open alluvial, volcanic ash-pumice desert
HOST PLANTS	Spurry Buckwheat (*Eriogonum spergulinum*)
NOTE	One of the most range-restricted, smallest, and endangered caterpillars
CONSERVATION STATUS	Not evaluated, but considered vulnerable and of concern

ADULT WINGSPAN
1¹¹⁄₁₆–³⁄₄ in (18–20 mm)

CATERPILLAR LENGTH
³⁄₈ in (10 mm)

PHILOTIELLA LEONA
LEONA'S LITTLE BLUE
HAMMOND & MCCORKLE, 2000

169

The tiny caterpillars of Leona's Little Blue develop from egg to chrysalis in just 10 to 12 days and feed only on the flowers and flower buds of Spurry Buckwheat. The vibrantly colored red-and-white larvae are cryptic on the red-and-white host plant. The adult flies from mid-June to late July, and most caterpillars pupate before the end of the flight period. The pupae oversummer and overwinter on the ground, withstanding temperatures ranging from 23°F (-5°C) to 154°F (68°C). Pupae sometimes take two to three years to produce butterflies.

First discovered in 1995, Leona's Little Blue is a highly specialized and range-restricted species, occupying a volcanic ash and pumice desert ecosystem and dependent on a similarly specialized host plant. Leona's Little Blue is closely related to the more widespread Small Blue (*Philotiella speciosa*), which occurs commonly in desert areas of California and Nevada.

Actual size

The Leona's Little Blue caterpillar is mostly white with vivid, bloodred markings in the form of an interrupted middorsal stripe and two stripes on either side. Tiny, short, white hairs densely cover the body. Underneath, the caterpillar is red, with yellow claspers and black legs.

FAMILY	Lycaenidae
DISTRIBUTION	Western North America, from British Columbia and Montana to southern California and New Mexico
HABITAT	Forest openings, shrub-steppe, subalpine meadows, and roadsides
HOST PLANTS	Lupine (*Lupinus* spp.)
NOTE	Caterpillar that is dormant for up to nine months
CONSERVATION STATUS	Not evaluated, but usually common, although some subspecies are threatened or endangered

ADULT WINGSPAN
1³⁄₁₆–1³⁄₈ in (30–35 mm)

CATERPILLAR LENGTH
³⁄₈–½ in (10–12 mm)

170

PLEBEJUS ICARIOIDES
BOISDUVAL BLUE
(BOISDUVAL, 1852)

Actual size

Boisduval Blue caterpillars hatch from pale, greenish-white eggs laid on lupines by the female butterfly some five to seven days earlier. First instar larvae feed for 14 days before molting to second instars. In most areas, second instars enter dormancy or diapause in midsummer and rest at the base of host plants. The larvae remain in this state through the fall and the winter, and then resume feeding the following spring on new plant growth. They feed on leaves initially, then focus on flowers and fruits. From this point, larval development to the fourth and final instar is rapid, with pupation on the host plant, in debris, or under stones occurring after 40 days. The pretty blue adult butterflies eclose in early April.

Caterpillars of this species are tended by ants, which provide protection from natural enemies such as wasps and predatory bugs. In turn, the caterpillars secrete a sugary substance on which the ants feed. Camouflage, diurnal concealment, and ant attendance are likely important features of defense.

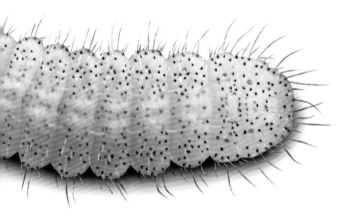

The Boisduval Blue caterpillar is green with a middorsal dark stripe and pale borders. The entire body is covered in tiny, black dots. The head is black but usually concealed. Long and short, pale setae adorn the body.

FAMILY	Lycaenidae
DISTRIBUTION	Europe, east into southern Russia, Turkey, Iraq, and Iran
HABITAT	Grassy slopes, especially on calcareous soil
HOST PLANTS	Horseshoe Vetch (*Hippocrepis comosa*), other *Hippocrepis* spp., *Coronilla* spp., *Lotus* spp., and *Securigera* spp.
NOTE	Green-and-yellow caterpillar that is tended by ants
CONSERVATION STATUS	Not evaluated

ADULT WINGSPAN
1³⁄₁₆–1⅜ in (30–35mm)

CATERPILLAR LENGTH
⁹⁄₁₆–¹¹⁄₁₆ in (15–18mm)

POLYOMMATUS BELLARGUS
ADONIS BLUE
(ROTTEMBURG, 1775)

171

The female Adonis Blue lays her eggs, singly, on the underside of leaves of the host plant, especially those growing in short grass. The caterpillars hatch and feed by day; in early instars, they consume the underside of leaves, leaving the upper surface intact. More mature larvae rest openly on leaves, moving to the base of the host plant, or on soil beside it, at dusk. They also pupate on the ground, usually in a small hollow.

There are two generations a year, with the first on the wing from May to June and the second from August to September. The caterpillars of the second generation overwinter and become active in spring. The larvae are attended by ants, especially those of the genera *Myrmicia*, *Lasius*, and *Tapinoma*. The ants feed on the caterpillar's sweet secretions and in return protect it from parasites and predators. The ants may even bury the overwintering caterpillar or pupa underground in a chamber linked to their nest.

Actual size

The Adonis Blue caterpillar has a sluglike shape and is dark green in color. There are two dorsal lines and one lateral line of broken yellow stripes running the length of the body. The caterpillar is covered in short setae.

FAMILY	Lycaenidae
DISTRIBUTION	North Africa, across Europe into Russia and the Middle East
HABITAT	Oak woodlands, parks, and sometimes gardens with oak trees
HOST PLANTS	Oak (*Quercus* spp.)
NOTE	Brown, sluglike caterpillar that feeds on young oak leaves
CONSERVATION STATUS	Not evaluated, but declining across much of its range

ADULT WINGSPAN
$^{15}/_{16}$–1$^{1}/_{8}$ in (24–28 mm)

CATERPILLAR LENGTH
$^{9}/_{16}$–$^{11}/_{16}$ in (15–18 mm)

172

QUERCUSIA QUERCUS
PURPLE HAIRSTREAK
(LINNAEUS, 1758)

Actual size

The eggs from which Purple Hairstreak caterpillars hatch are laid singly at the base of oak buds, where they remain for the winter. The larvae emerge in spring and burrow into the bud, out of sight of predators, to feed on the young leaves. The older caterpillar spins a loose silk web over a cluster of leaves and remains within it, feeding at night. When mature, the caterpillar drops to the ground to pupate. Here, it may be tended by ants that bury it in the leaf litter.

There is a single generation. The iridescent adult butterflies are on the wing high up in the oak canopy from June to August, so are often overlooked. The species depends on the oak tree to complete its life cycle, and numbers have declined as oak woodland has been lost. The species is also sometimes referred to as *Favonius quercus*.

The Purple Hairstreak caterpillar is brown and sluglike, with the body covered in short hairs. Dorsally there are a series of chevron-like shapes in pale brown running the length of the body and a broken lateral stripe in pale cream brown. At the end of the abdomen is a false head.

FAMILY	Lycaenidae
DISTRIBUTION	Western North America, from British Columbia to New Mexico
HABITAT	Oak-pine forest openings, canyons, riparian areas, and shrub-steppe
HOST PLANTS	Bitterbrush (*Purshia tridentata*)
NOTE	Well-camouflaged caterpillar that feeds on buds and leaves
CONSERVATION STATUS	Not evaluated, but periodically numbers fluctuate widely

ADULT WINGSPAN
1–1⅜ in (25–35 mm)

CATERPILLAR LENGTH
1¹¹⁄₁₆ in (17 mm)

SATYRIUM BEHRII

BEHR'S HAIRSTREAK
(W. H. EDWARDS, 1870)

173

Behr's Hairstreak caterpillars hatch from eggs laid on Bitterbrush stems, with which they contrast sharply. The eggs are not hidden in crevices but placed on exposed surfaces, where they overwinter, hatching in spring when the host plant starts sprouting new leaves. Initially, the caterpillars feed only on Bitterbrush buds, hollowing out the contents completely with their extendable necks. When half grown, they feed on leaves. Their coloring blends well with the pastel-green host plant, providing excellent camouflage and protection from enemies. The caterpillars develop through four instars and do not construct nests.

Mature larvae leave the host plant and pupate under cover about 24 days after hatching. The adults emerge two weeks later. Hairstreaks may be locally common but are subject to wide fluctuations in population numbers and may be scarce for many years at a time. The butterflies readily take nectar from flowers such as milkweed (*Asclepias* spp.), thistle (*Cirsium* spp.), and buckwheat (*Eriogonum* spp.), where they grow close to Bitterbrush.

Actual size

The Behr's Hairstreak caterpillar is dark forest green, highlighted with white markings. The longitudinal white line along the ventrolateral margin is prominent, bordered above and below with thin, dark green lines. Each segment has a distinct, dorsolateral, diagonal white line, bordered below with dark green and above with lighter green.

FAMILY	Lycaenidae
DISTRIBUTION	North America, south to Colombia and Venezuela
HABITAT	Most habitats, from urban areas to mountaintops, avoiding dense forest
HOST PLANTS	Very wide range, often pea (Fabaceae) and mallow (Malvaceae) families
NOTE	Versatile feeder that adopts the colors of its host plant
CONSERVATION STATUS	Not evaluated, but common except at the periphery of its range

ADULT WINGSPAN
1–1⅜ in (25–35 mm)

CATERPILLAR LENGTH
1¹¹⁄₁₆ in (17 mm)

174

STRYMON MELINUS
GRAY HAIRSTREAK
HÜBNER, 1818

The Gray Hairstreak caterpillar is a true generalist feeder that will develop on just about any plant—native or introduced. It hatches in about three days from eggs laid on buds, flowers, or leaves; pupation occurs 27 days later. The larvae prefer to feed on flowers and will adopt the color of the flower on which they are feeding. However, if flowers are not available, they will readily feed on leaves and other plant parts. Ant associations occur and help defend the caterpillars from enemies.

Pupation takes place in a sheltered location, such as a curled leaf, and the pupa usually overwinters. There are multiple generations annually, depending on latitude and elevation. Sometimes, the caterpillars cause problems for commercial bean production, but it is rare for large numbers of them to occur in small areas. The Gray Hairstreak is one of the first butterflies to emerge in spring and one of the last to stop flying in late fall.

Actual size

The Gray Hairstreak caterpillar is enormously variable in coloration. The larvae may be green, gray, tan, orange, olive, yellow, pink, or purple, depending on the host plant they are utilizing. There are few if any contrasting markings, although some forms have two indistinct dorsal lines. Its light colored setae are short and dense.

FAMILY	Lycaenidae
DISTRIBUTION	Parts of Africa, the Middle East, the Balkans, Iran, and much of India
HABITAT	Dry savannahs
HOST PLANTS	Christ's Thorn (*Paliurus spinus-christi*) and *Ziziphus* spp.
NOTE	Green caterpillar that lives in hot, dry grasslands
CONSERVATION STATUS	Least concern

ADULT WINGSPAN
$^{11}/_{16}$–$^7/_8$ in (18–22 mm)

CATERPILLAR LENGTH
$^3/_8$–$^1/_2$ in (10–12 mm)

TARUCUS BALKANICUS
LITTLE TIGER BLUE
(FREYER, 1844)

The female Little Blue Tiger, also known as the Balkan Pierrot, lays her eggs at the base of thorns on the host plant and occasionally on leaves. The caterpillars nibble away a groove on the underside of the leaf, leaving a characteristic transparent bar in the leaf surface. The species overwinters as a pupa. The fast-flying adults—the male much bluer in color than the browner female—eclose and are on the wing from April through to the end of summer, and there is a series of overlapping broods.

In the Balkans, the area of southeast Europe after which the species is named, there has been a considerable decline in the population, by as much as 30 percent in some areas, but numbers are holding up well in other parts of its range. There are several subspecies, including the Black-spotted Pierrot (*Tarucus balkanicus nigra*), which is found in India. The *Tarucus* genus comprises 23 species, commonly known as blue Pierrots.

Actual size

The Little Tiger Blue caterpillar is green in color with a distinctive, yellow-orange dorsal stripe running the length of the body. There are pale green spots laterally, and the whole body is covered in short, white hairs.

FAMILY	Lycaenidae
DISTRIBUTION	Southeastern Indonesia, Timor, mainland New Guinea, and Australia
HABITAT	Eucalypt woodlands and open forests; also semi-arid and arid scrublands
HOST PLANTS	Mainly wattle (*Acacia* spp.), but also small gum trees (*Eucalyptus* spp.) and occasionally species of Fabaceae, Sapindaceae, and Combretaceae
NOTE	One of the few butterfly caterpillars that feeds on *Eucalyptus*
CONSERVATION STATUS	Not evaluated, but common in northern areas of its range

ADULT WINGSPAN
⅞ in (22 mm)

CATERPILLAR LENGTH
½ in (13 mm)

176

THECLINESTHES MISKINI
WATTLE BLUE
(T. P. LUCAS, 1889)

Actual size

The Wattle Blue caterpillar is often attended by numerous ants from one of five genera. The ants may provide some protection from predators and parasitoids, and in return the caterpillars supply the ants with nutrient-rich secretions. The caterpillars feed openly on the young leaves and sometimes on the fleshy galls and flowers of their host plants—often small trees, and seedlings or suckers from larger trees. The larvae are sometimes found in ant nests or under leaf litter on the ground. In warmer areas, the caterpillars complete their development in four weeks or less.

The caterpillars generally pupate on the food plant and are attached by anal hooks and a central silk girdle. The species is common in central and northern Australia, breeding there throughout the year. The adult male butterflies will hilltop, defending territory from perches on the highest foliage. The genus comprises six species all restricted to Australasia.

The Wattle Blue caterpillar is prominently humped at the thorax and either green or dark purplish brown in color. It has a broad, reddish-brown or green dorsal band, often more pronounced on the thoracic segments and abdominal segments one to six, which is edged in white and has sparse marginal hairs.

FAMILY	Lycaenidae
DISTRIBUTION	Portugal, Spain, southern France, and North Africa
HABITAT	Dry grasslands and meadows up to 5,600 ft (1,700 m) elevation
HOST PLANTS	Various Fabaceae members, including *Anthyllis* spp., *Astragalus* spp., *Dorycnium* spp., *Lotus* spp., and *Medicago* spp.
NOTE	Olive-green caterpillar that is tended by ants
CONSERVATION STATUS	Not evaluated

ADULT WINGSPAN
1⅛–1³⁄₁₆ in (28–30 mm)

CATERPILLAR LENGTH
⁹⁄₁₆–¹¹⁄₁₆ in (15–18 mm)

TOMARES BALLUS

PROVENCE HAIRSTREAK

(FABRICIUS, 1787)

177

Provence Hairstreak caterpillars hatch from small, pale eggs laid by the female butterfly, who hides them under a leaf of the host plant. The young larvae burrow into flower buds, where they feed out of sight of predators, sometimes taking on the color of the flower to improve their camouflage. The more mature larvae emerge and feed on leaves. Ants of various genera, including *Plagiolepis*, attend the caterpillars, feeding on the sugary secretion (honeydew) produced by the larvae and in return protecting them from parasites and predators. The caterpillar is often transported by ants into their anthill to pupate, and the brown pupa will overwinter there.

There is a single generation annually, and the butterflies are seen on the wing from January to April. The species is endangered in southern France and Spain due to the move away from traditional grazing of grassland by livestock, a practice that kept the length of the sward down. *Tomares ballus* is more common in North Africa.

Actual size

The Provence Hairstreak caterpillar is olive green in color with a sluglike shape. There is a dark dorsal band with a series of oblique, yellow-brown marks either side and a pink-brown mark on each side of the thorax. The body is covered in short, white setae.

FAMILY	Nymphalidae
DISTRIBUTION	The Himalayas, northeast India, southern China, Chinese Taipei, and much of Southeast Asia
HABITAT	Open forests and disturbed areas where invasive host plants flourish
HOST PLANTS	Members of nettle family (Urticaceae), including *Boehmeria* spp., *Debregeasia* spp, *Elatostema* spp, *Urtica* spp., and *Pouzolzia* spp.; also *Buddleja* spp.
NOTE	Hungry, unpalatable, foul-smelling, spiny caterpillars that live en masse
CONSERVATION STATUS	Not evaluated, but locally common

ADULT WINGSPAN
2⅜–2¾ in (60–70 mm)

CATERPILLAR LENGTH
1⅜ in (35 mm)

ACRAEA ISSORIA
YELLOW COSTER
(HÜBNER, 1819)

178

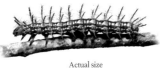

Actual size

The Yellow Coster caterpillar has a purplish-brown base color with broken white lines running the length of its body. On each body segment along these lines are rings of long spines with multiple branchlets over their length. The base and shaft of each spine are orange, and they are tipped with black.

Yellow Coster females lay several dozen eggs on the undersides of leaves, which take around 20 days to hatch. The caterpillars live gregariously and, as a result, frequently defoliate their host plants. Like the adult butterfly, the caterpillars are known to be unpalatable and malodorous to predators, but they can suffer heavy losses to parasitoid flies and wasps in the pupal stage. The chrysalis is white with yellow and black markings clearly defining the underlying anatomy. Pupation occurs on the host plant or nearby vegetation, and the butterfly emerges after 15 days. Depending on location, there are two (western China) to five (Chinese Taipei) generations annually. In Chinese Taipei, the butterfly overwinters as the larval stage.

The Yellow Coster butterfly is slow-flying and of variable appearance but with a yellowish-brown base color, black markings, and leathery, thinly scaled wings that can appear transparent with wear and tear. Populations can be concentrated to form "roosts" where the host plants grow, as the adults do not travel far.

FAMILY	Nymphalidae
DISTRIBUTION	Western United States, Mexico
HABITAT	Oak woodlands and riparian areas
HOST PLANTS	Oak (*Quercus* spp.)
NOTE	Caterpillar that has a variety of defense plans
CONSERVATION STATUS	Not evaluated, but common

ADELPHA CALIFORNICA
CALIFORNIA SISTER
(BUTLER, 1865)

ADULT WINGSPAN
3–3⅛ in (75–80 mm)

CATERPILLAR LENGTH
1⅜–1⁹⁄₁₆ in (35–40 mm)

179

California Sister caterpillars are masters of defense, employing camouflage, chemistry, and aggression to avoid predators. They feed only on oak leaves and adopt a variety of protective strategies that change during development. The newly hatched caterpillar feeds at the edge of a leaf, forming a pier from a vein or midrib. Frass is used to extend this pier, which the caterpillar rests upon when not feeding, and which somehow confers protection. Mid-stage caterpillars adopt a cryptic serpentine posture when at rest, while mature caterpillars are aggressive and will attempt to bite if disturbed.

Both the butterflies and caterpillars of this species are thought to be distasteful to bird and mammal predators. Some other butterflies mimic California Sisters in appearance to gain some protection by reputation. *Adelpha* is a genus of about 85 butterflies found in the western and southern United States, Mexico, and South America. They are commonly known as "sisters" due to the white markings on their wings, which resemble a nun's habit.

Actual size

The California Sister caterpillar is bright green with many tiny, white spots and nine pairs of orange, spiny horns of varying length. The spiny head is dark purplish brown with a pair of enlarged, black-tipped, short spines that resemble eyes. Prior to pupation, the caterpillar turns light brown.

FAMILY	Nymphalidae
DISTRIBUTION	Europe, much of Asia, east to Japan; also a small population recently discovered in New England, United States
HABITAT	Open woodlands, meadows, parks, gardens, and grasslands up to 8,200 ft (2,500 m) elevation
HOST PLANTS	Nettle (*Urtica* spp.) and hop (*Humulus* spp.)
NOTE	Jet-black, spiny caterpillar that builds a communal web
CONSERVATION STATUS	Not evaluated, but common

ADULT WINGSPAN
2⅛–2⅜ in (55–60 mm)

CATERPILLAR LENGTH
1⁹⁄₁₆–1¾ in (40–45 mm)

AGLAIS IO
PEACOCK
(LINNAEUS, 1758)

180

Peacock caterpillars hatch from green eggs laid in an untidy mass of up to 400 on the underside of a stinging nettle leaf. First instars build a communal web near the top of the plant, from which they emerge to bask and feed, and are usually highly conspicuous. Feeding may occur at any time of the day and night. As the caterpillars develop, they move to new plants, building new webs along the way. Webs are decorated with shed larval skins and frass, and are easily found.

The caterpillars have several strategies to avoid predation. If disturbed, a group of larvae will often jerk their bodies from side to side in unison—probably to appear larger—while a single caterpillar may regurgitate bitter, green fluid, curl up in a ball, and drop to the ground. There are five instars, and mature larvae leave the plant to pupate in nearby vegetation. The Peacock is a long-lived butterfly that overwinters as an adult and is one of the first butterflies to be seen in spring.

The Peacock caterpillar is jet black with numerous white dots all over its body. There are six rows of barbed spines, and the head is black and shiny. The true legs are black, and the prolegs are reddish orange.

Actual size

FAMILY	Nymphalidae
DISTRIBUTION	North America, except far north and far south
HABITAT	Wherever stinging nettles grow in the lowlands and mountains, especially along watercourses
HOST PLANTS	Nettle (*Urtica* spp.)
NOTE	Gregarious, spiny caterpillar
CONSERVATION STATUS	Not evaluated, but common

ADULT WINGSPAN
1⁹⁄₁₆–1¾ in (40–45 mm)

CATERPILLAR LENGTH
1–1³⁄₁₆ in (25–30 mm)

AGLAIS MILBERTI
FIRE-RIM TORTOISESHELL
(GODART, 1819)

181

Fire-rim Tortoiseshell caterpillars hatch from eggs laid in untidy masses of 20 to 900, often piled on top of each other, usually on the underside of a terminal nettle leaf. First instar larvae feed communally, skeletonizing and webbing leaves. Webbing increases in second and third instars, providing access and support for caterpillars between leaves and shoots. Fourth and fifth instars become solitary and live in folded nettle leaves tied with silk. Prepupal caterpillars wander, and most pupae are formed away from nettles.

Development is rapid, pupation occurring three weeks after egg hatch. Protection from enemies is based on aggregation in early instars and concealment later. Mature caterpillars also gain protection from being very spiny. Natural enemies include predatory bugs and parasitic wasps. Fire-rim Tortoiseshell butterflies may migrate from lowlands to high-elevation areas in summer to escape hot and dry conditions. The adults overwinter and may live for up to ten months.

The Fire-rim Tortoiseshell caterpillar is black, peppered with white dots dorsally, and has well-developed black and smaller white spines. Laterally, two intermittent wavy, creamy yellow lines border the spiracles, and the prolegs are white. The head is shiny and black with medium length, white setae.

Actual size

FAMILY	Nymphalidae
DISTRIBUTION	From Europe across Asia to the Pacific coast; also a small population recently discovered in New England, United States
HABITAT	Varied, from parks and gardens to farmlands, grasslands, and woodlands
HOST PLANTS	Nettle (*Urtica dioica* and *U. urens*)
NOTE	Caterpillar that, when young, clusters in silk webs on nettle
CONSERVATION STATUS	Not evaluated

ADULT WINGSPAN
1¾–2⁹⁄₁₆ in (45–65 mm)

CATERPILLAR LENGTH
¾–⅞ in (20–22 mm)

182

AGLAIS URTICAE
SMALL TORTOISESHELL
(LINNAEUS, 1758)

Small Tortoiseshell caterpillars hatch from eggs laid on the underside of leaves. The larvae are gregarious, clustering together in the safety of a silken web. They emerge from the web to feed and bask in the sun. As they move from plant to plant in search of food, they build a new web, leaving a trail of defoliated stems, silk threads, and droppings. When disturbed, a group of caterpillars will twist their bodies from side to side to deter a would-be predator. Sometimes, they roll up and drop to the ground. After the last molt, they disperse and pupate.

Typically, there are two generations a year, with the adults of the second generation overwintering in caves and buildings such as garages and sheds. In some parts of this species' range, particularly in Europe, there has been a steep decline in numbers. Some scientists suspect that this is linked to global warming and increasing numbers of the parasitic fly, *Sturma bella*.

The Small Tortoiseshell caterpillar has a black head and a mostly black body covered with tiny, creamy-white spots. There are broken yellow stripes and rows of yellow and black spines along the back and sides.

Actual size

FAMILY	Nymphalidae
DISTRIBUTION	From southern United States (San Francisco Bay area in the west and Virginia in the east) and the Caribbean, south to central Argentina
HABITAT	Open areas, disturbed habitats, and suburbs where its host plant occurs
HOST PLANTS	Passion vine (*Passiflora* spp.)
NOTE	Aposematic caterpillar protected against predators by toxic plant alkaloids
CONSERVATION STATUS	Not evaluated, but common

ADULT WINGSPAN
2⅜–3¾ in (60–95 mm)

CATERPILLAR LENGTH
1%₁₆ in (40 mm)

AGRAULIS VANILLAE

GULF FRITILLARY

(LINNAEUS, 1758)

183

Gulf Fritillary caterpillars hatch from yellow eggs laid on or near their food plant, where the larvae feed on all parts, sometimes defoliating the host. Although the caterpillars are protected from most predators by spines and black-and-orange coloration, warning of the toxicity conferred by plant alkaloids, in California the Greater Roadrunner (*Geococcyx californianus*) still finds a way to feed on them. Paper wasps, ants, lizards, predatory bugs, and praying mantids will also eat the larvae. Many are killed by parasitic tachinid flies—in Florida up to 90 percent can be parasitized in the fall.

Mature larvae pupate in a chrysalis resembling a dead leaf; in north Florida, larvae may hibernate through the winter when their host plant freezes until it sprouts fresh growth in the spring. There are multiple generations annually, uninterrupted by hibernation or diapause in the tropical parts of the range. To avoid cold temperatures and take advantage of the host plants' broad distribution, the butterflies move seasonally up and down the coast of the southeastern United States.

The Gulf Fritillary caterpillar is orange and black. Some larvae have only black spines, legs, and prolegs, while others have broad, black, longitudinal stripes, which sometimes dominate the general coloration. This makes the caterpillar appear black with orange stripes rather than orange with black stripes.

Actual size

FAMILY	Nymphalidae
DISTRIBUTION	The Andes of Venezuela, south through Bolivia
HABITAT	Subtemperate to upper subtropical forest edges and second growth
HOST PLANTS	*Erato* spp.
NOTE	Warningly colored caterpillars that are undoubtedly toxic to some predators
CONSERVATION STATUS	Not evaluated, but not considered threatened

ADULT WINGSPAN
2⅝–2⅞ in (66–73 mm)

CATERPILLAR LENGTH
1⁹⁄₁₆–1⅞ in (40–48 mm)

184

ALTINOTE DICAEUS
RED-BANDED ALTINOTE
(LATREILLE, [1817])

In many parts of its range, the Red-banded Altinote caterpillar is among the most commonly encountered and conspicuous of lepidopterous larvae. This is due, in part, to the fact that the eggs are laid in clusters of 50 to 110, with the hatching cohorts quickly devouring one host plant before munching down a second. Additionally, some time after the third instar, larvae disperse and crawl away to find food where there are fewer siblings to compete with, so are more likely to be seen as they wander farther afield. Pupation occurs after a period of wandering, often some distance away from the food plant, but with pupae always hanging in a sheltered cranny under a leaf or branch, or even the eaves of a house.

Adults are noticeably weak fliers and are often found post-mortem in roadways, having been run over while taking a meal from cow dung or mineral-enriched sand. The Latin name of the Red-banded Altinote is frequently misspelled, for instance, as *dice* or *diceus*, which has led to some taxonomic confusion over the two centuries since the species' original description.

The Red-banded Altinote caterpillar is split-toned: shiny black above and creamy yellow below. It is generally bare except for long, spiny scoli arranged in rows along the dorsum, subdorsum, and spiracular areas. The head is shiny black.

Actual size

FAMILY	Nymphalidae
DISTRIBUTION	The Andes, from Venezuela to at least central Peru
HABITAT	Cloud forests dominated by *Chusquea* bamboo around landslides and forest edges
HOST PLANTS	Bamboo (*Chusquea* spp.)
NOTE	Caterpillar that changes color but keeps its tails throughout development
CONSERVATION STATUS	Not evaluated, but not considered threatened

ADULT WINGSPAN
3¼–3⁷⁄₁₆ in (82–88 mm)

CATERPILLAR LENGTH
2¹⁄₁₆–2¼ in (52–57 mm)

ANTIRRHEA ADOPTIVA
ADOPTED MORPHET
(WEYMER, 1909)

185

Adopted Morphet caterpillars hatch from flattened, dome-shaped eggs laid in neat rows of two to seven on the underside of host leaves. First instar larvae feed in loose aggregations on the same leaf, but in the second and subsequent instars individuals disperse to adjacent leaves, with the distance between larvae increasing with age. They are frequently attacked by parasitic flies and wasps but have a small, pinkish, eversible neck gland that, presumably, helps deter their enemies. On the dorsum of the thorax, a pair of silvery-colored fissures mark the openings of "grooming glands" whose exact function is poorly known.

From oviposition to adult, the life cycle is quite variable in length, taking from 120 to 143 days. The butterflies are crepuscular, searching for mates and for oviposition sites in the early morning and just after sunset. They are almost exclusively found close to their bamboo host plants and frequently pause to feed on rotting fruit or animal dung.

The Adopted Morphet caterpillar is complexly patterned but overall chalky pink with irregular, rusty markings laterally and a bright yellow or whitish stripe middorsally that becomes light blue toward the rear. Laterally, it is marked with bright orange and black. It has a pair of long, whitish caudal tails, and the head is bright orange and roundly triangular in shape.

Actual size

FAMILY	Nymphalidae
DISTRIBUTION	From southern England and western France east across Europe and temperate Asia to China, Japan, and Korea
HABITAT	Deciduous, broad-leaved woodlands
HOST PLANTS	Willow (*Salix* spp.) and poplar (*Populus* spp.)
NOTE	Cryptic, sluglike caterpillar that has a pair of threatening horns
CONSERVATION STATUS	Not evaluated, but vulnerable in some countries, including England

ADULT WINGSPAN
2¾–3½ in (70–90 mm)

CATERPILLAR LENGTH
1¾–2⅛ in (45–55 mm)

186

APATURA IRIS
PURPLE EMPEROR
(LINNAEUS, 1758)

Purple Emperor caterpillars hatch from green eggs laid singly in late summer on the upper surfaces of host tree leaves. First instars lie perfectly camouflaged along the upper midrib of leaves and feed nocturnally. After reaching the second or third instar, the caterpillars enter hibernation resting on a silk pad spun on the upper surface of a withered leaf or twig, turning brown to blend in with their surroundings during the winter. Feeding resumes in spring, and the large, plump, green caterpillars pupate in June. The pupa, suspended from a leaf, is perfectly camouflaged, and the adult emerges after two weeks.

The Purple Emperor is an iconic butterfly, especially in England, where its relatively few well-known breeding sites often become tourist attractions, with people craning their necks to spot the emperors flying in the treetops. *Apatura iris* is single-brooded, with adults on the wing in July and August. The caterpillars can be found from August to June.

Actual size

The Purple Emperor caterpillar is bright green, speckled with tiny, yellow dots. There are diagonal yellow lines laterally and a pair of long, green-and-white anterior horns tipped in red. The head is brown and white.

FAMILY	Nymphalidae
DISTRIBUTION	Most of Europe and the Middle East, across central Asia to China and Japan
HABITAT	Sheltered places in damp woodlands, woodland margins and glades, and hedgerows
HOST PLANTS	Grasses, including Cock's-foot (*Dactylis glomerata*), meadow grass (*Poa* spp.), and Couch Grass (*Elytrigia repens*)
NOTE	Solitary caterpillar that emerges at night to feed
CONSERVATION STATUS	Not evaluated, but common in most parts of its range

ADULT WINGSPAN
1⅝–2⅟₁₆ in (42–52 mm)

CATERPILLAR LENGTH
1 in (25 mm)

APHANTOPUS HYPERANTUS
RINGLET
(LINNAEUS, 1758)

Ringlet caterpillars hatch from eggs scattered by the female while flying over grass. These solitary caterpillars are nocturnal, hiding during the day near the base of grass tussocks and emerging at night to feed on young leaves. If disturbed, they drop to the ground and remain still to avoid predation. The larvae hibernate, although they will feed during mild weather, becoming fully active again in spring. Pupation takes place in early summer when the mature caterpillars move to the base of the grass plant.

There is a single generation a year. The adults, which live for about two weeks, are on the wing during summer months, spending time resting on blades of grass and, unusually, flying on overcast days and even in the rain. *Aphantopus hyperantus* is not threatened—in fact, studies have shown it to be increasing in parts of its range, including the United Kingdom.

Actual size

The Ringlet caterpillar is pale reddish brown in color with many tiny, brown dots. The body is covered in short, brown setae. There is a cream lateral line and also a distinctive brown dorsal line that gets darker toward the rear. The head is dark brown with several pale, longitudinal stripes.

FAMILY	Nymphalidae
DISTRIBUTION	North Africa, Europe, Turkey, and across temperate Asia to China and Japan
HABITAT	Open woodlands with glades and rides, woodland margins, bracken-covered slopes, and subalpine meadows
HOST PLANTS	Violet (*Viola* spp.)
NOTE	Striking spiny caterpillar that basks in the sun
CONSERVATION STATUS	Not evaluated, but locally vulnerable

ADULT WINGSPAN
2⅛–2¹¹⁄₁₆ in (55–69 mm)

CATERPILLAR LENGTH
1½–1⅝ in (38–42mm)

188

ARGYNNIS ADIPPE
HIGH BROWN FRITILLARY
([DENIS & SCHIFFERMÜLLER], 1775)

The female High Brown Fritillary butterfly lays her pale pink eggs singly on dead leaves and stems either on or near the food plant. The eggs overwinter, becoming gray as the larvae develop within, and hatch in spring. The caterpillars feed along the edge of *Viola* leaves, leaving characteristic cutouts along the leaf margin. They are active during the day and can often be spotted basking in the sun. The caterpillars pupate on their host plant, suspended below a twig or leaf by a silk pad, and take on the appearance of a shriveled leaf.

The adult butterflies are on the wing in summer, with a single generation annually. In parts of its range, especially the United Kingdom, the species has declined significantly, mostly as a result of the shift away from traditional woodland management in which glades and rides were opened up to allow more light to reach the woodland floor.

The High Brown Fritillary caterpillar is either light or dark reddish brown with a prominent white line running along the length of its back and dividing a series of dorsal black spots. There are rows of backward-pointing brown spines, both dorsally and laterally.

Actual size

FAMILY	Nymphalidae
DISTRIBUTION	Northern United States, southern Canada
HABITAT	Open boggy areas within cool, boreal spruce fir habitats
HOST PLANTS	Violet (*Viola* spp.)
NOTE	Secretive, nocturnal, spiny caterpillar that is rarely seen
CONSERVATION STATUS	Not evaluated, but not common

ADULT WINGSPAN
2⅜–2⁹⁄₁₆ in (60–65 mm)

CATERPILLAR LENGTH
1⅜–1⁹⁄₁₆ in (35–40 mm)

ARGYNNIS ATLANTIS
ATLANTIS FRITILLARY
(W. H. EDWARDS, 1862)

189

Atlantis Fritillary caterpillars hatch from eggs laid singly in late summer and early fall on or near dried-up violets. They immediately overwinter under debris and rocks, without feeding, and only start to feed when violets produce growth in spring. The larvae, which are solitary with no nests, take about two months to reach the pupal stage. Most feeding occurs at night, and the caterpillars rest under leaves by day. Later instars have an eversible ventral "neck" gland that produces a musky odor when they are disturbed. The odor is thought to repel predators such as ground beetles and ants.

Pupae are formed in silked-together "leaf-tents" close to the ground. Adults fly during June to August, and males emerge before females. The males feed on mud and animal scat, as well as flowers such as thistles and yarrow. The species is closely related to other northern boreal fritillaries, including *Argynnis hesperis*. The Atlantis Fritillary is never common, and populations are sedentary with little movement away from breeding habitats.

The Atlantis Fritillary caterpillar is dark brown to black with a pair of distinct dorsal white stripes and a pattern of white lines reminiscent of crocodile skin. The black head has distinct brown markings. The spines are black, except for those on the sublateral row, which are bright orange.

Actual size

FAMILY	Nymphalidae
DISTRIBUTION	Western United States, from Washington State south to Arizona and southern California
HABITAT	Shrub-steppe, canyons, hillsides, and mountain meadows
HOST PLANTS	Violet (*Viola* spp.)
NOTE	Secretive, spiny caterpillar that feeds nocturnally on desert violets
CONSERVATION STATUS	Not evaluated, but common

ADULT WINGSPAN
2⅜–2¾ in (60–70 mm)

CATERPILLAR LENGTH
1⅜–1⁹⁄₁₆ in (35–40 mm)

190

ARGYNNIS CORONIS
CORONIS FRITILLARY
(BEHR, 1864)

Coronis Fritillary caterpillars hatch from eggs laid singly during fall in shrub-steppe among patches of dried-up violets. Initially they do not feed but, instead, they hide under rocks and plant debris for overwintering. The larvae begin feeding in spring as the violets develop new growth. The caterpillars are rarely seen as they are nocturnal, feeding by night and hiding under rocks by day. Young larvae prefer to feed on violet flowers and young leaves at first, moving to older leaves later. They have three modes of defense—concealment, their spines, and an eversible ventral gland that emits a bad odor.

Pupation occurs close to the ground within a few leaves silked together to provide a tent. Adults migrate up to 100 miles (160 km), leaving the hot, dry shrub-steppe to spend summer in cooler, flower-rich, high-elevation meadows. They return to the shrub-steppe for egg-laying in early fall. It is unknown how females find and recognize areas of dried-up violets for egg-laying but it is assumed that they are able, somehow, to "smell" the host plants.

Actual size

The Coronis Fritillary caterpillar is dark gray to black, mottled with white spots and patches, especially laterally. There is a prominent pair of middorsal, white stripes, and the black dorsal spines have orange bases. A row of lateral spines is orange. The head is black with very few or no orange markings.

FAMILY	Nymphalidae
DISTRIBUTION	North America, from southern Canada to central United States
HABITAT	Mid–high elevation, lightly forested, hilly, and mountainous areas
HOST PLANTS	Violet (*Viola* spp.)
NOTE	Spiny caterpillar that lives on the forest floor
CONSERVATION STATUS	Not evaluated, but common

ADULT WINGSPAN
2¾–3 in (70–75 mm)

CATERPILLAR LENGTH
1¾–2 in (45–50 mm)

ARGYNNIS CYBELE
GREAT SPANGLED FRITILLARY
(FABRICIUS, 1775)

191

Great Spangled Fritillary females lay their eggs singly in late summer and early fall on the forest floor where violets are present. The newly hatched caterpillars do not feed but overwinter in dormancy under plant debris, fallen branches, stones, and rocks. Feeding commences when violets begin to grow in spring. The larvae are mostly nocturnal, rarely seen, and feed chiefly at night. Second and later instar caterpillars have an eversible ventral "neck" gland that produces a musky odor when disturbed. This likely provides protection from ground-crawling natural enemies, such as ants and ground beetles.

Like other fritillaries, the caterpillars develop through six instars over a period of about two months. Their pupae are more wriggly than other fritillary species and are formed close to the ground under a protective tent made of silked-together leaves. Adults emerge in June, and the females are relatively inactive for a month or so, preferring to shelter in cool areas before they start to lay eggs.

Actual size

The Great Spangled Fritillary caterpillar is jet black with an orange head dorsally. Its spines are pale to bright orange with black tips or, in the two dorsal rows, all black with bright orange bases. The true legs are black, and the prolegs are brown.

FAMILY	Nymphalidae
DISTRIBUTION	Western North America, from Alaska and Manitoba to Arizona and Colorado
HABITAT	Moist alpine and subalpine meadows, watercourses, and roadsides
HOST PLANTS	Violet (*Viola* spp.)
NOTE	Variably colored, nocturnal, spiny caterpillar that is associated with violets
CONSERVATION STATUS	Not evaluated, but common

ADULT WINGSPAN
1¾–2 in (44–50 mm)

CATERPILLAR LENGTH
1⁵⁄₁₆–1⅜ in (30–35 mm)

192

ARGYNNIS MORMONIA
MORMON FRITILLARY
(BOISDUVAL, 1869)

Female Mormon Fritillaries crawl on the ground seeking out suitable oviposition sites among violets, where they lay their eggs singly. The eggs hatch after ten days, and first instar caterpillars seek refuge under leaves and rocks for overwintering. For optimum survival they require humid conditions. In spring, the larvae start feeding on violets as plant growth commences. The caterpillar's defense is based on spines, chemical protection from the bad odor produced by a ventral "neck" gland, and concealment. The coloration and markings of the larvae are influenced by geography and elevation.

Development to pupation takes about two months. Mature caterpillars silk leaves together as "pupation tents," and pupation occurs close to the ground; pre-pupae that fall to the ground are able to pupate successfully. Adults, produced after a further two or three weeks, fly from June to September. Dispersal of adults is limited, and males patrol for females. Both sexes feed on flower nectar, but males will also feed on animal scat, mud, and carrion.

Actual size

The Mormon Fritillary caterpillar is typically dark brown to black, with orangey-white spines and a dorsal, orangey-white stripe bisected by a dark line. Lateral pale markings are limited. The black head is covered dorsally with brown markings, dotted with black. Some forms are almost jet black with orange spines.

FAMILY	Nymphalidae
DISTRIBUTION	Southern and eastern United States, extending into Mexico
HABITAT	Woodland edges, riparian areas, and fields
HOST PLANTS	Hackberry (*Celtis* spp.)
NOTE	Thorny-headed, long-tailed, camouflaged caterpillar
CONSERVATION STATUS	Not evaluated, but common

ADULT WINGSPAN
1%₁₆–1⅞ in (40–47 mm)

CATERPILLAR LENGTH
1³⁄₁₆–1⅜ in (30–35 mm)

ASTEROCAMPA CELTIS

HACKBERRY EMPEROR
(BOISDUVAL & LECONTE, 1835)

193

Female Hackberry Emperor butterflies lay their white or pale yellow eggs singly or in small groups on the underside of Hackberry leaves, choosing trees with new growth. Young caterpillars rest there and are particularly easy to see at night if you shine a flashlight upward. Third instars turn brown and overwinter in rolled leaves, sometimes falling to the ground in the fall. In spring, the larvae climb back up the tree and resume feeding. Many different kinds of insect predators feed on Hackberry Emperor caterpillars, and some fly and wasp parasitoids destroy both caterpillars and pupae.

In northerly parts of the range, only one generation is produced annually, but two or three generations appear elsewhere. The adults have a very rapid flight, and males perch on foliage to await females or aggressively patrol territories. The females are less active than the males, but both sexes can be attracted to rotting fruit baits.

The Hackberry Emperor caterpillar is variably colored green with white dots. A pair of yellow dorsal lines extends along the body from the base of the spiny horns on the head, and there are also yellow lateral stripes. A pair of short tails extends from the posterior.

Actual size

FAMILY	Nymphalidae
DISTRIBUTION	Eastern North America into Mexico
HABITAT	Forest woodlands
HOST PLANTS	Hackberry (*Celtis* spp.)
NOTE	Gregarious caterpillar that becomes solitary as it matures
CONSERVATION STATUS	Not evaluated, but secure throughout most of its range

ADULT WINGSPAN
1⅝–2¾ in (42–70 mm)

CATERPILLAR LENGTH
1⁹⁄₁₆ in (40 mm)

ASTEROCAMPA CLYTON
TAWNY EMPEROR
BOISDUVAL & LECONTE, 1835

194

Actual size

The Tawny Emperor caterpillar is light green with some whitish stripes and markings, and it has two sharp tails and two multipronged antlers on the head. The similar Hackberry Emperor caterpillars have longer antlers and are usually solitary on hackberry leaves, even when young.

Young Tawny Emperor caterpillars feed and live together after hatching from a mass of several hundred eggs laid by the female butterfly on mature hackberry leaves. Half-grown larvae turn brown in the fall, when groups of about ten silk several leaves to a branch to make a nest for the winter. In spring, they become solitary and rest on the underside of a leaf curled downward with silk. The green pupa is flattened for its entire length against the underside of a leaf or twig and attached to a silked spot on the leaf by just a ⅛ in (3 mm) area of tiny hooks on the abdomen tip.

The larvae resemble those of the Hackberry Emperor (*Asterocampa celtis*), which usually feed on younger, tenderer hackberry leaves than *A. clyton*. Although Tawny Emperor butterflies are generally uncommon or rare, they sometimes explode in numbers locally. When this occurs, the subsequent caterpillars can almost defoliate some trees. Adults often glide with wings spread, and they prefer to feed on tree sap, rotten fruit, or even decaying carrion, rather than flowers.

FAMILY	Nymphalidae
DISTRIBUTION	The Himalayas, India, southern China, Chinese Taipei, and most of Southeast Asia
HABITAT	Open forests and surrounding secondary growth
HOST PLANTS	*Glochidion* spp. and *Phyllanthus* spp. (both Phyllanthaceae); also *Wendlandia* spp. (Rubiaceae)
NOTE	Caterpillar that uses perches and frass barriers to avoid predators
CONSERVATION STATUS	Not evaluated, but common in its main range

ADULT WINGSPAN
2⅛–2⁹⁄₁₆ in (55–65 mm)

CATERPILLAR LENGTH
1⅜–1⁹⁄₁₆ in (35–40 mm)

ATHYMA PERIUS
COMMON SERGEANT
(LINNAEUS, 1758)

195

Early larval stages of the Common Sergeant isolate themselves from potential threats by constructing perches, where they rest between meals. By consuming the leaf lamina either side of the midrib of the leaf tip and building a barrier of frass woven together with silk at its entrance, the caterpillar can deter hungry ants and other predators. Although the maturing caterpillar is adorned with multiple branched spines, they are purely ornamental and non-stinging. When it does feel threatened, the caterpillar adopts a defensive posture, arching the thorax and pressing the head flat against the leaf, presenting its spines as a physical barrier.

The adult Common Sergeant is a low-flying, territorial butterfly that patrols paths and clearing edges, often basking on the ground or on low vegetation. Typical of the sergeants, the topside is black with white dots and dashes; the underside is a light brown with white spots highlighted with black. The species occurs year-round within its main range.

Actual size

The Common Sergeant caterpillar is green with multiple rows of branched "antlers" running the length of its body. The trunks of these spines are red, the branches black and tipped with white points. The base of each spine is a deep purple. The head capsule is bordered with spines, and there are multiple contrasting conical nodules on the surface, resembling a face.

FAMILY	Nymphalidae
DISTRIBUTION	Tropical and subtropical Asia
HABITAT	Wooded areas, roadways, paths, and margins of clearings
HOST PLANTS	Plants of Rubiaceae family, including *Adina* spp., *Mussaenda* spp., and *Wendlandia* spp.
NOTE	Caterpillar that, when young, uses excrement as a defensive barrier
CONSERVATION STATUS	Not evaluated, but very common

ADULT WINGSPAN
2⅛–2%₁₆ in (55–65 mm)

CATERPILLAR LENGTH
1⅜ in (35 mm)

ATHYMA SELENOPHORA
STAFF SERGEANT
(KOLLAR, 1844)

The Staff Sergeant caterpillar leaves behind the sturdy midrib when it consumes its host plant leaf. In the early instars, this is where it isolates itself by building a barrier of frass at the leaf margin as a deterrent against intruders such as ants. As the caterpillar grows, the frass patch itself becomes the ideal hiding place as it perfectly matches its brown livery. By the time the leaf larder is exhausted and the caterpillar is too large to take advantage of the precarious perch, it molts into the mature larval green colors and rests on the topsides of leaves. The larval instars take 30 days, and the pupal period lasts 13 days.

The Staff Sergeant butterfly is gender dimorphic: males have fewer but more pronounced white markings on a black background. As many Southeast Asian butterflies were described and named in colonial times, often by collectors and laymen associated with the military, there is a preponderance of British military and peerage common names among them, hence the name "sergeant" (others include the commanders, dukes, earls, barons, and lascars).

Actual size

The Staff Sergeant caterpillar has branched red "antlers" on each segment. The body is green with pinpoint white spots throughout and a dark dorsal saddle mid-abdomen. The base of each proleg is red orange with additional small, white spines. There are white lateral patches on the first and last abdominal segments. The red head capsule has simple spines around its margins and white tubercles on the face.

FAMILY	Nymphalidae
DISTRIBUTION	North America, from Alaska and western Canada to Washington State and Montana
HABITAT	Arctic-alpine and high Arctic rockslides, ridges, and tundra
HOST PLANTS	Saxifrage (*Saxifraga* spp.)
NOTE	High-alpine caterpillar that takes two years to reach maturity
CONSERVATION STATUS	Not evaluated, but rare in parts of its range

ADULT WINGSPAN
2–2⅛ in (50–55 mm)

CATERPILLAR LENGTH
1–1⅜ in (25–35 mm)

BOLORIA ASTARTE

ASTARTE FRITILLARY
(DOUBLEDAY & HEWITSON, 1847)

197

The eggs of the Astarte Fritillary are laid singly on or near the host plant in midsummer. First instars hatch after eight days and do not consume their eggshells. Some development takes place in the first year, with caterpillars overwintering as first or second instars. Feeding then recommences in spring, and the caterpillar reaches the fourth or fifth instar by the end of the short summer. The mature caterpillar overwinters and pupates on or near the host plant in the following June. Adults emerge from late July to early August.

Astarte Fritillary caterpillars use camouflage, concealment, and chemicals produced from a ventral gland to defend themselves. The species, a true alpine specialist, generally occurring above 8,200 ft (2,500 m), may be vulnerable to climate warming or may adapt to longer summers. There is some evidence that the caterpillars can complete development in a single season if they are exposed to warm conditions.

Actual size

The Astarte Fritillary caterpillar is black with contrasting gray-white dorsal markings in complex but regular patterns of dashes and Vs. All spines and setae are black, with the bases of the two dorsal rows circled in yellow gold. White spots pepper the body laterally, and the spiracles are black and narrowly encircled in white.

FAMILY	Nymphalidae
DISTRIBUTION	Across Europe and northern Asia to Russian Far East
HABITAT	Woodland clearings, margins, and tracks; and recently coppiced wood with open ground and bracken
HOST PLANTS	Violet (*Viola* spp.)
NOTE	Dark, bristly caterpillar that feeds during the day
CONSERVATION STATUS	Not evaluated, but endangered in parts of its range

ADULT WINGSPAN
1½–1⅞ in (38–47 mm)

CATERPILLAR LENGTH
1 in (25 mm)

BOLORIA EUPHROSYNE
PEARL-BORDERED FRITILLARY
(LINNAEUS, 1758)

198

Actual size

Caterpillars of the Pearl-bordered Fritillary hatch from pale yellow eggs laid singly either on the host plant or on the surrounding leaf litter, especially bracken (*Pteridium aquilinum*). The larvae are active by day, feeding around the base of violet leaves and creating a characteristic pattern of leaf damage that gives away their presence. They molt three times and then overwinter in a dry, rolled-up leaf. Feeding resumes early in spring, before the caterpillars pupate in the leaf litter.

The adult butterflies, which get their name from the white, pearl-like spots on the underside of the hindwing, are active from April to June, while a second generation may be seen in August. The species is under threat from the loss of suitable habitat and the decline in traditional woodland management, especially coppicing, which creates the open, sunny conditions needed by the low-growing host plants. Once the ground becomes overgrown and dense, the butterfly disappears.

The Pearl-bordered Fritillary caterpillar is dark, almost black, in color. Sometimes there are pale brown hues along the ventral surface. Each segment has a ring of small, yellow tubercles, each bearing a tuft of black, bristly hairs.

FAMILY	Nymphalidae
DISTRIBUTION	Throughout much of northern North America, Asia, and Europe
HABITAT	Bogs, fens, and riparian habitats at mid-elevations
HOST PLANTS	Violet (*Viola* spp.)
NOTE	Spiny, cryptically colored, nocturnal caterpillar that is associated with violets
CONSERVATION STATUS	Not evaluated, but threatened in some locations due to declining habitat

ADULT WINGSPAN
1¾–2 in (45–50 mm)

CATERPILLAR LENGTH
1–1³⁄₁₆ in (25–30 mm)

BOLORIA SELENE
SILVER-BORDERED FRITILLARY
(DENIS & SCHIFFERMÜLLER, 1775)

199

Silver-bordered Fritillary caterpillars hatch within five to six days from eggs laid on violets. In late summer, the larvae enter dormancy and overwinter as second to fourth instars. Early instars feed on the undersides of violet leaves, while later instars eat large holes from the leaf edge inward. Their survival is based on three means of protection—concealment, spines, and a ventral gland that produces a musky odor to deter predators. There are five instars, and no nests are made. Development from first instar to pupation takes about 30 days, with adults emerging after a further 10 to 14 days.

There are one to three generations per season, depending on location. Adult males patrol conspicuously over grassy, boggy areas near violets, looking for females. After mating, females remain concealed in the vegetation. Grazing by deer or livestock is necessary to keep bog violets alive and healthy for *Boloria selene*. Without grazing, vegetative succession invariably causes extinction of both violets and Silver-bordered Fritillaries.

Actual size

The Silver-bordered Fritillary caterpillar is purplish gray, mottled with numerous black splotches and soft, yellow spines bearing many setae. The anterior three segments are black, and the horns on the first segment are long and black with yellow bases. The head is shiny black with dark setae.

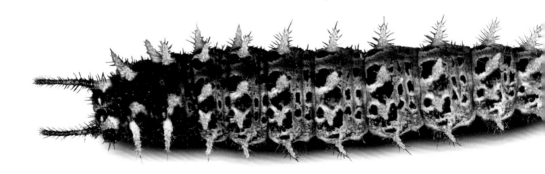

FAMILY	Nymphalidae
DISTRIBUTION	Across Europe and southern Scandinavia into southern Russia, central Asia, and northern China
HABITAT	Damp mountain meadows at 1,970–5,900 ft (600–1,800 m) elevation near to woodland
HOST PLANTS	*Polygonum bistorta* and violet (*Viola* spp.)
NOTE	Spiky, black caterpillar that is found in damp alpine meadows
CONSERVATION STATUS	Not evaluated, but classed as near threatened in Europe

ADULT WINGSPAN
1½–1⅞ in (38–48 mm)

CATERPILLAR LENGTH
¹¹⁄₁₆–⅞ in (18–22 mm)

200

BOLORIA TITANIA
TITANIA'S FRITILLARY
(ESPER, 1793)

Titania's Fritillary caterpillars hatch from eggs laid singly by the female on host plants or on nearby vegetation. The young caterpillars do not eat but go into diapause, overwintering on the host plant. In spring, the larvae, which are solitary, become active again and start to feed. They pupate on plants close to the host plant, where they remain for three weeks. The pupa is well camouflaged, with the appearance of a shriveled leaf.

There is a single generation annually. The fast-flying butterflies are on the wing from late June to August. Though still quite common in the Alps at higher altitudes, the species is in decline as a result of habitat loss, especially its favored meadows close to forests. Today there are scattered and relatively isolated colonies, which may account for the numerous subspecies that occur across the range. The species was formerly classified within the genus *Clossiana*, and many texts still refer to it as *Clossiana titania*.

The Titania's Fritillary caterpillar has a dark body that ranges from brown to black. There are rings of long tubercles, light brown in color, each bearing numerous short spines. The long, antennae-like appendages on the head are black.

Actual size

FAMILY	Nymphalidae
DISTRIBUTION	From northern Spain across central and southern Europe; also Russia, Turkmenistan, Mongolia, northern China, Korea, and Japan
HABITAT	Woodland clearings and scrub
HOST PLANTS	*Filipendula* spp. and *Rubus* spp.
NOTE	Striking caterpillar that feeds mostly on bramble species
CONSERVATION STATUS	Not evaluated, but increasingly rare in some areas

ADULT WINGSPAN
1¹¹⁄₃₂–1⁹⁄₁₆ in (26–40 mm)

CATERPILLAR LENGTH
1⅜ in (35 mm)

BRENTHIS DAPHNE
MARBLED FRITILLARY
BERGSTRÄSSER, 1780

201

The female Marbled Fritillary lays her ridged, conical-shaped, yellow-brown eggs singly on leaves and sometimes the flowers of the host plants. The embryonic larvae overwinter within the eggs and hatch the following spring. The caterpillars feed and then complete their development quickly over just a few weeks, before they pupate. The leaflike pupa, which has a ridge of spikes, is found hanging under stems and twigs.

There is a single generation annually. The adult butterflies are seen throughout the summer in warm woodland clearings and sunny slopes at altitudes of 245–5,750 ft (75–1,750 m). The species is declining in many areas due to loss of its scrubby woodland habitat, which is often cleared for agriculture and vineyards. It is one of four species within the genus *Brenthis* and is frequently confused with the Lesser Marbled Fritillary (*B. ino*), which is on the wing at the same time.

The Marbled Fritillary caterpillar has a pale brown head and body with two prominent, white dorsal stripes and a number of thin, dark brown stripes running the length of the body. There are rings of pale brown spikes bearing short, black spines around each segment.

Actual size

FAMILY	Nymphalidae
DISTRIBUTION	Hispaniola (Dominican Republic and Haiti)
HABITAT	Dry scrub, acacia-cactus woodlands, and pine forests
HOST PLANTS	Grasses, including *Poa* spp. and *Stenophrum* spp.
NOTE	Green-brown caterpillar that closely resembles related species
CONSERVATION STATUS	Not evaluated

ADULT WINGSPAN
½–⅝ in (12–16 mm)

CATERPILLAR LENGTH
¾ in (20 mm)

202

CALISTO OBSCURA
CALISTO OBSCURA
MICHENER, 1943

Calisto obscura caterpillars hatch from eggs laid on grasses. They are extremely well camouflaged on their host grasses and difficult to find, with the life cycle taking around 73 days to complete. The butterfly adults are on the wing in all seasons and possibly produce two generations a year. *Calisto obscura* is found across Hispaniola, occurring in a wide range of habitats, from sandy, coastal grassland to highland pine forest, and including very dry habitats, such as acacia-cactus woodlands.

The genus *Calisto* is endemic to the Caribbean, where there are 34 species and 17 subspecies. Most occur on the island of Hispaniola. Two species in particular, *C. confusa* and *C. batesi*, are very similar in appearance to *C. obscura*, and they often share the same habitat. There are, however, minor differences in the appearance of the caterpillars and their time of development. The *C. obscura* species has been identified conclusively through the use of DNA profiling.

The *Calisto obscura* caterpillar may have either a basic body color of olive brown or dark brown to black. The rings of tiny tubercles with short tufts of hairs around the body give a ribbed appearance. There is a pale dorsal stripe and a lateral row of pale, rhombus-shaped markings.

Actual size

FAMILY	Nymphalidae
DISTRIBUTION	Central America, from Honduras to Panama; South America, south to Peru, Bolivia, and southern Brazil
HABITAT	Humid, mid-elevation subtropical to temperate cloud forests
HOST PLANTS	*Alchornea* spp.
NOTE	Caterpillar with metallic highlights that make it resemble sparkling dew
CONSERVATION STATUS	Not evaluated, but unlikely to become endangered

ADULT WINGSPAN
2¾–3¾ in (70–95 mm)

CATERPILLAR LENGTH
2⁹⁄₁₆–3 in (65–75 mm)

CATONEPHELE CHROMIS
SISTER-SPOTTED BANNER
(DOUBLEDAY, [1848])

203

The eggs of the Sister-spotted Banner are laid singly on fresh leaves of the host plant, and young larvae build a frass chain on which to rest. Older larvae rest on the upper surface of the leaf with their long head horns pressed tightly to the leaf, waving them wildly back and forth over their body if disturbed. Such a defense is presumably effective at deterring the parasitic flies and wasps that commonly attack this species in some parts of its range. After the fifth, final instar, larvae pupate on the dorsal surface of a leaf, the emerald-green chrysalis resembling a leafy portion of the host plant.

As is the case in most other species of *Catonephele*, *C. chromis* adults are sexually dimorphic. The yellow-striped females are easily attracted to fruits placed on the ground and so are more common in collections. Males appear to prefer to fly in the canopy, guarding sunny perches near gaps or at the forest edge. A versatile species, the Sister-spotted Banner can be found from nearly sea level to above 7,875 ft (2,400 m) in some parts of the Andes.

The Sister-spotted Banner caterpillar is bright emerald green with three rows of small, caramel-orange spots along the dorsum, from each of which arises a three- to five-branched scolus. The bright yellow head is topped by two long, slender, black horns with yellow, clubbed tips, and greenish at their bases, and decorated with two or three whorls of sharp spines along their length.

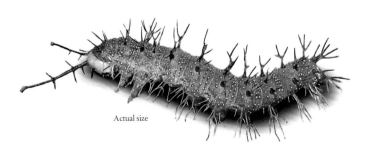

Actual size

FAMILY	Nymphalidae
DISTRIBUTION	North America (except northern Canada and the southwest United States), south to northern Mexico
HABITAT	Prairies, meadows, roadsides, parks, and forest openings
HOST PLANTS	Grasses, including *Festuca* spp., *Poa* spp., and *Avena* spp.
NOTE	Superbly camouflaged caterpillar that feeds on grass
CONSERVATION STATUS	Not evaluated, but common

ADULT WINGSPAN
1¾–2 in (45–50 mm)

CATERPILLAR LENGTH
1–1³⁄₁₆ in (25–30 mm)

CERCYONIS PEGALA
COMMON WOOD NYMPH
(FABRICIUS, 1775)

204

Common Wood Nymph caterpillars hatch from eggs laid singly nine to ten days earlier on grasses or nearby surfaces, and often tucked deep into a clump. The young caterpillars overwinter in dormancy and do not start feeding until the following spring, when grasses produce new growth. The larvae feed mostly at night on grass-blade edges, often spending the day at the base of the grass. No nests are made, and survival is based on camouflage. There are five or six instars, and pupation usually occurs on host grasses, with pupae suspended from a bent-over stem or blade, sometimes encircled by silk strands.

Development in spring to adulthood takes two to three months. A single brood of adults flies from May to September, becoming dormant during the heat of summer in many areas. During this time, females may be found in groups of 6 to 20, resting in the shade, neither feeding nor egg-laying. Both sexes feed on flower nectar and sap flows.

The Common Wood Nymph caterpillar is yellowish green, densely clothed with white spots and short, white setae, with a dark, middorsal stripe. There are two yellow stripes laterally, the lower one bolder. The head is green with white spots, and the posterior segment has two red-tipped tails.

Actual size

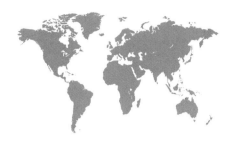

FAMILY	Nymphalidae
DISTRIBUTION	Java to Timor and northern Australia
HABITAT	Monsoon forests, especially along streams
HOST PLANTS	Lacewing Vine (*Adenia heterophylla*)
NOTE	Brightly colored, gregarious caterpillar
CONSERVATION STATUS	Not evaluated, but of no immediate conservation concern

ADULT WINGSPAN
2⅝ in (66 mm)

CATERPILLAR LENGTH
1¼ in (32 mm)

CETHOSIA PENTHESILEA
ORANGE LACEWING
(CRAMER, 1777)

205

Orange Lacewing caterpillars hatch from clusters of eggs usually laid on the tendrils of the food plant vine or on the upper surface of leaves. The caterpillars are gregarious and consume large amounts of leaf material and green stems; in captivity they will also consume fruit and woody stems. The larvae pupate together, suspended head down by the cremaster. They can be found throughout the year, although adult butterflies are most numerous early in the dry season (April to July).

The caterpillars have aposematic coloration and are distasteful to predators, as are many species in this subfamily of butterflies (Heliconiinae). Although most members of the subfamily occur in the Amazon basin of South America, the genus *Cethosia* has 12 species that range from India to Australia. The Orange Lacewing is often displayed in butterfly houses in Australia because of its ease of rearing and attractive coloration, and the flight behavior of the adults.

The Orange Lacewing caterpillar is orange brown with brown, intersegmental bands. The second, sixth, and eighth abdominal segments are largely white, the fourth segment is entirely white, and other segments and the thorax have white patches. Each of the body segments has six unbranched, black spines, except for the fourth abdominal segment, where the spines are white and black tipped. The head is black with two long, cylindrical, black spines.

Actual size

FAMILY	Nymphalidae
DISTRIBUTION	Across much of Africa and southern Europe
HABITAT	Dry grasslands, maquis, lightly wooded hillsides, parks, and gardens
HOST PLANTS	Various, including Strawberry Tree (*Arbutus unedo*), *Sorghum roxburghii*, *Lonchocarpus cyanescens*, and *Cassine* spp.
NOTE	Bizarre-looking, "dinosaur-head" caterpillar with four backward-pointing spines
CONSERVATION STATUS	Not evaluated

ADULT WINGSPAN
2⁹⁄₁₆–3 in (65–75 mm)

CATERPILLAR LENGTH
Up to 2 in (50 mm)

CHARAXES JASIUS

TWO-TAILED PASHA

(LINNAEUS, 1767)

206

Two-tailed Pasha caterpillars hatch from large eggs laid by the female butterfly on the upper surface of leaves of the host plant; she usually lays just one egg per leaf. Emerging, each caterpillar spins its own leaf tent of silken threads in which it shelters when not feeding. The caterpillars pupate away from the host plant, either in the leaf litter or suspended from a twig by a silken pad. The pupae are well camouflaged as they resemble small, ripening fruits. There are two generations a year, one in early summer and the second in late summer. The caterpillars of the second generation overwinter in a leaf tent and become active again the following spring.

The adults, whose hind wings terminate in the two short tails suggested by the species' common name, are attracted to ripe fruits and can be observed around a wide range of fruit-bearing species. They are also spotted close to cafés, where they are drawn to sugary drinks.

The Two-tailed Pasha caterpillar is distinctive, as its large head resembles that of a tiny dinosaur with four backward-pointing spines. The body is green with rings of tiny, white raised dots. There are two small eyespots on the dorsal surface and a yellow lateral line on both sides.

Actual size

FAMILY	Nymphalidae
DISTRIBUTION	Northern, eastern, and southern Australia
HABITAT	Wide range of lightly wooded habitats, including urban areas
HOST PLANTS	More than 45 hosts recorded; commonly wattles (*Acacia* spp.) but also plants from other families, including Caesalpiniaceae and Sterculiaceae
NOTE	Large, variably banded, and colorful caterpillar
CONSERVATION STATUS	Not evaluated, but in low densities throughout most of its range

ADULT WINGSPAN
3–3⅜ in (75–85 mm)

CATERPILLAR LENGTH
2⅛ in (55 mm)

CHARAXES SEMPRONIUS
TAILED EMPEROR
(FABRICIUS, 1793)

207

Tailed Emperor caterpillars hatch from round, green-and-brown eggs laid singly on host plants and feed openly on the leaves, mostly at night. During the day, and during molts, they rest on a silken pad spun on the upper side of a leaf. Even in early instars, the caterpillar's small head is adorned with two pairs of long horns. In the cooler, southern areas of the species' range there are two generations a year, with three to five generations occurring in northern regions, where the larvae can be found throughout the year.

The caterpillars pupate on the food plant, forming a shiny green, ovate chrysalis, suspended head down by the cremaster on a twig or the underside of a leaf. The male butterflies are strong, rapid fliers and commonly hilltop, while both sexes are attracted to fermenting fruit and tree sap. The subfamily of butterflies (Charaxinae) to which *Charaxes sempronius* belongs is well represented in the Afro-tropical and Oriental regions, and caterpillars of these species all have long horns.

The Tailed Emperor caterpillar is green or bluish green with a yellow ventrolateral line and transverse, yellow bands edged anteriorly blue on abdominal segments three and five but often also on other segments. The body is covered with tiny, white spots from which minute hairs arise. The head is green and edged yellow with four long, green-and-yellow, blue-tipped horns.

Actual size

FAMILY	Nymphaliade
DISTRIBUTION	North Africa, southern Europe into central Asia, Siberia, and across China to Korea
HABITAT	Dry, chalk grassland slopes with exposed rock outcrops grazed by sheep
HOST PLANTS	Grasses, particularly *Festuca ovina*, but also other members of Poaceae family
NOTE	Striped, brown caterpillar that is found on grazed grasslands
CONSERVATION STATUS	Not evaluated, but threatened in parts of Europe

ADULT WINGSPAN
1¾–2⅜ in (45–60 mm)

CATERPILLAR LENGTH
1⅜–1⁹⁄₁₆ in (35–40 mm)

208

CHAZARA BRISEIS
HERMIT
(LINNAEUS, 1764)

Hermit caterpillars hatch from conical-shaped, ridged eggs laid by the female butterfly on grasses, close to the ground, or sometimes on nearby moss and lichens. The larvae feed on grass, tending to prefer grazed or trampled tufts, and often they are active only at night. The caterpillars overwinter while still in an early instar and become active the following spring. They pupate in early summer, when their orange-brown pupae can be found at the base of the host plants.

The brown-and-white adults can be seen in late summer, flying from July to September. *Chazara briseis* is in sharp decline in central Europe due to its dependence on chalk grassland grazed short by sheep. This type of terrain is disappearing with the decrease in the traditional management of sheep by transhumance, in which sheep are moved up to higher slopes in summer, resulting in longer grass and colonization by shrub and tree species.

The Hermit caterpillar is brown in appearance with distinct, dark brown stripes running the length of the body and finer, paler lines in between. The stripes of brown are visible also on the head. The abdomen ends with two small, backward-pointing horns.

Actual size

FAMILY	Nymphalidae
DISTRIBUTION	Northeast India, northern Myanmar, Thailand, Laos, and southern Yunnan (China)
HABITAT	Open montane forests, at 3,300–6,600 ft (1,000–2,000 m) elevation
HOST PLANTS	Hackberry (*Celtis* spp.)
NOTE	Caterpillar whose head is heavily armored with horns and spines
CONSERVATION STATUS	Not evaluated, but common across its range, occurring in localized colonies

ADULT WINGSPAN
2⁷⁄₁₆–3⅜ in (65–85 mm)

CATERPILLAR LENGTH
2⅛ in (55 mm)

CHITORIA NAGA

NAGA EMPEROR
(TYTLER, 1915)

209

Naga Emperor caterpillars have a heavily horned head capsule with a pair of branched antlers and a trio of lengthy spikes either side of the head. Only when the larvae are mobile and feeding is the head held upright, otherwise it is pressed flush against the leaf surface. Early instars are gregarious. The larvae rest on a silken mat on the underside of host leaves and feed on adjacent foliage. The chrysalis is also formed on the underside of leaves attached to the midrib. It is green and streamlined, with a yellow-edged dorsal crest and wing margins, twin cranial horns, and blue spiracles.

The Apaturinae subfamily of nymphalid butterflies includes many species known as "emperors." The name "naga" refers to the mountainous Naga Hills district on the Indian–Myanmar border, where the species was described. Typical of many of the emperors, the butterfly will often be seen feeding from leaking tree sap or basking high in the canopy.

The Naga Emperor caterpillar is long and slender, its body base color bright green with multiple paler spots within broad, longitudinal stripes and with fine primary setae all over. The anal segment bears two black-tipped tail spikes. A pair of white processes project upward mid-abdomen. The head capsule has lengthy, arched horns with multiple branches and a halo of prominent spines around the margins of the faceplate.

Actual size

FAMILY	Nymphalidae
DISTRIBUTION	North America, from southwest and south central Canada south to Nebraska and New Mexico
HABITAT	Sage deserts, savannahs, washes, gulches, and canyons
HOST PLANTS	Rabbitbrush (*Chrysothamnus viscidiflorus*) and fleabane (*Erigeron* spp.)
NOTE	Spiny caterpillars that overwinter en masse underneath host plants
CONSERVATION STATUS	Not evaluated, but common

ADULT WINGSPAN
1⅜–1¾ in (35–45 mm)

CATERPILLAR LENGTH
1–1³⁄₁₆ in (25–30 mm)

210

CHLOSYNE ACASTUS
SAGEBRUSH CHECKERSPOT
(W. H. EDWARDS, 1874)

The Sagebrush Checkerspot caterpillar is black with numerous tiny, white spots, and each spiracle is encircled with a thin, white ring. Most segments bear lateral and dorsal, bright yellow-orange, elongate spots that fade with maturity. Setaceous black spines occur in clumps on each segment. The head is black and setaceous, increasingly so with maturity.

Female Sagebrush Checkerspots lay eggs in May in large clusters of 100 to 150 on the undersides of leaves at the base of Rabbitbrush plants. The young caterpillars hatch after six days and feed immediately, developing until the third instar, when they become dormant. The gregarious larvae seek refuge en masse at the base of the plant, where they oversummer and overwinter in dormancy. Feeding resumes in early spring, resulting in adults during April. Post-dormancy caterpillars are solitary, feed diurnally, and rest openly. Protection is based on gregariousness and sheltering in the early instars, whereas later instars depend on their spines and a ventral "neck" gland, which emits chemicals that appear to deter predators.

Sagebrush Checkerspot butterflies can be common or even abundant in some areas. At such locations, caterpillars can be found in early spring in large numbers sunning themselves on bare branches of Rabbitbrush. Both host plants have numerous narrow leaves, making it difficult to recognize larval feeding damage.

Actual size

FAMILY	Nymphalidae
DISTRIBUTION	Across Canada (except the far north), northern and western United States, Europe, and northern Asia
HABITAT	Low to high-elevation grassy habitats, prairies, steppes, roadsides, forest clearings, and subalpine meadows
HOST PLANTS	Grasses, including *Poa* spp., *Festuca* spp., *Stipa* spp., and *Bromus* spp.
NOTE	Caterpillar that blends in well on its host grasses
CONSERVATION STATUS	Not evaluated, but common

ADULT WINGSPAN
1³⁄₁₆–1³⁄₈ in (30–35 mm)

CATERPILLAR LENGTH
⅞–1 in (23–25 mm)

COENONYMPHA TULLIA

OCHRE RINGLET
(MÜLLER, 1764)

211

The cream eggs of the Ochre or Common Ringlet are laid singly on grass stems or blades, and the caterpillars hatch after six days. The larvae are extremely well camouflaged, with green, longitudinal stripes that blend perfectly with grasses. A small proportion of caterpillars in some populations are brown, which may occur as a result of drier conditions and frequently senescing grasses. The larvae develop through four or five instars. Most do not construct nests, but late summer caterpillars build loose silken shelters before dormancy in which they overwinter.

There are four or five instars, and development from egg to pupa takes between 40 and 60 days depending on temperature. An adult female may have one to three broods a year according to location. Males patrol all day for females, moving with a characteristic bouncing flight. Both sexes visit flowers for nectar. In warmer, drier areas, adults have a summer reproductive dormancy to avoid laying eggs in conditions that would be unsuitable for caterpillar survival.

The Ochre Ringlet caterpillar is green with a bluish cast and a lightly white-speckled appearance. Two tails at the tip of the abdomen are pale peach, and a white lateral stripe is prominent. There are several obscure to distinct, narrow, pale white dorsolateral stripes along the body.

Actual size

FAMILY	Nymphalidae
DISTRIBUTION	Mexico, Central America, most of the Caribbean, south to Colombia and Ecuador west of the Andes; east of the Andes, south from the Guianas and Venezuela to southeast Brazil
HABITAT	Humid and semi-humid forests, forest edges, and regenerating habitat
HOST PLANTS	*Cecropia* spp.
NOTE	Striking caterpillars that are gregarious and found in leaf shelters
CONSERVATION STATUS	Not evaluated, but unlikely to become endangered

ADULT WINGSPAN
2¾–3 in (70–75 mm)

CATERPILLAR LENGTH
1³⁄₁₆–1⅜ in (30–35 mm)

212

COLOBURA DIRCE
ZEBRA MOSAIC
(LINNAEUS, 1758)

When not feeding, young Zebra Mosaic caterpillars rest on frass chains protruding from the leaf margins, which provide them with a defense against marauding ants. For reasons that are not fully understood, ants seem unwilling to walk over the frass chains. As the larvae grow, they leave their frass chains and feed gregariously in groups of 5 to 20 caterpillars. When feeding, they usually bite through leaf veins and stems, helping to drain toxic plant compounds from the leaves and forming a loose shelter of drooping leaves around themselves. At the end of the fifth and final instar, larvae leave the group to pupate alone, either on or close to their host plant.

Although the original spelling of the Zebra Mosaic's Latin scientific name was *Papilio dirco*, Linnaeus, who first described the species more than 250 years ago, misspelled it *"dirce."* All subsequent authors followed suit until this incorrect spelling became the correct name, based on nomenclatural rules of general usage.

The Zebra Mosaic caterpillar has a shiny, black head bearing two short, white horns with brown tips, each horn armed with accessory setae. The body is velvet black with white thoracic scoli and pale yellow abdominal scoli.

Actual size

FAMILY	Nymphalidae
DISTRIBUTION	Andean regions of Venezuela, Colombia, and Ecuador (possibly Peru)
HABITAT	Cloud forests, forest edges, and landslides at mid to upper elevations
HOST PLANTS	Bamboo (*Chusquea* spp.)
NOTE	Rarely encountered caterpillar that is extremely cryptic on host plants
CONSERVATION STATUS	Not evaluated, but unlikely to become endangered

ADULT WINGSPAN
2⅜–2¾ in (60–70 mm)

CATERPILLAR LENGTH
2⅛–2⅜ in (55–60 mm)

CORADES CHELONIS
CORADES CHELONIS
HEWITSON, 1863

213

Unlike other species of its genus, *Corades chelonis* deposits only one to two eggs at a time, and larvae feed solitarily. When mature, the caterpillars are almost indistinguishable from an aging leaf of their host plant. Their forward and rearward projections make their shape very similar to the long, slender bamboo leaves, and their brown and yellowish markings are a perfect match to the aging patterns seen on older bamboo leaves. Pupation occurs on the host plant stem, often close to new growth. The pupa resembles, in shape and color, freshly emerging leaves.

Adults are rapid and active fliers, often descending to the ground to feed at urine or dung. They seem to be equally active on sunny and cloudy days, at least in eastern Ecuador. When ovipositing, females fly rapidly over areas of bamboo, dropping suddenly to land on a leaf and curling their abdomen under to deposit the egg below.

The *Corades chelonis* caterpillar is leaf green with various spots, stripes, and lines of yellows and browns that mimic leaf damage. Its triangular head bears two long horns that are fused to form an elongated cone, matched at the posterior end by two long, caudal tails, also held together to create one projection.

Actual size

FAMILY	Nymphalidae
DISTRIBUTION	The Himalayas, much of India through to western and southern China and Vietnam; isolated subspecies in Chinese Taipei and Okinawa, Japan
HABITAT	Primary and secondary forest, often near water
HOST PLANTS	Figs (*Ficus* spp.)
NOTE	Multi-horned caterpillar that resembles host plant new growth
CONSERVATION STATUS	Not evaluated, but relatively common in its range

ADULT WINGSPAN
2⅛–2¾ in (55–70 mm)

CATERPILLAR LENGTH
1⁹⁄₁₆ in (40 mm)

214

CYRESTIS THYODAMAS

COMMON MAP

(BOISDUVAL, 1836)

The eggs of the Common Map, also known as the Common Mapwing, are laid on new host plant shoots and take three days to hatch, the caterpillar forcing open a cap on the topside of the egg. The larvae are well camouflaged as their impressive heads and body horns look remarkably like the new shoots of the fig trees they live on. The caterpillar feeds for 15 days before pupating on the host plant or adjacent vegetation. The chrysalis hangs unsupported by its tail and is dark brown with a high ridge along the backline. The head end is produced into a long, curved snout, giving the pupa an elongated shape. The pupal period lasts seven days.

There are generally two generations annually between March and December. The wings of the Common Map are a stunning, marbled, cartographic pattern of fine lines and colored patches on a white background, almost always opened wide and flat and rarely seen held upright.

Actual size

The Common Map caterpillar is slender and smooth. The base color is green, brighter laterally than on the dorsum, broken only by deep-brown zones associated with the body horns; the underbelly is brown. The head bears a prominent pair of outwardly curving horns. Twin scimitar-shaped horns occur dorsally mid-body and at the rear. The cranial horn curves backward and the caudal horn forward. They are brown in color and covered in short spines, giving a serrated appearance.

FAMILY	Nymphalidae
DISTRIBUTION	Eastern slopes of the Andes in Ecuador
HABITAT	Mid-elevation cloud forests, especially in areas with heavy bamboo undergrowth
HOST PLANTS	Dwarf Bamboo (*Chusquea* cf. *scandens*)
NOTE	Caterpillar that feeds gregariously when young and is solitary later
CONSERVATION STATUS	Not evaluated, but may have a very restricted geographic range

ADULT WINGSPAN
2⅛–2⅜ in (55–60 mm)

CATERPILLAR LENGTH
1³⁄₁₆–1⅜ in (30–35 mm)

DAEDALMA RUBROREDUCTA

DAEDALMA RUBROREDUCTA

PYRCZ & WILLMOTT, 2011

215

Just after hatching, tiny *Daedalma rubroreducta* caterpillars huddle in a tightly packed group on the underside of a leaf. Later, prior to molt, they aggregate at the apex of the skeletonized food plant leaf, rearing back and regurgitating dark fluid when disturbed. As they grow older, they form lines along a narrow portion of a leaf, hanging in a vertical position under the larval weight. They later break into groups of two to five individuals, only dispersing as final instars.

Discovered only recently, *Daedalma rubroreducta* may have a broader range than is currently known, possibly extending into the eastern Andes of Colombia and Peru. The butterflies, when seen, are usually found feeding at animal droppings or other rotting organic matter. *Daedalma rubroreducta* is the only member of the genus to have its full life history described. However, related species are also known to feed on montane bamboos, and similarly to have caterpillars that resemble sticks and pupae that look like dead leaves.

Actual size

The *Daedalma rubroreducta* caterpillar is, overall, much like a mossy stick and roughly square in cross section. Its coloration is complex, consisting of various shades of brown with mossy green and black flecks and highlights. The caterpillar's resemblance to detritus is enhanced by its rough, tubercle-covered skin and short, fleshy, bifid tails. The head is roughly square, with small, conical "ears" on the top.

FAMILY	Nymphalidae
DISTRIBUTION	Africa, southern Europe, India, Sri Lanka, China, and areas of Southeast Asia
HABITAT	Open country, deserts, grasslands, and gardens up to 8,200 ft (2,500 m) elevation
HOST PLANTS	Milkweed (*Asclepias* spp.)
NOTE	Monarch-like caterpillar that has extra tentacles
CONSERVATION STATUS	Not evaluated, but common

ADULT WINGSPAN
2⅛–2⅜ in (55–60 mm)

CATERPILLAR LENGTH
1³⁄₁₆–1⅜ in (30–35 mm)

216

DANAUS CHRYSIPPUS
PLAIN TIGER
(LINNAEUS, 1758)

The female Plain Tiger perches on the upper side of a leaf and, curling its abdomen around the edge, lays an egg on the underside. Only one egg is laid per leaf to avoid overcrowding. After the caterpillar hatches, its first meal is the eggshell. It lives its entire life on the lower side of the leaves. Larvae defend themselves against some predators by sequestering poisons from their host plants. However, at least one parasitic wasp is specific to this species and can be responsible for up to 85 percent of mortality in an affected population.

Plain Tigers are usually encountered singly or in twos and threes. They have a slow, undulating flight, and both sexes patrol flowery areas, taking nectar. An alternative name for the species is African Monarch, and it is closely related to the well-known North American Monarch (*Danaus plexippus*). The caterpillars of the two species are very similar, but the Plain Tiger has an extra pair of tentacles, which it uses to try to avoid being parasitized by flies and wasps.

The Plain Tiger caterpillar is banded in black and white, interspersed with thick, yellow dorsolateral spots. It has three pairs of red-based, long, black, tentacle-like dorsal appendages on the third, sixth, and twelfth segments. The head is smooth and has alternating black and white semicircular bands. The true legs and prolegs are black.

Actual size

FAMILY	Nymphalidae
DISTRIBUTION	Southern United States, south to Argentina
HABITAT	Warm areas everywhere, except dense forest
HOST PLANTS	Milkweed and milkweed vine (*Asclepias* spp., *Sarcostemma* spp., *Cynanchum* spp., and *Matelea* spp.)
NOTE	Caterpillar and adult that are unpalatable to birds
CONSERVATION STATUS	Not evaluated, but most common in tropical parts of its range

DANAUS GILIPPUS

QUEEN

(CRAMER, 1775)

ADULT WINGSPAN
2⅝–3⅞ in (67–98 mm)

CATERPILLAR LENGTH
2⅛ in (55 mm)

217

Queen caterpillars eat plants in the milkweed family, ingesting the cardiac glycoside poisons they contain, which makes both larvae and adults poisonous to predating birds. Queen and related Monarch (*Danaus plexippus*) caterpillars are often found on the same milkweed plants. The larvae of both species chew through the midrib of leaves before feeding on them, in order to reduce the flow of the poisonous milky sap into the leaf they eat. The Queen caterpillar goes through six instars before it pupates, and the adult emerges within seven to ten days.

The principal difference between Queen and Monarch caterpillars is that Queens have three pairs of fleshy filaments (tentacles) while Monarchs have only two. Like Monarchs, Queens migrate north in spring and south in fall, but their migrations are much smaller and they remain common in Florida and Mexico during the summer. As the Queen uses the same pheromones as the African Milkweed Butterfly (*Danaus chrysippus*), the two could potentially mate if artificially introduced, though hybrids might be infertile.

Actual size

The Queen caterpillar is transversely banded white and black, often with partial yellow bands or spots in the black bands dorsally, often edged with maroon. The filaments are maroon, and the head is striped in black and white. The true legs and prolegs are black.

FAMILY	Nymphalidae
DISTRIBUTION	Southern Canada, United States, Bermuda, Mexico, the Canary Islands, Australia, and New Zealand
HABITAT	Almost any open habitat, especially riparian areas
HOST PLANTS	Milkweed (*Asclepias* spp.)
NOTE	Colorful caterpillar of arguably the world's best-known butterfly
CONSERVATION STATUS	Not evaluated, but numbers in steep decline in North America

ADULT WINGSPAN
3½–4 in (90–100 mm)

CATERPILLAR LENGTH
2⅛ in (55 mm)

218

DANAUS PLEXIPPUS
MONARCH
(LINNAEUS, 1758)

The Monarch caterpillar hatches from cream-colored eggs laid singly on the underside of young milkweed leaves. Feeding only on milkweeds, Monarch caterpillars, like other related milkweed-feeding species, advertise their distastefulness by their striking, banded coloration. The unpalatability of adults and larvae to birds and other predators is due to stored cardiac glycosides, or heart toxins, that are obtained from their host plant. The caterpillar develops rapidly through five instars, then forms a bright green, gold-spotted pupa, suspended from a silken pad. The complete life cycle from egg to adult takes only around 30 days.

Eastern North American Monarchs migrate up to 3,000 miles (4,800 km) from Canada to high-elevation forests in Mexico for overwintering. The species belongs to a family of about 300 milkweed butterflies that generally remain within tropical and subtropical regions. Numbers have declined in the past 20 years as a result of habitat destruction and a depletion of milkweeds.

The Monarch caterpillar is smooth and transversely banded white, yellow, and black, with the area occupied by the black bands greater under cool conditions. There are two pairs of fleshy filaments at the front and rear, which the caterpillar waves around when disturbed. The head is striped black and yellow.

Actual size

FAMILY	Nymphalidae
DISTRIBUTION	Guatemala, south to Bolivia, Paraguay, and southern Brazil
HABITAT	Humid lowland and foothill forest borders and mature second growth
HOST PLANTS	Various species of Ulmaceae, especially *Trema* spp.
NOTE	Caterpillar that, when young, is protected by its frass chain
CONSERVATION STATUS	Not evaluated, but not considered threatened

DIAETHRIA CLYMENA
WIDESPREAD EIGHTY-EIGHT
(CRAMER, 1775)

ADULT WINGSPAN
1½–1¾ in (38–45 mm)

CATERPILLAR LENGTH
1⅜–1¾ in (35–45 mm)

219

Caterpillars of the Widespread Eighty-eight hatch from pale green eggs shaped like truncated cones and laid singly on the very margin of host plant leaves. Immediately, the tiny larvae begin building a frass chain close to the oviposition site, resting in safety at its tip while not feeding. By the third instar, they abandon this perch and rest on the dorsal surface of the leaf with their now-sizeable head scoli pressed flat to the leaf surface. When disturbed, older caterpillars rear backward and lift their hind prolegs, clashing together the terminal scoli with those on the head.

The caterpillars move away from the host plant to pupate, forming a green chrysalis, attached to a leaf or stem. The adults are rapid fliers and frequently visit mud puddles and rotting fruit or dung. The species is so-named because of the undeniable similarity its underwing markings bear to the number 88 (sometimes more like 89).

Actual size

The Widespread Eighty-eight caterpillar is slender and nearly uniformly lime green, closely matching the color of its host leaves. The body bears several rows of short, stiff scoli, with those on the terminal abdominal segment the most formidable. The head, however, bears two very long scoli with several whorls of shorter spines emanating from its length. These are banded brown and pale yellow.

FAMILY	Nymphalidae
DISTRIBUTION	India, Sri Lanka, much of Southeast Asia, the Philippines, Papua New Guinea, northeast Australia, and adjacent southwest Pacific Islands
HABITAT	Lowland rain forests and adjacent areas
HOST PLANTS	Chinese Violet (*Asystasia gangetica*) and *Pseuderanthemum* spp.
NOTE	Aggressive and voracious caterpillar
CONSERVATION STATUS	Not evaluated, but common

ADULT WINGSPAN
2⁷⁄₁₆–2⁹⁄₁₆ in (62–65 mm)

CATERPILLAR LENGTH
2⅛–2⅜ in (55–60 mm)

220

DOLESCHALLIA BISALTIDE
LEAFWING
(CRAMER, [1777])

The Leafwing caterpillar, also known as the Autumn Leaf, occurs singly or in small numbers often on the widespread weed Chinese Violet, which in some parts of its range is the host plant of choice. It is a nocturnal feeder on young plants or regrowth on more mature plants, hiding by day in ground litter or under stones near the base of the food plant. The caterpillar is very active and can move rapidly. It can strip small plants and sometimes cannibalizes young caterpillars. Initially pale yellow in color, the caterpillar develops through five instars in as little as 12 days, though over a longer period in cooler conditions.

The caterpillar pupates head down, suspended by the cremaster attached to a silken pad on the underside of a leaf, usually some distance from the food plant. The hindwing of the adult butterfly is protruded into a short tail, giving a resting butterfly, with wings closed, the appearance of a dead leaf, and making it extremely difficult to detect.

The Leafwing caterpillar is black with cream subdorsal and lateral spots and further prominent lateral spots of blue and red. It is covered with numerous branched, black spines. The head is black with a bluish sheen and has a pair of branched spines.

Actual size

FAMILY	Nymphalidae
DISTRIBUTION	United States (Florida, Texas), Mexico, the Caribbean, Ecuador, Peru, Brazil, and Bolivia
HABITAT	Forest and woodland clearings and paths, and nearby fields
HOST PLANTS	Passion vine (*Passiflora* spp.)
NOTE	Long-spined, horned, multicolored caterpillar that can defoliate passion vines
CONSERVATION STATUS	Not evaluated, but common

ADULT WINGSPAN
3⅛–3½ in (80–90 mm)

CATERPILLAR LENGTH
1⁹⁄₁₆–1¾ in (40–45 mm)

DRYAS IULIA
JULIA BUTTERFLY
(FABRICIUS, 1775)

221

Julia Butterfly caterpillars hatch from eggs that are intially buff yellow when laid singly on the tendrils of passion vines, but become mottled just before the larvae emerge. The caterpillars, which develop through five instars, consume host plant leaves and utilize withered sections of leaves as resting perches. The larvae are unpalatable to birds and lizards because they contain varying amounts of cyanogenic glycosides, sequestered from the passion vine host plants. They are also protected by their many long spines and so feed openly.

Adult males congregate, sometimes in hundreds or thousands, on wet mud or sand, feeding on dissolved minerals; in Peru, males have also been observed imbibing the tears of turtles and alligators. It appears that a constant evolutionary war is being waged between the Julia Butterfly and passion vines. Some vines produce temporary stipules that attract egg-laying but then drop to the ground, exposing eggs to predation by ants. The life cycle from egg to adult takes about one month.

The Julia Butterfly caterpillar is blackish with variable amounts of white or cream, transverse striping dorsally, and usually with a row of white spots along the lower body. The spines are very long and black. The head is orange brown with black markings.

Actual size

FAMILY	Nymphalidae
DISTRIBUTION	India, Southeast Asia
HABITAT	Wooded areas; also parks, gardens, and plantations containing palms
HOST PLANTS	Palm, including Coconut Palm (*Cocos nucifera*), Oil Palm (*Elaeis guineensis*), Yellow Butterfly Palm (*Dypsis lutescens*), *Arenga* spp., *Calamus* spp., and *Phoenix* spp.
NOTE	Cryptic horned and tailed caterpillar
CONSERVATION STATUS	Not evaluated, but very common, particularly on cultivated land and around ornamental parks and gardens

ADULT WINGSPAN
2⅛–2¾ in (55–70 mm)

CATERPILLAR LENGTH
1¾ in (45 mm)

ELYMNIAS HYPERMNESTRA
COMMON PALMFLY
(LINNAEUS, 1763)

222

Common Palmfly caterpillars are crepuscular (feeding at dawn and dusk). Being long and thin, they blend in well with the host palm fronds. The round eggs, laid singly, usually on the underside of leaf blades, hatch in four days and are the caterpillar's first meal. The larvae develop through five instars over 19 days. Competitive cannibalism has been observed among these caterpillars as a natural control against overcrowding. The mature larvae wander and pupate on the underside of a frond. The pretty chrysalis, suspended from a silken pad, head downward, is of the same vibrant green as the larva, with red, yellow, and white highlights. The pupal period lasts seven days.

The Common Palmfly is a shade-loving butterfly with striking color on the topsides of its wings, although usually only the cryptic brown undersides are visible. The caterpillars of many species in the genus *Elymnias* feed on palm fronds and share the common name of palmfly.

Actual size

The Common Palmfly caterpillar is bright green and covered with short, stout bristles. A series of fine, yellow lines run longitudinally from head to tail. The wider, dorsolateral lines continue as yellow markings through to the head capsule and into a pair of pointed, pink, anal processes. Although variable, these lines can include orange and blue spots on some or all of the abdominal segments. The head capsule has two horns with spiked branches and a halo of spines around its margins.

FAMILY	Nymphalidae
DISTRIBUTION	Nepal, northeast India through to Myanmar, northern Thailand, southern Yunnan (China), and Vietnam
HABITAT	Forests and human habitations around host plant plantations
HOST PLANTS	Banana (*Musa* spp.) and palms (Arecaceae)
NOTE	Caterpillar whose food—bananas or palm leaves—determines its color
CONSERVATION STATUS	Not evaluated, but common within its fairly limited range

ADULT WINGSPAN
2⅜–3⅛ in (60–80 mm)

CATERPILLAR LENGTH
1⁹⁄₁₆ in (40 mm)

ELYMNIAS MALELAS
SPOTTED PALMFLY
(HEWITSON, 1863)

223

Although the immature stages of the Spotted Palmfly are yet to be scientifically described, the appearance and development of the larvae are likely to be similar to those of other *Elymnias* species. However, while many related larvae are exclusively palmivorous, *E. malelas* is an exception as banana leaves are an alternative food source. Consequently, larvae raised on banana leaves are yellow to match the leaf midrib on which they rest when not feeding at the leaf margins. Finding a caterpillar on a banana within its species range is therefore a strong indication of its identity. Differentiating species on palms, where a green base color predominates, requires closer examination of anatomical differences such as subtle head capsule features.

The male Spotted Palmfly has a strong purple iridescence on the topside of the forewings, which is present but reduced in the female and accompanied by more widespread white markings. The adult males and females mimic the similarly dimorphic genders of the *Euploea mulciber* species, which are known to be unpalatable to birds.

Actual size

The Spotted Palmfly caterpillar is long and spindle-shaped. Feeding on banana leaves, the base color is neon yellow with subtle longitudinal lines running the length of the body. Those larvae that feed on palm leaves are less yellow and tinged with green. There are twin pointed tails at the rear and a pair of branched horns on the head capsule. The entire body is covered in short, drumstick-like setae.

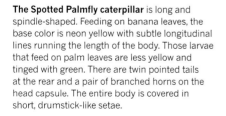

FAMILY	Nymphalidae
DISTRIBUTION	Mountainous regions of northwest Europe, across central Europe to the Urals, southern Siberia, Mongolia, and northeastern China
HABITAT	Forest margins and clearings, and damp grasslands up to 7,875 ft (2,400 m) elevation
HOST PLANTS	Grasses, particularly *Agrostis* spp., *Dactylis* spp., and *Poa* spp.
NOTE	Striped, brown caterpillar of upland grassland and forest margins
CONSERVATION STATUS	Not evaluated

ADULT WINGSPAN
1¾ in (45 mm)

CATERPILLAR LENGTH
1 in (25 mm)

EREBIA AETHIOPS
SCOTCH ARGUS
(ESPER, 1777)

224

Actual size

The Scotch Argus caterpillar is sluglike in appearance with a large head and tapered body. Many dark brown and pale brown stripes run along its length, and bands of small, raised tubercles each bear a short hair.

The female Scotch Argus lays her spherical egg singly on short grass leaves and seed heads. The caterpillar is slow growing, often feeding at night and resting near the base of the grass during the day. If disturbed, it falls to the ground and "plays dead." The young caterpillar overwinters in leaf litter near the base of the food plant and becomes active again the following April. Mature larvae move to the ground to pupate in a loose cocoon near the base of the food plant, often in moss and lichen.

There is one generation a year, with the adults flying late in summer, usually in July and August. *Erebia aethiops* has experienced a decline in its distribution due to loss of habitat and lack of habitat management, but numbers are increasing again in some parts of its range. In favorable habitats, the species may exist in colonies of hundreds or even thousands of individuals.

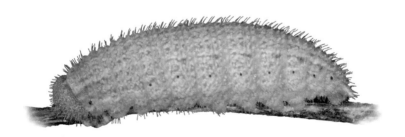

FAMILY	Nymphalidae
DISTRIBUTION	Western North America, from Alaska to New Mexico
HABITAT	Moist mountain meadows, bogs, fields, and canyons
HOST PLANTS	Grasses, including *Poa* spp. and *Setaria* spp.
NOTE	Rarely seen, nocturnally feeding, mountain grass caterpillar
CONSERVATION STATUS	Not evaluated, and though usually common may be rare in parts of its range

ADULT WINGSPAN
1¾–2 in (45–50 mm)

CATERPILLAR LENGTH
1³⁄₁₆–1⅜ in (30–35 mm)

EREBIA EPIPSODEA
COMMON ALPINE
BUTLER, 1868

225

In June, the female Common Alpine selects an egg-laying site by walking through grasses, then lays her eggs singly. After eight to ten days, the caterpillars hatch and feed for 20 to 30 days before entering dormancy in the third or fourth instar. Feeding recommences after overwintering, and the larvae may complete development, pupate, and produce adults in the summer. In some locations, overwintering occurs for a second time before pupation. The caterpillars are solitary, feed nocturnally, and rest at the base of grasses during the day.

Camouflage appears to be the primary defense of these caterpillars, which can be attacked by spiders and predatory mites. To survive overwintering, they require high humidity, but the same conditions are detrimental during spring. Adults visit flowers for nectar, and males also imbibe from mud and dung. Males emerge before females and patrol circular routes, while females tend to remain hidden in grass.

Actual size

The Common Alpine caterpillar is pinkish tan with a prominent, middorsal, black stripe. The less distinct, lateral, white stripes are bordered below in black. Profuse short, pale setae arise from tiny, white spots, giving an overall granulated appearance, and the abdominal tails are short. The setaceous head is greenish tan.

FAMILY	Nymphalidae
DISTRIBUTION	Eastern slopes of the Andes, from northern Ecuador to northern Peru
HABITAT	Montane cloud forests and forest edges
HOST PLANTS	Bamboo (*Chusquea* spp.)
NOTE	Recently described species first noticed for its distinctive caterpillar
CONSERVATION STATUS	Not evaluated, but not likely to become endangered

ADULT WINGSPAN
5¼–5¹¹⁄₁₆ in (135–145 mm)

CATERPILLAR LENGTH
4¼–4½ in (110–115 mm)

ERYPHANIS GREENEYI
ERYPHANIS GREENEYI
PENZ & DEVRIES, 2008

The caterpillar of *Eryphanis greeneyi* goes through three very distinctive "morphs" as it develops, each corresponding with a shift in behavior and crypsis. When first hatched, the bulbous-headed, intricately patterned first instars are easily mistaken for a small irregularity in the host plant leaf. As they grow, third and fourth instars are countershaded beige and green to perfectly match the natural patterns of mature leaves. Finally, when too heavy to rest on the leaves, last instar caterpillars resemble rotting, moldy portions of the host plant stem.

The bamboo host plants of *Eryphanis greeneyi* are likely not very nutritious, and their cloud forest homes are rather cool. As a result, it takes a long time for the caterpillars to build up the weight needed for successful pupation. The life cycle from egg to adult can take six to seven months, with as many as 20 days in the egg stage alone, and parasitoids are a significant threat at all life stages. Adults are crepuscular, most frequently seen guarding perches, mating, and laying eggs at dusk and dawn.

The *Eryphanis greeneyi* caterpillar has a helmetlike head with a corona of short, conical projections. The ground color of the body is dull orange brown, and there are complex, diffuse patterns of black, orange, ocher, pale blue, and white, which create a moldy-stick appearance. The dorsum bears several fleshy projections, and the terminal segment has a pair of long caudae that are covered in pointed tubercles.

Actual size

FAMILY	Nymphalidae
DISTRIBUTION	Western North America, from British Columbia to Washington State, Idaho, Oregon, Nevada, and California
HABITAT	Mountains, foothills, upper shrub-steppe, open forests, meadows, and roadsides
HOST PLANTS	Common Snowberry (*Symphoricarpos albus*), Shrubby Penstemon (*Penstemon fruticosus*), Common Mullein (*Verbascum thapsus*), honeysuckle (*Lonicera* spp.), and Indian paintbrush (*Castilleja* spp.)
NOTE	Caterpillar that adapts to conditions, prolonging or shortening its development
CONSERVATION STATUS	Not evaluated, but common in many locations

ADULT WINGSPAN
2–2⅛ in (50–55 mm)

CATERPILLAR LENGTH
1³⁄₁₆–1⅜ in (30–35 mm)

EUPHYDRYAS COLON
SNOWBERRY CHECKERSPOT
(W. H. EDWARDS, 1881)

227

Snowberry Checkerspot females lay batches of 50 to 200 eggs on host plant leaves in June, which hatch into first instar caterpillars within two weeks. The first three instars are gregarious, consuming leaves in messy silk nests, and expanding or moving nests when the food supply is exhausted. The caterpillars overwinter in the second or third instar, often in nests on upper parts of the host plant. They recommence feeding in spring, but if food is in short supply or of poor quality they will reenter dormancy. In challenging conditions, caterpillars may pass through up to seven instars before pupation.

Caterpillars of *Euphydryas colon* likely sequester iridoid alkaloids from host plants, making them unpalatable to some predators. Their spines also confer protection. There are a number of *Euphydryas* checkerspot species in western North America, all with similar caterpillars and life cycles. The adults are a common sight in mountainous habitats, nectaring on alpine flowers.

Actual size

The Snowberry Checkerspot caterpillar is variably colored but usually some combination of black, orange, and white. Three large, orange spots are present on each segment, and large, branching setae impart a spiny appearance. The black head bears long, white setae.

FAMILY	Nymphalidae
DISTRIBUTION	India, Sri Lanka, southern China, Sumatra, Java, Bali, and northern and eastern Australia
HABITAT	Open forests and woodlands, riparian areas, gullies, and gardens
HOST PLANTS	Wide range in Asclepiadaceae (milkweeds), Moraceae (figs), and Apocynaceae (dogbanes)
NOTE	Smooth-bodied, vividly colored caterpillar that has ornate tentacles
CONSERVATION STATUS	Least concern and common in its range

ADULT WINGSPAN
2¼–3 in (70–75 mm)

CATERPILLAR LENGTH
1¾–2⅛ in (45–55 mm)

228

EUPLOEA CORE
COMMON CROW
(CRAMER, 1780)

Common Crow caterpillars hatch from eggs laid singly on the undersides of young leaves and flowers of host plants. The larvae feed on soft new growth and usually pupate on the underside of a leaf or some other part of the host plant. Larval development varies in duration from three to ten weeks, according to host plant species, temperature, and the seasonal quality of new shoots. Caterpillar survival rates are poor at temperatures below about 68°F (20°C).

The caterpillars feed on generally poisonous plants and have evolved feeding strategies to minimize exposure to such toxins, while using some of them for defense. For example, if disturbed, the larvae will exude fluid containing toxins from the mouth to deter a predator. Adults form large, nonreproductive aggregations during the dry season in sheltered habitats, often near creeks. Occasionally, migratory flights will occur, usually to more humid areas. Many subspecies occur throughout the Common Crow's range.

Actual size

The Common Crow caterpillar is orange or orange brown, each segment with several narrow, transverse black bands, partly edged with white, and a black-and-white ventrolateral band. There are four pairs of long, black, fleshy tentacles on the third, fourth, sixth, and twelfth segments. The head is shiny and smooth, with alternating black and white, semicircular bands.

FAMILY	Nymphalidae
DISTRIBUTION	Southern India to southern China, Chinese Taipei, and Southeast Asia, including the Philippines
HABITAT	Open forests, and often seen in urban environments due to ornamental uses of host plants
HOST PLANTS	Fig (*Ficus* spp.), Oleander (*Nerium oleander*), *Gymnanthera* spp., *Oxystelma* spp., *Toxocarpus* spp., and *Aristolochia* spp.
NOTE	Caterpillars that cut their food before eating it
CONSERVATION STATUS	Not evaluated, but not uncommon

ADULT WINGSPAN
3⅛–3½ in (80–90 mm)

CATERPILLAR LENGTH
1¾–2 in (45–50 mm)

EUPLOEA MULCIBER
STRIPED BLUE CROW
(CRAMER, 1777)

229

The yellow eggs of the Striped Blue Crow are laid singly on the underside of leaves on a range of host plants and become the caterpillar's first meal when it hatches. The larvae develop through five instars over 14 days. Most of the growth occurs during the final, longest instar, when the caterpillar doubles in size. Curiously, it feeds by first disconnecting the leaf at its stalk or midrib and securing it in place with silk before consuming the severed portion. Early instars feed on younger leaves, while older instars eat more mature foliage.

The caterpillar pupates on the underside of a leaf, spinning a silk pad on the midrib, from which the chrysalis is suspended. The chrysalis changes color from orange brown to mirrored silver to black during the week-long pupation. The adult Striped Blue Crow, most commonly seen feeding at flowers, is known to be distasteful to predators and is mimicked by other butterflies and moths as an evolved defensive strategy.

Actual size

The Striped Blue Crow caterpillar has four pairs of reddish-brown, black-tipped, tendril-like processes—three on the anterior segments and one pair at the rear. The smooth, cylindrical body is striped with white, black, and reddish-brown bands with variable yellow or orange lateral spots. The head capsule has distinctive white outlines on a black or reddish-brown base color.

FAMILY	Nymphalidae
DISTRIBUTION	North America, south to Argentina
HABITAT	Open areas everywhere
HOST PLANTS	Wide range, including flax (*Linum* spp.), violet (*Viola* spp.), passion vine (*Passiflora* spp.), and stonecrop (*Sedum* spp.)
NOTE	Caterpillar whose bright markings advertise its unpalatability to predators
CONSERVATION STATUS	Not evaluated, but considered secure within its range

ADULT WINGSPAN
1¾–3⅛ in (45–80 mm)

CATERPILLAR LENGTH
1¾ in (45 mm)

EUPTOIETA CLAUDIA
VARIEGATED FRITILLARY
(CRAMER, 1775)

230

The Variegated Fritillary caterpillar is beautifully striped in orangey red, black, and white. The body is adorned with numerous shiny, black spines, and behind the black head are two long, hornlike projections, which are twice as long as the spines. The two white stripes are strongly developed and divided vertically into three to five pieces per segment.

The striking and solitary Variegated Fritillary caterpillar emerges from pale green or cream-colored eggs laid on the host plant and feeds on leaves and flowers. Its pupa is an equally attractive light blue green with gold cones, yellow antennae, and orange eyes. The adults and caterpillars cannot survive severe freezes so in winter are found only in warm or tropical areas. However, adults migrate north every spring as far as southern Canada, where they produce several generations.

The Variegated Fritillary is a "connecting link" to the Heliconiinae subfamily, known as the heliconians, especially the Gulf Fritillary (*Agraulis vanillae*), whose caterpillar, which is distasteful to birds, it mimics as a defense mechanism. The Gulf Fritillary caterpillar is likewise striped red and black (but sometimes without the white stripes). The caterpillars of both butterflies often eat passion flower leaves, and adults are similar in size and color, both with pointed forewings, although the Variegated Fritillary lacks the silver spots of the Gulf Fritillary.

Actual size

FAMILY	Nymphalidae
DISTRIBUTION	Southeastern Mexico, south to the Amazon basin, including Trinidad, as far south as southeastern Peru and northern Bolivia
HABITAT	Forest borders and secondary growth of humid and semi-humid tropical forests, usually below 3,600 ft (1,100 m) elevation
HOST PLANTS	*Croton* spp.
NOTE	Intricately patterned caterpillar that hides within rolled leaves
CONSERVATION STATUS	Not evaluated, but not considered threatened

ADULT WINGSPAN
2–2⅛ in (50–55 mm)

CATERPILLAR LENGTH
1¾–2⅛ in (45–55 mm)

FOUNTAINEA RYPHEA
FLAMINGO LEAFWING
(CRAMER, [1775])

231

Flamingo Leafwing caterpillars hatch from almost perfectly round, smooth, and yellowish-white eggs, which are laid singly on the underside of leaves. Like other, related species, early instars build and rest upon small lines of their own frass that are silked into chains extending from the margins of their host plant leaves. As they grow, the larvae switch to living within a tubular shelter, which they build by curling up one of the leaves, usually remaining inside almost constantly, reaching out to feed on nearby leaf tissue or on the leaf material of the shelter itself. The larvae may pupate on or off their host plant, hanging their emerald green and yellow-edged pupa from a thin branch or the bottom of a leaf.

With its bright colors and intricate patterning, the Flamingo Leafwing caterpillar lives up to its name—as does the flashy butterfly. The adults are rapid and powerful fliers, zipping across the treetops, where males often guard small territories, and frequently descending to the ground to feed on organic material, especially rotting carcasses or dung.

The Flamingo Leafwing caterpillar is stout, nearly cylindrical, and has a bulbous head with small, conical, black or yellow bumps. It is complexly patterned with green, yellow, brown, red, and black. The body has minute setae sparsely scattered across it, though they are visible only due to the small, bright white protrusions from which they emerge.

Actual size

FAMILY	Nymphalidae
DISTRIBUTION	North and South America, from southern United States and the Caribbean south to northern Peru
HABITAT	Forests and meadows, including disturbed habitats
HOST PLANTS	Passion vine (*Passiflora* spp.)
NOTE	Spiny caterpillar that is unpalatable
CONSERVATION STATUS	Not evaluated, but common

ADULT WINGSPAN
2–4 in (50–100 mm)

CATERPILLAR LENGTH
1¹⁄₁₆–2 in (40–50 mm)

232

HELICONIUS CHARITHONIA
ZEBRA LONGWING
(LINNAEUS, 1767)

Zebra Longwing caterpillars are found on the stems or underside of leaves of various species of passion vine. The plants are toxic, so ingesting them makes the larvae distasteful, providing a chemical defense against predators. Their conspicuous, contrasting black-and-white patterning advertises their unpalatable taste. Additionally, the caterpillars are protected by sharp, black spines. There are thought to be five larval instars, and development from egg to adult takes around a month.

Pupation occurs on or near the host plant, and the pupae are visited by Zebra Longwing males. They frequently sit on the pupae, competing with each other to insert their abdomen and mate before a female butterfly ecloses; pheromones determine the adult's sex at later stages of pupal development. Mated females then disperse long distances to lay eggs on as many passion vines as possible. Like all *Heliconius* species, Zebra Longwing adults feed not only on nectar but also on pollen, which allows them to live longer than other butterflies.

Actual size

The Zebra Longwing caterpillar is white, with a white head and long, black spines. There are also two black dots on the front of the head, two on the sides, and several rows of dots on the body. This coloration is clearly aposematic, making the caterpillar stand out to naive predators.

FAMILY	Nymphalidae
DISTRIBUTION	The Andes, from southern Colombia to Bolivia
HABITAT	Humid and semi-humid cloud forests and forest borders at 2,625–6,600 ft (800–2000 m) elevation
HOST PLANTS	Passion vine (*Passiflora* spp.)
NOTE	Caterpillar whose natural history, however, is little known
CONSERVATION STATUS	Not evaluated, but not considered threatened

ADULT WINGSPAN
3¹⁄₁₆–3⁵⁄₁₆ in (78–84 mm)

CATERPILLAR LENGTH
2⅛–2⅜ in (55–60 mm)

HELICONIUS TELESIPHE
TELESIPHE LONGWING
(DOUBLEDAY, 1847)

233

The Telesiphe Longwing caterpillar hatches from an egg laid singly on the newly opening leaves or young tendrils of the host plant; the female may, however, return multiple times to the same plant, leaving behind several eggs. First instar larvae consume most of their eggshell before beginning to feed on nearby leaf tissue. While the caterpillars are generally solitary at all instars, the adults sometimes gather in small groups to spend the night hanging together in loose clusters from the tip of a thin branch or vine. Prior to eclosion of a female pupa, males will also congregate in the area, vying for the chance to copulate with the newly emerged female.

While ovipositing, the fluttery, hovering flight of Telesiphe Longwing females is distinctive, and watching for this characteristic behavior is often the best way to find host plants and caterpillars. With their paired, facial, black spots on the front of the head, the black-spotted larvae are typical of other, related species of *Heliconius* caterpillars, all of which also feed on *Passiflora* plants.

The Telesiphe Longwing caterpillar is creamy white to yellow with irregularly shaped, black spots along the sides, which create broad stripes, and smaller, paler, blackish spots across the dorsum. Each segment bears several long, unbranched, black scoli that, along with the long, slightly curved pair of scoli on the head, give the caterpillar a spiky look.

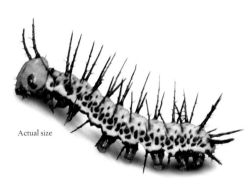

Actual size

FAMILY	Nymphalidae
DISTRIBUTION	Southeastern and southern Australia
HABITAT	Woodlands and eucalypt forests with grassy understory, from alpine to semi-arid and urban environments
HOST PLANTS	Native and introduced grasses, including *Cynodon* spp., *Poa* spp., *Themeda* spp., *Microlaena* spp., *Bromus* spp., and *Ehrharta* spp.
NOTE	Long-lived caterpillar that develops in the cooler months
CONSERVATION STATUS	Least concern, and common in southern areas of its range

ADULT WINGSPAN
2³⁄₁₆–2½ in (56–64 mm)

CATERPILLAR LENGTH
1⁷⁄₁₆ in (36 mm)

234

HETERONYMPHA MEROPE
COMMON BROWN
(FABRICIUS, 1775)

Hatching of Common Brown caterpillars is timed to coincide with the start of the fall and winter rains, which produces soft growth of the grasses. The caterpillars occur singly or in small groups on a range of grasses. The larvae feed at night on the leaf blades and hide under litter during the day. Caterpillar development takes five to six months and is slow during the cooler months with rapid growth in the spring.

Pupation occurs in early spring, the pupae lying loose on the ground. This sexually dimorphic species flies low in grassy areas, and males will hilltop. Males emerge in the spring before females, with mating occurring in the spring. Females become dormant, unless disturbed, during the hot, dry summer, while males die in summer. The females become active in the early fall, laying eggs as the days shorten and rainfall increases.

The Common Brown caterpillar is variably colored brown or green, mottled with darker brown, with a brown, broken middorsal line and pale-colored, wavy, or broken lateral and sublateral longitudinal lines. There are numerous short setae, and the anal segment has a forked posterior projection.

Actual size

FAMILY	Nymphalidae
DISTRIBUTION	Southern United States, south through Mexico, Central America, and some Caribbean islands, and South America to northern Argentina and Uruguay
HABITAT	Humid lowland and foothill forests
HOST PLANTS	*Cecropia* spp.
NOTE	Distinctive, large, spiky caterpillar whose pupa wriggles if disturbed
CONSERVATION STATUS	Not evaluated, but not likely to become endangered

HISTORIS ODIUS
STINKY LEAFWING
(FABRICIUS, 1775)

ADULT WINGSPAN
4¼–4⅝ in (110–120 mm)

CATERPILLAR LENGTH
2¾–2¹³⁄₁₆ in (70–72 mm)

235

Early instar Stinky Leafwing larvae make frass chains at the margins of their host plant leaf, presumably to avoid the stinging ants that often reside within the stems of *Cecropia* trees. Larger larvae rest along the terminal meristem and are apparently not bothered by the ants. If touched, the large, dead-leaf colored pupa wriggles vigorously about, like a fish out of water. This behavior has led to the local name of *pescadillo*, meaning "little fish," in some areas of Costa Rica.

Like the adults of the only other member of the genus, *Historis acheronta*, adult Stinky Leafwings are extremely powerful flyers and spend most of their time in the forest canopy feeding on overripe and damaged fruit. They are, however, familiar to anyone who has camped or stayed outdoors in the American Tropics, as adults quickly descend to procure minerals from backpack straps, sweaty socks, and muddy boots left out to dry.

The Stinky Leafwing caterpillar has a somewhat square head capsule, which is prominently lobed, each lobe bearing a stout, spiny horn ending in a rosette of spines. The head is mostly dark brown, except for an orange area around the base of the horns. The body is pale tannish brown with whitish-yellow transverse stripes and bears clusters of spines—orange dorsally and yellowish laterally.

Actual size

FAMILY	Nymphalidae
DISTRIBUTION	Extreme southeastern United States, through Central America and Trinidad; also the Andes from Venezuela to Bolivia, southeast through Paraguay to southern Brazil and Uruguay
HABITAT	Disturbed or regenerating humid and semi-humid foothill and montane forest, occasionally tropical deciduous forest
HOST PLANTS	Nettle (*Boehmeria* spp., *Phenax* spp., *Urera* spp.) and hackberry (*Celtis* spp., *Sponia* spp., *Trema* spp.)
NOTE	Spiny caterpillar that creates leaf shelters
CONSERVATION STATUS	Not evaluated, but generally widespread and common

ADULT WINGSPAN
2⅜–2⁹⁄₁₆ in (60–65 mm)

CATERPILLAR LENGTH
1⅜–1⁹⁄₁₆ in (35–40 mm)

236

HYPANARTIA LETHE
ORANGE MAPWING
(FABRICIUS, 1793)

Newly hatched Orange Mapwing caterpillars make a silk-lined tube from the host plant leaf and rest inside, feeding on the epidermis within the tube and on adjacent leaves. Older larvae form a pouch-like tent out of leaves, resting inside when not feeding. Unlike the larvae of many other shelter-building species, such as skippers (Hesperiidae), the caterpillars appear to lack the ability to fling their frass, a tactic thought to deter predators, and instead simply drop it outside the shelter. Pupation may occur within the shelter or occasionally on an adjacent plant. When alarmed, larvae move their head rapidly back and forth against the leaf surface, making a surprisingly loud noise.

Orange Mapwings are usually found between 985 ft (300 m) and 4,920 ft (1,500 m), but in eastern Ecuador they appear to make altitudinal migrations every several years, breeding at elevations as high as 7,545 ft (2,300 m) for one or two generations before disappearing again. Adults commonly feed on fruits, at puddles, or at water seeps, but both sexes occasionally visit flowers, particularly large asters such as *Baccaris* species.

The Orange Mapwing caterpillar is predominantly dull, creamy white to yellow with, short, black, branched spines and lime green or bluish highlights on the body. The mature larva in eastern Ecuador has a striped, green head, though larvae in Costa Rica reportedly have dull orange heads, suggesting that more than one species may be involved.

Actual size

FAMILY	Nymphalidae
DISTRIBUTION	Mexico, south to the Amazon basin
HABITAT	Humid lowland rain forests and adjacent second growth
HOST PLANTS	*Croton* spp.
NOTE	Caterpillar that is widespread but rarely seen
CONSERVATION STATUS	Not evaluated, but not likely to become endangered

ADULT WINGSPAN
3½–4 in (90–100 mm)

CATERPILLAR LENGTH
2⅜–2¾ in (60–70 mm)

HYPNA CLYTEMNESTRA
JAZZY LEAFWING
(CRAMER, 1777)

237

The oddly shaped caterpillar of the Jazzy Leafwing resembles a piece of dead leaf, a bird dropping, or some other inedible piece of forest detritus, making it unlikely to be found by hungry predators such as birds. While not feeding, larvae often rest near a damaged portion of the host plant to further enhance their crypsis. Early instars rest out on the ends of fragile frass chains, which they begin to build immediately after hatching. At later instars, too, the caterpillars are one of the few species of their tribe Anaenini that do not rest inside rolled-up host plant leaves.

Jazzy Leafwings are among the largest members of Anaenini. The fast-flying adults, which feed predominantly on rotting fruit, are widely distributed but rarely encountered. The caterpillar's host plants are in the genus *Croton*, among them *C. lechleri*, known in some areas of eastern Ecuador as *Sangre de Drago*, or Dragon's Blood, for its thick, bloodred sap, which is harvested and used for myriad medicinal purposes, from treating minor cuts to curing ulcers.

Actual size

The Jazzy Leafwing caterpillar has a pale brown head, which is ringed with a corona of eight bristled tubercles and bears whitish "warts" across the front. The body is tapered behind the head, greatly enlarged at the posterior thorax, and quickly tapered toward the rear. It is overall chocolate brown, darker along the back, and has numerous reddish tubercles along the dorsum, from which arise long, hairlike bristles.

FAMILY	Nymphalidae
DISTRIBUTION	Northeastern and eastern Australia
HABITAT	Tall, open forests and the margins of rain forests
HOST PLANTS	Tropical and temperate grass species in the Poaceae family
NOTE	Nocturnal, long-lived, cryptic caterpillar
CONSERVATION STATUS	Not evaluated, but locally common

ADULT WINGSPAN
1³⁄₁₆ in (30 mm)

CATERPILLAR LENGTH
⅞ in (22 mm)

238

HYPOCYSTA METIRIUS
BROWN RINGLET
BUTLER, 1875

The Brown Ringlet caterpillar feeds nocturnally on grass, resting at the base of a tuft during the day, although early instars stay on the leaf on which they are feeding. Young caterpillars feed down the leaf blade but only across to the midvein, forming scars down one side of the leaf. Mature caterpillars consume the full leaf as they move down toward the base. The larvae feed and complete development over a period of between five weeks and six months, depending on the location of the population and the time of year. In the warmer northern areas, caterpillars are found throughout the year.

The caterpillar pupates on a grass stem, the angular brown pupa resembling a small, curled, dead leaf hanging from the stem. The adult butterflies are weak fliers and fly close to the ground, although males will often fly to hilltops. The genus contains 12 species restricted to Australia, New Guinea, and the Aru Islands.

The Brown Ringlet caterpillar is green or brown and covered in minute, white dots, with thin lateral lines, some dark and some light, running along the body. The head has two, short, crimson-tipped horns, edged laterally with a white line. The anal segment has a forked projection edged in white.

Actual size

FAMILY	Nymphalidae
DISTRIBUTION	Madagascar, India, Southeast Asia, Chinese Taipei, southern Japan, New Guinea, Australia, and islands of the South Pacific, east to French Polynesia and Easter Island
HABITAT	Savannahs and open woodlands, particularly in tropical and subtropical areas
HOST PLANTS	Wide range from at least nine families, but commonly joyweed (*Alternanthera* spp.) and Common Asystasia (*Asystasia gangetica*)
NOTE	Caterpillar that is gregarious in the early instars
CONSERVATION STATUS	Not evaluated, but common in tropical and subtropical areas

ADULT WINGSPAN
3–3⅜ in (76–86 mm)

CATERPILLAR LENGTH
2¹⁄₁₆ in (53 mm)

HYPOLIMNAS BOLINA
VARIED EGGFLY
(LINNAEUS, 1758)

239

Varied Eggfly caterpillars hatch from eggs that are laid in batches on the underside of leaves of the food plant. The caterpillars feed gregariously when young but singly when mature. Feeding occurs at night, and caterpillars may hide some distance from their food plant during the day. In tropical areas, caterpillars can be found throughout the year, and the larval stage can last as little as three weeks. The pupa is attached to a silk pad and suspended by the cremaster on or near the food plant.

Some populations of the Varied Eggfly, also known as the Great Eggfly or Blue Moon, have been known to produce mainly female offspring because of the presence of *Wolbachia* bacteria, which kill male caterpillars in the egg. Although the species is not migratory, specimens are occasionally recorded well outside their normal breeding areas. The adult butterflies are found in moist gullies and urban gardens. The males are territorial and will guard a particular perch even after being disturbed by other intruding butterflies.

The Varied Eggfly caterpillar is brown or black, with a yellowish-orange ventrolateral line, and each segment has several brownish-orange, branched spines. The head is orange with two long, black, bristly spines and a lateral black spot near the eyes.

Actual size

FAMILY	Nymphalidae
DISTRIBUTION	The Andes of Colombia, south to southern Peru
HABITAT	Regenerating montane habitat, especially at forest margins and landslides, from around 6,600 ft (2,000 m) elevation to the treeline
HOST PLANTS	Bamboo (*Chusquea* spp.)
NOTE	Caterpillar that near perfectly mimics a dying host plant leaf
CONSERVATION STATUS	Not evaluated, but not considered threatened

ADULT WINGSPAN
2⅜–2¾ in (60–70 mm)

CATERPILLAR LENGTH
2⅛–2%₁₆ in (55–65 mm)

240

JUNEA DORINDA

DORINDA SATYR

(FELDER & FELDER, 1862)

Dorinda Satyr caterpillars hatch from round, yellow-white eggs, laid singly on fresh leaves of their host plants. Like many other satyrine larvae feeding on bamboo, they are wonderfully cryptic when at rest on a specific portion of the plant. The yellowish ground color and pointed head and rear of the mature caterpillar make it nearly impossible to see when it rests on a dying (though not fully dried) bamboo leaf. Younger caterpillars are greenish and generally rest upon green leaves, often along the skeletonized vein of a partially consumed leaf.

Adults are extremely fast and erratic fliers, almost impossible to identify on the wing. However, the unique, complexly patterned undersides of their wings easily distinguish them from other species in the cloud forest when they rest on the ground, frequently feeding on rotting fruit, carrion, or dung. Although *Junea dorinda* is predominantly a species of bamboo patches, its butterflies can occasionally be seen zipping across open paramo (high, treeless plateaus) above the treeline.

The Dorinda Satyr caterpillar is almost entirely, including the head, a dull yellowish or orangish-brown color, looking much like a yellowed and dying leaf of its host plant. Irregular, small, brownish markings add to its crypsis. The body ends with a bifid tail that is held together, and the head is conically projected upward into a similar point.

Actual size

FAMILY	Nymphalidae
DISTRIBUTION	North America, from southwestern and southeastern Canada to northern Mexico
HABITAT	Disturbed open, weedy habitats, including roadsides, watercourses, and fields
HOST PLANTS	Plantain (*Musa* spp.), figwort (*Scrophularia* spp.), penstemon (*Penstemon* spp.), acanthus (*Acanthus* spp.), and verbena (*Verbena* spp.)
NOTE	Colorful, spiny caterpillar that is prone to oral regurgitation
CONSERVATION STATUS	Not evaluated, but common

ADULT WINGSPAN
2⅛–2⅜ in (55–60 mm)

CATERPILLAR LENGTH
1½–1⁹⁄₁₆ in (38–40 mm)

JUNONIA COENIA
COMMON BUCKEYE
HÜBNER, 1822

241

Female Common Buckeye butterflies lay eggs singly on lower surfaces of host plant leaves or on terminal shoots. The caterpillars hatch within three days and consume the eggshells before feeding on leaves. All of the food plants used by this species contain iridoid glycosides, which encourage the larvae to feed as well as supplying them with defensive compounds. First to third instars feed on the upper surface of leaves, producing clear patches or spots. Older instars feed openly, consuming leaves from the edges. During the day, the larvae often wander off host plants and return at night to feed.

These caterpillars are prone to much oral regurgitation caused by interactions with other larvae. This may provide an overall defense benefit by repelling predators. Development from egg hatch to pupation takes just 15 days in warm temperatures. Pupation usually occurs on host plants, sometimes under lightly silked leaf shelters or on other surfaces. Adults emerge after a week, and there are numerous generations during spring to fall.

Actual size

The Common Buckeye caterpillar is black dorsally with two orange stripes broken into small spots. There are numerous tiny, white spots, and the large dorsal spines are iridescent blue. Laterally the caterpillar is strongly marked in orange and white, and the head is orange (dorsally) and black (ventrally).

FAMILY	Nymphalidae
DISTRIBUTION	Australia, mainland New Guinea, New Zealand, and islands of the southwest Pacific
HABITAT	From woodlands to grasslands and urban gardens
HOST PLANTS	Herbaceous plants from several families, including Acanthaceae, Asteraceae, Convolvulaceae, Gentianaceae, Goodeniaceae, and Plantaginaceae
NOTE	Black caterpillar with branched spines that has numerous food plants
CONSERVATION STATUS	Not evaluated, but common and widespread in many habitats

ADULT WINGSPAN
1⁹⁄₁₆–1¹¹⁄₁₆ in (40–43 mm)

CATERPILLAR LENGTH
1⁷⁄₁₆–1⁹⁄₁₆ in (37–40 mm)

242

JUNONIA VILLIDA
MEADOW ARGUS
(FABRICIUS, 1787)

The Meadow Argus caterpillar is found singly and feeds during the day and at night. When not feeding, it rests under litter at the base of the plant. In the hot tropical conditions of the wet season, the caterpillar completes growth in two weeks, but breeding during the winter dry season appears to be limited. Several generations are completed annually in tropical areas, but perhaps only two in temperate regions. In cooler areas, when adults are not present, it is uncertain how overwintering occurs.

The caterpillars often leave their food plant to pupate; attached by the cremaster to a silken pad, they hang head down on a rock or fence. Sporadic migration of the adult butterflies occurs but not in all locations or in all years. Caterpillars developing in the cool, short days of winter produce small adults preadapted to spring migration. The genus contains 30 to 35 species, commonly called "buckeyes," which occur throughout the world.

The Meadow Argus caterpillar has a black body with blue-based, black, short-branched dorsal spines and yellow, short-branched lateral and sublateral spines. Numerous fine, white hairs emerge from tiny, white spots within black dorsolateral lines. The head is black with short hairs, and the prothorax is orange.

Actual size

FAMILY	Nymphalidae
DISTRIBUTION	Temperate and tropical Asia, as far north as southeast Siberia
HABITAT	Forests, usually at higher elevations
HOST PLANTS	*Smilax* spp. and *Heterosmilax* spp.
NOTE	Fierce-looking caterpillar that becomes a fierce-acting butterfly
CONSERVATION STATUS	Not evaluated, but locally common

ADULT WINGSPAN
2⅜–2¾ in (60–70 mm)

CATERPILLAR LENGTH
1⁹⁄₁₆ in (40 mm)

KANISKA CANACE
BLUE ADMIRAL
(LINNAEUS, 1763)

243

The eggs of the Blue Admiral are laid on the leaves of host plants singly, enabling a wider distribution of the progeny. The larvae feed and develop through five instars over 23 days. In the later stages, the caterpillars rest on the undersides of leaves, adopting a characteristic U-shape when they are not feeding or when they are disturbed. Despite having a formidable set of spines, the caterpillar is harmless. When mature, it pupates in a spiky, ocherous chrysalis on the host plant stem or adjacent branches, and the adult ecloses after 12 days. There can be multiple generations throughout the year.

The Blue Admiral, a powerful and acrobatic flyer, is insanely territorial and spends most of its time patrolling a defined area chasing off other butterflies—and even challenging people by flapping in their faces. *Kaniska canace* is the only species of its genus, but there are numerous subspecies, some of which are colored differently in their larval stages.

Actual size

The Blue Admiral caterpillar is a striking orange or yellow color. The colored segments have multiple black spots. These alternate with black-streaked white segments. Each orange band bears a ring of seven white, branching, black-tipped, non-stinging spines. The head capsule is orange and black with numerous long setae.

FAMILY	Nymphalidae
DISTRIBUTION	Southeastern Colombia, through the Andes to Bolivia and northern Argentina
HABITAT	Montane forest edges, stream banks, and landslides
HOST PLANTS	Bamboo (*Chusquea* spp.)
NOTE	Caterpillar that mimics twiggy portions of its host plant
CONSERVATION STATUS	Not evaluated, but not considered threatened

ADULT WINGSPAN
2¹⁄₁₆–2⁵⁄₁₆ in (52–58 mm)

CATERPILLAR LENGTH
2–2⅜ in (50–60 mm)

244

LASIOPHILA ORBIFERA
FIERY SATYR
BUTLER, 1868

Fiery Satyr caterpillars, like other members of the subtribe Pronophilina, hatch from round, yellowish eggs laid singly on their host plant. The pale, brownish appearance of the larvae is unremarkable, but that coloration, together with their shape, provides wonderful camouflage, making them a near perfect match for the dried leaves of their host. When not feeding, the caterpillars rest on the numerous dead leaves that often remain attached to, or caught in bunches on, the living portions of bamboo. In this position, the larvae are all but invisible, and only a trained eye can see them.

Adults are commonly seen in rapid, erratic flight along roadsides and streams, almost exclusively in association with bamboo. While feeding at rotting fruit, dung, or carrion, they sit with wings held closed and the forewings tucked rearward between the hindwings, the somewhat complex, leaflike patterning of which makes the adults difficult to spot.

The Fiery Satyr caterpillar is pale brown with variable amounts of dark or reddish-brown striping and spotting across the dorsum. The head is similarly colored and bears two long, rounded head horns, while the terminal abdominal segment is similarly divided into two long tails, which are almost always held together to further mimic the pointed end of a leaf.

Actual size

FAMILY	Nymphalidae
DISTRIBUTION	Northeast India, southern China, and Southeast Asia
HABITAT	Clearings and tracks in tropical forests
HOST PLANTS	*Cratoxylum formosum* and *Cratoxylum cochinchinense*
NOTE	Striking caterpillar that has lengthy protuberances bearing spines
CONSERVATION STATUS	Not evaluated

ADULT WINGSPAN
3½–6 in (90–150 mm)

CATERPILLAR LENGTH
2 in (50 mm)

LEXIAS PARDALIS
COMMON ARCHDUKE
(MOORE, 1878)

245

On the underside of host plant leaves, the female Common Archduke lays green, dome-shaped eggs that are unusually pitted with honeycomb-shaped depressions and covered in tiny, hairlike spines. The caterpillars hatch and first consume their egghell before feeding on older and more mature leaves. Their unique appearance, with long protuberances bearing feather-like spines, provides excellent camouflage against the foliage of its tropical forest habitat. As a defense, the larva also curls up, tucking its head under its spines. The mature caterpillar spins a silk mound on the underside of a leaf to which it attaches and then pupates. The pupa is smooth and green, with tapering ends, and is also well camouflaged.

After about ten days, the pupa darkens, and a day later the butterfly ecloses. The large and fast-flying adults feed on rotting fruits on the floor of tropical forests and are present year-round. The *Lexias* genus has 17 species, including several that are farmed for butterfly houses, all commonly known as Archdukes.

The Common Archduke caterpillar has a green body arrayed with a series of long, spectacular, protuberances along both sides, each ending in a blue and orange tip. Each protuberance bears two rows of spines, which give it the appearance of a feather.

Actual size

FAMILY	Nymphalidae
DISTRIBUTION	North America, from southern Canada to northern Mexico
HABITAT	Low-elevation riparian habitats, usually along watercourses
HOST PLANTS	Willow (*Salix* spp.), poplar (*Populus* spp.), plum (*Prunus* spp.), and apple (*Malus* spp.)
NOTE	Horned caterpillar that has "bird dropping" markings
CONSERVATION STATUS	Not evaluated, but locally common

ADULT WINGSPAN
3–3⅛ in (75–80 mm)

CATERPILLAR LENGTH
1⅜–1½ in (35–38 mm)

246

LIMENITIS ARCHIPPUS
VICEROY
(CRAMER, 1776)

Female Viceroys lay eggs singly on the upper side tips or edges of host plant leaves. After six days, the caterpillars hatch and begin to feed, mostly at night. First instars feed at the tip of a leaf, leaving the midrib exposed. On this, the caterpillar spins a silk mat and rests there, adding frass pellets to the tip of the midrib to extend the pier. The larvae use piers until the third instar, likely to provide protection from predators, which may have an aversion to frass. For defense, older instars use crypsis and aggression—waving their spiked horns when disturbed.

The third instar overwinters in a shelter constructed from host plant leaves. Pupation usually occurs on the host plant, and development from egg to adults takes as little as 40 days. The adult Viceroy is a mimic of the Monarch (*Danaus plexippus*), gaining protection from the latter's distastefulness. Recent research indicates that, like Monarch caterpillars, Viceroy larvae also sequester toxins from their host plants for defense.

The Viceroy caterpillar is smooth and reddish brown with a white saddle and white posterior ventral markings. It also displays tiny, blue spots and long, spiked horns. The head is orange and flat with knobby protuberances. In some populations, the ground color is green or dark brown.

Actual size

FAMILY	Nymphalidae
DISTRIBUTION	Europe, across Asia to China and Japan
HABITAT	Deciduous forests and forest edges near streams
HOST PLANTS	Aspen (*Populus tremula*) and Black Poplar (*Populus nigra*)
NOTE	Well-camouflaged caterpillar that feeds on young leaves
CONSERVATION STATUS	Not evaluated, but threatened in some parts of its range

ADULT WINGSPAN
2⁹⁄₁₆–3⅛ in (65–80 mm)

CATERPILLAR LENGTH
1³⁄₁₆ in (30 mm)

LIMENITIS POPULI

POPLAR ADMIRAL
(LINNAEUS, 1758)

247

Female Poplar Admirals can be seen gliding around treetops before they lay their large eggs singly on leaves. After seven days, the caterpillars hatch and feed on leaf buds and young leaves. The larvae have a particular feeding pattern, consuming the leaf from its tip and avoiding the midrib on which they rest. By silking together frass pellets, the caterpillars may extend the midrib, a habit that is thought to deter ants and other predators. While still quite young, the larvae spin a loose cocoon within a rolled leaf in which they spend winter, emerging in spring to complete their growth and pupate on a leaf, webbing it so that the edges curl and protect the chrysalis.

The large and distinctive *Limenitis populi* butterflies eclose and are on the wing from May to August, the actual months varying across the range. Despite being widespread, the species is not very common, the main reason being the loss of its forest habitat as poplar trees now have little commercial value. There are a number of subspecies, some differing slightly in appearance.

The Poplar Admiral caterpillar has a brown head and predominantly green body, with some areas of brown and even black. There are four protuberances behind the head, the first two being longer, brown-tipped, and covered in short spines. There are also two short horns at the end of the abdomen. The body is covered in many white, raised dots, some with hairs.

Actual size

FAMILY	Nymphalidae
DISTRIBUTION	Eastern Andean slopes, from Colombia to Bolivia
HABITAT	Interior of humid, montane cloud forests, generally at 3,950–8,900 ft (1,200-2,700 m) elevation
HOST PLANTS	*Drymonia* spp.
NOTE	Simply patterned caterpillars that have been reared few times
CONSERVATION STATUS	Not evaluated, but not considered threatened

ADULT WINGSPAN
3⅜–3¾ in (85–95 mm)

CATERPILLAR LENGTH
1⁹⁄₁₆–2 in (40–50 mm)

248

MEGOLERIA ORESTILLA
MEGOLERIA ORESTILLA
(HEWITSON, 1867)

Megoleria orestilla caterpillars hatch from yellowish-white, ribbed eggs, shaped like an elongated barrel, that are laid singly or in small groups of five to eight, usually on the underside of host plant leaves. The larvae hatch together and feed in a characteristic manner, tending to eat channels and small holes near the end of the leaf, causing it to droop, which helps conceal them. When fully developed, the caterpillar forms a mottled, pale green pupa, oddly bent in shape and suspended from a leaf. From oviposition to eclosion takes around 70 to 80 days.

Although it is not known with certainty, these slow-moving caterpillars are likely chemically defended against attacks by vertebrate predators. Nevertheless, rearing projects in eastern Ecuador have found them to frequently fall prey to several types of parasitic wasps and flies. Like most other related species, adults of *Megoleria orestilla* are slow, floppy fliers—another indicator that they are likely chemically defended. The genus *Megoleria* belongs to the subfamily Ithomiinae of so-called clearwing butterflies.

Actual size

The *Megoleria orestilla* caterpillar is dark green, but grayish green dorsally, and has a broad, dirty-yellow stripe laterally spanning the spiracular area. The anterior thoracic segments are washed with grayish white and the true legs are shiny black. The pale reddish-brown head capsule is round and shiny with black stemmata.

FAMILY	Nymphalidae
DISTRIBUTION	Most of Africa, South and Southeast Asia, extending to Australia and New Zealand
HABITAT	Meadows and forest edges as well as disturbed habitat; also in rice fields and among other grassy crops
HOST PLANTS	Grasses and bamboo, including *Poa* spp. and *Oryza* spp.
NOTE	Widely occurring caterpillar that can be a pest on rice
CONSERVATION STATUS	Not evaluated, but common

ADULT WINGSPAN
2–3⅛ in (50–80 mm)

CATERPILLAR LENGTH
2–2¾ in (50–70 mm)

MELANITIS LEDA
COMMON EVENING BROWN
(LINNAEUS, 1758)

249

The Common Evening Brown caterpillar is typical of its subfamily Satyrinae, with cryptic coloration that matches the pattern and shape of the long grass blades it feeds on. While living on such an abundant resource offers many advantages, it has shortcomings, too, as grasses are a tough, low-nutrient food. As a result, it usually takes a long time for the caterpillar to develop, and the ingested chemicals do not offer protection from predators or parasitoids. Hence the larvae have to rely on being undetected on the matching background.

The Satyrinae subfamily of butterflies encompasses more than 2,400 species, whose caterpillars—with a few exceptions—feed on monocot plants such as grasses and bamboos and are quite similar in appearance. Like some other satyrine butterflies, *Melanitis leda* adults have a distinct wet season form with numerous eyespots on their underside—thought to deflect attacks from predators—and various dry season forms that mimic diverse fallen leaf patterns.

Actual size

The Common Evening Brown caterpillar is cryptically colored green with longitudinal stripes: one median, dorsal, darker stripe and several lighter stripes. The head can be either green with dark horns from which two vertical, dark stripes originate, or it can also be colored entirely black. Both horns and body are covered with numerous short, thin setae.

FAMILY	Nymphalidae
DISTRIBUTION	North Africa, across Europe, Middle East, and northern Asia (Russia and Mongolia)
HABITAT	Open grassland on thin, rocky soils, and subalpine meadows
HOST PLANTS	Ribwort Plantain (*Plantago lanceolata*) and speedwell (*Veronica* spp.)
NOTE	Spiny, black caterpillar that lives in a silken web
CONSERVATION STATUS	Not evaluated, but locally endangered

ADULT WINGSPAN
1½–1⅞ in (38–47 mm)

CATERPILLAR LENGTH
1 in (25 mm)

250

MELITAEA CINXIA
GLANVILLE FRITILLARY
(LINNAEUS, 1758)

Actual size

The Glanville Fritillary caterpillar is easily recognized, with its red head and prolegs and black, spiny body. The body has rings of black tubercles bearing many black spines, alternating with rings of white spots.

Glanville Fritillary caterpillars hatch from clusters of between 50 and 200 yellow eggs laid on the underside of a leaf of the host plant. The caterpillars are gregarious, staying together within a silk web that they spin over the host plant and in which they feed. They are often spotted basking on the surface on sunny days. The larvae spin a silk tent within tall grass, where they overwinter, becoming active again in spring. The mature caterpillars are solitary and when disturbed roll up into a ball and drop to the ground. They pupate on the host plant, attached to a stem, or they drop to the ground around the plant and pupate within the leaf litter.

The orange, black, and white chequer-patterned adults are on the wing between May and July. There is usually one generation in the northern part of the Glanville Fritillary's range, but two generations appear in the south. The species, named for Lady Eleanor Glanville (ca. 1654–1709), an English entomologist, is in decline across much of its range due to loss of habitat. In the United Kingdom, it is now restricted to the Isle of Wight.

FAMILY	Nymphalidae
DISTRIBUTION	Europe, across central Asia, southern Siberia, Mongolia, northeast China, Korea, and Japan
HABITAT	Various, including pine woodlands, coppiced woodlands, alpine meadows, grasslands, wetlands, and fens
HOST PLANTS	Mostly *Valeriana* spp.; also cow wheat (*Melampyrum* spp.)
NOTE	Spiny caterpillar that overwinters within curled-up dead leaves
CONSERVATION STATUS	Not evaluated, but locally threatened

ADULT WINGSPAN
1¼–1⅝ in (32–42 mm)

CATERPILLAR LENGTH
¹¹⁄₁₆ in (18 mm)

MELITAEA DIAMINA
FALSE HEATH FRITILLARY
(LANG, 1789)

251

False Heath Fritillary caterpillars hatch from pale yellow eggs laid by the female in batches of about 100 on the underside of leaves of the host plant. The larvae are gregarious, living and feeding together in a communal web spun from silk threads. On cool but sunny days they may be seen basking on the surface. The caterpillars overwinter inside dead, curled leaves beneath the host plant and become active again from April. As they mature, they become increasingly solitary. The larvae pupate, hanging from stems of the host plant. The pupa is creamy white with brown-black marks.

The adults, strongly marked in a chequered pattern of orange, black, and white, are on the wing from May to September, depending on altitude. There is usually a single generation annually. The species is in decline across much of its range, although it continues to be common on damp alpine meadows and fenland.

Actual size

The False Heath Fritillary caterpillar is dark brown with gray-white dots and a single, dark dorsal stripe. There are bands of yellow-brown spines, often gray at the tips. The head is black with black hairs.

FAMILY	Nymphalidae
DISTRIBUTION	Eastern Colombia to southeast Ecuador, but possibly farther south
HABITAT	Intact tracts of subtropical cloud forests, forest edges, and light gaps, at 5,600–7,200 ft (1,700–2,200 m) elevation
HOST PLANTS	*Nectandra* spp. and *Ocotea* spp.
NOTE	Caterpillar that hides inside a rolled-up leaf when mature
CONSERVATION STATUS	Not evaluated, but not considered threatened

ADULT WINGSPAN
2^{11}⁄$_{16}$–3 in (68–75 mm)

CATERPILLAR LENGTH
2⅛–2^{9}⁄$_{16}$ in (55–65 mm)

252

MEMPHIS LORNA

LORNA LEAFWING

(DRUCE, 1877)

From observations in northeastern Ecuador, the rare, brownish, tubular-shaped, green-speckled Lorna Leafwing caterpillar looks remarkably like a moldy, mossy twig. As first to third instars, the larvae rest on chains of frass projecting from the edge of host plant leaves, then create a tube-shaped shelter out of a rolled host plant leaf during the fourth and fifth instars. The spiky nature of the caterpillar's head, in conjunction with its unusual thickness, create a protective "plug" to fill the entrance to the shelter, preventing predator access to the larva's more vulnerable parts.

Memphis lorna is confined to a narrow elevational range in eastern Colombia and Ecuador, but may also occur east of the Andes in Peru and Bolivia. Adults are rarely seen, but males can be locally common in some areas, usually observed while feeding at rotting flesh, ripe fruit, or dung, their extremely realistic, leaf-patterned, ventral wings making them almost invisible on the forest floor. A formal description of the species has not yet been published.

The Lorna Leafwing caterpillar has a bulbous, black head patterned with creamy vertical stripes and washed across the front with deep crimson. Across the top, it bears several short, conical protrusions. The body is dark brown, blackish posteriorly, finely flecked with bright green on the abdomen, and striped with green on the thorax. It is sparsely covered in thin, crooked setae.

Actual size

FAMILY	Nymphalidae
DISTRIBUTION	South and Southeast Asia
HABITAT	Glades and clearings in heavily forested regions with heavy rainfall
HOST PLANTS	Rubiaceae, including *Cinchona* spp. and *Wendlandia* spp.; also Capparaceae
NOTE	Camouflaged caterpillar that uses its frass to deter predators
CONSERVATION STATUS	Not evaluated, but most common of its genus

ADULT WINGSPAN
2⅜–3 in (60–75 mm)

CATERPILLAR LENGTH
¾–⅞ in (20–22 mm)

MODUZA PROCRIS
COMMANDER
(CRAMER, 1777)

253

Commander larvae emerge from spiny, green eggs—resembling tiny sea urchins—which the butterfly lays on the underside of leaves near the shoot tip up to four days earlier. The fast-growing and bizarre-looking caterpillar has an unusual defensive strategy. It partially consumes a leaf, and then combines its excreted frass with bits of chewed leaf to create a long chain held together with silk thread. The chain, together with scattered frass, which is believed to contain toxins, acts as a barrier to deter ants and other predatory insects from reaching the caterpillar while it rests.

Mature caterpillars may move some distance from the host plant to pupate on the ground in the leaf litter. The brown pupa is camouflaged with lines and markings to resemble a rolled-up dead leaf. The colorful, red, brown, and white butterflies are most commonly seen after the monsoon and in winter. The Commander is the most common and widespread of the nine *Moduza* species.

Actual size

The Commander caterpillar has an unusual appearance that provides effective camouflage. It is chestnut brown in color with darker spots, and Its body is covered in thick tubercles that bear many spiny processes, giving it a spiky outline. This helps to disrupt its shape when resting and may reduce predation.

FAMILY	Nymphalidae
DISTRIBUTION	Central America and northern part of South America
HABITAT	Tropical rain forests
HOST PLANTS	Leguminosae, including *Arachis hypogaea*, *Lonchocarpus* spp., *Inga* spp., *Medicago sativa*, and *Pithecellobium* spp.; also at least one member of Bignoniaceae, *Paragonia pyramidata*
NOTE	Tropical caterpillar that apparently relies mostly on crypsis for defense
CONSERVATION STATUS	Not evaluated, but common, although, like all tropical forest species, affected by habitat loss

ADULT WINGSPAN
4–6 in (100–150 mm)

CATERPILLAR LENGTH
4–5 in (100–130 mm)

254

MORPHO PELEIDES
BLUE MORPHO
KOLLAR, 1850

The Blue Morpho caterpillar is unusual because its appearance is neither clearly cryptic nor clearly aposematic. Under the dark, dull conditions of the tropical forest, its brown burgundy coloring is likely to blend in, concealing it from predators, while its neon-yellow or green patterning when viewed from above makes it look like an unpalatable insect. The peripheral hairs along the abdomen, for instance, resemble spider legs. Patterning with such a dual cryptic-aposematic function is quite common in the tropical forest. The pupa is less colorful but an equally cryptic green.

Some consider *Morpho peleides* a subspecies of *M. helenor*, which has a broader range, from Mexico southward. The genus *Morpho* contains some of the largest, most extravagant butterfly species in the world, famous for their structural iridescent coloration. Many of them are also highly monophagous and rare. The fact that *M. peleides* is oligophagous, feeding on relatively common hosts, is responsible for its success as a species.

The Blue Morpho caterpillar is burgundy with neon-yellow or green, rhomboid patterning on its back and sides. It has dorsal and lateral, finlike tufts of hair as well as tufts of hair on the first thoracic segment ornamenting the head. The elaborate pattern vanishes when the caterpillar suspends itself upside down to pupate; the prepupa and pupa are green.

Actual size

FAMILY	Nymphalidae
DISTRIBUTION	From the coastal Andean range of Venezuela, south to Colombia and northern Ecuador on both Andean slopes
HABITAT	Cloud forests and bamboo thickets at middle and upper elevations
HOST PLANTS	Bamboo (*Chusquea* spp.)
NOTE	Distinctive caterpillar, but rarely seen
CONSERVATION STATUS	Not evaluated, but unlikely to be endangered

MYGONA IRMINA
MYGONA IRMINA
(DOUBLEDAY, [1849])

ADULT WINGSPAN
2⁹⁄₁₆–2¾ in (65–70 mm)

CATERPILLAR LENGTH
1⅛–1¼ in (28–32 mm)

255

Mygona irmina caterpillars hatch from eggs laid singly on the underside of mature bamboo leaves. Initially, they rest near the tip of mature leaves, feeding in a way that leaves one margin of the leaf apex intact. Against this thinned portion of the leaf, they are very cryptic. During the middle stage of development, larvae rest on the dorsal surface of leaves, often traveling to adjacent leaves to feed. Immediately after molting to the final instar, the caterpillars are various shades of brown, green, and pale blue dorsally and dorsolaterally, with shades of white and pink spiracularly to ventrolaterally. Within a day, they become darker overall, turning to shades of brown.

To date, fifth instars have not been observed in the field, but their distinctive, dark coloration suggests they rest somewhere other than on host plant leaves or stems. The full cycle from egg to butterfly takes between 102 and 109 days. Adults fly rapidly over bamboo-growing areas, usually on sunny days, and also feed at mammal droppings.

The *Mygona irmina* caterpillar has a dark brown to black head and well developed but rounded and curved scoli. Its body shape is slightly flattened (trapezoidal in cross section). The body ground color is predominantly brown, with darker areas forming distinct dorsal chevrons, each highlighted with small, green flecking.

Actual size

FAMILY	Nymphalidae
DISTRIBUTION	Mainland New Guinea, northeastern and eastern Australia
HABITAT	Mainly lowland rain forest along streams, but occurs up to 2,625 ft (800 m) elevation
HOST PLANTS	Stinging trees in the Urticaceae family, such as *Dendrocnide* spp., and Native Mulberry (*Pipturus argenteus*)
NOTE	Caterpillar that moves and feeds gregariously
CONSERVATION STATUS	Not evaluated, but can be locally common

ADULT WINGSPAN
2–2¼ in (50–57 mm)

CATERPILLAR LENGTH
1⁹⁄₁₆–1¾ in (40–45 mm)

256

MYNES GEOFFROYI
JEZEBEL NYMPH
(GUÉRIN-MÉNEVILLE, [1830])

Jezebel Nymph caterpillars hatch in clusters of up to 50 individuals. They are initially orange with black hairs, well camouflaged and gregarious, staying in clusters while feeding on the underside of leaves. They move en masse to new leaves when they have devoured the entire leaf. The larval stage lasts six to seven weeks in the cooler southern locations, and breeding can occur throughout the year, with several generations completed.

The caterpillars pupate together or close by, often with ten or more under a single leaf, suspended head down and attached to a silken pad by the cremaster. The pupae thrash wildly for several seconds if disturbed. Butterflies from the same cohort all emerge within a day. Male butterflies are very territorial and have corridors that they patrol, chasing any other butterflies entering their territory. The genus *Mynes* contains 12 species, which occur only in Australia, New Guinea, or Indonesia.

The Jezebel Nymph caterpillar is dark brown or black, with numerous white dots and rows of large, branched, pinkish or pale blue spines. The head is black or grayish brown with two small, black branched spines.

Actual size

FAMILY	Nymphalidae
DISTRIBUTION	Northwest Himalayas, northern India, southern China, Chinese Taipei, Myanmar, Thailand, Malay Peninsula, and Sumatra
HABITAT	Subtropical and tropical evergreen forests
HOST PLANTS	Hall Crab Apple (*Malus halliana*) and Loquat (*Eriobotrya japonica*)
NOTE	Trapeze artist of the caterpillar world
CONSERVATION STATUS	Not evaluated, but occurs in localized colonies so prone to threats

ADULT WINGSPAN
2⅛–2⁹⁄₁₆ in (55–65 mm)

CATERPILLAR LENGTH
1³⁄₁₆ in (30 mm)

NEPTIS SANKARA
BROAD-BANDED SAILER
(KOLLAR, 1844)

257

Like many nymphalid caterpillars, the Broad-banded Sailer incorporates frass chains and perches during its development. By creating a safe place on the food plant and building barriers of silk and frass, otherwise defenseless larvae can isolate themselves from predators and accidents. The species persists with the strategy into later instars by "trapezing." Having stripped the leaf down to the midrib, the caterpillar lines it with silk and leaf fragments, then spends much of the day hanging there, resembling debris. It migrates to other leaves to feed and pupates on foliage within a curved chrysalis. There are two generations annually.

Typical of its genus, the Broad-banded Sailer butterfly is disruptively marked in black and white on its topside with cryptic brown-and-white patterns on the underside. This species and fellow "sailers" of *Neptis* are named for the way the butterflies rotate between perches in a very similar gliding or sailing fashion. *Neptis* larvae share the broad, horned head and fleshy thoracic tubercles of *N. sankara*, although each species has its own variations. Populations of the Broad-banded Sailer are localized and often isolated.

Actual size

The Broad-banded Sailer caterpillar is dark brown with tiny, irregular flecks of green across the head capsule and along the length of the body laterally. Mid-abdomen, there are elongated, reddish-brown, wavy blotches incorporating another green spot within a larger white zone. The disproportionately broad head is topped with a pair of small horns. The third thoracic segment bears a pair of prominent, forward-pointing, soft-tipped, fleshy processes, and the second segment has a single ridge-like process.

FAMILY	Nymphalidae
DISTRIBUTION	Central Europe, through central Asia and southern Russia to Japan, and south to Southeast Asia
HABITAT	Damp, temperate woodlands and tropical rain forests up to 3,950 ft (1,200 m) elevation
HOST PLANTS	Pea (*Lathyrus* spp.) and Black Locust (*Robinia pseudacacia*)
NOTE	Oddly shaped caterpillar that has disruptive coloring for camouflage
CONSERVATION STATUS	Not evaluated, but common across its range

ADULT WINGSPAN
1⁹⁄₁₆–1⅞ in (40–48 mm)

CATERPILLAR LENGTH
1 in (25 mm)

NEPTIS SAPPHO
COMMON GLIDER
(PALLAS, 1771)

258

Common Glider caterpillars hatch from round eggs laid by the female in shady places on the upper surface of leaves of the host plant. Each egg is covered with a network of hexagonally shaped ridges, which bear tiny hairs. The young caterpillars shelter within curled-up leaves and emerge to feed on the leaf, leaving the midrib, along which they rest. Their body has a disruptive outline, which provides excellent camouflage among dead leaves. The larvae overwinter and pupate the following spring. The pupa has the appearance of a dried-up leaf suspended from a stem.

The Common Glider completes its development in just five to six weeks, so there are as many as four generations a year, and the adults fly from April to September. Also known as the Pallas Sailor, the species gets its common names from the butterfly's powerful gliding or sailing style of flight.

Actual size

The Common Glider caterpillar has an unusual shape and coloration, with bands of olive green and brown that provide a disruptive shape, especially when the abdomen is raised. The body is covered in short, white hairs, and there are spines on the thorax and abdomen.

FAMILY	Nymphalidae
DISTRIBUTION	North America, Europe, and Asia
HABITAT	Riparian corridors, glades, groves, parks, and yards
HOST PLANTS	Many, including willow (*Salix* spp.), poplar (*Populus* spp.), birch (*Betula* spp.), apple (*Malus* spp.), and alder (*Alnus* spp.)
NOTE	Spiny, gregarious caterpillar that is often seen on backyard willows
CONSERVATION STATUS	Not evaluated, but common

ADULT WINGSPAN
3–3⅛ in (75–80 mm)

CATERPILLAR LENGTH
2–2⅛ in (50–55 mm)

NYMPHALIS ANTIOPA
MOURNING CLOAK
(LINNAEUS, 1758)

259

Female Mourning Cloak butterflies lay eggs in "collars" of 100 to 200 eggs on the branches of the host plants. The caterpillars hatch after five to nine days and are highly gregarious, feeding and moving in groups throughout development. They react in unison when disturbed, rearing their heads and waving them about. Such synchronous head-jerking, as well as their spines and likely emission of repellent chemicals from a ventral "neck" gland, are their main forms of defense.

Mature caterpillars, which have bright orange-red patches, signaling distastefulness, leave the host plant before pupating and wander on paths or roads. Development of this species is rapid, as the larvae pupate as little as two weeks after egg hatch. Adults emerge after another two weeks. Populations of Mourning Cloaks are subject to "boom and bust" cycles, thought to be caused by disease or natural enemy pressures. En masse, they can be destructive: a large group can denude a small willow tree.

Actual size

The Mourning Cloak caterpillar is black with black, branched spines, and many tiny, white spots in broken, transverse lines. The bold dorsal patches may be orange or red. The numerous setae are short and white, imparting a shaggy appearance. The head is black and shiny with short, white hairs.

FAMILY	Nymphalidae
DISTRIBUTION	Western North America, from British Columbia to southern California
HABITAT	Mountain slopes, canyons, watercourses, parks, and gardens
HOST PLANTS	Mountain Balm (*Ceanothus velutinus*), Deerbrush (*Ceanothus integerrimus*), and Redstem Ceanothus (*Ceanothus sanguineus*)
NOTE	Spiny, gregarious caterpillars that reportedly produce "millions of shimmering pupae"
CONSERVATION STATUS	Not evaluated, but common

ADULT WINGSPAN
2⅜–2⁹⁄₁₆ in (60–65 mm)

CATERPILLAR LENGTH
1¾–2 in (45–50 mm)

NYMPHALIS CALIFORNICA
CALIFORNIA TORTOISESHELL
(BOISDUVAL, 1852)

260

California Tortoiseshell caterpillars hatch from eggs laid in clusters of up to 250 four to five days earlier on the upper and lower surfaces of host plant leaves. The first instars partially consume the eggshell, then feed on the host plant. Early instars feed only on new growth and also use silk to cover and join leaves together. The caterpillars feed and rest openly and are gregarious in the first three instars, dispersing in the fourth and fifth instars. Adults emerge about five weeks after eggs are laid.

When disturbed, larvae of the first three instars jerk their heads in unison to intimidate predators. The heavily sclerotized posterior segments in third and fourth instars resemble the head capsule, giving the caterpillar a two-headed appearance, which may divert the attention of birds. Reports of acres of defoliated Mountain Balm and "millions of shimmering pupae" suggest that caterpillar survival can be excellent. The species is subject to periodic population explosions, and adults are long lived as well as migratory.

The California Tortoiseshell caterpillar is black with profuse long, white setae on the body and head. Spines are either orange or black with bulbous, orange or black bases. A pair of interrupted, dorsal, white or yellow lines may be present. The posterior two segments are heavily sclerotized.

Actual size

FAMILY	Nymphalidae
DISTRIBUTION	North America, from Alaska to Quebec and south to Wisconsin
HABITAT	Mid-elevation pine forests and grasslands
HOST PLANTS	Various sedges and grasses, including *Carex* spp. and *Festuca* spp.
NOTE	Caterpillar that may be vulnerable in a warming climate
CONSERVATION STATUS	Not evaluated, but considered secure in most of its range

ADULT WINGSPAN
1¾–2⅛ in (45–54 mm)

CATERPILLAR LENGTH
1⅜ in (35 mm)

OENEIS CHRYXUS
CHRYXUS ARCTIC
(DOUBLEDAY, [1849])

261

Chryxus Arctic caterpillars are biennial, overwintering twice, and have five instars. The female finds sedges growing in a turf-like mat under a pine tree and lays an egg on a dead branch just above the sedge. The tiny caterpillar then drops onto the sedge, feeds for several months, and overwinters. It feeds during the next brief summer and then overwinters once more as a nearly mature caterpillar. The following spring, after feeding briefly, it turns into a brown-orange pupa in a slight silken cocoon in the litter, emerging as an adult in June.

Adults suck the nectar from a number of flower species but also sip moisture from wet soil. Several dozen *Oeneis* species occur in high latitude areas in North America and Eurasia. All have similar caterpillars, but most live on grassland or tundra. As the species are well adapted to living in harsh environments, global warming could reduce numbers.

The Chryxus Arctic caterpillar is pinkish to tan to dark brown with many paler and darker stripes. The head bears six distinct, brown to black, vertical stripes. There is a prominent black dorsal stripe and other brown stripes covering the sides, which make the larvae resemble grass. At the posterior there is a pair of short tails.

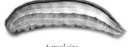

Actual size

FAMILY	Nymphalidae
DISTRIBUTION	Eastern Ecuador
HABITAT	Humid montane cloud forest understory and forest edges
HOST PLANTS	*Brugmansia aurea*
NOTE	Caterpillar that snips off leaves to eat at ground level
CONSERVATION STATUS	Not evaluated, but not considered threatened

ADULT WINGSPAN
2⅛–2⁹⁄₁₆ in (55–65 mm)

CATERPILLAR LENGTH
¾–1³⁄₁₆ in (19–21 mm)

OLERIA BAIZANA
BAEZA GLASSWING
(HAENSCH, 1903)

262

Actual size

The Baeza Glasswing caterpillar is very simply patterned and tubular in shape. Its head is round and shiny black, as are its true legs. The body is entirely ochraceous olive to dark greenish black with only a few wavy, indistinct, white markings laterally.

Baeza Glasswing caterpillars hatch from eggs laid singly, off the host plant in the leaf litter. All instars are rather dull and similar in appearance to the mature caterpillar. They are slow moving and reluctant to react, even when touched, generally doing nothing more than curling into a tight ball. However, during the night, the larvae climb a food plant seedling and sever a leaf petiole, parachuting with the leaf to the ground, where they remain while feeding. There are five larval instars, and individuals take 75 to 80 days to mature from oviposition to eclosion.

Larvae pupate under curled, dead leaves on the forest floor. The tiny, rounded pupa is a subtle, translucent, dark yellow with black markings, well camouflaged in its leaf-litter habitat. The cycle recommences as the adults eclose and mate. Females are most frequently encountered searching for oviposition sites in the deep shade of the cloud forest understory.

FAMILY	Nymphalidae
DISTRIBUTION	The Andes of Colombia and Ecuador
HABITAT	Mid-elevation cloud forests, forest edges, and bamboo-dominated gaps
HOST PLANTS	Bamboo (*Chusquea* spp.)
NOTE	Extremely cryptic caterpillar that is common but difficult to find
CONSERVATION STATUS	Not evaluated, but unlikely to be endangered

ADULT WINGSPAN
1⅞–2⅛ in (48–54 mm)

CATERPILLAR LENGTH
1½–1⅝ in (38–42 mm)

PEDALIODES PEUCESTAS
PEDALIODES PEUCESTAS
(HEWITSON, 1862)

263

Eggs of *Pedaliodes peucestas* are laid singly, occasionally in pairs, on the underside of host plant leaves. The newly hatched caterpillar is tiny and pale white, with a bulbous, brown head. Its first meal is its own eggshell, but it quickly moves on to leaves, soon taking on the green coloration of its host plant and becoming nearly impossible to see. The cycle from egg-laying, through all five larval stages to the emergence of the adult can take more than 110 days. The pupal stage alone may last more than 25 days.

This widespread species is often one of the most common adult butterflies within its habitat. Adults, which feed on animal droppings and carrion, are seen flitting almost ceaselessly along roadsides and over large tracts of their bamboo host plant. They fly in all weathers, except for heavy rain, although they are quick to resume activity as soon as the storm has passed.

Actual size

The *Pedaliodes peucestas* caterpillar is overall extremely cryptically colored, resembling a decaying mossy stick or portion of a bamboo leaf petiole. Its complex mottling of various shades of brown is here and there highlighted with just the right amount of green to enhance the resemblance to a twig. The head capsule is generally squared-off, with a slightly cat-eared appearance.

FAMILY	Nymphalidae
DISTRIBUTION	The Andes of Venezuela, Colombia, Ecuador, and northern Peru
HABITAT	Mid-elevation cloud forests and forest edges
HOST PLANTS	*Paullinia* spp.
NOTE	Rarely encountered caterpillar that has only recently been described
CONSERVATION STATUS	Not evaluated, but not likely to be endangered

ADULT WINGSPAN
1¾–2 in (45–50 mm)

CATERPILLAR LENGTH
¾–⅞ in (20–22 mm)

264

PERISAMA OPPELII
CITRON PERISAMA
(LATREILLE, [1809])

Young Citron Perisama caterpillars rest on the skeletonized midveins of leaf tips that have been extended by a frass chain. Molting occurs near the tips of these safe havens. Later instars rest on the dorsal surface of leaves with their head tipped forward, scoli pressed flat, and body held either straight or in a slight S-curve. When disturbed, larvae thrash their head and abdomen, attempting to brush the offending object away with the head scoli. They drop from the plant only reluctantly, and must be strongly provoked before thrashing.

The emerald-green pupa is attached by the cremaster to the dorsal surface of a leaf, and adults emerge about 20 days after pupation. Male Citron Perisama butterflies visit wet sand enriched with urine or feces, and they often feed on the sides of buildings or on dirty clothes, periodically curling their abdomen under their body and exuding a droplet of liquid that is then re-ingested.

Actual size

The Citron Perisama caterpillar is green with small, yellow granulations and several faint, whitish lines laterally. The terminal segment has a pair of short, green scoli, topped with a rosette of dark spines. The head is mottled brownish and white and bears two long scoli with several whorls of spikes, including a rosette at the end that has five to six points.

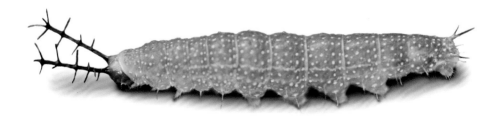

FAMILY	Nymphalidae
DISTRIBUTION	Mainland New Guinea, and northeastern and eastern Australia
HABITAT	Coastal rain forest edges, especially along creeks, gullies, and urban gardens
HOST PLANTS	Wide range, including species from Bombacaceae, Boraginaceae, Fabaceae, Sterculiaceae, Tiliaceae, and Ulmaceae
NOTE	Caterpillar that creates dead-leaf camouflage on its food plant
CONSERVATION STATUS	Not evaluated, but locally common

ADULT WINGSPAN
2³⁄₁₆ in (56 mm)

CATERPILLAR LENGTH
1¹⁄₃₂ in (26 mm)

PHAEDYMA SHEPHERDI

WHITE-BANDED PLANE

(MOORE, 1858)

265

The White-banded Plane caterpillar hatches from pale yellow, pitted eggs laid on the shoots and young leaves of its food plant. The caterpillar cuts out small pieces of leaf and hangs them from the leaf edge on which it is feeding. These decaying leaf pieces provide an excellent camouflage for the larva as it rests along the midrib of the leaf. Such dead-leaf fragments are a strong indicator of the presence of this species. The caterpillars develop slowly but are present throughout the year in the tropical areas of their range.

Pupation occurs on the underside of a nearby uneaten leaf, the brownish pupa hanging head downward and attached by the cremaster to a silken pad. The adult butterflies have a distinctive gliding flight. The males will defend territories in sunny patches in the rain forest, frequently returning to the same leaf to survey their own territory.

Actual size

The White-banded Plane caterpillar is light brown with oblique, darker brown bands on the abdomen and green-brown patches at the end of the abdomen. It has a pair of yellow lateral spots on segments eight and nine, while its mesothorax, metathorax, and segments two and eight have a pair of branched, spiny dorsolateral processes. The head has two short, spiny dorsolateral projections.

FAMILY	Nymphalidae
DISTRIBUTION	Areas of sub-Saharan Africa, and southern and Southeast Asia to northern Australia
HABITAT	Tropical, riparian, monsoon forests from sea level to above 4,920 ft (1,500 m) elevation
HOST PLANTS	*Flacourtia* spp. and species from Acanthaceae, Compositae, Primulaceae, Salicaceae, Rubiaceae, and Violaceae
NOTE	Most common and widespread caterpillar of its genus
CONSERVATION STATUS	Not evaluated, but locally common and widely distributed

ADULT WINGSPAN
1¹³⁄₁₆–2⅛ in (46–55 mm)

CATERPILLAR LENGTH
1 in (25 mm)

PHALANTA PHALANTHA
COMMON LEOPARD
(DRURY, [1773])

266

The Common Leopard caterpillar, also known as the Spotted Rustic, feeds openly on the foliage of its food plant, where it is very active and develops fast, taking about seven days to mature. The larvae are most abundant during the late dry season and at the end of the wet season following new growth of the plants on which they feed. When the caterpillar stops feeding, it wanders to the underside of a leaf where it spins a silk pad from which it hangs vertically to pupate. The pupa is a spectacular green with red and silver tubercles.

Several generations of *Phalanta phalantha* are completed each season. The fast-flying adult is a sun-loving species, commonly seen flying around flowering shrubs taking nectar, and at times exhibiting puddling behavior to imbibe salts from damp ground. The Common Leopard is the most common and widespread member of its genus, which contains six species.

The Common Leopard caterpillar is orange brown, changing to bright green just before pupation, with black, branched, dorsolateral and lateral spines. A white line joins the base of the subspiracular, black-and-white, branched spines. A narrow, dark band runs dorsally on the body. The head is orange brown dorsally and black near the mouth parts.

Actual size

FAMILY	Nymphalidae
DISTRIBUTION	Western North America, from British Columbia to Arizona
HABITAT	Dry foothill or shrub-steppe gullies, streambeds, and hillsides
HOST PLANTS	Thistle (*Cirsium* spp.)
NOTE	Spiny, communal caterpillar that has many defenses
CONSERVATION STATUS	Not evaluated, but common

ADULT WINGSPAN
1¾–2 in (45–50 mm)

CATERPILLAR LENGTH
1–1³⁄₁₆ in (25–30 mm)

PHYCIODES PALLIDA
PALE CRESCENT
(W. H. EDWARDS, 1864)

267

Pale Crescent caterpillars hatch from an ordered mass of around 90 eggs, laid eight to nine days earlier on the underside of thistle leaves at the plant's mid-height. First-year thistles are preferred, and mature, flowering thistles are avoided. First and second instars live communally in loosely woven silk nests, usually in a fold of the host leaf. The survival strategy of this species includes group behavior (synchronous reactions to disturbances), the physical protection of its silken webs, concealment inside or under leaves, avoidance (coiling up and dropping to the ground when disturbed), and camouflage.

The larvae usually develop to the fourth instar in about 18 days. In most locations, the third or fourth instar enters dormancy and overwinters, resuming feeding and development in the following spring. Adults fly from the middle of May to July. Males perch conspicuously and chase females, which spend much of their time low in the vegetation, avoiding males.

The Pale Crescent caterpillar is dark brown to black with numerous white spots and several broken, longitudinal, white stripes. In some populations, the white markings coalesce, making the caterpillar largely white. The spines are black and clumped, and have pronounced orange bases. The head is shiny black, sometimes with orange patches dorsally.

Actual size

FAMILY	Nymphalidae
DISTRIBUTION	Southwestern United States into Mexico
HABITAT	Lower mountain hills
HOST PLANTS	Beardtongue (*Penstemon* spp.)
NOTE	Caterpillar that is protected by the toxic chemicals it ingests
CONSERVATION STATUS	Not evaluated, but generally common if occasionally rare at the periphery of its range

ADULT WINGSPAN
1¼–1¾ in (32–45 mm)

CATERPILLAR LENGTH
1⅜ in (35 mm)

POLADRYAS ARACHNE

ARACHNE CHECKERSPOT
(W. H. EDWARDS, 1869)

268

Actual size

Young Arachne Checkerspot caterpillars are gregarious, hatching from clusters of about 40 eggs laid on the underside of young green beardtongue leaves, which the caterpillars eat. The larvae hibernate half grown and become solitary feeders in spring, but may reenter diapause if the host plant or weather conditions are unsuitable. Caterpillars and adults are poisonous to birds and mice because of iridoid glycoside chemicals they ingest from *Penstemon* leaves. The pupa is white with black marks and orange bumps.

There are several generations of adults during the summer, and the bright orangish adults can be seen on hilltops, where males wait for females to arrive for mating. Several hundred other species of Checkerspots occur in the northern hemisphere and American tropics. Young caterpillars of all species are gregarious on various herbs or bushes, and all but the youngest caterpillars are adorned with numerous spines that help protect them from mice and birds.

The Arachne Checkerspot caterpillar is boldly striped in black and white, and is covered with branching, orange-and-black spines. The top row of spines is black. The head is orange and pubescent, the true legs are black, and the prolegs are orange.

FAMILY	Nymphalidae
DISTRIBUTION	North Africa, Europe, central Asia, Siberia, northern India, eastern China, and Japan
HABITAT	Woodlands, especially glades, rides, and margins; also hedgerows and gardens
HOST PLANTS	Various, including nettle (*Urtica* spp.), currants (*Ribes* spp.), and sometimes willow (*Salix* spp.)
NOTE	Caterpillar that looks like a bird dropping in later instars
CONSERVATION STATUS	Not evaluated, but common and increasingly widespread

ADULT WINGSPAN
1⅝–1⅞ in (42–47 mm)

CATERPILLAR LENGTH
1⅜ in (35 mm)

POLYGONIA C-ALBUM

COMMA
(LINNAEUS, 1758)

269

Comma caterpillars hatch from eggs laid singly, near the edge of the upper surface of leaves. Over the two or three weeks before they hatch, the eggs turn from green to yellow to gray. The newly emerged larvae move to the underside of the leaf, where they start feeding, and then the older caterpillars move back to the upper surface, protected to some extent by their bird-dropping appearance. The caterpillars pupate on the host plant or nearby vegetation, suspended beneath a twig or stem; the pupae darken to resemble withered leaves.

The adults, which eclose within three weeks, can be seen almost any time of year as overwintering adults may become active on warm winter days. The first generation appears in early summer, while in late summer there is a second generation of adults much darker in color. In the southernmost part of the Comma's range, there may be a third generation. This is one of the few species that has an expanding range as a result of a warming climate.

Actual size

The Comma caterpillar is dark brown to black, with orange markings on the thorax and along the sides, and a large patch of white on the dorsal surface. There are bands of long tubercles that bear spines.

FAMILY	Nymphalidae
DISTRIBUTION	Western North America, from Alaska to New Mexico
HABITAT	Mountain areas above 3,300 ft (1,000 m) elevation, including meadows, streams, roads, and trails
HOST PLANTS	Currant (*Ribes* spp.) and elm (*Ulnus* spp.)
NOTE	Caterpillar protected by spines and often by spiny host plants
CONSERVATION STATUS	Not evaluated, but common

ADULT WINGSPAN
1¾–2 in (45–50 mm)

CATERPILLAR LENGTH
1⅜–1½ in (35–38 mm)

POLYGONIA GRACILIS
HOARY COMMA
(GROTE & ROBINSON, 1867)

270

Hoary Comma caterpillars hatch from eggs laid singly, or in groups of three or four, on the underside of host plant leaves four or five days earlier. Feeding from the leaf edges, they create jagged holes. The caterpillars rest on stems or leaves of the host, usually on the undersides, where they are hidden from aerial predators. Concealment and camouflage afford the larvae some protection, as do the spines of some of their host plants. Chemicals emitted from a small ventral gland near the head may also repel some attackers. Shelters are not constructed. In later instars, caterpillars disperse, with only one or two per shrub.

The caterpillars go through five instars, each taking about five days to complete, and pupation occurs on or near the host plant; the pupal period lasts about nine days. The butterfly adults are long-lived (up to 12 months), and there is only one generation a year. The flight period extends from mid-March to October, and adults overwinter.

The Hoary Comma caterpillar is black anteriorly with orange or mustard spines, and the posterior area is black with nearly solid, white frosting dorsally. The black sides have rusty-orange, wavy lines resembling links in a chain. The head is shiny black with two conical "horns." Some caterpillars become bright rusty orange anteriorly prior to pupation.

Actual size

FAMILY	Nymphalidae
DISTRIBUTION	Canada (southern Yukon, Northwest Territories, and British Columbia), and northern and western United States
HABITAT	Many habitats, from sea level up to 8,200 ft (2,500 m) elevation, including canyons, open deciduous woodlands, watercourses, parks, and gardens
HOST PLANTS	Nettle (*Urtica* spp.) and hop (*Humulus* spp.)
NOTE	Spiny, white caterpillar that builds an open-ended leaf nest
CONSERVATION STATUS	Not evaluated, but common

ADULT WINGSPAN
2–2⅛ in (50–55 mm)

CATERPILLAR LENGTH
1³⁄₁₆–1⅜ in (30–35 mm)

POLYGONIA SATYRUS
SATYR COMMA
(W. H. EDWARDS, 1869)

271

Satyr Comma eggs are laid on the underside of nettle leaves—sometimes up to seven eggs may hang down in a string from one leaf. The young caterpillars emerge after five to seven days, and early instars are usually solitary, resting in the open on the underside of a leaf. The larvae build individual nests by folding the edges of a nettle leaf and loosely silking them together. The ends are left partially open, so the caterpillar is visible inside. Caterpillars feed either from the leaf edge or at midleaf, making deep, jagged holes.

Concealment is this species' principal means of defense, although natural enemies such as small predatory bugs—for example, Anthocoridae—may enter the nests. Caterpillar development is rapid, taking just 23 days to reach pupation. Only nine days are spent as a pupa, and the adults overwinter and may live for up to 12 months. Both sexes visit flowers for nectar, but males may also feed on animal scat and mud.

Actual size

The Satyr Comma caterpillar is almost entirely white dorsally, and most of the spines are white. The anterior horns are small, black, and antler-like. Black chevrons interrupt the white dorsum, one per segment. Laterally, the caterpillar is black, and the black spiracles are encircled in white.

FAMILY	Nymphalidae
DISTRIBUTION	Nepal, northern India, central and southern China, Japan, Chinese Taipei, and mainland Southeast Asia
HABITAT	Elevated evergreen forests
HOST PLANTS	*Rhamnella franguloides*, *Celtis boninensis*, and *Albizia* spp.
NOTE	"Dragonhead" caterpillar that has an impressive set of horns
CONSERVATION STATUS	Not evaluated, but not uncommon, although some subspecies are geographically isolated

ADULT WINGSPAN
3½–4⅝ in (90–120 mm)

CATERPILLAR LENGTH
2⅜ in (60 mm)

POLYURA EUDAMIPPUS
GREAT NAWAB
(DOUBLEDAY, 1843)

272

By the time the Great Nawab caterpillar has devoured its own eggshell at hatching, its full set of horns has inflated and hardened. This is a characteristic of the Charaxinae "dragonhead" caterpillars not shared by related subfamilies of horned larvae, which develop horns only by the mid-instar stages. The caterpillars build a silk pad on a leaf tip, expanding on this "base camp" for their entire cycle and wandering away at night to feed on adjoining foliage. During the day, they make no attempt to conceal themselves but remain stationary and exposed, often with the anterior half of the body raised. Pupation occurs on twigs (rather than the leaves) of adjacent non-host plants. The chrysalis is glossy green, round, and smooth.

The name "Nawab" is an honorary title once bestowed on regional rulers or officials in South Asia. The Great Nawab is a large, powerful canopy flier but can be observed closely (and often very stubbornly) on the ground, mud-puddling or feeding from animal feces. There are numerous subspecies, which vary quite dramatically in appearance at both larval and adult stages.

The Great Nawab caterpillar has a partially flattened, green body, widest around the fifth segment and tapering to a flattened, rectangular anal plate. The body is covered with lightly colored blunt points, densest and largest laterally, giving the appearance of a fringe. The head capsule bears two pairs of lengthy, serrated, and nodular horns, between which is a small, beak-like pair of conical horns. Body markings are variable and can include a crescent-shaped saddle, or two saddles, on the backline.

Actual size

FAMILY	Nymphalidae
DISTRIBUTION	The Andes of Colombia, Ecuador, Peru, and Bolivia
HABITAT	Landslides and second growth within Andean cloud forests
HOST PLANTS	Bamboo (*Chusquea* spp.)
NOTE	Caterpillar that near-perfectly mimics dead bamboo leaves
CONSERVATION STATUS	Not evaluated, but not considered threatened

ADULT WINGSPAN
2¾–3 in (70–75 mm)

CATERPILLAR LENGTH
2⁷⁄₁₆–2¹¹⁄₁₆ in (62–68 mm)

PRONOPHILA ORCUS
ORCUS GREAT-SATYR
(LATREILLE, 1813)

273

Orcus Great-satyr caterpillars hatch from white, spherical eggs, laid in clusters of two to five on the underside of fresh host plant leaves. The larvae, however, are not gregarious and quickly disperse to their own leaves. Their complex patterning of shades of brown, along with their elongate, tapering shape, make them extremely cryptic on their host plant. They are nearly identical in shape and color to a dead bamboo leaf, and even the white spots on the thorax are part of this mimicry, being very similar to a type of white fungus that frequently attacks dead leaves. The pupa, hung from the underside of a cluster of dead bamboo leaves, is similarly colored and also difficult to spot.

The Orcus Great-satyr is among the largest and fastest-flying of the numerous species of Andean bamboo-feeding satyrs, known collectively as the tribe Pronophilini. The name "Orcus" refers to the god of the underworld in Roman mythology, the rough equivalent of Hades in Greek mythology.

The Orcus Great-satyr caterpillar is elongate, broadest just behind the thorax, and tapers rearward to a point created by two short, conical caudal tails that are held tightly together. The head, bearing two short horns, and the entire body are dull brownish with complex patterns of various shades of brown and white. A thin, white middorsal line bordered by brown on the thorax, along with a pair of white spots and a pair of brown crescents on the abdomen, are its most prominent markings.

Actual size

FAMILY	Nymphalidae
DISTRIBUTION	Northwest Himalayas through to central China, Thailand, and northern Vietnam
HABITAT	Hill forests and streams
HOST PLANTS	*Debregeasia* spp.
NOTE	Horned, green caterpillar that blends seamlessly with its food plant
CONSERVATION STATUS	Not evaluated, but less common in its Himalayan range, although not threatened

ADULT WINGSPAN
2–2⅛ in (50–55 mm)

CATERPILLAR LENGTH
1¾ in (45 mm)

PSEUDERGOLIS WEDAH
TABBY
(KOLLAR, 1844)

274

The early instar Tabby caterpillar constructs a resting perch from the lateral leaf margins, protected by a wall of silk-bound frass. By the third instar, the caterpillar starts to develop the horns that are so prominent in later instars. As the larvae mature, they also take on the color and texture of the deeply corrugated and hairy leaves, progressing from the green of fresh young foliage to the yellower dappled appearance of older leaves, which renders them almost invisible when at rest. The caterpillar more than doubles in length from the fourth to the fifth (final) instar, at which time it rests flat against the leaf upper surface on a bed of silk.

The Tabby's chrysalis is also very cryptic and has a broad keel that curves downward from the thorax to meet another upwardly curved protuberance from the abdomen. The genus *Pseudergolis* includes only two species of butterfly. They are sun-baskers, particularly near water, and can be very territorial.

Actual size

The Tabby caterpillar matches the texture and color of the host plant precisely—dappled green with a dense coverage of small, white tubercles, each topped with a fine hair. The head bears a lengthy pair of curved, branched horns. A wartlike bump protrudes about one-third of the way along the topside, and there is a pair of sharp, black spines at the tail.

FAMILY	Nymphalidae
DISTRIBUTION	Tropical Mexico to Venezuela and south along the Andes to central Peru
HABITAT	Humid subtemperate and upper tropical montane forests and forest borders
HOST PLANTS	*Urera* spp.
NOTE	Caterpillar that, when young, rests on frass suspended from leaves
CONSERVATION STATUS	Not evaluated, but not considered threatened

ADULT WINGSPAN
$3\frac{7}{16}$–$3\frac{3}{4}$ in (88–95 mm)

CATERPILLAR LENGTH
$2\frac{3}{8}$–3 in (60–75 mm)

PYCINA ZAMBA
CLOUD-FOREST BEAUTY
DOUBLEDAY, [1849]

275

The Cloud-forest Beauty caterpillar, which is, unlike many species, as deserving of its name as the adult, hatches from an egg laid singly on the host plant. Ovipositing females dash at breakneck speeds through the canopy, pausing momentarily on the upper surfaces of large leaves to check for suitability, sometimes returning multiple times to the same plant or even the same leaf. Young larvae build and rest upon frass chains, which, unlike those of most other chain-building species, are constructed carefully to hang below the leaf rather than projecting laterally from its margin. From oviposition to eclosion, the life cycle (at least in Central America) lasts 43 to 45 days, with males apparently taking longer to complete metamorphosis than females.

When disturbed, mature Cloud-forest Beauty larvae will rear back and thrash at the offending intruder with their quite formidable head scoli, and will attempt to bite with their powerful mandibles. Although the spines of the caterpillar do not deliver an urticating sting, the host plant often does, something that likely also helps protect the caterpillar.

The Cloud-forest Beauty caterpillar is complexly patterned with swirls and stripes of brown on a background of pale yellow and white, and with splashes of crimson. It bears several rows of long, branched scoli along its body, with several additional rows on the thorax. The head is shiny black and crowned with two short, stout scoli.

Actual size

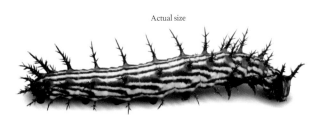

FAMILY	Nymphalidae
DISTRIBUTION	From Mexico through Central America and northern South America, south as far as Bolivia
HABITAT	Lowland humid forests
HOST PLANTS	*Serjania* spp.
NOTE	Newly described caterpillar that has a leaflike pupa
CONSERVATION STATUS	Not evaluated, but not likely to be endangered

ADULT WINGSPAN
2⅛–2⁹⁄₁₆ in (55–65 mm)

CATERPILLAR LENGTH
1⁵⁄₁₆–1⁷⁄₁₆ in (33–37 mm)

276

PYRRHOGYRA OTOLAIS
DOUBLE-BANDED BANNER
BATES, 1864

Double-banded Banner larvae hatch from eggs laid singly on the bright red, newly expanding leaves of their host plants. Young caterpillars rest near the ends of frass chains, while larger instars rest on top of the host leaf, with their face pressed tightly to the leaf surface. When disturbed, older larvae rear backward to thrash wildly about with their long scoli, often lifting their terminal abdominal segments simultaneously, bringing these scoli to bear as well. Their leaflike, emerald-green pupae are somewhat oddly attached, being affixed by the cremaster to the dorsal surface of the leaf and standing upright, overall resembling a piece of green leaf that has curled upward.

Despite the Double-banded Banner's wide geographic distribution, and its formal recognition as a species more than a century and a half ago, its larval stages and host plant preferences were described only at the beginning of this century. The adults are very fast fliers and feed on fallen fruit, rotting fungi, dung, and carrion.

The Double-banded Banner caterpillar has an orange head with long, recurved scoli, each bearing several smaller chalazae. While its body is predominantly white with black spotting and striping, irregularly shaped markings in yellow and blue form longitudinal stripes laterally and subdorsally. All thoracic and most abdominal segments bear prominent, multipronged dorsal scoli, with small scoli protruding laterally above the prolegs.

Actual size

FAMILY	Nymphalidae
DISTRIBUTION	From southern United States through the Caribbean and Central America to the Amazon basin
HABITAT	Humid, semi-humid, and deciduous forests, both tropical and subtemperate
HOST PLANTS	Mostly *Blechum* spp., *Justicia* spp., and *Ruellia* spp.; also *Calliandra* spp., *Salvia* spp., and *Plantago* spp.
NOTE	Sluggish, inconspicuous caterpillar that is hidden on ground-cover plants
CONSERVATION STATUS	Not evaluated, but unlikely to be endangered

ADULT WINGSPAN
2–2⅜ in (50–60 mm)

CATERPILLAR LENGTH
2–2⅛ in (50–55 mm)

SIPROETA STELENES
MALACHITE
(LINNAEUS, 1758)

277

Malachite larvae of all ages are somewhat sluggish and reluctant to move, even when prodded. They generally stay hidden on the underside of host plant leaves and, despite their relatively bright coloration, can be difficult to find, especially when feeding on low-growing, densely leaved succulents. Given the Malachite's very large geographic range, it is not surprising that the species is found in so many habitats, and that its caterpillars are known to feed on so many types of plants.

The adults are nimble, wary flyers, often seen perching on the upper surface of leaves, wings partially open, some 3–6 ft (1–2 m) above the ground. They are quickly distinguished from the very similar, but much longer-winged, Scarce Bamboo Page (*Philaethria dido*) by the irregular-shaped margins of both wings. In many areas where the species inhabits fairly seasonal forests, local or regional migrations are suspected, but these are poorly documented.

The Malachite caterpillar has a shiny, black head bearing two long, recurved horns, knobbed at the end. The body is velvety greenish black, with three pairs of branched scoli per segment. The two lateral pairs are blackish, and the dorsal pairs are reddish yellow. The terminal segments are often purplish black.

Actual size

FAMILY	Nymphalidae
DISTRIBUTION	The Himalayas, India, Myanmar, southwestern China (Yunnan), and mainland Southeast Asia
HABITAT	Moist, shaded deciduous forests
HOST PLANTS	Wild Guava (*Careya arborea*) and *Melastoma malabathricum*
NOTE	Caterpillar that has a flattened, feathered skirt of spines
CONSERVATION STATUS	Not evaluated, but common

ADULT WINGSPAN
2⁹⁄₁₆–3 in (65–75 mm)

CATERPILLAR LENGTH
1⁹⁄₁₆ in (40 mm)

278

TANAECIA LEPIDEA
GREY COUNT
(BUTLER, 1868)

Grey Count caterpillars hatch from eggs resembling tiny, hairy half golf balls that are laid singly on host plant leaves. By the mid-instar stages, the caterpillar starts to develop its skirt of flexible spines, employing them to mute its body outline and blend into the leaf surface. In its usual resting posture on the topside of leaves aligned with the midrib, the caterpillar is practically invisible. Pupation occurs on the host plant or on adjoining vegetation, in which case the caterpillar uses a silk line to lower itself to the ground in search of secluded locations. The broad, angular chrysalis is ¹¹⁄₁₆ in (18 mm) long and bright green with highlights of yellow and orange.

The Grey Count butterfly is often spotted on the forest floor or in clearings, effortlessly gliding short distances between patches of sunlight. When fresh, the silvery, moon-shaped crescent on its wing margins is eye-catching. The immature life history and behavior of *Tanaecia lepidea*, while similar to other genus members, has not been documented in detail, and its full range of host plants is not known.

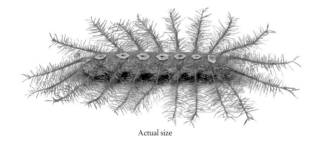

Actual size

The Grey Count caterpillar looks substantially bigger than it really is. Each segment has a pair of long, soft, heavily feathered lateral spines held flat against the leaf surface, completely obscuring the caterpillar's body outline. At the extremities of the spines, the feathering darkens and has white tips. Dorsally, each body segment also bears a hexagonal spot with a blue center.

FAMILY	Nymphalidae
DISTRIBUTION	Indonesia, the Philippines, New Guinea, and northern and eastern Australia
HABITAT	Monsoon vine thickets and littoral rain forests
HOST PLANTS	Usually Corky Milk Vine (*Secamone elliptica*); also other Apocynaceae spp.
NOTE	Colorful caterpillar that can defoliate its host plant vine
CONSERVATION STATUS	Not evaluated, but common near breeding areas

ADULT WINGSPAN
2¹³⁄₁₆ in (72 mm)

CATERPILLAR LENGTH
2 in (50 mm)

TIRUMALA HAMATA

BLUE TIGER
(W. S. MACLEAY, 1826)

279

The Blue Tiger caterpillar is generally solitary on the food plant, but it is a voracious feeder, and when populations are high, large numbers of larvae can strip milk vines of all foliage. Caterpillars feed exposed on the plants, developing rapidly to complete their growth in two weeks or less. The larvae accumulate chemicals that provide protection from bird predation and also enable them to produce pheromones, which are released from the hair pencil tufted pheromone signaling organs of the male butterflies.

Caterpillars will often leave the food plant and pupate on nearby foliage, suspended head down by the cremaster. During the dry season of winter, the milk vines have no fresh foliage and are unsuitable for caterpillar development. Adult butterflies congregate in shady creek banks to overwinter, often with other species of milkweed butterflies, and then disperse in the spring to commence reproduction. Large migration flights often occur during the breeding season.

The Blue Tiger caterpillar is greenish gray with an orange-brown lateral line and a white ventrolateral line; each segment has transverse black bands enclosing white and narrow gray bands. There is a dorsolateral pair of black, fleshy filaments on the mesothorax and eighth abdominal segment. The head is black with two white, transverse bands.

Actual size

FAMILY	Nymphalidae
DISTRIBUTION	North America, Europe, and much of Asia
HABITAT	Many habitats, including parks, gardens, woodlands, meadows, orchards, and riparian areas
HOST PLANTS	Nettle (*Urtica* spp.), pellitory (*Parietaria* spp.), and hop (*Humulus* spp.)
NOTE	Variably colored, nettle-feeding, spiny caterpillar
CONSERVATION STATUS	Not evaluated, but common

ADULT WINGSPAN
2–2⅛ in (50–55 mm)

CATERPILLAR LENGTH
1³⁄₁₆–1⅜ in (30–35 mm)

VANESSA ATALANTA
RED ADMIRAL
(LINNAEUS, 1758)

280

Actual size

Red Admiral caterpillars hatch from eggs laid singly on the underside of nettle leaves, typically on a leaf vein. Early instars rasp leaf surfaces with their mandibles, creating holes, and produce small amounts of silk to cover themselves, forming loosely silked nests. Later instars form shelters by folding a leaf over or silking a group of leaves together. Feeding occurs inside or outside nests at any time of day or night. Defense is based mainly on concealment within nests, but the caterpillars also possess a ventral gland near the head that likely secretes chemicals to deter predators.

The caterpillars go through five instars and pupate about three weeks after hatching, usually on the host plant. The Red Admiral adult, a familiar sight in home gardens, is one of the northern hemisphere's best known and most charismatic butterflies. If the garden also has a nettle patch, then Red Admiral caterpillars are likely to be found there.

The Red Admiral caterpillar is black, peppered with white dots and short, white setae. The spines (black or pale) are prominent, as are the creamy white subspiracular dashes. The prolegs are brown, and the head is black with white dots and short, pale setae. Caterpillars vary considerably in ground color from black to gray to brown to white.

FAMILY	Nymphalidae
DISTRIBUTION	Every continent, except Antarctica
HABITAT	Most habitats from urban to mountaintops
HOST PLANTS	Wide variety, including thistle (*Cirsium* spp.), nettle (*Urtica* spp.), mallows (Malvaceae), and legumes (Fabaceae)
NOTE	Well-known species often reared to teach students about metamorphosis
CONSERVATION STATUS	Not evaluated, but widespread and common

ADULT WINGSPAN
2¾–3 in (70–75 mm)

CATERPILLAR LENGTH
1¼ in (32 mm)

VANESSA CARDUI
PAINTED LADY
(LINNAEUS, 1758)

281

Painted Lady caterpillars hatch from green, inconspicuous eggs laid singly on host plants. Their development is rapid, taking three weeks from egg hatch to pupation. Adults eclose ten days later. The caterpillars feed on leaves primarily, and all instars build protective nests made out of a web of silk; in later instars these incorporate leaves that are curled or silked together. First instars feed and rest on upper leaf surfaces, covering themselves with a few strands of silk. Nests become more complex as the caterpillars mature, but there is only one caterpillar per nest. A considerable amount of frass collects at the bottom of the nest.

As the caterpillars wander before pupating, the pupae are usually formed in sheltered areas away from the host plant. Immature stages do not diapause, but adults migrate toward lower latitude areas in the fall and overwinter in climates with mild winters. Northerly migrations with enormous numbers of individuals sometimes occur in spring in North America and Europe.

Actual size

The Painted Lady caterpillar is highly variable in coloration, especially in later instars. The ground color is usually black with variable red, yellow, and white markings. Branched spines occur along the body and are yellow or white with dark tips. Ventrolaterally, the body is pale or white, and the head is black.

FAMILY	Nymphalidae
DISTRIBUTION	Most of United States
HABITAT	Fields, parks, gardens, and canyons
HOST PLANTS	Pearly Everlasting (*Anaphalis margaritaceae*), cudweed (*Gnaphalium* spp.), and pussytoe (*Antennaria* spp.)
NOTE	Spiny caterpillar that builds complex silk nests
CONSERVATION STATUS	Not evaluated, but common

ADULT WINGSPAN
2–2⅛ in (50–55 mm)

CATERPILLAR LENGTH
1½–1¹¹⁄₁₆ in (38–43 mm)

282

VANESSA VIRGINIENSIS
AMERICAN LADY
(DRURY, 1773)

American Lady females lay eggs singly or a few at a time on their low-growing host plants. The eggs hatch after six to seven days, and first instars immediately excavate lightly silked nests under leaf pubescence. Most first instar caterpillars stay between leaf membranes, feeding and creating "windowpane" areas. Older instars move to the outer leaf surface and form increasingly complex nests by silking leaves together. Feeding occurs inside or outside the nests, mostly by night. Final (fifth) instar caterpillars leave their nests and rest exposed on stems and leaves.

The caterpillar's defense is based mostly on concealment in nests until the fifth instar, when prominent spines and bold color patterns may deter predators. Pupation occurs mainly on the host plant and lasts for less than a week. Development from egg-laying to adult eclosion takes about a month during summer. Adults fly fast and erratically, usually close to the ground, and visit numerous types of flowers for nectar.

Actual size

The American Lady caterpillar is spiny and transversely banded in black and yellowish white, with prominent orange spots and white spots. The pale intersegmental areas comprise five to six indistinct white bands on a black background. The head is black and bears numerous long, white setae.

FAMILY	Nymphalidae
DISTRIBUTION	Eastern Indonesia, Papua New Guinea, Australia, and the Solomon Islands
HABITAT	Lowland rain forests and monsoon forests
HOST PLANTS	Passion vine (*Adenia heterophylla*, *Passiflora aurantia*, and *Hollrungia* spp.)
NOTE	Extremely fast-moving and active caterpillar
CONSERVATION STATUS	Not evaluated, but not uncommon

ADULT WINGSPAN
3–3¼ in (75–82 mm)

CATERPILLAR LENGTH
1⁹⁄₁₆ in (40 mm)

VINDULA ARSINOE
CRUISER
(CRAMER, [1777])

283

Cruiser caterpillars hatch from eggs laid by the female butterfly on the tendrils of the food plant. In the early instars, the caterpillar rests on the tendril during the day and feeds on the nearby leaves during the night. Later instars rest on leaves during non-feeding periods. The caterpillars are very active and sensitive to touch, often dropping to the ground if disturbed. Unlike those of other passion vine-feeding butterflies, in this species the larvae are usually solitary. The larval stage lasts about 16 days in the wet season and 25 days in the winter dry season.

Pupation takes place with the pupa attached to a silken pad on the stem of the food plant and suspended head down from the cremaster. The pupa resembles a dead leaf and twitches if disturbed, an action that probably provides some protection from parasitoids and predators. The adults fly in patchy sunlight and shady areas in rain forests and have a gliding fight.

Actual size

The Cruiser caterpillar has a greenish or yellow body with broad, blackish-green, and yellow-spotted dorsolateral bands. Each segment has a pair of black, branched dorsolateral and lateral spines and a pair of white, branched ventrolateral spines. The head is black with two curved, long, thick, black, branched spines.

MOTH
CATERPILLARS

Most Lepidoptera are moth species, with larvae that are highly diverse in size, appearance, and habitat. This chapter includes caterpillars from 31 moth families, many of which evolved long before the butterfly superfamily, Papilionoidea.

The earliest family featured here is Psychidae, whose caterpillars are extraordinary "bagworms," spending their entire lives in protective cases constructed from materials within their habitat. Almost as ancient is the Tineidae family, of which only a small number feed on plants; most consume fungi, lichens, and dead organic material. Among its members is the Case-bearing Clothes Moth (*Tinea pellionella*), which feeds on household fabrics. Larvae of the Cossidae family are tree and root borers, and some are notoriously bad smelling. Limacodidae caterpillars are distinguished by their sluglike gait; they have suckers instead of prolegs and secrete a lubricant to facilitate movement.

Later in the evolutionary order come the stout caterpillars of the giant silkmoths, royal moths, and emperor moths of Saturniidae, the hawkmoth larvae of Sphingidae, and the inchworms of Geometridae. The chapter ends with larvae from five of six families in the superfamily Noctuoidea—Notodontidae, Erebidae, Euteliidae, Nolidae, and Noctuidae—including cutworms, and owlet, puss, and tiger moth caterpillars.

FAMILY	Psychidae
DISTRIBUTION	Eastern United States, west to New Mexico, and south to the Caribbean
HABITAT	Forests and also urban trees
HOST PLANTS	At least 50 families of deciduous and evergreen trees and shrubs, including juniper and cedar (*Juniperus* spp.), oak (*Quercus* spp.), willow (*Salix* spp.), maple (*Acer* spp.), and pine (*Pinus* spp.)
NOTE	Caterpillar that lives inside a silk bag decorated with vegetation
CONSERVATION STATUS	Not evaluated, but common

ADULT WINGSPAN
¾–1³⁄₁₆ in (20–30 mm)

CATERPILLAR LENGTH
1⁵⁄₁₆–1¼ in (24–32 mm)

THYRIDOPTERYX EPHEMERAEFORMIS
EVERGREEN BAGWORM
(HAWORTH, 1803)

286

Actual size

The Evergreen Bagworm caterpillar has a pigmented and sclerotized head. Its front end, only ever visible as it feeds, partially emerging from the bag's turtleneck opening, is usually white or beige with extensive black markings. Past the abdomen, the larva is a nondescript black brown. The bag's appearance varies according to the host plant.

Evergreen Bagworm caterpillars hatch from eggs laid by the wingless adult female in the bag that she, as a caterpillar, constructed. The young larvae are dispersed through the air, attached to silk threads—the only effective method of dispersal over greater distances. Once they land and start feeding on the new host, the caterpillars construct their own silk bags around themselves, continuing to enlarge them and attaching fragments of foliage throughout their growth. Excrement is expelled from a hole in the bag's lower end. Female caterpillars gradually ascend to the crown of a tree, while males stay at the same level throughout their development.

The larvae develop through seven instars over about three months, then pupate in the bag. When the adults eclose, only the winged males start flying—looking for bags containing wingless females. The male mates by inserting his abdomen into the lower opening of the female bag. The fertilized female then lays a large number of eggs inside her bag, where the eggs spend the winter.

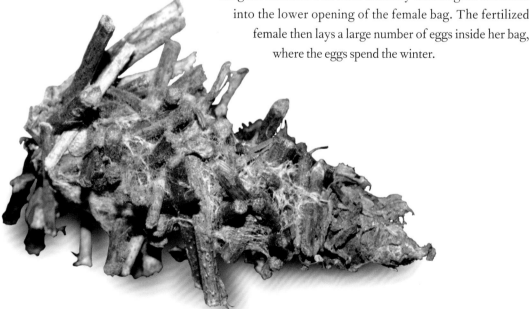

FAMILY	Tineidae
DISTRIBUTION	Widespread, recorded across the Americas, Africa, Europe, Asia, and Australasia
HABITAT	Mainly in buildings and outbuildings, occasionally outdoors in birds' nests
HOSTS	Animal fibers, including fur, hair, feathers, clothing, and carpets; also stored vegetable products, owl pellets, other debris, and even wallpaper
NOTE	Worldwide caterpillar pest of economic importance, mainly damaging natural fibers
CONSERVATION STATUS	Not evaluated, but widespread in many parts of the world

ADULT WINGSPAN
⅜–⅝ in (10–16 mm)

CATERPILLAR LENGTH
⅜ in (10 mm)

TINEA PELLIONELLA
CASE-BEARING CLOTHES MOTH
LINNAEUS, 1758

287

The Case-bearing Clothes Moth caterpillar hatches from up to 100 eggs laid in or near the host materials, which it prefers soiled, rather than clean. In the first instar, it spins a tubular silken case covered in tiny fragments of debris and lives within it for its entire larval life. At each end, the case is flattened and open. Using its true legs, the caterpillar crawls around in this case, into which it can fully retract itself. Pupation also occurs within the case, after the caterpillar has moved well away from the food source. The caterpillars can be found at any time of year.

The rather plain, brown adults may be found throughout the year but more frequently in the warmer months, in one or more broods. Several clothes moths of the genus *Tinea* have very similar caterpillars that live in silken tubes; this is one of the most widespread and abundant of such species in temperate or subtropical regions. The specific name *pellionella* comes from the Latin *pellionis*, meaning "a furrier."

Actual size

The Case-bearing Clothes Moth caterpillar is mainly plain, smooth, and whitish, apart from the head, which is dark brown. There are two dark brown plates on the prothorax on the upper side behind the head. The cuticle is quite translucent so that the gut contents may be visible as a dark central line along the body.

FAMILY	Yponomeutidae
DISTRIBUTION	Europe, Siberia, India, and eastern Asia to Japan
HABITAT	Woodlands, hedgerows, parks, and gardens
HOST PLANTS	Bird Cherry (*Prunus padus*)
NOTE	Small, gregarious caterpillar that can defoliate whole trees
CONSERVATION STATUS	Not evaluated, but widespread and common

ADULT WINGSPAN
⅝–1 in (16–25 mm)

CATERPILLAR LENGTH
¾ in (19 mm)

288

YPONOMEUTA EVONYMELLUS
BIRD-CHERRY ERMINE MOTH
(LINNAEUS, 1758)

Actual size

Bird-cherry Ermine Moth caterpillars hatch from eggs laid on the host plant and develop slowly, overwintering there and becoming active again the following spring, when they spin huge, communal, silken webs over branches of the host tree, giving it a frost-like covering. The gregarious caterpillars feed on the leaves beneath it, the web giving them protection from predators. They also pupate in the trees, spinning white, opaque cocoons that are suspended within the web.

The small, day-flying adults emerge and fly in late summer. While the caterpillars can reach pest proportions, defoliating many trees, in the past their silken webs were put to a unique artistic use. In the sixteenth century, in the Austrian Tyrol, monks made canvases from spiders' webs and the silken webs of ermine moth caterpillars, and produced miniature so-called "cobweb paintings." One example, depicting the Virgin and Child and made from the web of *Yponomeuta evonymellus*, can be seen in Chester Cathedral, England.

The Bird-cherry Ermine Moth caterpillar
has a yellow-brown body. There are two rows of brown-black spots running from the head to the end of the abdomen. The head and legs are brown.

FAMILY	Lyonetiidae
DISTRIBUTION	Florida and adjacent areas of southeast United States
HABITAT	Open areas, such as forest edges and upland pine
HOST PLANTS	Coral Bean (*Erythrina herbacea*)
NOTE	Minute caterpillar that lives within a leaf until pupation
CONSERVATION STATUS	Not evaluated, but common where host plant is present

ADULT WINGSPAN
¹⁄₁₆–⅛ in (2–3 mm)

CATERPILLAR LENGTH
⅛–³⁄₁₆ in (3–4 mm)

LEUCOPTERA ERYTHRINELLA
ERYTHRINA LEAFMINER
BUSCK, 1900

289

The eggs of the Erythrina Leafminer, one of the smallest moth species, are only about ten times the size of leaf epidermal cells. The caterpillar feeds inside the leaf, consuming the mesophyll but not making holes in the leaf epidermis until it is ready to pupate; the leaf discoloration its feeding causes can be mistaken for disease. In each leaf, there is usually a single larva, and one leaf seems to be sufficient for its entire development. Before pupation, the larva emerges to the ventral surface of the leaf and spins a silk structure resembling a miniature hammock. Then it makes a cocoon beneath it, from which it emerges about two weeks later.

Despite their tiny size, the moths are sexually dimorphic and fully functional. The adult does not appear to venture far from the host plant and, when flying, looks like a white fly or a snowflake. The genus *Erythrinella* includes more than 60 species; some are agricultural pests, but this species is of marginal economic importance.

Actual size

The Erythrina Leafminer caterpillar has ten prolegs. The head is cream colored and sclerotized, but the caterpillar is otherwise colorless, translucent white with very pronounced, bulging segments, superficially resembling the larva of a beetle or a fly. Although concealed until pupation, its outline can be seen in the mines if a leaf is held in front of a light source.

FAMILY	Elachistidae
DISTRIBUTION	Most of Europe, east to the Urals and Iran, northeastern United States, and eastern Canada (introduced in the 1960s and slowly spreading)
HABITAT	Open, disturbed, calcareous or sandy places
HOST PLANTS	Viper's Bugloss (*Echium vulgare*) and other *Echium* spp., including *E. tuberculatum*; also Hound's Tongue (*Cynoglossum officinale*), alkanet (*Anchusa* spp.), and gromwell (*Lithospermum* spp.)
NOTE	Caterpillar that wriggles vigorously when disturbed—behavior typical of microlarvae
CONSERVATION STATUS	Not evaluated, but widespread in suitable habitat throughout most of its range

ADULT WINGSPAN
¾–1⅛ in (20–28 mm)

CATERPILLAR LENGTH
¾ in (20 mm)

290

ETHMIA BIPUNCTELLA
VIPER'S BUGLOSS MOTH
(FABRICIUS, 1775)

Actual size

The Viper's Bugloss Moth caterpillar lives among the flowers or leaves of its food plant, sheltering under a slight web. When disturbed, like many "micromoth" larvae, it wriggles vigorously backward. The eggs are laid singly on the underside of leaves, hatching after about ten days. There are two broods annually, and the caterpillar can be found in June and July, then again in September. It leaves the food plant to pupate in a cocoon among leaf litter on the ground and passes the winter as a pupa.

The black-and-white patterned adults fly in the spring and again in the late summer. The species is a member of the small subfamily Ethmiinae (sometimes classified as the family Ethmiidae), comprising about 300 species of fairly small moths, found in most parts of the world. Many of the larvae are colorful, and some are gregarious. A high proportion feed on plants in the Boraginaceae family, often living in a silken web but in some cases openly.

The Viper's Bugloss Moth caterpillar is whitish, heavily marked with lines of black spots and large, black blotches. It has a line of alternating black and orange-yellow spots along the back and an irregular but more or less complete orange-yellow stripe along the sides. The head is black with a conspicuous white, triangular mark in the center.

FAMILY	Elachistidae
DISTRIBUTION	Europe from Spain and southern United Kingdom to southern Scandinavia and eastern Russia, Asia Minor, northern Iran, and east Asia from Korea north to eastern Siberia
HABITAT	Woodland rides and edges, and wooded fenland
HOST PLANTS	Common Gromwell (*Lithospermum officinale*)
NOTE	Spotted caterpillar whose striking moth resembles "ermine" moths (*Yponomeuta* spp.)
CONSERVATION STATUS	Not evaluated

ADULT WINGSPAN
$^{11}/_{16}$–$^{3}/_{4}$ in (18–21 mm)

CATERPILLAR LENGTH
$^{11}/_{16}$ in (18 mm)

ETHMIA DODECEA
DOTTED ERMEL
(HAWORTH, 1828)

291

The Dotted Ermel caterpillar lives singly or in groups under the leaves of its host plant, sheltering beneath a slight silken web. The eggs are undescribed but are probably laid on the leaves in small groups or singly, during the late spring and summer, and hatch within one or two weeks. There is a single brood each year, and the caterpillar can be found in August and September. It leaves the food plant to pupate in a cocoon among plant litter on the ground and overwinters as a pupa.

The black-spotted, grayish-white adults fly from May to August. The species is one of many fairly small moths in the subfamily Ethmiinae that produce brightly colored caterpillars that specialize in feeding on plants in the family Boraginaceae. Ethmiinae comprises about 300 species, found in most parts of the world.

Actual size

The Dotted Ermel caterpillar is slender, with long, scattered, white bristles. It is pale yellow but largely white along the back with a fine, black, broken central line and alternating large and small, black blotches. It also has small, black spots and blotches on the lower sides. The black head has a short, white frontal band.

FAMILY	Pterophoridae
DISTRIBUTION	Europe
HABITAT	Grasslands, sand dunes, and shingle
HOST PLANTS	Restharrow (*Ononis* spp.)
NOTE	Green caterpillar that is well camouflaged on its food plant
CONSERVATION STATUS	Not evaluated, but locally rare

ADULT WINGSPAN
$1^{11}/_{16}$–$^7/_8$ in (18–22 mm)

CATERPILLAR LENGTH
$^1/_4$–$^5/_{16}$ in (6–8 mm)

292

MARASMARCHA LUNAEDACTYLA
CRESCENT MOON
(HAWORTH, 1811)

Crescent Moon caterpillars hatch from eggs laid on restharrows—tough weeds so-named because in the past they obstructed harrows from breaking up the soil. The emerging hairy, green caterpillars are well camouflaged among the leaves and shoots of their host plant as they feed. When fully developed, final instar larvae also pupate on the host plant. The pupae are found attached to the underside of a leaf or stem. There is a single generation each year.

The tiny moths eclose and are active from June to August, appearing at dusk. Adults of the Pterophoridae family are called "plume moths" for their modified forewings, often consisting of just a few feathery plumes. The species gets its common name from the distinctive pale crescent around the cleft in the forewing. At rest, the moths roll up their wings, which resemble dead grass, and hold them out at right angles to form a T-shape.

Actual size

The Crescent Moon caterpillar is hairy and green. The body is uniformly green with a faint dorsal line. The hairs are long and white but not dense, arising in rings of tubercles around the segments. The head is shiny black.

FAMILY	Tortricidae
DISTRIBUTION	Large areas of both northern and southern hemispheres at latitudes between 30 degrees and 60 degrees
HABITAT	Horticultural and urban areas where host plants occur
HOST PLANTS	Pome fruit (Rosaceae): apple (*Malus* spp.), pear (*Pyrus* spp.), and Quince (*Cydonia oblonga*); occasionally stone fruits: plum and apricot (*Prunus* spp.) and walnut (*Juglans* spp.)
NOTE	Caterpillar that is a pest of apple and pear orchards
CONSERVATION STATUS	Not evaluated, but a widespread and common pest species

ADULT WINGSPAN
1¹¹⁄₁₆ in (17 mm)

CATERPILLAR LENGTH
⁹⁄₁₆–¾ in (15–19 mm)

CYDIA POMONELLA
CODLING MOTH
(LINNAEUS, 1758)

293

Codling Moth caterpillars hatch from eggs laid on the fruit surface. They burrow into the fruit and tunnel to the core to feed on the seeds. Pupation occurs in a cocoon, usually under bark or in crevices of the host tree. There are one to three generations a year, with the caterpillar overwintering in a state of diapause. Day length, temperature, and food quality are the main factors influencing induction of diapause. If the caterpillars hatch under short day lengths and cooler weather, when feeding is complete they will enter diapause in their cocoon.

Diapause is broken only after an extended period of cold weather—below 50°F (10°C)—followed by warmer weather, which results in pupation, then moth emergence in the spring. Control of Codling Moth caterpillar damage in commercial pome fruit orchards underpins pest control throughout most orchards worldwide, although the species is not an apple pest in Japan or most of China, despite the climatic suitability.

Actual size

The Coding Moth caterpillar is moderately stout and usually creamy white but turns slightly pink dorsally when mature. The head and prothoracic shield are yellow brown, often overlaid with a darker brown pattern. The anal plate is yellow with a moderately distinct pattern of pale brown spots and specks. Small, sclerotized plates on the body are gray, with short, white setae arising from them.

FAMILY	Cossidae
DISTRIBUTION	Europe, the Middle East, central Asia, Russia, and China
HABITAT	Woodlands, parks, and gardens
HOST PLANTS	Deciduous trees such as oak (*Quercus* spp.), ash (*Fraxinus* spp.), and apple (*Malus* spp.)
NOTE	Wood-eating and tunneling caterpillar that is rarely seen
CONSERVATION STATUS	Not evaluated, but widespread though declining in parts of its range

ADULT WINGSPAN
2¹¹⁄₁₆–3¾ in (68–96mm)

CATERPILLAR LENGTH
4 in (100 mm)

COSSUS COSSUS
GOAT MOTH
(LINNAEUS, 1758)

294

The young Goat Moth caterpillar, named for the peculiar, strong, goatlike odor it gives off, burrows into the trunk of the host tree and gnaws out a chamber in which to overwinter. Wood is difficult to digest, so the caterpillar grows slowly and may spend up to five years in the larval stage. Eventually, it abandons the tree and moves to the ground or an old tree stump, where it spins a silk cocoon and pupates, incorporating particles of soil to disguise the cocoon. In more northerly areas, the caterpillar pupates inside the tree.

The large, grayish-brown adult emerges and flies between April, in more southerly locations, and August, in the north of its range. The caterpillar is unusual for its habit of feeding on wood, which, together with its tunneling, damages the host tree and causes economic harm to orchards and olive groves. Pest control has resulted in a steep decline in numbers since the 1960s.

The Goat Moth caterpillar has a distinctive, black head and a large, black mark on the thorax. The black mandibles are particularly large and serrated. It is a deep red-purple color along the back, with lighter and more orange tones along the lateral and ventral sides. Scattered over the body are long, pale hairs.

Actual size

FAMILY	Castniidae
DISTRIBUTION	Southeastern Australia
HABITAT	Native grasslands
HOST PLANTS	Wallaby grasses (*Rytidosperma* spp.), spear grasses (*Austrostipa* spp.), and Chilean Needle Grass (*Nassella neesiana*)
NOTE	Subterranean caterpillar that feeds on grass roots
CONSERVATION STATUS	Not evaluated, but listed as critically endangered under Commonwealth of Australia legislation

SYNEMON PLANA

GOLDEN SUN MOTH

WALKER, 1854

ADULT WINGSPAN
1¼–1⁵⁄₁₆ in (31–34 mm)

CATERPILLAR LENGTH
1⅛ in (28 mm)

295

The Golden Sun Moth caterpillar hatches from an egg laid at the base of a grass clump. A female may lay up to 200 eggs over many clumps. The caterpillar tunnels underground, where it remains feeding on grass roots for up to two years, although the details of the development of caterpillars is not well documented. Caterpillars seem to be able to survive well in areas invaded by the weedy Chilean Needle Grass. After completing growth, the caterpillar digs a vertical tunnel to the surface before pupating.

The pupal period is six weeks, and, after eclosion, the empty pupal casing protrudes from the soil surface. Male moths fly rapidly, during the day, in a zigzag flight pattern about 3 ft (1 m) above the grass, searching for females that sit on the ground and rarely fly. *Synemon plana* has become a flagship species for remnant native grasslands, and environmental impact statements are required before construction projects can occur on its habitat.

Actual size

The Golden Sun Moth caterpillar is white, tinged with brown. Its thoracic segments are larger than the abdominal segments, and the abdomen tapers slightly toward the posterior end. The prothoracic plate is large, and the legs are small. The body is without prolegs, and there are a few secondary setae. The head is brown and has long, tactile setae on its anterior part.

FAMILY	Limacodidae
DISTRIBUTION	From Mexico to Peru
HABITAT	Medium to high elevation rain forests, at 650–2,625 ft (200–800 m) elevation
HOST PLANTS	Trees, including South American Holly (*Ilex paraguariensis*), oil palm (*Elaeis* spp.), Avocado (*Persea americana*), plum (*Prunus* spp.), and *Citrus* spp.
NOTE	Spiny caterpillar that has stinging hairs
CONSERVATION STATUS	Not evaluated, but likely secure

ADULT WINGSPAN
1⁹⁄₁₆–2³⁄₈ in (40–60 mm)

CATERPILLAR LENGTH
1 in (25 mm)

ACHARIA NESEA
STOLL'S CUP MOTH
(STOLL, 1780)

296

Actual size

Caterpillars of the Limacodidae family are extraordinarily spiny, and the Stoll's Cup Moth caterpillar is no exception. It hatches among eggs that are dorsoventrally flat and thin, and quite transparent, and bears fleshy horns carrying groups of stinging hairs. These hairs can break off and cause serious pain to anyone who touches them. Like all limacodids, the Stoll's Cup Moth caterpillar has suckers instead of prolegs, which help it attach securely to substrates.

The Stoll's Cup Moth caterpillar is usually light to dark brown, or grayish. A distinctive thick, white-lined saddle runs dorsally between five pairs of black, fleshy horns, which carry tan-colored, stinging, spiny hairs. Ventrally, the caterpillar is pinkish. The black spiracles are bordered with orangish, spiny hairs.

The host plants always have smooth leaves, which the first instars skeletonize and older instars consume entirely. The pupal period lasts about a month, inside a tough, fibrous cocoon constructed by the caterpillar, with a built-on line of weakness that acts as an escape hatch for the emerging adult—a moth notable for its huge abdomen and outsized hairy legs. When reared in captivity, Stoll's Cup Moth caterpillars often fail to survive to adulthood.

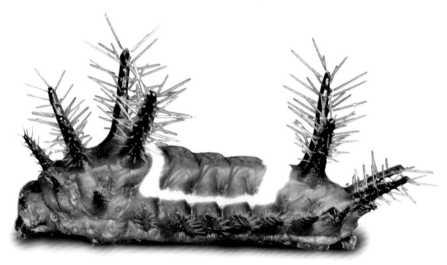

FAMILY	Limacodidae
DISTRIBUTION	Eastern North America
HABITAT	Mixed deciduous forests
HOST PLANTS	Wide variety, both native and exotic, including Manila Palm (*Adonidia merrillii*); *Aster* spp. and *Helianthus* spp. (Asteraceae); *Celtis* spp. (Cannabaceae); and dogwood (*Cornus* spp.)
NOTE	Aposematically colored caterpillar that stings
CONSERVATION STATUS	Not evaluated, but common

ADULT WINGSPAN
1–1¹¹⁄₁₆ in (25–43 mm)

CATERPILLAR LENGTH
¾ in (20 mm)

ACHARIA STIMULEA
SADDLEBACK
(CLEMENS, 1860)

297

Saddleback moth caterpillars hatch from batches of 30 to 50 eggs, laid by the female moth on the upper side of host plant leaves. The caterpillars are truncated and sluglike, having prolegs with suction cups (instead of crochets, as in most caterpillars) on the ventral surface of the body. They produce a semifluid silk from their ventral pores as they move, which provides an adhesive bond to smooth leaf surfaces. Young larvae feed gregariously. The second instars develop bright green saddle markings and urticating spines on the fleshy tubercles of the body, which effectively deter vertebrate and invertebrate predators. Contact with the caterpillar can cause a painful, swollen rash.

After feeding for four or five months, the caterpillar excretes white frass pellets and spins a compact cocoon, creating a thinner ring within its end, from which the adult will eclose. To harden the cocoon, the larva excretes calcium oxalate and also breaks off some of its spines and weaves them into the cocoon wall as additional protection for the pupa.

Actual size

The Saddleback caterpillar is dark brown at each end, its posterior end also bearing three bright, neon spots, giving it a face-like appearance. A contrasting bright green midsection has a brown saddle bordered with white. The fleshy tubercles extending from both anterior and posterior ends are covered in long, urticating spines and setae. Urticating setae are also located on shorter projections that line the caterpillar laterally.

FAMILY	Limacodidae
DISTRIBUTION	North America, from southern Canada to Florida, west to Missouri
HABITAT	Deciduous woodlands and forests
HOST PLANTS	Oak (*Quercus* spp.) and beech (*Fagus* spp.)
NOTE	Covert slug caterpillar that is rarely seen
CONSERVATION STATUS	Not evaluated

ADULT WINGSPAN
¾–1³⁄₁₆ in (19–30 mm)

CATERPILLAR LENGTH
Up to ¾ in (20 mm)

APODA BIGUTTATA
SHAGREENED SLUG
(PACKARD, 1864)

The Shagreened Slug Moth caterpillar gets its common name from the coarse, "shagreen leather" texture of its skin and its Limacodidae family resemblance to slugs. First instar caterpillars are non-feeding, but later instars can be found by searching the underside of leaves of their favorite host tree, the White Oak (*Quercus alba*). The larvae prefer to feed on leaves growing among the low-lying branches of young oaks rather than those of more mature trees. In the south of the caterpillar's range, multiple generations are possible from May onward. In the northern range, a single generation occurs between July and early September. Adult moths typically fly from March throughout the summer months.

The Shagreened Slug is similar in appearance to the more common Yellow-shouldered Slug (*Lithacodes fasciola*) but sports a distinguishing pair of "racing stripes" along the subdorsal area. Avid caterpillar hunters consider *Apoda biguttata* a welcome find, as the larvae are fairly uncommon within their range.

Actual size

The Shagreened Slug caterpillar is oval shaped and slightly elongated with a short, squared-off tail section. The overall body color can be light green or blue green. Two broad, white or cream-colored stripes traverse the entire dorsal length, bordered along the inside by thin, black lines. The larva's coarse, leathery skin is free of spines, unlike many other Limacodidae species.

FAMILY	Limacodidae
DISTRIBUTION	From southern United Kingdom, northern Spain, and southern Scandinavia east to Asia Minor and the Caucasus
HABITAT	Forests, smaller woodlands, and parklands
HOST PLANTS	Oak (*Quercus* spp.), hornbeam (*Carpinus* spp.), and European Beech (*Fagus sylvatica*)
NOTE	Caterpillar whose species name derives from the *Limax* slug genus
CONSERVATION STATUS	Not evaluated, but widespread throughout most of its range

ADULT WINGSPAN
1⁵⁄₁₆–1¼ in (24–32 mm)

CATERPILLAR LENGTH
½–⅝ in (13–16 mm)

APODA LIMACODES
FESTOON
(HUFNAGEL, 1766)

299

The Festoon caterpillar lives among tree foliage on which it feeds and can be found from July to October. The flattened oval, yellowish eggs are laid singly or in small groups on the underside of a leaf and hatch after two weeks. The caterpillar is well camouflaged and hard to find, often living high in the tree and clinging on very tightly. Pupation occurs in a tough, brownish oval cocoon (with a circular escape hatch), spun on the upper side of a leaf, which falls to the ground in fall.

The rusty brown, rather broad-winged adults fly in June and July. Limacodids are mainly tropical moths. Their caterpillars are highly varied, and some are remarkable, brightly colored creatures with poisonous, protective, stinging barbs and strange cactus-like growths and spines, or starfish-like "arms." Other limacodid larvae are smooth and, like the Festoon, may resemble those of lycaenid butterflies. However, lycaenid caterpillars are covered in fine, very short hairs and the head extends out of the hood when feeding.

Actual size

The Festoon caterpillar is bright green and sluglike. The retractile head is concealed under a mainly yellowish hood, and two yellowish ridges extend along the back. The true legs are reduced, and the prolegs are replaced by small suckers; a silky liquid produced by the larva further aids adhesion. Younger larvae have numerous warts, each bearing a long spine.

FAMILY	Limacodidae
DISTRIBUTION	Nepal, northeast India, Myanmar, southern China, Thailand, Laos, Vietnam, Borneo, Chinese Taipei, and southern Japan
HABITAT	Lowland and montane tropical and subtropical forests
HOST PLANTS	Many, including Rock Oak (*Lithocarpus konishii*), lychee (*Litchi* spp.), and *Liquidambar* spp.
NOTE	Caterpillar that has a geometric, tentlike shape
CONSERVATION STATUS	Not evaluated, but common

ADULT WINGSPAN
1 in (25 mm)

CATERPILLAR LENGTH
⅜ in (10 mm)

300

DEMONAROSA RUFOTESSELLATA
DEMONAROSA RUFOTESSELLATA
(MOORE, 1879)

Actual size

The *Demonarosa rufotessellata* caterpillar is one of the non-stinging species of limacodid larvae. It has a strange, uncaterpillar-like tent shape with no obvious head or tail and can be difficult to detect on the shaded underside of leaves. Typical of cup moth larvae (also called slug moths or skiff moths), the caterpillars move slowly in a smooth, sluglike fashion. They do not travel far and, before moving on, will consume the same leaf until there is nothing left. The caterpillars pupate in a hardened ball cocoon usually sandwiched between two leaves. Those with a temperate distribution overwinter as pupae.

These caterpillars can occur in large numbers but are heavily parasitized by species of braconid parasitic wasps. The wasp larva developing inside a caterpillar eventually "mummifies" its host into a hardened protective shell, within which it completes its life cycle. The adult moth is particularly fluffy and colorful and easy to recognize.

The *Demonarosa rufotessellata* caterpillar is a smooth, non-stinging limacodid species that has an unusual, peaked, angular shape with an ambiguous rounded front end, pointed tail end, and no distinct walking legs. It moves like a slug with an adhesive muscular underbelly. It is green, with the dorsal peaks outlined in brown and intricate, armor-plating markings across its top and sides.

FAMILY	Limacodidae
DISTRIBUTION	Eastern United States, from Maine to Florida, and west to Texas and Missouri
HABITAT	Woodlands, parks, and gardens
HOST PLANTS	Woody trees, such as oak (*Quercus* spp.), ash (*Fraxinus* spp.), apple (*Malus* spp.), beech (*Fagus* spp.), and cherry (*Prunus* spp.)
NOTE	Covert, stinging caterpillar that is rarely seen
CONSERVATION STATUS	Not evaluated, but not considered threatened

ADULT WINGSPAN
¾–1¼ in (19–31 mm)

CATERPILLAR LENGTH
¾ in (20 mm)

EUCLEA DELPHINII
SPINY OAK-SLUG
(BOISDUVAL, 1832)

301

Spiny Oak-slug caterpillars hatch from single eggs or small clusters, laid on the underside of leaves. Like most slug caterpillars, they prefer to feed and rest beneath older leaves, usually around the leaf edges. During the heat of the day, they can sometimes be found hiding between two leaves. The larvae are present throughout the summer and early fall, but the peak for mature caterpillars is from late August through September. Prepupal caterpillars overwinter within a cocoon of brown silk. A single generation of larvae is typical throughout most of the species' range, with multiple broods more likely in southern areas. Adult moths fly from May to August, or later in the northern fringe.

Eclea delphinii ranks among the more festive and charismatic of the Limacodidae species encountered within its range. Its unique design is unmistakable, although the coloration of individuals can be highly variable. Numerous spines across the caterpillar's body are poisonous to the touch. Even though the sting is considered mild in comparison with that of the Saddleback (*Acharia stimulea*) and other slug species, cases of severe allergic reaction have been reported.

Actual size

The Spiny Oak-slug caterpillar has a mottled and variable base color, usually green, pink, red, or tan, although green is the most common form. Two dorsal stripes spanning the length of the body can also vary in coloration and can include four or more red, rectangular spots. Extended lobes of stinging spines encompass the entire body.

FAMILY	Limacodidae
DISTRIBUTION	Eastern United States, southern Ontario
HABITAT	Deciduous forest
HOST PLANTS	Various trees, including oak (*Quercus* spp.), cherry (*Prunus* spp.), maple (*Acer* spp.), basswood (*Tilia* spp.), elm (*Ulmus* spp.), and beech (*Fagus* spp.)
NOTE	Deceptively attractive caterpillar whose sting can cause a severe reaction
CONSERVATION STATUS	Not evaluated, but common

ADULT WINGSPAN
1¹⁄₁₆–1 in (17–25 mm)

CATERPILLAR LENGTH
⁹⁄₁₆–1 in (15–25 mm)

302

ISA TEXTULA
CROWNED SLUG
(HERRICH-SCHÄFFER, [1854])

Crowned Slug Moth caterpillars, frequently found on the underside of oak leaves in the fall, can be both cryptic and aposematic. When small, they blend in with the surface and pattern of the leaf, but, as they develop, their memorable coloring is an apt warning to predators that, if disturbed, they can deliver a powerful sting. When mature, the caterpillar constructs a silk cocoon, strengthened by infusions of calcium oxalate that it produces. The pupa within forces open an escape hatch before the moth emerges.

Limacodidae slug caterpillars move differently from other moth or butterfly larvae because they have suckers instead of prolegs. The larval head is usually concealed under folds of skin, a feature shared with Blue and Hairstreak caterpillars of the Lycaenidae family, which, however, have normal prolegs and are covered in minute hairs. While slug caterpillars have a spectacular and diverse appearance, the resultant adult moths are mostly brown or gray.

Actual size

The Crowned Slug caterpillar is green and flat, with a uniform pair of lobes for each abdominal segment. In the thoracic segments, the lobes develop into red spines with black tips. Regardless of their shape, all of these projections are equipped with stinging spines. The back ridge can have a pair of yellow or red stripes, and the front edge of the body is edged with orange or red. The head of the caterpillar is not visible from above.

FAMILY	Limacodidae
DISTRIBUTION	Eastern United States, from New York State west to Texas, and one area in Florida
HABITAT	Woodlands, parks, and field and roadside edges
HOST PLANTS	Swamp Oak (*Quercus bicolor*) and other *Quercus* spp.
NOTE	Well-defended caterpillar typically found on the underside of leaves
CONSERVATION STATUS	Not evaluated, and uncommon

ISOCHAETES BEUTENMUELLERI

BEUTENMUELLER'S SLUG MOTH

(HY. EDWARDS, 1889)

ADULT WINGSPAN
¾–1⁵⁄₁₆ in (19–24mm)

CATERPILLAR LENGTH
⅜–⁹⁄₁₆ in (10–15 mm)

303

The Beutenmueller's Slug Moth caterpillar is covert and solitary in nature, seldom seen, and an uncommon species throughout most of its normal range. The tiny larva is nearly transparent and sports multiple appendages armed with glass-like "hairs." Physical contact with the caterpillar should be avoided, as it can cause dermatitis and other skin irritations. During later instars, these poisonous filaments encompass the caterpillar's entire body, hence its more common name, Spun Glass Slug.

During the eighth (final) instar, the larva sheds its formidable weaponry prior to metamorphosis within a brown, silken cocoon. A circular hatch at one end of the cocoon allows the transformed moth to exit. Adults—the females larger than the males—fly from June through August. Like all Limacodidae caterpillar species, these crystalline creatures glide across the leaf with a sluglike motion, leaving a shiny trail wherever they go. They are active at night and dormant throughout most of the daylight hours.

The Beutenmueller's Slug Moth caterpillar is transparent light green, with multiple, hairy appendages spanning the length of the body in an elongated starlike pattern. Its internal organs are visible through the dorsal area as a dark stripe along the length of its back. Translucent, knobby protrusions contain stinging spines.

Actual size

FAMILY	Limacodidae
DISTRIBUTION	Eastern United States, from Missouri to the Atlantic coast
HABITAT	Forests
HOST PLANTS	Beech (*Fagus* spp.), hickory (*Carya* spp.), oak (*Quercus* spp.), chestnut (*Castanea* spp.), and hornbeam (*Carpinus* spp.)
NOTE	Caterpillar with a sting that can cause mild skin irritation
CONSERVATION STATUS	Not evaluated, but common

ADULT WINGSPAN
⅝–1⅛ in (16–29 mm)

CATERPILLAR LENGTH
¾–1³⁄₁₆ in (20–30 mm)

NATADA NASONI
NASON'S SLUG MOTH
(GROTE, 1876)

304

Actual size

The Nason's Slug Moth caterpillar hatches from an egg laid individually on the host plant and is initially non-feeding, molting into its second instar without growth. Larger larvae are probably well defended from vertebrate predators by their toxins. In humans, inadvertent contact with the caterpillar causes a sting, which, while of relatively low intensity, causes pain and can produce a rash, or blistering. Their toxins, however, do not make the larvae immune to attacks by parasitoids, such as the braconid *Triraphis discoideus*. In the late stages of development, the caterpillar makes a dense cocoon inside which it diapauses until spring, when it pupates. Adults fly in midsummer.

Slug caterpillars, including those of the Nason's Slug Moth, have unusual locomotion in which their highly elastic underbelly moves in wavelike pulses, aided by a semifluid silk that sticks to smooth-leaved food plants. This may be why, while being generalist feeders, *Natada nasoni* larvae are not often found on plants that have hairs on the leaf surface. There are about 1,500 species of slug moths worldwide.

The Nason's Slug Moth caterpillar is green, with a thin, yellow subdorsal stripe on both sides, each with a row of orange verrucae equipped with stinging spines. The latter are larger in the thoracic and final abdominal segments. Another row of verrucae defend the caterpillar laterally. The caterpillar is short and stout, flat ventrally and convex dorsally, and has vestigial prolegs.

FAMILY	Limacodidae
DISTRIBUTION	United States, from southern New England to Florida, west to Texas
HABITAT	Barrens and woodland edges
HOST PLANTS	Woody trees, including oak (*Quercus* spp.), apple (*Malus* spp.), and elm (*Ulmus* spp.)
NOTE	Leaf-feeding slug caterpillar
CONSERVATION STATUS	Not evaluated, but common

ADULT WINGSPAN
1¹¹⁄₁₆–1¹⁄₁₆ in (18–27 mm)

CATERPILLAR LENGTH
¾ in (20 mm)

PARASA CHLORIS

SMALLER PARASA
(HERRICH-SCHÄFFER, [1854])

305

The Smaller Parasa caterpillar has an elliptical, sluglike appearance typical of Limacodidae species, and glides rather than walks using a muscular pad on its underside. The caterpillars occur from August into October. Mid-September is the peak time for mature caterpillars, when they are most readily found on the leaves of oak, elm, and other deciduous trees. Like most Limacodidae larvae, the caterpillar prefers the underside of leaves while feeding or at rest. Resting Smaller Parasa caterpillars will generally wake up and move around when disturbed.

Adult moths fly from May until August. A single generation is the norm, with a second generation possible in the southern fringes of the species' range. Young larvae are often confused with early-instar *Euclea* species until further growth defines their distinct characteristics. The Smaller Parasa is a mildly venomous caterpillar that can partially retract its arsenal of weaponry while feeding and at rest. When alarmed or threatened, the caterpillar's stinging spines become fully exposed.

Actual size

The Smaller Parasa caterpillar has a distinctive, humpbacked, sluglike form. Spiny "warts" are present on the upper thorax and posterior, and along the subspiracular region. The overall body color can be tan, pink, or orange, with a brighter orange occasionally developing on the posterior in mature caterpillars. There are wavy lines laterally along the abdomen, and the spiracles are elongated.

FAMILY	Limacodidae
DISTRIBUTION	Eastern Siberia, Japan, Korea, eastern China, and Chinese Taipei
HABITAT	Forests, agricultural land, parks, and gardens
HOST PLANTS	Many, including willow (*Salix* spp.), poplar (*Populus* spp.), chestnut (*Castanea* spp.), persimmon (*Diospyros* spp.), *Citrus* spp., and many other fruit trees
NOTE	Caterpillar that spends almost a year in the prepupal stage
CONSERVATION STATUS	Not evaluated, but extremely common

ADULT WINGSPAN
1⅜–1⁹⁄₁₆ in (35–40 mm)

CATERPILLAR LENGTH
1–1⅛ in (25–28 mm)

306

PARASA CONSOCIA
PARASA CONSOCIA
WALKER, 1865

Actual size

The *Parasa consocia* caterpillar has a bright blue dorsal stripe with a broken outline of darker green. There are longitudinal rows of tubercles bearing clusters of stinging spines laterally and dorsolaterally. The third and largest dorsolateral pair of tubercles includes modified, thickened spines tipped in black. The lateral clusters have a central thickened orange seta. A pair of black decoy eyespots are found on a fleshy apron protecting the head and duplicated at the rear as four bulbous, black patches.

Parasa consocia caterpillars hatch from up to 150 eggs laid in small batches over a seven-day period. For the first couple of instars, the larvae are gregarious, grazing only the surface of leaves. They then disperse for the remainder of the eight or nine instars (totaling 27 to 37 days), devouring the entire leaf. Food consumption in the final instar constitutes 80 percent of the total. The caterpillars overwinter inside a dark brown, oval cocoon, wedged into tree bark or in the soil, often in aggregations, for an average 300 days.

Generally, there is one generation annually, with adult moths flying June to July, and larvae feeding through till September. However, particularly in the southern range, there can be a second complete generation produced in only 40 days from late August to October. Populations experience heavy losses from parasitic ichneumonid wasps and flies in both larval and pupal stages. Direct contact with the caterpillar or cocoon, or with loose, shed spines, can cause dermatitis in humans.

FAMILY	Limacodidae
DISTRIBUTION	Oriental tropics, from western India southeast to Borneo, east to Chinese Taipei
HABITAT	Low to medium altitude forests
HOST PLANTS	Many, including banana (*Musa* spp.), Wood-oil Tree (*Aleurites cordata*), Green Tea (*Camellia sinensis*), Teak (*Tectona grandis*), and Chinese Tallow (*Triadica sebifera*)
NOTE	"Stinging nettle" slug caterpillar that has a lavender streak
CONSERVATION STATUS	Not evaluated, but common

ADULT WINGSPAN
1⅜–1⅝ in (35–42 mm)

CATERPILLAR LENGTH
1 in (25 mm)

PARASA PASTORALIS

PARASA PASTORALIS
BUTLER, 1885

307

Like other cup moth (or slug moth) caterpillars, *Parasa pastoralis* larvae characteristically move in a fluid, sluglike, peristaltic fashion, as they do not have the defined prolegs seen in most other caterpillars. Instead, they rely on an adhesive, muscular underbelly. Although they are polyphagous, their mode of locomotion restricts their choice of host plants to those with smooth-surfaced foliage. As early instars, caterpillars graze only from the outer layer of the leaf but progress to full-leaf consumption in middle to late instars. At this time, the larva feeds from the leaf margin with its head retracted beneath fleshy thoracic folds, concealing the movement of mouthparts from potential predators.

As a "stinging nettle" species of limacodid caterpillar, the larva has both sharp, hollow spines, which inject a toxin from glands at the spine base, and urticating hairs and needlelike spicules, which break off on contact. *Parasa pastoralis* overwinters in a cocoon, and adults fly from March to November. The moths have the distinctive green forewings of the *Parasa* genus, with a brown basal spot and margins.

Actual size

The *Parasa pastoralis* caterpillar has four rows of spiny tubercles running the length of its green body. The anterior dorsal pair is the largest cluster. Its spines are orange, and the central few are markedly thicker and black-tipped. There are four black patches on the posterior segments, a vivid purple stripe outlined in black along the back, and wave-shaped markings along the sides.

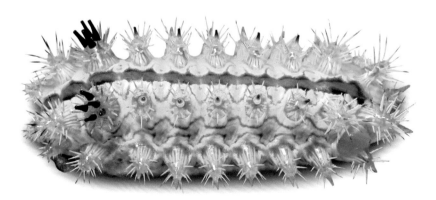

FAMILY	Limacodidae
DISTRIBUTION	Eastern North America, from southeastern Canada to Florida, west to Texas
HABITAT	Forests
HOST PLANTS	Deciduous trees and shrubs, including apple (*Malus* spp.), ash (*Fraxinus* spp.), cherry (*Prunus* spp.), dogwood (*Cornus* spp.), oak (*Quercus* spp.), and willow (*Salix* spp.)
NOTE	Unusually shaped caterpillar with projections that bear stinging hairs
CONSERVATION STATUS	Not evaluated, but common

ADULT WINGSPAN
1³⁄₁₆ in (30 mm)

CATERPILLAR LENGTH
1 in (25 mm)

308

PHOBETRON PITHECIUM
MONKEY SLUG
(J. E. SMITH & ABBOT, 1797)

The Monkey Slug caterpillar is initially black, becoming brown in later instars. It is highly unusual in shape, with nine pairs of fleshy lobes of variable length that project laterally, making it look flat and similar to lichen growth. Every second lobe is longer. Ventrally, the caterpillar is translucent yellow, including its legs, which are barely noticeable but for slightly more sclerotized claws at the tips.

The Monkey Slug, or Hag Moth, caterpillar feeds solitarily on the underside of leaves. With its curly, hairy projections it can easily be mistaken for a spider's shed skin or a leaf deformation, such as a gall or lichen growth. Like other limacodid larvae, the caterpillars move very slowly in a sluglike manner, using suction-like structures instead of prolegs. When fully developed, the caterpillar pupates in a hard, dark, round cocoon, hidden beneath a second cocoon made of softer silk threads that matches the dead leaf background and is shaped like a flipped-over cup. The species produces a single generation annually in the north of its range but two or more in the south.

The *Phobetron pithecium* caterpillar is clearly defended by at least two means: it is cryptic and also memorable to predators if attacked, as its shape is unique and the sting is potent. For humans, the stings are immediately painful, but the effects are not long lasting and anaphylaxis is rare. Akin to a lizard's tail, the projections carrying harmful setae can fall off without harming the caterpillar.

Actual size

FAMILY	Limacodidae
DISTRIBUTION	Southeast Asia, notably Malay Peninsula and Indonesia
HABITAT	Forests, but adapted to intensive plantation industries
HOST PLANTS	Many, including coconut palm (*Cocos* spp.), Oil Palm (*Elaeis guineensis*), banana (*Musa* spp.), citrus (*Citrus* spp.), coffee (*Coffea* spp.), and tea (*Camellia* spp.)
NOTE	Stinging caterpillar that has a heavy impact on palm plantations
CONSERVATION STATUS	Not evaluated, but very common with sporadic outbreaks

ADULT WINGSPAN
1³⁄₁₆–1⅜ in (30–35 mm)

CATERPILLAR LENGTH
1⅜–1⁹⁄₁₆ in (35–40 mm)

SETORA NITENS
SETORA NITENS
WALKER, 1855

309

Setora nitens caterpillars hatch from between 250 and 350 eggs laid in chains on palm fronds. The early instar larvae strip the surface of the leaves only, creating translucent windows in the fronds and leaving them vulnerable to secondary viral and fungal infections. The later instars completely strip the mature fronds before moving to newer growth, causing substantial yield reductions. There are five instars over three to seven weeks before the larvae descend to the trunk base or nearby vegetation to pupate; the pupal period lasts from two-and-a-half to four weeks.

Populations are not constant and can surge in occasional devastating outbreaks during summer months. Like many limacodids, the caterpillars are vulnerable to parasitic wasps and flies, as well as predatory pentatomid bugs and assassin bugs (Reduviidae), and fungal and viral infections, all of which are employed in biocontrol. Because the larvae have adapted to modern agricultural practices and emerged as a major defoliator of palms, affecting the oil palm industry, the species is well documented. The intense stinging capacity of the caterpillars can also have an impact on plantation workers.

The *Setora nitens* caterpillar is block-shaped and deeper than it is broad. The base color is green with a brown dorsal line broken by segmental spots in blue and yellow. There are corresponding, similarly colored, oblique gash-like markings on the sides. The lateral spines are small, but the front and rear dorsolateral spine clusters are large and banded in black. These spines are normally collapsed like a wet paintbrush but splayed open like a pompom when the caterpillar is threatened.

Actual size

FAMILY	Limacodidae
DISTRIBUTION	Southern China, Chinese Taipei, Thailand, and Indochina
HABITAT	Forests
HOST PLANTS	Polyphagous, including nutmeg (*Myristica* spp.)
NOTE	Caterpillar that is sluglike and festooned with stinging spines
CONSERVATION STATUS	Not evaluated, but common

ADULT WINGSPAN
1–1⅜ in (25–35 mm)

CATERPILLAR LENGTH
1⅜ in (35 mm)

SUSICA SINENSIS
SUSICA SINENSIS
(WALKER, 1856)

Susica sinensis caterpillars, like many other lepidopteran larvae, consume their freshly shed skins after molting from one instar to the next. Cup moth (or slug moth) caterpillars of the stinging variety, including *S. sinensis*, typically consume everything, spines and all. This species undergoes a dramatic transformation in terms of size, morphology, and color when it molts into the final instar. At earlier stages, the larvae display vivid blue markings on a green body, with the anterior and posterior dorsolateral scoli much longer than those in between. At maturity, the body is glossy white with green markings and all spiny scoli of uniform size.

The stinging spines of limacodid slug caterpillars, including those of *Susica sinensis*, are capable of causing painful injuries if in contact with human skin, introducing toxins produced from glands at the base of the hollow spines. This species is sometimes informally referred to as the Statuesque Cup Moth for its erect cartoon character-like postures, often while inverted.

Actual size

The *Susica sinensis* caterpillar is pearly white, capsule-shaped, and has three broad, green stripes with darker outlines. The outlines of the dorsal stripe converge over the thoracic segments to produce a pair of X marks. The lateral stripes are heavily waved to circumvent the bases of the heavily spined scoli and incorporate the spiracles. There are four longitudinal rows of scoli forming a uniform halo of defensive, green, black-tipped spines.

FAMILY	Limacodidae
DISTRIBUTION	United States, from New England south to Mississippi, west to Missouri
HABITAT	Forests, woodlands, and field edges
HOST PLANTS	Oak (*Quercus* spp.), beech (*Fagus* spp.), willow (*Salix* spp.), and other deciduous trees
NOTE	Slug caterpillar often mistakenly dismissed as a leaf abnormality
CONSERVATION STATUS	Not evaluated

ADULT WINGSPAN
⁹⁄₁₆–1¹¹⁄₁₆ in (15–43 mm)

CATERPILLAR LENGTH
⅜ in (10 mm)

TORTRICIDIA PALLIDA
RED-CROSSED BUTTON SLUG
(HERRICH-SCHAFFER, 1854)

311

The Red-crossed Button Slug and other button slug caterpillars appear to the naked eye as colorful oval (or round) "buttons" on the undersides of leaves, usually near the leaf edges. Solitary in nature, they are rarely observed occupying the same leaf as other caterpillars. While some have been seen "rocking in place" at three to five-second intervals, the majority are inactive when encountered and easily mistaken for leaf abnormalities. Early instars typically begin to appear in early July and feed by skeletonizing leaf surfaces. Mature caterpillars can be found feeding along leaf edges during August and September. There is one generation in the north of the range, two in the south.

Observation of the genus suggests that button slugs are localized within their range, as one small section of forest can yield multiple encounters, while trees a short distance away may be button slug-free. *Tortricidia pallida* is difficult to differentiate from the Abbreviated Button Slug (*T. flexuosa*), particularly during early instars. Some experts believe that they could be the same species.

Actual size

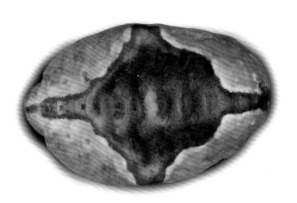

The Red-crossed Button Slug caterpillar is oval shaped and lime green, with a distinctive brownish-red saddle across the dorsum. Considerable variation in color and markings occurs, fully manifested during later instars. Side peripheral markings are noticeably wider than on the Abbreviated Button Slug (*Tortricidia flexuosa*). Thin, yellow, and brighter red outlines border the saddle on mature caterpillars.

FAMILY	Zygaenidae
DISTRIBUTION	Southwestern China
HABITAT	Elevated forests and urban ornamental plantings
HOST PLANTS	*Prunus cerasoides* and *P. majestica*
NOTE	Defoliating, distasteful caterpillar of a day-flying moth
CONSERVATION STATUS	Not evaluated, but can be very common in parts of its range and elsewhere rarely seen

ADULT WINGSPAN
3⅜ in (85 mm)

CATERPILLAR LENGTH
1³⁄₁₆ in (30 mm)

312

ACHELURA YUNNANENSIS
ACHELURA YUNNANENSIS
HORIE & XUE, 1999

Achelura yunnanensis caterpillars can occur in huge numbers on host plants. They have only one generation annually, with the late instar caterpillars leaving the host plant to pupate in late summer. The silk-wrapped pupa is formed in adjoining evergreen foliage or in the leaf litter to overwinter, as the host plant is deciduous but often already defoliated by the larvae. The moths emerge the following summer en masse.

Typical of most caterpillars of the subfamily Chalcosiinae, *Achelura yunnanensis*, when threatened, secretes a transparent, viscous liquid from glands associated with its numerous body verrucae. These droplets contain cyanoglucosides, which act as a taste deterrent against natural enemies; the distastefulness is passed on to the adult moth. The moths are high-flying, day-flying butterfly mimics and do not travel far from the host plant, so are rarely seen or identified. Members of the genus *Achelura*, including *A. yunnanensis*, have only recently been described, with new discoveries still to be made.

Actual size

The *Achelura yunnanensis* caterpillar is a striking, aposematic, yellow-and-black caterpillar. Each segment bears a rectangular black zone either side of the midline that includes two, raised, conical structures bearing twin spinelike setae. These are the source of the defensive cyanide secretions, from visible cuticular cavities. A skirt of white plumose hairs surrounds the body.

FAMILY	Zygaenidae
DISTRIBUTION	Southern Europe
HABITAT	Dry, south-facing slopes with meadow and scree, up to 6,600 ft (2,000 m) elevation
HOST PLANTS	Dock (*Rumex* spp., especially *Rumex scutatus*)
NOTE	Sluglike, hairy caterpillar that is found on alpine meadows
CONSERVATION STATUS	Not evaluated, but is at risk in parts of its range

ADULT WINGSPAN
¾–1³⁄₁₆ in (20–30 mm)

CATERPILLAR LENGTH
¾ in (20 mm)

ADSCITA ALPINA

ADSCITA ALPINA
(ALBERTI, 1937)

313

Adscita alpina caterpillars hatch in summer or early fall from oval-shaped eggs laid on the underside of leaves of various dock species; the eggs, at first pale yellow, turn blue green in color just before the larvae emerge. The young caterpillars mine the leaves, while the older ones feed on the outside of the leaf. The caterpillars then overwinter on or near the ground and become active again the following spring. They pupate on or near the host plant, in a loose cocoon of white silk thread. There is a single generation annually.

The eye-catching, metallic blue-green adults are on the wing over summer, usually from June to August, but they may be seen as late as September. They can often be spotted feeding on the flowers of thistles. The species is under threat from agricultural intensification, tourist development, and the lack of traditional management of its meadow habitat, particularly scrub management, which has led to trees and shrubs colonizing the meadows upon which *Adscita alpina* relies.

The *Adscita alpina* caterpillar has a plump, sluglike shape and is dark brown in color. There are rings of distinctive, large, caramel-yellow tubercles, which bear tufts of short hairs. The head is black.

Actual size

FAMILY	Zygaenidae
DISTRIBUTION	The Himalayas, northern India, southern China, and Southeast Asia
HABITAT	Shaded understory of elevated forests
HOST PLANTS	*Aporosa* spp.
NOTE	Caterpillar that mimics particular butterfly larvae
CONSERVATION STATUS	Not evaluated, but not uncommon

ADULT WINGSPAN
2⅜–2¾ in (60–70 mm)

CATERPILLAR LENGTH
1³⁄₁₆ in (30 mm)

314

CYCLOSIA MIDAMA
CYCLOSIA MIDAMA
(HERRICH-SCHÄFFER, 1853)

Cyclosia midama caterpillars hatch from eggs laid individually, usually on twigs and branches of the host plant or adjoining vegetation, and tend to position themselves on the underside of leaves during development. When under threat, they produce distasteful cyanoglucosides, expressed in a halo of clear droplets, from the tips of their body tubercles; if not dislodged from the secretory pore, these can be reabsorbed once the threat has passed. Associated spines are also urticating on contact. There are two generations during the summer months; the caterpillars feed in June and July, and in October and November.

It has been proposed that both the caterpillars and adults of *Cyclosia midama* mimic separate species of butterflies (*Troidini papilionid* swallowtail larvae and *Euploea mulciber* adults). All are unpalatable at every life stage, so this would be a case of Müllerian mimicry (where different species develop similar characteristics as a protective measure). At least seven subspecies are described, but obvious differences in size, wing shape, markings, and adult genitalia may indicate there is actually more than one species involved.

Actual size

The *Cyclosia midama* caterpillar is covered in evenly spaced, crimson protuberances on a pale pink base interspersed with black markings. Each of these protuberances is topped with one or two sharp spines as well as a cyanoglucoside-secreting cuticular cavity. The central two abdominal segments are glossy porcelain white bordered by black. The head is almost always kept concealed beneath a membranous hood.

FAMILY	Zygaenidae
DISTRIBUTION	Southern India and Sri Lanka, Southeast Asia, and southern China
HABITAT	Tropical rain forests and humid deciduous forests to 3,300 ft (1,000 m) elevation
HOST PLANTS	*Aporusa dioica* and *Dipterocarpus tuberculatus*
NOTE	Caterpillar with candy-like appearance that creates its own poison
CONSERVATION STATUS	Not evaluated

ADULT WINGSPAN
2⅜–2¾ in (60–70 mm)

CATERPILLAR LENGTH
1³⁄₁₆ in (30 mm)

CYCLOSIA PAPILIONARIS
DRURY'S JEWEL
(DRURY, 1773)

315

The Drury's Jewel caterpillar is brightly colored, the yellow and red warning colors alerting predators to its toxic nature. As they develop, the larvae feed on the leaves of tall, high-altitude, tropical trees in primary and secondary forest, and then pupate in a cocoon attached to the upper surface of a leaf. When alarmed, the caterpillar exudes droplets of cyanide from the tip of each of its tubercles. The caterpillars make the cyanide within their body, unlike many other distasteful species that acquire the poison from their food plant. However, the toxin is no protection against frequent paratisitism by nematode worms—often tiny juveniles that larvae accidentally ingest.

The caterpillar passes its toxin to the adult, a brightly colored, day-flying moth, which is also poisonous. It is related to the burnet moths, also of the Zygaenidae family, but has a butterfly-like appearance, reflected in its species name—*papilionaris*. The genus *Cyclosia* comprises 100 or so species. There are eight subspecies of the Drury's Jewel found across Southeast Asia.

Actual size

The Drury's Jewel caterpillar is quite striking, with its blunt shape and rows of raised, yellow tubercles, each of which bears one or two short, black hairs. There are six red tubercles. The body is gray white, and the dorsal tubercles are edged in white, while the lateral tubercles are edged in black. The prolegs are yellow.

FAMILY	Zygaenidae
DISTRIBUTION	Northeast India, Indochina, southern China, Chinese Taipei, and Japan
HABITAT	Forested mountain slopes and valleys
HOST PLANTS	*Helicia* spp., *Buddleja* spp., *Coffea* spp., *Hibiscus* spp., and tallow (*Triadica* spp.)
NOTE	Caterpillar that advertises its noxiousness with classic aposematic colors
CONSERVATION STATUS	Not evaluated, but not uncommon

ADULT WINGSPAN
2¾–3⅛ in (70–80 mm)

CATERPILLAR LENGTH
1³⁄₁₆–1⅜ in (30–35 mm)

316

ERASMIA PULCHELLA
ERASMIA PULCHELLA
HOPE, 1841

Erasmia pulchella caterpillars are not cryptic and make no effort to conceal themselves from potential threats while feeding on the topsides of leaves. Their defensive strategy is one of aposematic red and yellow warning coloration, distasteful cyanogenic body secretions, urticating hairs and spines, and ambiguous head and tail anatomy. The main natural threat they face is from parasitoids, particularly braconid wasps, against which they have no defenses. The specific parasitoids of the Chalcosiine subfamily to which *E. pulchella* belongs have been able to adapt to detoxify hydrocyanic acids, thus bypassing the group's main defensive mechanism.

Pupation occurs beneath a parchment-like silken sheet formed over a concavity on the host plant leaf or adjoining vegetation or in leaf litter. Adult moths are stunningly colored and iridescent and highly prized in collections. They fly over the summer months and inherit their distastefulness from the larval stage. In the adult, this is expressed as a toxic froth, which bubbles from thoracic glands when the moths are threatened.

Actual size

The *Erasmia pulchella* caterpillar is black, with each segment ringed by six protuberances topped with setae. The lateral, cranial, and caudal-most protuberances are bright red with long, white, flexible setae. The dorsolateral protuberances are white and the dorsal pairs yellow, with shorter black setae. There is a canary-yellow saddle mid-body.

FAMILY	Zygaenidae
DISTRIBUTION	Sri Lanka, southern India, northeast India and the Himalayas, Thailand, Myanmar, Indochina, China, Chinese Taipei, and Japan
HABITAT	Montane forests
HOST PLANTS	Many, including tea (*Camellia* spp.), *Aporosa* spp., wild peony (*Melastoma* spp.), *Eurya* spp., *Buddleja* spp., and *Rhododendron* spp.
NOTE	Caterpillar that is sluglike and a noted pest of tea
CONSERVATION STATUS	Not evaluated, but generally very common, although some Japanese subspecies are specific to individual islands and locations

ADULT WINGSPAN
2⅜–2¾ in (60–70 mm)

CATERPILLAR LENGTH
1³⁄₁₆ in (30 mm)

ETERUSIA AEDEA
ETERUSIA AEDEA
(LINNAEUS, 1763)

Eterusia aedea caterpillars, also known as red slug caterpillars, hatch from up to 300 ovoid, yellow eggs laid by the female moth in stacked batches. New hatchlings are gregarious and graze the leaf surface but later disperse, preferring to feed on mature foliage. When threatened, like related moth larvae of the Chalcosiinae subfamily, *E. aedea* larvae secrete droplets of distasteful cyanic fluid from their many tubercles. There are five larval instars over three to four weeks. Pupation, lasting three to four weeks, occurs on the midrib of the topside of leaves with the leaf folded over the silken cocoon. In populations associated with tea plantations in temperate China, there are two generations annually—June to August and October to November. The second generation overwinters as a pupa.

Eterusia aedea is easily the most widely distributed Asian species of the Chalcosiinae subfamily. It includes 13 subspecies (eight from the islands of Japan) with varied wing markings and colors, sizes, flight times, life histories, and host plant preferences. Several subspecies have been reported as pests of tea plantations, which means their life cycles are generally better known.

Actual size

The *Eterusia aedea* caterpillar is slow-moving, sluglike, and brick red in color, bearing six, longitudinal rows of tubercles. These are tipped with pairs of setae that are spinelike, except for those around the margins, where the tubercles are a brighter orange and bear longer, softer hairs. The body color deepens toward the dorsum with a lighter saddle mid-abdomen.

FAMILY	Zygaenidae
DISTRIBUTION	Southeastern United States
HABITAT	Forests, forest edges, and disturbed habitats
HOST PLANTS	Carolina Laurelcherry (*Prunus caroliniana*)
NOTE	Toxic, sluglike caterpillar that defoliates its hosts
CONSERVATION STATUS	Not evaluated, but common

ADULT WINGSPAN
⁹⁄₁₆–¹³⁄₁₆ in (15–21 mm)

CATERPILLAR LENGTH
½ in (13 mm)

NEOPROCRIS FLORIDANA
LAURELCHERRY SMOKY MOTH
TARMANN, 1984

318

Actual size

The Laurelcherry Smoky Moth caterpillar is yellow and white with black lines. It is covered with verrucae carrying venomous setae. The head retracts into the body and can be seen more clearly if the larva is viewed from underneath or when feeding.

Laurelcherry Smoky Moth caterpillars feed on the underside of leaves. As young caterpillars they are found in groups, as eggs are laid in clusters. Later they disperse to feed alone or in pairs. There are usually numerous larvae on each plant, so the plants' tough leaves are often completely skeletonized. When that happens, the caterpillars suspend themselves on silk threads to find a new host plant. The larvae behave similarly if disturbed, so dropping off the host plant appears also to be a defense against predators. Larger caterpillars are aposematically colored, probably indicating their toxicity, as they feed on cyanogenic, glycoside-rich host plant leaves. Contact with a full-grown caterpillar can cause mild skin irritation.

The Laurelcherry Smoky Moth produces three generations a year and diapauses as a pupa. The adults, though dark colored, look like wasps and have an iridescent abdomen suggesting that the caterpillars sequester the toxic chemicals and pass on the chemical defense to the moths. The species may be a part of a mimicry complex that includes wasps and other insects.

FAMILY	Zygaenidae
DISTRIBUTION	Japan, Korean peninsula, Chinese Taipei, and southern China (Hong Kong)
HABITAT	Low to medium-altitude forests
HOST PLANTS	*Eurya japonica* and *E. emarginata*, and *Euonymus japonicus*
NOTE	Noxious yellow-and-black caterpillar
CONSERVATION STATUS	Not evaluated, but very common

ADULT WINGSPAN
1¾–2⅜ in (45–60 mm)

CATERPILLAR LENGTH
1–1¹/₁₆ in (25–27 mm)

PIDORUS ATRATUS
PIDORUS ATRATUS
BUTLER, 1877

319

Pidorus atratus caterpillars hatch from eggs laid on the host plant in bark crevices or developing flower buds. As a result, the larvae are evenly distributed over the food plant rather than on isolated leaves and often in large numbers. When disturbed, the caterpillars produce and secrete droplets of cyanide compounds (mainly linamarin and lotaustralin), which is typical of the Chalcosiinae subfamily of moths. This acts as a taste deterrent against natural predators. Contact with human skin can also cause a delayed urticating effect due to breakaway fragments of irritating hairs. With the larva's head disguised beneath a fleshy hood, the symmetrical body markings make it difficult to tell the front end from the rear, unless the caterpillar is seen in motion.

There are two generations annually, with caterpillars feeding from April to June and August to September. Pupation occurs in a silken cocoon on host leaves or branches. The adult moths have distinctive red heads and a white V-shaped blaze on the wings at rest, and they are day-flyers during July and September.

Actual size

The *Pidorus atratus* caterpillar is yellow with broad, black stripes on the sides and bands across the cranial and caudal segments. A lighter, almost gray, stripe runs the length of the dorsal midline. Single black setae occur on the protuberances within these black zones, while those in the dorsal yellow zone bear two setae. Long, thick, white setae form a sparse fringe around the entire body.

FAMILY	Zygaenidae
DISTRIBUTION	Across Europe and Russia to Russian Far East and Japan
HABITAT	Forests, scrub, and heathlands
HOST PLANTS	Bog-rosemary (*Andromeda polifolia*), bilberry (*Vaccinium* spp.), Heather (*Calluna vulgaris*), and *Prunus* spp.
NOTE	Small, slug-shaped caterpillar that has tufts of white hairs
CONSERVATION STATUS	Not evaluated, but classed as endangered in parts of its range

ADULT WINGSPAN
⅞–1½₂ in (22–26 mm)

CATERPILLAR LENGTH
¾–1 in (20–25 mm)

320

RHAGADES PRUNI

BLACKTHORN AURORA MOTH

(DENIS & SCHIFFERMÜLLER, 1775)

The caterpillars of the Blackthorn Aurora Moth hatch from yellow, conical-shaped eggs laid either singly or in small groups on the host plants. The larvae feed on leaves and buds and overwinter in the vegetation, becoming active again in spring. After a period of further growth, the caterpillars pupate on the host plant, spinning a white, boat-shaped cocoon.

The day-flying moths are butterfly-like with an eye-catching metallic blue body and dark wings. They are seen in June and July, and there is a single generation. The female moths stay close to the place where they eclosed, while the males fly over greater distances. The adults do not feed. Unusually, *Rhagades pruni* is found in very different habitats, ranging from heathland and bogs, dry bushy slopes, and grasslands to woodland and hedgerow. They are under threat from the loss of scrub, drainage of bogs, reforestation, and general development.

Actual size

The Blackthorn Aurora Moth caterpillar is small and slug-shaped. It is dark brown with orange stripes running the length of the body. There are rings of tubercles, each bearing a tuft of white hairs.

FAMILY	Zygaenidae
DISTRIBUTION	Across southern Europe into the Balkans, southern Russia, and the Caucasus, Turkey, North Africa, and the Middle East
HABITAT	Dry, scrubby woodlands, grassy slopes, verges, and gardens
HOST PLANTS	Vines such as *Vitis* spp. and *Parthenocissus* spp.
NOTE	Caterpillar that has been a vineyard pest since ancient times
CONSERVATION STATUS	Not evaluated

ADULT WINGSPAN
¾–1 in (20–25 mm)

CATERPILLAR LENGTH
¾–1 in (20–25 mm)

THERESIMIMA AMPELLOPHAGA
VINE BUD MOTH
(BAYLE-BARELLE, 1808)

321

The female Vine Bud Moth lays clusters of creamy-white eggs on the underside of leaves. The caterpillars feed on vine leaves, and their dark brown color has given the moth its alternative name of Dark Brown Vine Moth. The mature caterpillar spins a loose cocoon attached to the underside of leaves in which it overwinters. The pupa is light brown in color.

The Vine Bud Moth is on the wing during the day, showing off its distinctive metallic blue body and brown wings. Across much of its range, the moth flies from May to July with a single generation, but in the south there is a second generation from the end of July to September. *Theresimima ampellophaga* is an important pest of vineyards, recognized as such since Roman times, as it damages the leaves of the grapevines and reduces yields. The species can be controlled by the use of pheromone traps that attract the males, and the use of these traps has shown that the moth is more widely distributed than was once thought.

The Vine Bud Moth caterpillar has a dark brown, sluglike body. Each segment has a ring of large, light brown tubercles bearing tufts of long, brown and white hairs.

Actual size

FAMILY	Zygaenidae
DISTRIBUTION	Across Europe to western Asia
HABITAT	Woodland glades, meadows, and coastal cliffs
HOST PLANTS	Legumes (Fabaceae), including *Lathyrus* spp., *Lotus* spp., and *Trifolium* spp.
NOTE	Caterpillar whose black and yellow colors warn of its toxicity
CONSERVATION STATUS	Not evaluated, but common throughout its range

ADULT WINGSPAN
1³⁄₁₆–1½ in (30–38 mm)

CATERPILLAR LENGTH
¾–⅞ in (20–22 mm)

322

ZYGAENA FILIPENDULAE
SIX-SPOT BURNET
(LINNAEUS, 1758)

The female Six-spot Burnet lays her eggs in small groups on the leaves of the host plant. The emerging caterpillars feed and then overwinter, becoming active the following spring. Occasionally they may overwinter twice. From the host plant they feed on, the caterpillars acquire cyanogenic compounds and can also create them. These compounds are protective, making the larvae, which release cyanide if attacked, highly distasteful to predators. The mature caterpillars pupate in a yellow, boat-shaped, papery cocoon attached to old grass stalks.

The brightly colored, day-flying moths, which are found living in colonies, emerge and are active in midsummer. Their distinctive black forewings with red spots and crimson hindwings warn would-be predators that, like the caterpillars, they are distasteful because of the presence of cyanogenic glucosides. The chemicals are passed from the caterpillars onto the adults to give them protection; the eggs laid by the female are also laced with the toxic compounds.

The Six-spot Burnet caterpillar has a dumpy shape with tapered ends. It is pale yellow green with rows of black spots, which gives it a speckled appearance. The disruptive coloration provides camouflage among the leaves of its host plant. The body is covered in tufts of short, white hairs.

Actual size

FAMILY	Zygaenidae
DISTRIBUTION	Across Europe to western Turkey and the Caucasus, and as far east as the Urals
HABITAT	Dry slopes, cliff tops, dry grasslands, and alpine meadows up to 6,600 ft (2,000 m) elevation
HOST PLANTS	Low-growing plants, including *Lathyrus* spp., clover (*Trifolium* spp.), and vetch (*Vicia* spp.)
NOTE	Hairy caterpillar that has distinctive rows of black spots
CONSERVATION STATUS	Not evaluated, but regionally rare

ADULT WINGSPAN
1¾₆–1¹³⁄₁₆ in (30–46 mm)

CATERPILLAR LENGTH
1³⁄₁₆ in (30 mm)

ZYGAENA LONICERAE
NARROW-BORDERED FIVE-SPOT BURNET
(SCHEVEN, 1777)

323

The female Narrow-bordered Five-spot Burnet moth lays her creamy-yellow and slightly oval eggs in large clusters around stems and under leaves. The hatched caterpillars feed through summer and then overwinter, emerging to feed again in spring and pupate in May. Like other *Zygaena* species, a proportion of the caterpillars remain in diapause for the second or even third winter to reduce the risk of an entire generation being wiped out by adverse weather. The mature larvae pupate on the vegetation, spinning a creamy-yellow, elongated cocoon. The pupa is dark brown.

The day-flying moths, with their bright, red-spotted wings, are often mistaken for butterflies. They have a variable appearance, and there are a number of subspecies. The moths are on the wing in June and July with a single generation. *Zygaena lonicerae* is often confused with the Five-spot Burnet (*Z. trifolii*), which flies at the same time in similar habitats.

The Narrow-bordered Five-spot Burnet caterpillar has a pale yellow body with rows of rectangular black spots. There are transverse bands of yellow and tufts of long, white hairs, which are longer than those on the larvae of other burnet moths.

Actual size

FAMILY	Zygaenidae
DISTRIBUTION	Across Mediterranean Europe and the Balkans, into southern Russia and the Caucasus
HABITAT	Forest margins, dry scrubby grasslands, and embankments up to 3,950 ft (1,200 m) elevation
HOST PLANTS	Low-growing plants, including *Lathyrus* spp. and vetch (*Vicia* spp.)
NOTE	Plump caterpillar that is seen feeding on low-growing plants
CONSERVATION STATUS	Not evaluated, but some subspecies are under threat

ADULT WINGSPAN
1⅛–1⅜ in (28–35 mm)

CATERPILLAR LENGTH
1³⁄₁₆ in (30 mm)

ZYGAENA ROMEO
RETICENT BURNET
(DUPONCHEL, 1835)

324

The caterpillars of the Reticent Burnet moth hatch from oval, cream-colored eggs laid in small clusters on the underside of leaves. The larvae feed during the summer, when they are often seen basking in the sun, then overwinter and resume feeding and growing in spring. As a precaution against potentially decimating weather conditions, a few individuals will then overwinter once more to further the survival chances of the species in their locality. This strategy is also believed to reduce inbreeding. In late spring, the caterpillars spin a cocoon on the host plant and pupate.

The striking red-spotted and black day-flying moth is on the wing in June and July with a single generation. There are a number of subspecies across the range, each with a slightly different appearance, both for adult and caterpillar. Often, *Zygaena romeo* is confused with *Z. osterodensis*, which has very similar coloration, in both caterpillar and adult moth, and occurs in the same area.

The Reticent Burnet caterpillar has a sluglike shape with a pale yellow body. There are two rows of triangular-shaped black spots, transverse bands of yellow, and tufts of white hairs. The spiracles are black.

Actual size

FAMILY	Thyrididae
DISTRIBUTION	Across central and southern Europe, into Russia as far as the Urals
HABITAT	Woodland margins, scrub, grassy slopes, verges, and gardens
HOST PLANTS	Burdock (*Arctium* spp.), Old Man's Beard (*Clematis vitalba*), and Elder (*Sambucus nigra*)
NOTE	Plump, orange caterpillar that builds a distinctive leaf shelter
CONSERVATION STATUS	Not evaluated, but locally endangered

ADULT WINGSPAN
⁹⁄₁₆–¾ in (15–20 mm)

CATERPILLAR LENGTH
⅜–½ in (10–12 mm)

THYRIS FENESTRELLA
PYGMY
(SCOPOLI, 1763)

325

The caterpillars of the Pygmy moth hatch from brown eggs laid up to 80 at a time, widely spaced on leaves. The larvae lead solitary lives, each one building a leaf shelter by rolling the leaf tip and securing it with silk. A new shelter is built after each molt, each shelter getting bigger, until, eventually, a whole leaf is used. The caterpillar hides in its shelter during the day and emerges at night to feed. It crawls to the ground to pupate, spinning a cocoon among dead leaves. The pupa is red brown in color.

The day-flying moths are on the wing from May to August, and there are two generations. The pupae of the second generation overwinter. *Thyris fenestrella* is endangered in parts of its range due to the clearance of scrub and verges favored by its host plants. In New Zealand, the Pygmy caterpillar is one of several insects being evaluated as a biological agent to control *Clematis vitalba*, an invasive plant.

The Pygmy caterpillar is plump and orange in color. There are rings of raised, dark brown tubercles on each segment, each tubercle bearing a tuft of short hairs. The head and legs are a shiny dark brown.

Actual size

FAMILY	Tortricidae
DISTRIBUTION	Every continent, except Antarctica
HABITAT	Stored grain or nut facilities, households, warehouses; can survive and breed outdoors among nut trees
HOSTS	Foodstuffs, including flour, cereals, dried fruits, nuts, and chocolate
NOTE	Common household pest that gives caterpillars a bad name
CONSERVATION STATUS	Not evaluated, but abundant

ADULT WINGSPAN
½–¾ in (13–20 mm)

CATERPILLAR LENGTH
⅜–½ in (10–12 mm)

PLODIA INTERPUNCTELLA
INDIAN MEAL MOTH
(HÜBNER, [1813])

326

Actual size

Indian Meal Moth caterpillars can be found year-round indoors, and in food processing and storage facilities. The life cycle can be completed in three weeks under optimal temperatures— 86–95°F (30–35°C)—so there may be more than 12 generations annually. Females lay 60 to 400 pinhead-sized, white, sticky eggs on food surfaces, which hatch in 2 to 14 days, depending on temperature. The caterpillars feed on the surface of grain, nuts, and flour, trailing silken threads that bind food, frass, and cast skins together. There are five to seven instars, and last instars pupate in a thin, white cocoon from which adult moths emerge in about seven days.

Adult moths fly at night, are non-feeding, and live only seven to ten days. After emergence, females produce a pheromone to attract males for mating, usually at dusk. Last instar caterpillars are capable of entering diapause, allowing them to survive the winter in unheated conditions.

The Indian Meal Moth caterpillar is generally white with a pinkish or greenish tinge. Its setae are sparse, pale, and relatively long. The setal bases are dark and sometimes appear as distinct spots. The head and true legs are reddish brown, as are the dorsal collar on segment one and the small dorsal shield on the posterior segment.

FAMILY	Crambidae
DISTRIBUTION	Southern United States, Central America, and the Caribbean
HABITAT	Various habitats, in close association with host plants
HOST PLANTS	Coral trees (*Erythrina* spp.)
NOTE	Caterpillar that folds leaves to create a shelter
CONSERVATION STATUS	Not evaluated, but not common, though not considered endangered

ADULT WINGSPAN
¹⁵⁄₁₆–1³⁄₁₆ in (24–30 mm)

CATERPILLAR LENGTH
1³⁄₁₆–1⁹⁄₁₆ in (30–40 mm)

AGATHODES MONSTRALIS
ERYTHRINA LEAF-ROLLER
GUENÉE, 1854

327

The Erythrina Leaf-roller caterpillar is so-called because it makes shelters from a single leaf, where it remains when not feeding. In spring, the larvae prefer flowers, while summer and fall generations feed on leaves and develop more slowly than the earlier generation. In north Florida, where the host plants freeze to the ground in November and December, resprouting from the roots in April, diapause occurs in a prepupal stage inside cocoons made of a double layer of silk. In the warmer parts of the moth's range, there may be no diapause. There are four generations of *Agathodes monstralis* between May and September in Florida, each generation taking about a month to develop.

Erythrina Leaf-rollers can have an economic impact as they attack a group of popular ornamental and medicinal plants in the genus *Erythrina*. Although both North American and South American moth populations were known as *Agathodes designalis*, the most recent genetic studies suggest two separate species, and so the name *A. monstralis* now applies to the North American populations and *A. designalis* to the South American ones. *Agathodes* moths are found throughout the subtropical and tropical regions, forming a complex of 15 species, three of them in the New World.

Actual size

The Erythrina Leaf-roller caterpillar when young is translucent and green (if feeding on leaves) or orange (if feeding on flowers), with six rows of short, black, sclerotized tubercles. The later instar larvae develop cream-colored, longitudinal stripes, and the black tubercles become more prominent on the background color, while the head is bright red. Before pupation, a caterpillar can turn orange or pink, especially if dieting on flowers.

FAMILY	Crambidae
DISTRIBUTION	Temperate and subtropical east Asia, including China, Japan, and Korea; accidentally introduced into Germany in 2007 and now in many other European countries
HABITAT	Scrub, woodlands, hedgerows, parks, and gardens
HOST PLANTS	Box (*Buxus* spp.)
NOTE	Caterpillar that in large numbers creates webbing and extensive defoliation
CONSERVATION STATUS	Not evaluated; widespread but rather local, although increasing in Europe

ADULT WINGSPAN
1⅜–1⁹⁄₁₆ in (35–40 mm)

CATERPILLAR LENGTH
1⁹⁄₁₆ in (40 mm)

328

CYDALIMA PERSPECTALIS
BOX TREE MOTH
(WALKER, 1859)

Actual size

The Box Tree Moth caterpillar takes its common name from its host plant. Clusters of 5 to 20 eggs are laid in a flat sheet on a leaf, and the caterpillars live in groups. They create a loose, silken web with frass pellets suspended within it and graze the leaf surface, causing conspicuous pale patches of leaf death or defoliation. In addition, the larvae sometimes eat the green bark of the young twigs. Pupation occurs in a cocoon between two leaves. Later broods of larvae overwinter on the food plant.

The slightly iridescent, blackish and white (or almost entirely brownish black) adults fly in two to three broods, mainly in summer. In Europe, in the absence of its usual natural enemies, the species has become a pest, threatening both native and ornamental populations of box. In some places, larvae have caused repeated heavy defoliation, which, along with stripping of bark, can result in the death of the plant.

The Box Tree Moth caterpillar is smooth, rather slender, and pale green in color with a black head. It has two irregular, yellow lines along the back and another low down along the sides. Laterally, there are interrupted black-and-white stripes and lines of raised black spots bearing fine, white bristles. Young caterpillars are yellowish and plainer.

FAMILY	Crambidae
DISTRIBUTION	Europe, North Africa, and the Middle East to western China
HABITAT	Woodlands, parks, grasslands, and heaths
HOST PLANTS	Mosses, particularly *Hypnum* spp.
NOTE	Small caterpillar that feeds and pupates on moss
CONSERVATION STATUS	Not evaluated, but quite common

ADULT WINGSPAN
⅝–¾ in (16–19 mm)

CATERPILLAR LENGTH
¼–⅜ in (7–10 mm)

EUDONIA MERCURELLA
EUDONIA MERCURELLA
(LINNAEUS, 1758)

329

The female *Eudonia mercurella* moths lay their eggs on the various mosses that are host plants for the caterpillars when they hatch. Such mosses may be growing on tree trunks, rocks, walls, and other structures. The young larvae feed and then overwinter, becoming active and feeding again in spring. As they feed, the caterpillars spin a silken tube through the moss and later pupate in a silken cocoon within the moss, usually during May and June. There is a single generation, with the adult moths flying at night from June to September.

The large and widespread *Eudonia* genus includes around 250 species of micro moths, most of whose larvae feed on moss. The moths are often referred to as grass moths because of their habit of resting on grass stems during the day. At night they are commonly attracted to lights. Identification is difficult as there are few distinguishing features.

Actual size

The *Eudonia mercurella* caterpillar is yellowish green or cream colored with a faint, dorsal, brown line and brown spots. The head is a shiny brown, as is the first abdominal segment. Its hairs are short and sparse.

FAMILY	Crambidae
DISTRIBUTION	Eastern United States, from South Carolina to Florida, west to California, and south to Argentina
HABITAT	Various habitats in association with *Erythrina* species
HOST PLANTS	Coral Bean (*Erythrina herbacea*) and other coral trees
NOTE	Caterpillar that lives and feeds inside the host plant
CONSERVATION STATUS	Not evaluated, but can be common where host plant is present

ADULT WINGSPAN
1³⁄₁₆–1⁹⁄₁₆ in (30–40 mm)

CATERPILLAR LENGTH
1–1⁹⁄₁₆ in (25–40 mm)

330

TERASTIA METICULOSALIS
ERYTHRINA BORER
GUENÉE, 1854

Erythrina Borer caterpillars are found inside the stems or bean pods of their host plants, creating tunnels from which they expel frass. Their feeding produces a characteristic dying-off of the tip of the stem. In Florida, the spring generation feeds on flowers, and mature larvae destroy the bean pods of the Coral Bean plant. Later in the season, they feed inside the stem, which is less nutritious than the beans or flowers, and colorless. As a result, the summer and fall generation larvae are paler than spring individuals, and the resulting adults are smaller.

Because it is a borer, the caterpillar is difficult to reach with insecticides and is also well protected from natural enemies, making it a formidable pest. It is capable of destroying the nursery stock of ornamental *Erythrina* species. The long-lived moths are strong fliers and lay eggs singly. *Terastia meticulosalis* is a New World representative of a complex of almost identical pantropical species with similar biology, found from Africa to Australia and on many tropical islands.

The Erythrina Borer caterpillar is pale reddish brown to pink. It has a few short setae, and a black head and prothoracic plate that are both heavily sclerotized. In older larvae, the prothoracic plate becomes lighter and only slightly darker than the rest of the body, which is otherwise translucent and cream colored but can turn pinkish red toward pupation.

Actual size

FAMILY	Cimeliidae
DISTRIBUTION	Southern Greece
HABITAT	Hot, dry slopes with sparse vegetation
HOST PLANTS	*Euphorbia* spp.
NOTE	Little-known green caterpillar that is found in Greece
CONSERVATION STATUS	Not evaluated

ADULT WINGSPAN
1¹⁄₁₆–1⁵⁄₁₆ in (27–33 mm)

CATERPILLAR LENGTH
⁹⁄₁₆–¾ in (15–20 mm)

AXIA NESIOTA
AXIA NESIOTA
REISSER, 1962

Axia nesiota caterpillars hatch from eggs laid on the host plants, which are found growing in dry, rocky, often hot habitats. The larvae are active from fall, all through winter, and into early spring, pupating from February to May. The pupa estivates during the hot summer months and ecloses in the cooler fall temperatures.

The adult moths are on the wing from September to November, with peak numbers seen during October. Although very little is known about *Axia nesiota*, it is present in many of the Greek islands, with Samos the most easterly, and its range may extend even farther into Turkey. The genus *Axia* comprises just five species of large, brightly colored, night-flying moths found only in southern Europe. *Axia* is classified within the family Cimeliidae, known as the gold moths. A distinctive feature of the adult is a pair of pocket-like organs on the seventh abdominal spiracle that may be linked to sound reception.

The *Axia nesiota* caterpillar has a plump, sluglike shape and is bright green in color. Tiny, white dots are scattered across the body. There is a prominent, yellow lateral stripe at the level of the spiracles. The head and true legs are pink brown.

Actual size

FAMILY	Drepanidae
DISTRIBUTION	Southern Greece, southern Balkans, Ukraine, and Asia Minor to Pakistan
HABITAT	Warm, dry, open scrub
HOST PLANTS	Hawthorn (*Crataegus* spp.), Blackthorn (*Prunus spinosa*), plum (*Prunus* spp.), and apple (*Malus* spp.)
NOTE	Caterpillar of a recently recognized species resembling others of its genus
CONSERVATION STATUS	Not evaluated

ADULT WINGSPAN
¾–1⁵⁄₁₆ in (19–24 mm)

CATERPILLAR LENGTH
⁹⁄₁₆–⅝ in (14–16 mm)

332

CILIX ASIATICA
EASTERN CHINESE CHARACTER
O. BANG-HAAS, 1907

Actual size

The Eastern Chinese Character caterpillar is generally light brown with little contrast and a distinct fine, complex, vein-like pattern. The tail spike is quite long, and the white markings near the tail end are not especially distinct.

The Eastern Chinese Character caterpillar hatches and feeds on much the same range of host plants as the very similar Chinese Character (*Cilix glaucata*), of which, until 1987, it was thought to be an eastern subspecies. However, the Eastern Chinese Character is restricted to dry, hot places, and so has a more southerly distribution than the Chinese Character. *Cilix asiatica* larvae can be found from May to late October, and there are likely to be three broods annually in most places. Like that of the Chinese Character, the pupation site is probably leaf litter or a bark crevice, with the species overwintering in this stage.

The Eastern Chinese Character species was first found in Europe (in the Crimea) in the late 1990s. The adults closely resemble Chinese Character moths but with four gray spots near the edge of the forewing instead of the Chinese Character's six. Further very similar species (at both caterpillar and adult stages), *Cilix hispanica* and *C. algirica*, occur farther west in southern Europe.

FAMILY	Drepanidae
DISTRIBUTION	Across Europe, east to the Urals and Asia Minor
HABITAT	Scrub, woodlands, hedgerows, and gardens
HOST PLANTS	Mainly hawthorn (*Crataegus* spp.) and Blackthorn (*Prunus spinosa*); also plum (*Prunus* spp.) and other woody Rosaceae, including apple (*Malus* spp.), Bramble (*Rubus fruticosus*), and Rowan (*Sorbus aucuparia*)
NOTE	Solitary caterpillar that mimics a shriveled leaf
CONSERVATION STATUS	Not evaluated, but common throughout most of its range

ADULT WINGSPAN
$^{11}\!/_{16}$–$1^{1}\!/_{16}$ in (17–27 mm)

CATERPILLAR LENGTH
$^{9}\!/_{16}$–$1^{1}\!/_{16}$ in (15–18 mm)

CILIX GLAUCATA
CHINESE CHARACTER
(SCOPOLI, 1763)

333

The Chinese Character caterpillar can be found from May to October in a wide variety of habitats where its food plants occur. The roughly oval eggs are laid singly on the leaves, where the solitary caterpillars will then feed openly and rest with their spiky tail end raised. They pupate in a strong, brown, silken cocoon spun among the leaves or in a bark crevice on the food plant. Like that of other drepanine moths, the pupa is covered in a pale, waxy bloom. Pupae of the later broods overwinter.

Cilix glaucata is the most widespread of the four *Cilix* species recognized in Europe and the Middle East. The adults, which resemble the fecal sac of a small juvenile bird when at rest, fly in two or three broods from April to September and are named for the elaborate silver marking in the center of the forewing. The species is closely related to hook-tip moths (so-called for the shape of their forewings), whose caterpillars share a similar, characteristic form, resembling a shriveled piece of leaf or a bird dropping.

Actual size

The Chinese Character caterpillar is dark brown to tawny brown, with a paler saddle mark along the back, and resembles a piece of shriveled leaf. The head is deeply divided, and the swollen, peaked frontal segments have two pairs of dorsal tubercles. The raised hind end tapers to a short, fleshy spike with white markings beside it.

FAMILY	Drepanidae
DISTRIBUTION	Europe, east to the Urals, and northeast Asia
HABITAT	Forests, heathland, scrub, and other wooded areas, including gardens
HOST PLANTS	Birch (*Betula* spp.) and alder (*Alnus* spp.)
NOTE	Common hook-tip caterpillar that is found in many habitats
CONSERVATION STATUS	Not evaluated, but common throughout most of its range

ADULT WINGSPAN
1³⁄₁₆–1⁹⁄₁₆ in (30–40 mm)

CATERPILLAR LENGTH
¹¹⁄₁₆–¹⁵⁄₁₆ in (18–24 mm)

DREPANA FALCATARIA
PEBBLE HOOK-TIP
(LINNAEUS, 1758)

334

Actual size

Pebble Hook-tip caterpillars hatch from eggs laid in short chains on the leaves of their host plant. They are solitary, each making a slight silken shelter when small, but live openly on the leaves when larger. The larvae often rest with the hind end raised and the front end arched so that the tubercles protrude. They can be found in almost any habitat within their range, from late May to October. Pupation occurs in a tough, brown cocoon constructed either in a rolled-up leaf or between two leaves joined together, and this is the overwintering stage.

The species is named for the pebble mark on the forewing of the adults, which fly from April to September (or May to July in the north of the range). Hook-tips are members of the subfamily Drepaninae, whose characteristic caterpillars have a single fleshy spike that has evolved from the hind pair of prolegs. The Dusky Hook-tip caterpillar (*Drepana curvatula*) is similar in appearance to *D. falcataria* and also feeds on birch and alder, but it has shorter tubercles and is generally slightly darker in color.

The Pebble Hook-tip caterpillar has many quite long, stiff hairs. It is blackish when small, with pale green markings. When larger, it is plump, bright green on the sides, and brown on the back with four pairs of large, raised tubercles on the front half. Its body tapers to the hind end, which has a short, blunt spike.

FAMILY	Drepanidae
DISTRIBUTION	North America, from southern Canada to southern United States
HABITAT	Moist forests, woodlands, and shrublands
HOST PLANTS	Alder (*Alnus* spp.), birch (*Betula* spp.), oak (*Quercus* spp.), poplar (*Populus* spp.), and willow (*Salix* spp.)
NOTE	Wrinkly caterpillar that is found in silked-up leaf shelters
CONSERVATION STATUS	Not evaluated

ADULT WINGSPAN
1⅝–1¹¹⁄₁₆ in (41–43 mm)

CATERPILLAR LENGTH
1¼–1⅜ in (32–35 mm)

PSEUDOTHYATIRA CYMATOPHOROIDES

TUFTED THYATIRIN

(GUENÉE, 1852)

335

Tufted Thyatirin caterpillars hatch from eggs laid by the female moth on a wide range of deciduous trees. The larvae are solitary and feed alone, usually by night, on a range of hardwood shrubs and trees; those in the Pacific northwest are said to favor members of the rose (Rosaceae) family. The caterpillars form shelters by tying together the edges of one or more adjacent leaves and rest concealed within them. When disturbed, the caterpillars drop to the ground and curl up, sometimes emitting clear fluid from the mandibles as a defensive ploy. The species overwinters as a pupa within a cocoon spun in leaf litter.

The adults are nocturnal and their flight period extends from early June to September. The species is generally single-brooded but may have a partial second brood in the south. Flies (tachinids) frequently parasitize this caterpillar, with maggots developing internally. The Tufted Thyatirin is the only member of its genus, *Pseudothyatira*.

Actual size

The Tufted Thyatirin caterpillar is yellow to orange brown with fine, reticulate mottling dorsally and dark transverse lines. The overall appearance is wrinkly, and the thorax is swollen. There is usually a white spot over the spiracle on abdominal segment one. The head is orange with fine, paler markings.

FAMILY	Drepanidae
DISTRIBUTION	From western Europe and North Africa (Algeria) across temperate Asia, east to Japan, and south to Borneo and Sumatra
HABITAT	Forests, scrub, hedgerows, and gardens
HOST PLANTS	Bramble (*Rubus fruticosus*), Dewberry (*Rubus caesius*), and Raspberry (*Rubus idaeus*)
NOTE	Seldom seen caterpillar of a familiar moth
CONSERVATION STATUS	Not evaluated, but common throughout most of its range

ADULT WINGSPAN
1⅜–1¾ in (35–44 mm)

CATERPILLAR LENGTH
1⅛–1⁵⁄₁₆ in (28–33 mm)

THYATIRA BATIS
PEACH BLOSSOM
(LINNAEUS, 1758)

336

Actual size

Peach Blossom caterpillars hatch from eggs laid either singly or in small groups on the leaves of its host plant. When young, the larva resembles a bird dropping and, so disguised, lives on the upper side of the leaves. When the caterpillar is larger, the markings are different and it can look more like a piece of dead leaf. At this point, the larva hides among leaf litter by day and only ascends the food plant at night, feeding from the leaf edge. It sometimes rests with the hind end raised. Pupation occurs in the ground in a silken cocoon, and this is the overwintering stage.

The species is named for the petal-like markings of the adults, which fly in one or two broods from April to September, depending on climate. It is one of a group of moths known as lutestrings for the lines on their forewings; their subfamily, Thyatirinae, is sometimes classified as a distinct family. Unlike that of the hook-tips, the hind end of these caterpillars is not tapered, or usually raised, and has a pair of normal prolegs. Many species live between spun leaves.

The Peach Blossom caterpillar is partly whitish at the front end when small, becoming dark to pale brown or greenish as it develops. The head is notched with two double, raised bumps behind it, the second pair being larger. It has five ridged peaks along the back, the darker slopes of which create a pale diamond pattern, and one ridge at the hind end.

FAMILY	Drepanidae
DISTRIBUTION	From Europe to the Urals, Asia Minor, and the Caspian Sea
HABITAT	Forests, hedgerows, parks, and gardens
HOST PLANTS	Oak, including Pedunculate Oak (*Quercus robur*), Sessile Oak (*Quercus petraea*), and Downy Oak (*Quercus pubescens*)
NOTE	Caterpillar that adopts a characteristic hook-tip, arched pose
CONSERVATION STATUS	Not evaluated, but common throughout most of its range

ADULT WINGSPAN
⅞–1⅜ in (22–35 mm)

CATERPILLAR LENGTH
⅞–1 in (22–25 mm)

WATSONALLA BINARIA
OAK HOOK-TIP
(HUFNAGEL, 1767)

337

The Oak Hook-tip caterpillar ecloses from oval eggs that are initially green but turn red before hatching and are laid on the leaf edge. It lives openly and rests with its head and its tail end raised in the usual hook-tip manner, often with the head held high, the front end arched, and the tail spike pointing upward. The caterpillars are present from June to October. The pale brown pupa, which has a waxy bloom, is formed in a whitish, mesh-like cocoon within a folded leaf or leaves drawn together and overwinters there.

The species has two broods annually, with adults flying in May and June—mainly at night, although males also fly by day. Several related species have caterpillars similar to the Oak Hook-tip. The Spiny Hook-tip caterpillar (*Watsonalla uncinula*), which occurs in southern Europe where the two species may overlap, is almost identical, and the adults are so similar that reliable identification is only possible by examination of the genitalia.

Actual size

The Oak Hook-tip caterpillar is light brown, light orange brown, or darker brown, with a double-pointed tubercle on the back near the peaked, swollen front end. It has a paler, often whitish but sometimes yellowish, elongated saddle-mark along the back, which extends as two whitish, sometimes dark-edged diverging lines toward both the head and hind end, which is formed into a spike.

FAMILY	Drepanidae
DISTRIBUTION	United Kingdom and western Europe, east to the Urals and south to the Caspian Sea
HABITAT	Forests, hedgerows, parklands, and gardens
HOST PLANTS	European Beech (*Fagus sylvatica*)
NOTE	Hook-tip caterpillar that is found on beech trees
CONSERVATION STATUS	Not evaluated, but probably common throughout most of its range

ADULT WINGSPAN
⅞–1⅜ in (22–35 mm)

CATERPILLAR LENGTH
¾–⅞ in (19–22 mm)

338

WATSONALLA CULTRARIA
BARRED HOOK-TIP
(FABRICIUS, 1775)

Actual size

The Barred Hook-tip caterpillar hatches on the leaf of its host plant from oval eggs that are yellowish green when first laid but later turn reddish. Like other hook-tips, it lives openly, being well disguised as a piece of dead leaf or bird dropping. The caterpillars are present throughout their range between June and October. The blue-gray pupa, which has a waxy bloom, is formed in a pale, quite dense cocoon within a folded leaf or leaves drawn together and overwinters there.

The caterpillar of this species is very similar to that of the Oak Hook-tip (*Watsonalla binaria*), but it is slightly slimmer and more inclined to be reddish in color (although there are several different color forms), and the double-pointed tubercle on the back of the peaked, swollen front section is smaller. The geographical ranges of the two species are also much the same. However, as their food plants are different, the two are unlikely to be confused in the wild.

The Barred Hook-tip caterpillar is light reddish brown, light orange brown, or darker brown, with a double-pointed tubercle on its back on the peaked, swollen front section. It has a paler, often whitish, elongated saddle-mark along the back, which extends as two whitish, diverging lines toward both the head and hind end, which is formed into a spike.

FAMILY	Lasiocampidae
DISTRIBUTION	European Alps, Scandinavia
HABITAT	Upland coniferous forests and moors, and scree slopes with low-growing shrubs
HOST PLANTS	Various, including alder (*Alnus* spp.), birch (*Betula* spp.), *Vaccinium* spp., and willow (*Salix* spp.)
NOTE	Brown, hairy caterpillar of limited distribution
CONSERVATION STATUS	Not evaluated

ADULT WINGSPAN
1¼–1¹¹⁄₁₆ in (31–43 mm)

CATERPILLAR LENGTH
¾–⅞ in (20–22 mm)

ERIOGASTER ARBUSCULAE
DWARF BIRCH SPINNER
FREYER, 1849

339

Dwarf Birch Spinner caterpillars hatch from eggs laid by the female moth in large clusters on the host plants. The larvae stay together, spinning a communal silken tent over the leaves and stems of the plant during June and July. They feed in the safety of the web, but on sunny days they can be seen basking in the sun. After their final molt, the caterpillars become solitary. To pupate, they drop to the ground and spin a tough, pale yellow cocoon that lies in the leaf litter or just beneath the soil surface. They overwinter in their cocoons.

The adult moths eclose and are on the wing in May and June, usually the following year with a single generation. If the conditions are unfavorable, however, it is not unusual for the species to overwinter for more than one year to increase its chances of survival. While the silken webs of the caterpillars are a relatively common sight within their range, the night-flying moths are rarely observed.

Actual size

The Dwarf Birch Spinner caterpillar is dark brown to black in color and covered in hairs of varying length, some white and others orange brown. There is a dorsal row of pale yellow spots and a lateral line of smaller, creamy spots. The head is black and the prolegs are brown.

FAMILY	Lasiocampidae
DISTRIBUTION	Parts of southern Europe and central Europe, from Spain to the Balkans, and into southern Russia
HABITAT	Limestone grasslands, hedgerows, and forest margins
HOST PLANTS	Various trees and shrubs, especially Blackthorn (*Prunus spinosa*); also on birch (*Betula* spp.), hawthorn (*Crateagus* spp.), poplar (*Populus* spp.), and oak (*Quercus* spp.)
NOTE	Hairy caterpillar that lives with others on a silken web
CONSERVATION STATUS	Data deficient, formerly endangered

ADULT WINGSPAN
1¹⁄₁₆–1⅜ in (27–35 mm)

CATERPILLAR LENGTH
2–2⅛ in (50–55 mm)

340

ERIOGASTER CATAX
EASTERN EGGAR
(LINNAEUS, 1758)

The caterpillars of the Eastern Eggar moth hatch from eggs laid in batches of around 150 to 200 on branches and then covered in gray hairs from the adult female's abdomen. The larvae overwinter and hatch the following spring, living together gregariously on a silken, gray web, jerking their heads when threatened to deter would-be predators. Older caterpillars move away, become solitary, and pupate in a cocoon.

The moths are on the wing from September until November, and there is a single generation. The species was once classed as endangered but is no longer at risk across the eastern part of its range, although still rare elsewhere. The adults are seldom seen, but the caterpillars are easily spotted. Much of the threat to *Eriogaster catax* comes from the loss of its habitat through agricultural intensification and clearance of hedgerows, plus the use of insecticides on trees with an infestation of hairy caterpillars. It also suffers from high levels of parasitism.

The Eastern Eggar caterpillar is very hairy. Tufts of long, black, and orange-brown hairs run the length of the dorsal surface, from the head to the end of the abdomen, together with a broken white line. The sides are covered in tufts of gray and orange-brown hairs.

Actual size

FAMILY	Lasiocampidae
DISTRIBUTION	Across Europe, extending east across Asia to China's eastern coast
HABITAT	Hedgerows
HOST PLANTS	Deciduous trees, including Blackthorn (*Prunus spinosa*), hawthorn (*Crataegus* spp.), birch (*Betula* spp.), and willow (*Salix* spp.)
NOTE	Social caterpillar that is covered in tufts of long hairs
CONSERVATION STATUS	Not evaluated, but numbers are declining

ADULT WINGSPAN
1³⁄₁₆–1⁹⁄₁₆ in (30–40 mm)

CATERPILLAR LENGTH
2 in (50 mm)

ERIOGASTER LANESTRIS
SMALL EGGAR
(LINNAEUS, 1758)

Small Eggar caterpillars hatch from eggs laid near the tip of branches, and they can be seen from late spring to mid summer. The larvae are gregarious, living together in a large, tent-shaped, silken web, which they spin between branches of a host tree. They leave the web to forage, following trails marked by returning caterpillars. On sunny days the caterpillars bask on the surface of the web. The largest webs are home to hundreds of caterpillars, and their weight can cause the branches to bend downward.

The full-grown caterpillars leave the web and crawl down the tree to find a place to pupate. They spin a brown cocoon, in which they remain for up to a year before emerging. Studies have found that some remain in the pupal stage for as long as ten years. The adult moths are on the wing in early spring, although the females are weak fliers and do not travel far.

The Small Eggar caterpillar is dark brown with two distinctive lines of tufts along its back made up of white, yellow, and orange hairs sprouting from a ginger base. Along its sides, it has more long hairs, and also white dots and lines, which create a series of U-shaped marks.

Actual size

FAMILY	Lasiocampidae
DISTRIBUTION	From United Kingdom and western Europe to Russia, Siberia, and Japan
HABITAT	Mostly damp places, including marshes, fens, open areas in forests, and on moorlands
HOST PLANTS	Coarse grasses (Poaceae), sedges (*Carex* spp.), and Common Reed (*Phragmites australis*)
NOTE	Striking, large, hairy caterpillar that is fond of drinking dew
CONSERVATION STATUS	Not evaluated, but probably common within most of its range

ADULT WINGSPAN
2–2¾ in (50–70 mm)

CATERPILLAR LENGTH
2⅜–2¾ in (60–70 mm)

EUTHRIX POTATORIA
DRINKER
(LINNAEUS, 1758)

342

The Drinker caterpillar lives in tall, mainly damp grassland and is named for its habit of drinking drops of dew or rainwater. The whitish, gray-marked eggs are laid in small groups attached to grass stems. The caterpillar, while small, hibernates near the ground in dense grasses and feeds, mainly at night, the following spring. It is often seen resting on grasses by day, and contact should be avoided as the hairs can cause skin rashes. The caterpillar pupates in a yellowish, papery, tapered cocoon low down among grass stems.

The large, furry, chestnut-brown or orange-brown adults fly from June to August in a single brood. The Lasiocampidae (often known as eggars and lappets) are medium-sized to large moths found in most parts of the world. The caterpillars all have a dense covering of short hairs and tufts of longer hairs. Many are conspicuous by day but, due to the hairs, are not eaten by most birds, except cuckoos.

Actual size

The Drinker caterpillar is dark gray and intricately patterned, with a distinctive long, slender, black or brown hair tuft near the front and back end, irregular yellow lines along the back and sides, and white and rusty hair tufts along the lower sides. It also has extensive short, rusty, and black hair tufts and longer, pale hairs.

FAMILY	Lasiocampidae
DISTRIBUTION	Europe to eastern Siberia, Russian Far East, China, and Japan
HABITAT	Forests, woodland edges, scrub, and hedgerows
HOST PLANTS	Blackthorn (*Prunus spinosa*), hawthorn (*Crataegus* spp.), buckthorn (*Rhamnus* spp.), oak (*Quercus* spp.), willow (*Salix* spp.), and other broadleaved trees
NOTE	Large, hairy caterpillar that rests camouflaged along a stem
CONSERVATION STATUS	Not evaluated, but declining in parts of Europe, including the United Kingdom

ADULT WINGSPAN
2⅜–3½ in (60–90 mm)

CATERPILLAR LENGTH
3–3½ in (75–90 mm)

GASTROPACHA QUERCIFOLIA
LAPPET
(LINNAEUS, 1758)

343

The Lappet caterpillar hatches in late summer from a white, green-marbled, oval egg, laid in small groups attached to twigs. It hibernates while still small, uncovered but well concealed on or often underneath small stems, low down close to the ground. In the spring, the larva ascends the bush to feed at night and is fully fed in May or June. The dark gray pupa is formed inside a tough, gray-brown cocoon attached low down on the food plant.

The species' common name comes from the hairy protuberances, or lappets, on the lower sides of the caterpillar. These, along with the color and shape, make the larva difficult to detect when it rests along a stem. The caterpillars of several related species are similar in form. The large, furry adults fly in a single brood from late June to August. The decline of this species in Europe is in part due to loss of suitable scrub and hedgerow habitats.

The Lappet caterpillar is dark gray to reddish brown, sometimes variegated white. It has blue-black bands between the frontal segments and sometimes pairs of orange spots along the back. The underside is flattened, and there are a series of fleshy protuberances along the sides, with long, downward-pointing hairs. It has a backward-pointing protuberance, resembling a small bud, on the back near the hind end.

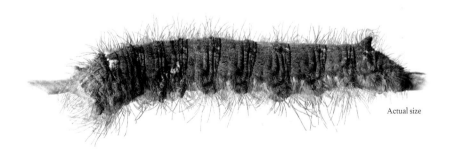

Actual size

FAMILY	Lasiocampidae
DISTRIBUTION	Europe, Asia Minor, and east to northeast Asia
HABITAT	Open places with low scrub, including mountain moorlands, lowland acid heathlands, downlands, coastal dunes, woodland edges, and hedgerows
HOST PLANTS	Mainly on shrubs, including Bilberry (*Vaccinium myrtillus*), Heather (*Calluna vulgaris*), heath (*Erica* spp.), Bramble (*Rubus fruticosus*), and willow (*Salix* spp.); also herbaceous plants and rushes (*Juncus* spp)
NOTE	Furry caterpillar that often basks and feeds in sunshine
CONSERVATION STATUS	Not evaluated, but common throughout most of its range

ADULT WINGSPAN
2⅜–3½ in (60–90 mm)

CATERPILLAR LENGTH
2⁹⁄₁₆–3⅛ in (65–80 mm)

LASIOCAMPA QUERCUS
OAK EGGAR
(LINNAEUS, 1758)

344

The Oak Eggar caterpillar is brown when young, sometimes partly blue gray, with transverse, triangular, orange marks along the back and sometimes longitudinal, white spots. When larger, it is blackish gray, with straw-colored or brown hairs on the back in broad bands alternating with narrower black bands. Rusty-brown hairs predominate along the sides, which have a fine white line.

The life cycle of the Oak Eggar caterpillar varies according to climate. In warmer regions, the adults fly and mate in July and August. Eggs are laid and larvae hatch soon after, overwintering when well developed and pupating the following spring in a tough, compact acorn-shaped cocoon. In cooler climates, the adults fly in May and June, and the larvae hatch in July, with their first winter spent as a small, early instar caterpillar and the second as a pupa. In both warm and cooler climates, the light brown eggs are dropped by the female as she flies low at dusk over suitable habitat.

Male Oak Eggars fly by day, searching for the nocturnal females, which release their sex pheromone from a resting place in the vegetation. This moth is not associated with oak but is thought to have been named after the acorn-like cocoons, with "Eggar" having been adopted for related species with similar, somewhat egg-shaped cocoons. Populations with a two-year cycle are often known as Northern Eggar—the subspecies *Lasiocampa quercus callunae*.

Actual size

FAMILY	Lasiocampidae
DISTRIBUTION	Europe and North Africa, east to Central Asia
HABITAT	Woodlands, heathlands, and sand dunes
HOST PLANTS	Various, including Heather (*Calluna vulgaris*), trefoil (*Trifolium* spp.), grasses, and trees such as oak (*Quercus* spp.) and poplar (*Populus* spp.)
NOTE	Hairy caterpillar found on a variety of food plants
CONSERVATION STATUS	Not evaluated, but locally scarce

ADULT WINGSPAN
1⁹⁄₁₆–2⅛ in (40–55 mm)

CATERPILLAR LENGTH
2–2⅜ in (50–60 mm)

LASIOCAMPA TRIFOLII
GRASS EGGAR
DENIS & SCHIFFERMÜLLER, 1775

345

Female Grass Eggar moths lay their gray-brown, oval eggs in a cluster, usually on the underside of leaves. The eggs overwinter and hatch the following spring. The caterpillars feed on a range of plants by night and seek shelter during the day. When mature, the larvae crawl down to the ground to pupate, often burrowing under leaf litter or into the soil. The adults emerge a few weeks later. The night-flying moths are active through summer, from June to September, with a single generation.

The caterpillars are covered in urticating hairs, which help deter predators. The larvae also incorporate the protective hairs as they spin their silk to create their brown cocoon. The emergent female moth is much larger than the male. This species is endangered in parts of its range as a result of loss of habitat and changes in farming practices. The species is sometimes classified as *Pachygastria trifolii*.

The Grass Eggar caterpillar is covered in dense tufts of caramel-brown and cream-colored hairs. Underneath, the body is dark brown to black with several rows of broken white lines. The head is brown with a central white band.

Actual size

FAMILY	Lasiocampidae
DISTRIBUTION	Western Europe, the Middle East, and east to central and northeast Asia
HABITAT	Acid moorlands, lowland heaths, calcareous grasslands, sand dunes, and open woodlands
HOST PLANTS	Many shrubs, including Bramble (*Rubus fruticosus*), Heather (*Calluna vulgaris*), Bilberry (*Vaccinium myrtillus*), Sea-buckthorn (*Hippophae rhamnoides*), and willow (*Salix* spp.); also herbaceous plants, such as clovers (*Trifolium* spp.)
NOTE	Large, very furry caterpillar that basks in spring sunshine
CONSERVATION STATUS	Not evaluated

ADULT WINGSPAN
2⅛–3⅛ in (55–80 mm)

CATERPILLAR LENGTH
2⅜–2¾ in (60–70 mm)

346

MACROTHYLACIA RUBI
FOX MOTH
(LINNAEUS, 1758)

Fox Moth caterpillars hatch in late June or July from light gray, brown-marbled eggs laid in clusters, often on plant stems, fences, or rocks. The larva feeds until fall and hibernates fully fed, on or slightly under the ground, unprotected but curled up under dense moss or other litter. It emerges briefly in spring and can then be found basking in sunshine, before it pupates low down in the vegetation in a thin, elongated, grayish cocoon.

This caterpillar is most often found in fairly extensive areas of open, rough, agriculturally unimproved mosaics of grassland and scrub rather than in shady forests, farmland, or urban habitats. The common name relates to the color of the hairs on the back of the caterpillar, and also the color of the adult moths, especially the male. The adults fly in May and June in a single brood, the males by day and the females at dusk.

The Fox Moth caterpillar has a black body. The narrow, yellow rings present at the early instars are lost at the final molt, when the back becomes densely covered with tawny-brown hair, interrupted by black bands. The body is also covered with long, brown, or blackish hairs, with tufts of whitish hairs along the sides.

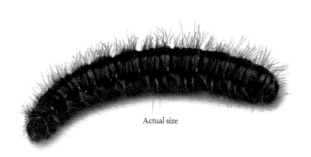

Actual size

FAMILY	Lasiocampidae
DISTRIBUTION	Across Europe, North Africa, and Asia (except for the far north and south)
HABITAT	Salt marsh, marshes, heaths, and woodlands
HOST PLANTS	Various, including Cypress Spurge (*Euphorbia cyparissias*), Golden Samphire (*Inula crithmoides*), Salad Burnet (*Sanguisorba minor*), and Common Sea-lavender (*Limonium vulgare*)
NOTE	Gregarious tent caterpillar that spins a silken web
CONSERVATION STATUS	Not evaluated, and although widespread may be locally scarce

ADULT WINGSPAN
1¼–1⅝ in (31–41 mm)

CATERPILLAR LENGTH
1⁹⁄₁₆–2 in (40–50 mm)

MALACOSOMA CASTRENSIS
GROUND LACKEY
(LINNAEUS, 1758)

347

Ground Lackey caterpillars emerge in spring and early summer from overwintering eggs laid in a ring around the stem of a host plant and fixed so that they remain firmly in place despite—in the case of salt marsh plants—inundations from the sea. The caterpillars feed on almost any plant and mass together as they forage, leaving a silken trail back to their web. Their long, urticating hairs deter predators. Once mature, the caterpillars disperse and pupate in grass in a slightly transparent cocoon.

The moth ecloses and flies in late summer. The Ground Lackey is one of the tent caterpillar species, so named because the larvae stay together and spin a silken web near the ground, often extending across a number of plants. On sunny days, they bask on the surface of their web, retreating inside when skies are overcast. In bumper years, caterpillar numbers are so great that they can overrun their habitat in their search for food.

The Ground Lackey caterpillar is brown black in color. Along the back are four broken, reddish lines and a central, blue line, with a broader lateral, blue stripe and, below this, black speckles. The body is covered with long, yellow-brown hairs. The head is gray black with no spots.

Actual size

FAMILY	Lasiocampidae
DISTRIBUTION	Across Canada and United States
HABITAT	Deciduous woodlands
HOST PLANTS	Deciduous trees, including oak (*Quercus* spp.), Maple (*Acer saccharum*), aspen (*Populus* spp.), and Paper Birch (*Betula papyrifera*)
NOTE	Widely distributed tent caterpillar that causes defoliation of hardwood trees
CONSERVATION STATUS	Not evaluated, but common

ADULT WINGSPAN
1–1¼ in (25–45 mm)

CATERPILLAR LENGTH
2–2⁹⁄₁₆ in (50–65 mm)

MALACOSOMA DISSTRIA
FOREST TENT CATERPILLAR
HÜBNER, 1820

348

Unlike other tent caterpillars, the Forest Tent caterpillar does not spin a tent. Instead, the gregarious caterpillars, which hatch in late winter from overwintering eggs laid in a band around a twig, spin silken sheets on trunks and branches. Here they gather to rest and molt, basking together to increase their body temperature, which accelerates development. While young, they feed together, too, traveling en masse along strands of silk from branch to branch, following a scent trail laid by returning caterpillars. Older caterpillars wander in search of food and a place to pupate.

The pale yellow cocoons of the Forest Tent caterpillar can often be found among leaves and in cracks in bark. Pupation takes around two weeks; the adult moth emerges and flies in summer. Because of the damage they cause hardwood trees, the caterpillars are an economic pest. In some years, outbreaks result in widespread defoliation.

The Forest Tent caterpillar is dark brown to black, with faint blue and yellow stripes running the length of the body. Each abdominal segment bears a dorsal, white, keyhole-shaped spot. Fine white hairs are distributed across the body.

Actual size

FAMILY	Lasiocampidae
DISTRIBUTION	Europe, North Africa, and across Asia (except for the far north and south) to Japan
HABITAT	Deciduous forests up to 5,250 ft (1,600 m) elevation, hedgerows, verges, grasslands, and farmlands
HOST PLANTS	Deciduous trees, including apple (*Malus* spp.), hawthorn (*Crataegus* spp.), willow (*Salix* spp.), and oak (*Quercus* spp.)
NOTE	Gregarious caterpillar that feeds within a vast silken web
CONSERVATION STATUS	Not evaluated, but decreasing locally

ADULT WINGSPAN
1–1⅜ in (25–35 mm)

CATERPILLAR LENGTH
1¾–2 in (45–50 mm)

MALACOSOMA NEUSTRIA

LACKEY MOTH

(LINNAEUS, 1758)

349

Lackey Moth caterpillars hatch en masse in spring from a ring of overwintering eggs laid around twigs. The gregarious larvae live together in a silken web, spun across branches of the host tree. The web helps them regulate their temperature; on sunny days the larvae bask on the surface to raise their body temperature so they can become active. It also gives them protection and deters natural enemies such as parasitic wasps and birds. Mature caterpillars disperse into the surrounding vegetation in search of food and a place to pupate.

The sparse cocoons can be found among leaves under trees where the larvae have fed. Lackey Moth caterpillars are voracious eaters, feeding on leaves of many different deciduous tree species as well as fruit trees, causing localized defoliation. Population explosions can occur, and the caterpillars are considered a pest, especially of orchards. The adult moths are on the wing in summer.

The Lackey Moth caterpillar is brown with blue, orange, and white stripes down the back. The body is covered with long, orange-brown hairs. The head is gray with dark spots.

Actual size

FAMILY	Lasiocampidae
DISTRIBUTION	Europe, Asia Minor, Russia, China, and Japan
HABITAT	Woodlands, orchards, and forests at low and high elevations up to 9,850 ft (3,000 m)
HOST PLANTS	Plum (*Prunus* spp.), pear (*Pyrus* spp.), hawthorn (*Crataegus* spp.), willow (*Salix* spp.), elm (*Ulmus* spp.), and oak (*Quercus* spp.)
NOTE	Extremely cryptic caterpillar that is sometimes found on fruit trees
CONSERVATION STATUS	Not evaluated

ADULT WINGSPAN
1³⁄₁₆–2³⁄₈ in (30–60 mm)

CATERPILLAR LENGTH
1¾–2⅛ in (45–55 mm)

ODONESTIS PRUNI
PLUM LAPPET
(LINNAEUS, 1758)

350

Plum Lappet caterpillars hatch in late summer or early fall from spherical white eggs, laid singly or in small groups on the leaves of host plants. The larvae develop slowly through five instars and overwinter as mid-instars, completing development the following spring. In most areas there are two generations, although there may only be one in northern ranges. The caterpillars are usually solitary, gaining protection from remarkable camouflage coloring that renders them hard to spot on twigs and branches. When mature, they spin a dense silken cocoon, from which the adult moth emerges within two to three weeks.

The *Odonestis pruni* caterpillar has been reported as an economic pest of fruit trees in parts of Europe, particularly on cherries and plums, as reflected in the first part of the common name. The second part of the name was inspired by the flaps covering the caterpillar's prolegs, like a fold, or "lappet," in a garment.

Actual size

The Plum Lappet caterpillar is light to dark brown with intricate thin, wavy, yellow lines, especially dorsally. On each of the six middle segments there is, posteriorly, a pair of indistinct white triangles. Segment three has a red transverse line bordered by yellow orange and a pair of white patches posteriorly. The body is covered with fine white setae, giving a slightly fuzzy appearance.

FAMILY	Apatelodidae
DISTRIBUTION	North America, from southern Canada to Texas and Florida
HABITAT	Deciduous forests
HOST PLANTS	Trees, such as maple (*Acer* spp.), ash (*Fraxinus* spp.), cherry (*Prunus* spp.), and oak (*Quercus* spp.)
NOTE	Conspicuous caterpillar that is covered in long, fluffy hairs
CONSERVATION STATUS	Not evaluated

ADULT WINGSPAN
1¼–1⅝ in (32–42 mm)

CATERPILLAR LENGTH
2–2⅜ in (50–60 mm)

APATELODES TORREFACTA
SPOTTED APATELODES
(SMITH, 1797)

351

The distinctive Spotted Apatelodes caterpillar is conspicuous for its shaggy covering of long, fluffy hairs. In some species variants the hairs are pure white, while in others they are gray-white or yellow. These hairs, unlike those of many other hairy caterpillars, are not irritants. Instead, research suggests that they are a result of mimicry, allowing the caterpillar to resemble more harmful larvae that do use irritant hairs as a defense. This is known as Batesian mimicry, reflecting the butterfly studies carried out in the nineteenth century by the English naturalist and explorer Henry Walter Bates (1825–92) in the rain forests of Brazil.

The caterpillars leave the food plant and pupate in the ground in a silken cocoon, where they overwinter. The adult moths are on the wing from May to August. There are usually two generations in the south of its range, but only one in the north. Members of the Apatelodidae family are sometimes known as American silkworm moths and are found only in the Americas—mostly in Neotropical regions.

The Spotted Apatelodes caterpillar is covered with either white, gray-white, or yellow hairs. Under the hairs, the body is creamy white with large, dorsal, black spots and smaller lateral black spots. There is a row of black, middorsal tufts, and two long, black pencils of hair arise from the second and third thoracic segments and the eighth abdominal segment.

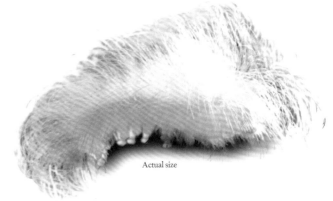

Actual size

FAMILY	Brahmaeidae
DISTRIBUTION	Russian Far East, Mongolia, China, and Korean peninsula
HABITAT	Forests and woodlands
HOST PLANTS	Privet (*Ligustrum* spp.), ash (*Fraxinus* spp.), and lilac (*Syringa* spp.)
NOTE	Spectacularly tentacled caterpillar
CONSERVATION STATUS	Not evaluated

ADULT WINGSPAN
4⅝–6 in (120–150 mm)

CATERPILLAR LENGTH
2¾–3⅛ in (70–80 mm)

BRAHMAEA CERTHIA
SINO-KOREAN OWL MOTH
(FABRICIUS, 1793)

352

The Sino-Korean Owl Moth caterpillar is
black dorsally with black tentacles and spines.
Laterally, each segment is light orangey brown
tending to white anteriorly. The prolegs and true
legs are black. The head is black and white and
appears skull-like viewed from the front.

Sino-Korean Owl Moth caterpillars emerge from large, creamy-white spherical eggs laid in clusters on host plant stems and trunks some seven to ten days earlier. First instar caterpillars, already with tentacles, hatch by eating a hole through the eggshell. The larvae grow rapidly, eating leaves from the edges and consuming a great amount of vegetation; they take less than three weeks to complete their development. The caterpillars are gregarious at first but then become solitary. Final instars lose their tentacles and head toward the ground, where they pupate in an earthen cell. The pupa is dark brown to black and overwinters.

Adult moths emerge in spring and are nocturnal, mostly active during the first part of the night. They live only a week or two and are well camouflaged when resting by day on tree trunks. *Brahmaea certhia* is the type species of the genus *Brahmaea*, that is the species on which the genus is based; the Brahmaeidae was only separated from the Bombycidae family in the early 1990s.

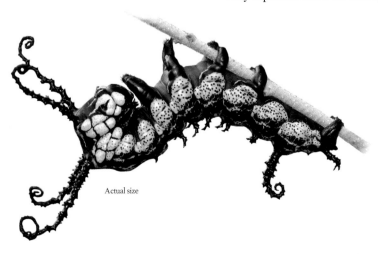

Actual size

FAMILY	Brahmaeidae
DISTRIBUTION	Japan
HABITAT	Humid deciduous forests
HOST PLANTS	Ash (*Fraxinus* spp.), privet (*Ligustrum* spp.), and *Osmanthus* spp.
NOTE	Large, brightly colored caterpillar that is found only in Japan
CONSERVATION STATUS	Not evaluated

ADULT WINGSPAN
3⅛–4½ in (80–115 mm)

CATERPILLAR LENGTH
2¾ in (70 mm)

BRAHMAEA JAPONICA

JAPANESE OWL MOTH
(BUTLER, 1873)

353

Japanese Owl Moth caterpillars hatch from smooth, brown, dome-shaped eggs laid in small clusters on the leaves of the host plant. The young and mature caterpillar look very different, with the young caterpillar white in color with markings of black and yellow and sporting four long, black thoracic filaments and three anal filaments. These filaments grow longer with each molt but are then replaced by small tubercles in the final molt. The mature caterpillars move to the ground, where they pupate under logs or stones and overwinter. The pupa is black.

The night-flying adult moths rest on tree trunks during the day and are seen in March to April, when they are attracted to light. They get their name from the resemblance of their upperside wing markings to the face of an owl. There is some confusion over the species' classification, as a number of authorities describe the moth as a subspecies of *Brahmaea wallichii*.

The Japanese Owl Moth caterpillar is large and brightly colored. The body is creamy white with a lateral black stripe between two broken yellow stripes and scattered markings in black and yellow. The head is black with yellow marks and white marks. There are small tubercles on the thorax and at the tail end of the abdomen.

Actual size

FAMILY	Brahmaeidae
DISTRIBUTION	Southern Siberia and Russian Far East, Mongolia, China, Korea, and south to Indonesia
HABITAT	Forests
HOST PLANTS	Privet (*Ligustrum* spp.)
NOTE	Striking caterpillar that displays false eyes to scare off predators
CONSERVATION STATUS	Not evaluated

ADULT WINGSPAN
4–4⅞ in (100–120 mm)

CATERPILLAR LENGTH
3½–4 in (90–100 mm)

354

BRAHMAEA TANCREI
SIBERIAN OWL MOTH
AUSTAUT, 1896

Siberian Owl Moth caterpillars hatch from round, white eggs laid on the underside of leaves, eating their eggshells before feeding on the leaves. They have four long, black filaments on their thorax and a further three at the tail end of their abdomen. These filaments are lost after the third molt and replaced by false eyespots on the thorax, which normally remain hidden. The species does not spin a cocoon but pupates underground or in the leaf litter, where it overwinters.

The adult moths are on the wing in April, and there is just a single generation each year. The fast-growing caterpillars are active from May to June and, when threatened, they rear up and curl their head down to reveal their eyespots, waving their body from side to side to deter the predator. To further put off predators, they may also make a squeaky noise.

The Siberian Owl Moth caterpillar is large with striking markings. The body is brown with caramel and black oblique stripes along the sides. The abdomen is covered with many small, white dots, some ringed in brown. The spiracles are ringed in white. The head and thorax are brown and black, while the legs and prolegs are black.

Actual size

FAMILY	Brahmaeidae
DISTRIBUTION	Nepal, Bhutan, southwest China, northern India, northern Myanmar, Thailand, and Japan
HABITAT	Temperate and tropical forests
HOST PLANTS	Ash (*Fraxinus* spp.) and privet (*Ligustrum* spp.)
NOTE	Large and distinctive caterpillar that rears up when threatened
CONSERVATION STATUS	Not evaluated

ADULT WINGSPAN
3½–6⅜ in (90–160 mm)

CATERPILLAR LENGTH
3½ in (90 mm)

BRAHMAEA WALLICHII
OWL MOTH
GRAY, 1831

355

The caterpillars of the Owl Moth hatch from large, creamy-white eggs between 10 and 14 days after they are laid on the underside of leaves of the host plant. The young larvae have four long, black filaments on the thorax and three on the tail of the abdomen. These all disappear after the third molt. When threatened, a caterpillar rears up into a threatening posture to deter would-be predators. Pupation takes place when the caterpillars move to the ground, creeping under stones or fallen leaves and into other moist places. The black pupa overwinters and ecloses in spring.

Brahmaea wallichii is one of the largest of the owl moths. A night flyer, it rests during the day on tree trunks or on the ground, its wing pattern of dark and light browns providing perfect camouflage. Two generations of moths appear each year, the first in April to May and the second in August.

The Owl Moth caterpillar has eye-catching coloration. The body is white with lateral stripes of black and yellow and spiracles ringed in black and white. The head and thorax have a reticulate pattern of yellow, green, and black with white spots. The legs and prolegs are black and blue.

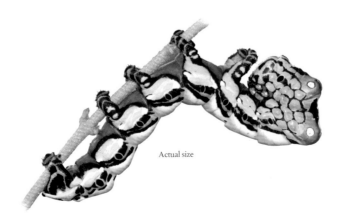

Actual size

FAMILY	Brahmaeidae
DISTRIBUTION	Europe, Asia Minor, and Russia to the Ural Mountains
HABITAT	Farmlands, orchards, grasslands, wet meadows, and fens
HOST PLANTS	Yarrow (*Achillea* spp.), hawkweed (*Hieracium* spp.), and dandelion (*Taraxacum* spp.)
NOTE	Hairy caterpillar of a day-flying moth
CONSERVATION STATUS	Not evaluated, but locally vulnerable

ADULT WINGSPAN
1¾–2⁹⁄₁₆ in (45–65 mm)

CATERPILLAR LENGTH
1⁹⁄₁₆ in (40 mm)

356

LEMONIA DUMI
LEMONIA DUMI
(LINNAEUS, 1761)

Female *Lemonia dumi* moths lay their brown eggs in clusters around the stems of the food plant. The eggs overwinter and the caterpillars hatch in early summer. The larvae feed on a range of food plants belonging to the Asteraceae family, including weeds such as dandelion. After the final instar, they move to the ground to pupate, crawling into underground chambers. The day-flying moths emerge late in the year, and are seen on warm days in October and November.

This species has suffered a sharp decline in recent decades due to loss of its habitat through farm intensification, the use of artificial fertilizers, land drainage, and the loss of traditional orchards. The genus *Lemonia* comprises about 12 species, found in Europe and temperate parts of Asia. Once considered to be in the family Lemoniidae, it is now, thanks to recent DNA analysis, placed in the family Brahmaeidae.

The *Lemonia dumi* caterpillar is dark brown black in color and covered in short, orange-brown tufts of long hairs. It also has pairs of pale white and black marks along the back.

Actual size

FAMILY	Anthelidae
DISTRIBUTION	East coast of Australia, Tasmania
HABITAT	Woodlands, scrub, and coastal scrub
HOST PLANTS	Wattle (*Acacia* spp.)
NOTE	Slender and hairy caterpillar that has stinging hairs
CONSERVATION STATUS	Not evaluated

ADULT WINGSPAN
1³⁄₁₆–1⁹⁄₁₆ in (30–40 mm)

CATERPILLAR LENGTH
2 in (50 mm)

NATAXA FLAVESCENS

YELLOW-HEADED ANTHELID
(WALKER, 1855)

The female Yellow-headed Anthelid moth lays her creamy, round eggs in a line along the edge of a leaf or stem of the host plant. The eggs overwinter, and the caterpillars hatch in spring. The fast-moving larva is often found on the trunk of the host plant, and its hairs can irritate if a person or a predator picks it up. The mature caterpillar spins a cocoon that it positions in a crevice or under bark, and the adult moth ecloses a few weeks later.

Yellow-headed Anthelid moths are nocturnal and attracted to light, and are on the wing in late summer and fall. The sexes look very different—the female is larger with dark gray and white wings, and the male is smaller with orange, brown, and cream wings. Anthelidae is a family of lappet moths found only in Australia and New Guinea. There are 74 named species, and the caterpillars are typically hairy.

The Yellow-headed Anthelid caterpillar is slender and gray, with a black dorsal stripe covered in gray hairs and edged in white and yellow. There are two tufts of black hairs on the thorax and another at the tail end, red tubercles behind the head, and tufts of long, gray hairs the length of the body. The head is brown.

Actual size

FAMILY	Endromidae
DISTRIBUTION	United Kingdom (only Scottish Highlands and possibly Worcestershire), Europe, across Asia to Siberia and northern China
HABITAT	Deciduous woodlands dominated by birch (*Betula* spp.), coppiced woodlands, moorlands, and bog edge woodlands
HOST PLANTS	Deciduous trees, including alder (*Alnus* spp.), birch (*Betula* spp.), hazel (*Corylus* spp.), and European Hornbeam (*Carpinus betulus*)
NOTE	Species with a pupal period lasting up to three years
CONSERVATION STATUS	Not evaluated, but scarce in parts of its range

ADULT WINGSPAN
2–2¾ in (50–70mm)

CATERPILLAR LENGTH
2 in (50 mm)

ENDROMIS VERSICOLORA
KENTISH GLORY
(LINNAEUS, 1758)

358

Kentish Glory females lay up to 250 yellow-brown eggs in rows on thin branches of the host trees. The caterpillars, initially colored black, hatch 10 to 14 days later and stay together at first in groups of up to 30. Later, they move apart and feed alone at night, descending to the ground after the final molt to pupate, spinning a cocoon about an inch deep in moss. They may remain in the cocoon for up to three years before emerging as adults.

The day-flying adults are among the earliest moths to be on the wing, emerging in late winter. Named for the English county of Kent, where it was once abundant, the increasingly scarce Kentish Glory moth is no longer found in England—except possibly for an isolated community in the county of Worcestershire—and is declining across the rest of its range. This reduction in numbers is due primarily to habitat loss, as the open, birch-dominated woodland the species prefers is cleared.

The Kentish Glory caterpillar is bright green, with a dorsal green line, lateral cream, oblique stripes, and white spiracles. The body is covered with tiny, black spots. The horned tail is similar to that of a hawkmoth caterpillar but much smaller. Just before pupating, the caterpillar becomes darker in color.

Actual size

FAMILY	Bombycidae
DISTRIBUTION	No longer found in the wild, but now bred worldwide; historically, north India, China, Korea, and Japan
HABITAT	Historically, woodlands and parks
HOST PLANTS	Mulberry (*Morus* spp.), especially White Mulberry (*Morus alba*)
NOTE	Renowned caterpillar bred for its cocoon's extensive silken thread
CONSERVATION STATUS	Not evaluated as commonly bred

ADULT WINGSPAN
1⅟₁₆ in (40 mm)

CATERPILLAR LENGTH
Up to 3 in (75 mm)

BOMBYX MORI

MULBERRY SILKWORM
(LINNAEUS, 1758)

359

Mulberry Silkworms hatch from several hundred eggs laid by the Mulberry Silkmoth around two weeks earlier; the silkmoth dies after laying her eggs. The caterpillars have huge appetites, eating continually and growing rapidly up to the end of the fifth instar. They then lose as much as a third of their length just prior to pupation—a distinctive characteristic of this species. The larvae pupate in large silk cocoons. *Bombyx mori* is closely related to the Wild Silkmoth (*B. mandarina*), and the two species are able to hybridize.

The silkworm was domesticated more than 5,000 years ago in China to create that country's silk industry. During the domestication process, the adults lost the ability to fly and to cling to their food plant. As a result, the species—once found across Asia—can no longer survive in the wild. The cocoon is also much larger than that of the wild silkworm and consists of a very fine, single raw silk thread up to 3,000 ft (900 m) in length. Commercially, the pupa is killed to prevent the adult damaging the thread as it emerges from the cocoon.

The Mulberry Silkworm is large and creamy white to buff brown in color with irregular, brown spots and brown-ringed spiracles. There is a short horn on the posterior.

Actual size

FAMILY	Saturniidae
DISTRIBUTION	Southeastern Canada and eastern United States, south to central Florida
HABITAT	Deciduous and mixed wood forests and woodlands
HOST PLANTS	Birch (*Betula* spp.), alder (*Alnus* spp.), persimmon (*Diospyros* spp.), Sweetgum, (*Liquidambar styracifluar*), hickory (*Carya* spp.), and walnut (*Juglans* spp.)
NOTE	Cryptic caterpillar that can strip a branch in a day
CONSERVATION STATUS	Not evaluated, but common

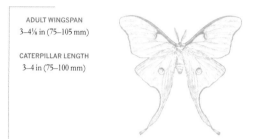

ADULT WINGSPAN
3–4⅛ in (75–105 mm)

CATERPILLAR LENGTH
3–4 in (75–100 mm)

ACTIAS LUNA
LUNA MOTH
(LINNAEUS, 1758)

360

The Luna Moth caterpillar is green with a thin, yellow lateral stripe. Its head, legs, and prolegs are rust colored, and there are rows of orange or pink verrucae—sclerotized structures, six on each segment, all bearing setae. In the thoracic segments and the last abdominal segments, the verrucae are larger and clearly defensive as they would be unpleasant for a predating bird or mammal to swallow.

Luna Moth caterpillars feed in groups during the first two to three instars and are solitary thereafter. The amount of food consumed grows exponentially; by the last instar, a caterpillar can defoliate a branch in a day. There are five instars, and each instar takes five to seven days to complete. At maturity, the caterpillar, like other silkmoth species, creates a tight, thick, silver-colored cocoon among the leaves of its host plant.

Within its large, diverse family that includes some of the world's biggest Lepidoptera species, *Actias* is a genus of long-tailed adults, some of which, such as *A. artemis* and *A. selena* in Asia, are very similar to the Luna Moth. *Actias luna* adults have recently been the subject of pioneering research to investigate how their tail helps them to survive bat attacks: via ultrasound reflection the bat detects the tail, attacking that rather than other vital parts.

Actual size

FAMILY	Saturniidae
DISTRIBUTION	Eastern Africa, from Ethiopia south to KwaZulu-Natal (South Africa)
HABITAT	Tropical forests and savannahs
HOST PLANTS	*Commihora* spp., walnut (*Juglans* spp.), *Sclerocarya* spp., and *Spirostachys* spp.
NOTE	Plump caterpillar that is often collected for human consumption
CONSERVATION STATUS	Not evaluated

ADULT WINGSPAN
4⅝ in (120 mm)

CATERPILLAR LENGTH
3⅛ in (80 mm)

ACTIAS MIMOSAE
AFRICAN MOON MOTH
(BOISDUVAL, 1847)

361

The African Moon Moth caterpillar usually hatches with two or three siblings but is completely independent of them. As it develops through its five instars it becomes very plump and firm, and as with many other large African moth species, this caterpillar is regularly collected for human consumption. When it has finished feeding, the larva discharges its gut contents and becomes smaller, then spins a silvery cocoon. There are two broods yearly in most African populations, but only one in the far south.

Actias mimosae was previously classified in the *Argema* genus of moon moths, whose members were recently transferred to the *Actias* genus. *Actias* now contains 26 moon moth species worldwide, including *A. luna*, the American Luna Moth. The genus probably originated in Eurasia, and only one ancestral species spread to North America, from which *A. luna* and two other Mexican and Central American species have derived. *Graellsia isabella* is a third related moon moth, found in and around the mountains of Spain.

The African Moon Moth caterpillar is green, shading to blue and yellow between the segments. It has a pair of very long, tubercular scoli on the dorsum of each segment, each tipped with a crown of small, black, harmless spines and studded with long, white, rumpled hairs. The head and thoracic legs are reddish brown, and the prolegs are yellow and black with curved white bristles.

Actual size

FAMILY	Saturniidae
DISTRIBUTION	South Asia, China, Japan, and parts of Southeast Asia
HABITAT	Temperate forests, scrub, and gardens
HOST PLANTS	Various, including *Hibiscus* spp., apple (*Malus* spp.), and wild pear (*Pyrus* spp.)
NOTE	Apple-green silkmoth caterpillar that "clicks" to scare predators
CONSERVATION STATUS	Not evaluated, but widespread

ADULT WINGSPAN
3⅛–4⅝ in (80–120 mm)

CATERPILLAR LENGTH
4 in (100 mm)

362

ACTIAS SELENE
INDIAN MOON MOTH
(HÜBNER, 1806)

Indian Moon Moth caterpillars emerge from around 100 pale brown eggs laid some two weeks earlier on the host plant. The larvae have an incredibly strong grip, attaching themselves firmly to branches, which makes it difficult for predators, such as birds, to pull them off. If disturbed, they defend themselves by twisting around and making a clicking noise with their jaws to scare their attackers. As the caterpillar approaches pupation, it becomes paler in color and spins a silk cocoon in which to pupate. The adult emerges about six weeks later.

There are usually two generations a year, but in the more southerly regions the moth breeds all year round. The species is easy to rear, and the lovely, night-flying adult, with large, pale green wings and a long tail, is, unsurprisingly, kept by entomologists worldwide. Although the Indian Moon Moth is a member of the silkmoth family, its silk is not used commercially.

The Indian Moon Moth caterpillar is mostly red with a black saddle at the first instar stage. It changes appearance as it molts, becoming bright apple green in the third instar. The head and legs are dark brown. Each segment, apart from the last, bears large, orange-yellow, spiny warts.

Actual size

FAMILY	Saturniidae
DISTRIBUTION	Southern Mexico to Brazil, Bolivia, and eastern Peru
HABITAT	Tropical forests at lower elevations
HOST PLANTS	Inga tree (*Inga* spp.); in captivity has fed on oak (*Quercus* spp.)
NOTE	Stocky silkmoth caterpillar that is the largest of its genus
CONSERVATION STATUS	Not evaluated, but quite common

ADULT WINGSPAN
2½–4¾ in (63–123 mm)

CATERPILLAR LENGTH
2⁹⁄₁₆ in (65 mm)

ADELONEIVAIA JASON

ADELONEIVAIA JASON
(BOISDUVAL, 1872)

363

The *Adeloneivaia jason* silkmoth caterpillars group together on a leaf after hatching and eating their eggshells. At first, the caterpillars are black but later turn green. At the end of the fifth instar they stop feeding, release fluids, and burrow into the ground, forming a chamber and wrapping themselves with a few strands of silk. In about ten days the pupa forms and remains in the ground for six weeks or more until the adult moth emerges, crawling out of its chamber before expanding its wings.

The *Adeloneivaia jason* caterpillar belongs to a very large subfamily of silkmoths—the Ceratocampinae—found only in the Americas. The adults of most members are plump-bodied and shaped like jet airplanes, and their caterpillars are stocky with little or no hairs. All species whose habits are known pupate in the soil. Of the more than 15 species of *Adeloneivaia*, *A. jason* is the largest.

The *Adeloneivaia jason* caterpillar is bright green with three large, white markings, shaped like bird heads with the blue spiracles for eyes. The "beak" is silver on the outer side and orange on the inner. Numerous thick, green horns angle backward along the body. The head is green and the feet are dark orange and black.

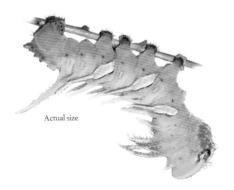

Actual size

FAMILY	Saturniidae
DISTRIBUTION	North America, from Arizona, Colorado, New Mexico, and west Texas to Mexico
HABITAT	Oak woodlands
HOST PLANTS	Coffee Berry (*Rhamnus californica*) and willow (*Salix* spp.)
NOTE	Caterpillar of a species that is distinctive within its genus
CONSERVATION STATUS	Not evaluated, but common within its range

ADULT WINGSPAN
2¹⁵⁄₁₆–3¹¹⁄₁₆ in (74–94 mm)

CATERPILLAR LENGTH
2³⁄₁₆–2⁹⁄₁₆ in (55–65 mm)

AGAPEMA HOMOGENA
ROCKY MOUNTAIN AGAPEMA
DYAR, 1908

The Rocky Mountain Agapema caterpillar hatches from a cluster of 45 to 160 ivory-colored, oval eggs in the summer. Initially, it is all black with white hairs but in the second and remaining instars it becomes black and yellow. It is gregarious until the fourth and final instar, when it becomes solitary. The mature caterpillar leaves the host plant and spins a fluffy, tan-colored cocoon in a crevice of rock or tree trunk in the fall, and the pupa spends the winter often covered by snow.

There are seven species of *Agapema* in the western United States and Mexico. All are similar as caterpillars and adults, except the Rocky Mountain Agapema, which is larger, more attractive, and lives in a forested environment at a higher altitude. The other six are desert species. Many of the larvae are parasitized by tachinid flies, which lay their eggs on the leaves that the caterpillars feed on, eventually killing a larva after it has made a cocoon.

Actual size

The Rocky Mountain Agapema caterpillar
is black, overlaid with a bold, yellow pattern repeated on each segment. There are many short, white hairs and fewer very long ones on the body and medium-long, curved ones on the abdominal legs and prolegs. The head and legs are black, and there is a white patch on each abdominal leg.

FAMILY	Saturniidae
DISTRIBUTION	Across Europe and Asia, excluding southern Asia
HABITAT	Beech forests, mixed conifer-deciduous forests, and among deciduous trees along rivers within otherwise coniferous forests
HOST PLANTS	European Beech (*Fagus sylvatica*), birch (*Betula* spp.), Alder (*Alnus glutinosa*), Goat Willow (*Salix caprea*), Mountain Ash (*Sorbus aucuparia*), and oak (*Quercus* spp.)
NOTE	Caterpillar with striped projections when young but cryptic when mature
CONSERVATION STATUS	Not evaluated, but common

ADULT WINGSPAN
2⅜–3⁵⁄₁₆ (60–84 mm)

CATERPILLAR LENGTH
2⅜ in (60 mm)

AGLIA TAU
TAU EMPEROR
(LINNAEUS, 1758)

365

Tau Emperor caterpillars hatch from eggs that the female moth lays at different strata of the forest, dispersing them by night in small clusters on the leaves and shoots of host trees. While in the early instars, the green caterpillars have several long, red-and-white dorsal projections that probably protect them to some extent from being swallowed by birds; the later instars are cryptic. The larvae have only four instars under good conditions but will molt into additional instars if conditions are not optimal. There is a single generation per year, and the pupa overwinters inside a cocoon.

Tau Emperor males fly fast and erratically when the sun is out in early spring, looking for freshly emerged females, which are, unlike males, nocturnal. Mating occurs during the day, when males, using pheromone trails and visual cues, find a newly eclosed female still sitting near her cocoon. Currently, four species in the genus *Aglia* are recognized, all of which are quite similar. *Aglia tau* has by far the widest distribution.

The Tau Emperor caterpillar is green with a white subspiracular stripe and beige spiracles. Its wavy shape dorsally is formed by pronounced segmentation, resembling the edge of a leaf. The minute, short, white spines and hairs throughout the body add to the leaf resemblance, as do the white, parallel, vein-like lines that lead off the white, longitudinal subspiracular line at a 45-degree angle.

Actual size

FAMILY	Saturniidae
DISTRIBUTION	Central and eastern North America
HABITAT	Deciduous forests
HOST PLANTS	Deciduous hardwood trees, in particular oak (*Quercus* spp.)
NOTE	Gregarious, defoliating caterpillar active in late summer and fall
CONSERVATION STATUS	Not evaluated, but among the more common in its family

ADULT WINGSPAN
1³⁄₁₆–2 in (30–50 mm)

CATERPILLAR LENGTH
2 in (50 mm)

ANISOTA SENATORIA
ORANGE-STRIPED OAKWORM
(J. E. SMITH, 1797)

Orange-striped Oakworms hatch from eggs laid 10 to 14 days earlier in large clusters of up to 500 eggs on the underside of leaves. The dark green young larvae are gregarious but disperse when they are older. The mature caterpillars tend to eat the entire leaf except for the midrib. They then drop to the forest floor and wander in search of suitable pupation sites underground. They overwinter as pupae. There is one generation a year, with the emerging orange-yellow moths active in early to mid summer.

The caterpillar is a pest of hardwood trees, especially the Red Oak (*Quercus rubra*) and other oak species—hence its common name. Trees can tolerate some defoliation as it occurs at the end of the growing season, but repeated infestations over several years can seriously damage them, especially if combined with an attack earlier in the season from other defoliating species, such as the Gypsy Moth caterpillar (*Lymantria dispar*).

The Orange-striped Oakworm caterpillar has a black body with yellow-orange stripes that run its entire length. There are two black horns on the second thoracic segment, tiny, backward-facing tubercles on the other segments, and several short, black spines at the anal end.

Actual size

FAMILY	Saturniidae
DISTRIBUTION	From southern Canada (southeast Ontario to Nova Scotia), south through United States to Florida, and west to Texas and Iowa
HABITAT	Deciduous forests and leafy suburbs
HOST PLANTS	Oak (*Quercus* spp.)
NOTE	Cryptically colored caterpillar that has two distinctive, black, thoracic projections
CONSERVATION STATUS	Not evaluated, but common, although in sharp decline

ADULT WINGSPAN
1⅝–2⅝ in (42–66 mm)

CATERPILLAR LENGTH
1¹¹⁄₁₆–2⅜ in (40–60 mm)

ANISOTA VIRGINIENSIS
PINK-STRIPED OAKWORM
(DRURY, 1773)

367

The Pink-striped Oakworm hatches from clutches of round, yellow eggs laid on oak leaves. The caterpillars are gregarious in the early instars, skeletonizing leaves of their host. Later instars are solitary and consume the entire leaf, except for the middle vein. The larvae descend to the ground to pupate in shallow burrows, and the pupae overwinter. There is one brood in the north of the species' range and two or more in the south. While *Anisota virginiensis* does not normally become a pest, it occasionally defoliates oak trees. During one outbreak in Manitoba in the late 1980s, the larvae stripped 95 percent of the trees.

The species, once very common, now appears to be in drastic decline—possibly because of the rise of *Compsilura concinnata*, a tachinid fly introduced from Europe to control the Gypsy Moth (*Lymantria dispar*). Three species of tachinid flies and two species of ichneumon wasps parasitize the larvae, in addition to predation by birds. Of 13 *Anisota* species on the North American continent, six quite similar species also occur in eastern North America.

The Pink-striped Oakworm caterpillar is cryptically colored, blending in with the background of leaves and twigs. It has two black thoracic projections, the head and prolegs are green, and the body color is greenish with alternating black and pink longitudinal stripes. Each segment also carries short spines that, while not dangerous to humans, may deter vertebrate predators.

Actual size

FAMILY	Saturniidae
DISTRIBUTION	Spain, India, China, and Japan
HABITAT	Oak woodlands
HOST PLANTS	Oak (*Quercus* spp.)
NOTE	Caterpillar that has produced Tussah silk since 200 BCE
CONSERVATION STATUS	Not evaluated

ADULT WINGSPAN
4¼–6 in (110–150 mm)

CATERPILLAR LENGTH
3⅜ in (85 mm)

368

ANTHERAEA PERNYI
CHINESE TUSSAH SILKMOTH
(GUÉRIN-MÉNEVILLE, 1855)

The Chinese Tussah Silkmoth caterpillar is
green and covered with tiny, white granules.
A pale yellowish stripe on each side ends
in a dark triangle on the claspers of the tail.
Its spiracles are black ovals with yellow centers,
and its sparse long hairs are mostly yellow.
The head is tan.

The Chinese Tussah Silkmoth caterpillar feeds on oak leaves and is farmed in China for its Tussah silk. It completes five instars before spinning its cocoon, which is wrapped in a leaf. Tussah is often produced by wild silkmoths farmed on guarded forest trees or indoors. Although various other species have been used over the centuries, the Chinese Tussah Silkmoth has been bred for 2,000 years. Recent genetic studies suggest the wild Royle Silkmoth (*Antheraea roylei*) as the original ancestor of *A. pernyi*.

Unlike the unrelated silkworm moth that produces most commercial silk, the domesticated Chinese Tussah Silkmoth can fly, and various introduced populations survive in the wild in Asia and in Europe. The majority of the 80 listed species of *Antheraea* occur mostly in Asia, with only four evolving in the Americas, including the Polyphemus Moth (*A. polyphemus*) in North America and *A. godmani*, with a range from central Mexico to northern Colombia, which is as far south in the Americas as oak trees occur.

Actual size

FAMILY	Saturniidae
DISTRIBUTION	United States and southern Canada
HABITAT	Deciduous woodlands, although common wherever food plants are found
HOST PLANTS	Decidous trees, including apple (*Malus* spp.), chokecherry (*Prunus* spp.), chestnut (*Castanea* spp.), oak (*Quercus* spp.), sycamore (*Platanus* spp.), and willow (*Salix* spp.)
NOTE	Caterpillar that regurgitates its food if threatened
CONSERVATION STATUS	Not evaluated, but common throughout most of its range

ADULT WINGSPAN
4–6¹⁄₁₆ in (102–152 mm)

CATERPILLAR LENGTH
4 in (102 mm)

ANTHERAEA POLYPHEMUS
POLYPHEMUS MOTH
(CRAMER, 1776)

369

Polyphemus Moth caterpillars hatch from large, flattened, round eggs that have a band of brown along the outside edge. They are solitary feeders and rarely found in close proximity to each other. When threatened, the larvae regurgitate their food, covering their bodies in a greenish-brown liquid in a bid to deter predators. Though their green coloration acts as an effective camouflage, the large size of fifth instar larvae makes them relatively easy to find.

The caterpillars' hard, large, egg-shaped cocoons are usually spun up among the leaves of the host plant and then drop to the ground in the fall. Since there is no escape valve on the cocoon, the adult moth secretes an enzyme that breaks down one end, allowing it to escape and pump up its wings. The Polyphemus Moth is one of North America's largest moths, sometimes rivaling the size of the Cecropia Moth (*Hyalophora cecropia*). It is most famous for the large eyespots on its hind wings, highlighted in yellow, blue, and black, which have given rise to the common name, from Polyphemus, one of the Cyclopes, a race of single-eyed giants in ancient Greek mythology.

The Polyphemus Moth caterpillar is bright green with yellowish, vertical lines on each of the abdominal segments, and pinkish protuberances at each end. The caterpillar has a pronounced, tan "face" and is sparsely covered with non-urticating hairs.

Actual size

FAMILY	Saturniidae
DISTRIBUTION	Madagascar and the Comoros Islands
HABITAT	Wet and dry forests
HOST PLANTS	Many trees and shrubs, including Oleander (*Nerium oleander*), privet (*Ligustrum* spp.), willow (*Salix* spp.), and beech (*Fagus* spp.)
NOTE	Variably colored silkmoth caterpillar that forms an edible pupa
CONSERVATION STATUS	Not evaluated, but generally common

ADULT WINGSPAN
4–5 in (100–130 mm)

CATERPILLAR LENGTH
2⅜–2¾ in (60–70 mm)

ANTHERINA SURAKA
SURAKA SILKMOTH
(BOISDUVAL, 1833)

370

The colors of Suraka Silkmoth caterpillars vary considerably, ranging from black with orange spines to green with pink spines. The larvae consume a large amount of vegetation from many different host plants and develop rapidly, taking only about three to four weeks from first instar to pupation, depending on temperature, with many caterpillars dying at low temperatures. Pupation takes place in a strongly built cocoon on the ground under the host plant.

In Madagascar, cocoons from the Suraka Silkmoth are used for silk production. Although the silk is less strong than that obtained from the Mulberry Silkworm (*Bombyx mori*), it is still of significant value. The pupae of *Antherina suraka* are edible and used increasingly in Madagascar as a source of protein. These silk and food properties have led to the species becoming a focus of conservation efforts to restore forests on the vast island, with local cooperatives beginning to grow forest trees for the cultivation of *A. suraka*.

The Suraka Silkmoth caterpillar is variably colored but often bright green or yellow green with pink spines. The spines are reduced laterally but prominent dorsally. Laterally on each segment there is a distinctive, small, yellow, triangle-shaped patch, strongly edged posteriorly in black. A few black spots occur below the yellow triangle, and the prolegs and true legs are black.

Actual size

FAMILY	Saturniidae
DISTRIBUTION	From Nepal southeast through Myanmar and Thailand to Vietnam, and south to Malaysia
HABITAT	Mountain forests
HOST PLANTS	Many, including *Ailanthus* spp. and Kashi Holly (*Ilex chinensis*)
NOTE	Large, polyphagous caterpillar of the largest moth in Asia
CONSERVATION STATUS	Not evaluated

ADULT WINGSPAN
Up to 9⅞ in (250 mm)

CATERPILLAR LENGTH
6 in (150 mm)

ARCHAEOATTACUS EDWARDSI
EDWARDS ATLAS SILKMOTH
WHITE, 1859

The Edwards Atlas Silkmoth caterpillar grows large, with the ability to feed on a variety of plants in the wild. In captivity, it has fed on ailanthus, privet (*Ligustrum* spp.), poplar (*Populus* spp.), willow (*Salix* spp.), lilac (*Syringa* spp.), and other plants. The first instar is white with narrow black bands. Subsequent instars are white and thickly covered in powdery wax, but the greenish fifth and last instar is more lightly covered. Recent genetic evidence suggests there are three species of *Archaeoattacus* instead of two as previously believed, all from Asia.

Currently, the Edwards Atlas Silkmoth is placed evolutionarily between the Ailanthus Silkmoth (*Samia cynthia*) and the Atlas Moth (*Attacus atlas*) because of its wing pattern. Moths from the *Rothschildia* genus of the American tropics are smaller but similar, and, in Africa, related species are represented by the genus *Epiphora*. The caterpillars of these genera all have similar characteristics and spin papery silk cocoons.

The Edwards Atlas Silkmoth caterpillar is dull green and blue with darker, tiny, rounded spots, and partially covered with white, powdery wax. A large, reddish-brown, rounded triangle decorates each clasper. Long, fleshy, blue horns are bent toward the rear. The larva's head is dull green and its feet are blue.

Actual size

FAMILY	Saturniidae
DISTRIBUTION	Eastern Mexico, south to Bolivia and southern Brazil
HABITAT	Tropical forests
HOST PLANTS	*Ceiba* spp. and other trees of the Bombacaceae family
NOTE	Highly social caterpillar of giant silkmoth family
CONSERVATION STATUS	Not evaluated

ADULT WINGSPAN
4⅜–6¾ in (115–170 mm)

CATERPILLAR LENGTH
4¼ in (110 mm)

372

ARSENURA ARMIDA
ARSENURA ARMIDA
CRAMER, 1779

The *Arsenura armida* caterpillar is brightly banded black and white with long horns for most of its stages, but is hornless and less contrasting in the final instar. It is a very social caterpillar; large numbers often congregate on the base of a tree by day to ascend in procession to the branches to feed at night, following a pheromone (scent) trail deposited by each member to assist grouping. After dispersing in the canopy to feed, the caterpillars then descend together around dawn, again using the pheromone trail as a guide.

Arsenura silkmoth species caterpillars belong to the subfamily Arsenurinae, adult members of which mostly resemble one another, except for size, and are found only in the Central and South American tropics. All pupate in the ground. The moths are very large and variously patterned in subdued browns, grays, reddish brown, or black and white. Although *A. armida* caterpillars are toxic enough to kill some birds, they are cooked and eaten by ethnic communities in southern Mexico.

Actual size

The *Arsenura armida* caterpillar is black and smooth on the dorsal half of its body with a band of orange separating each segment. The lower half is covered with fine, dense, cream-colored filigree, the spiracles are black ovals, and the head and claspers of the tail are rich brown. The feet are black.

FAMILY	Saturniidae
DISTRIBUTION	Central and South America, including Costa Rica, Colombia, and Ecuador
HABITAT	Tropical rain forests
HOST PLANTS	Mostly members of Malvaceae, including *Apeiba* spp., *Luehea* spp., and *Pithecellobiun* spp.
NOTE	Large, unusual-looking caterpillar that is camouflaged by disruptive coloring
CONSERVATION STATUS	Not evaluated

ADULT WINGSPAN
6⅛–6½ in (155–165 mm)

CATERPILLAR LENGTH
Up to 4 in (100 mm)

ARSENURA BATESII

ARSENURA BATESII
(R. FELDER & ROGENHOFER, 1874)

373

The female *Arsenura batesii* moth lays several hundred creamy-white eggs, depositing them singly on the upper surface of leaves, so they are scattered through the crowns of the food plant trees. The caterpillars hatch and consume their eggshell before eating leaves. Although little is known of their feeding habits, one of two subspecies, *A. batesii* Druce, reportedly feeds by day on leaves in the tree crown during early instars, resting on foliage. In the penultimate instar, however, it is said to rest on the trunk by day, returning to the canopy at night, and simultaneously changes color from mottled brown green to brown black.

The unusual patterning of *Arsenura batesii* larvae is thought to be disruptive, helping to break up their outline and making it difficult for predators to spot them. By the final instar, larvae resemble dead sticks, losing the long tentacles they have at earlier stages. The species belongs to the subfamily Arsenurinae, which consists of large Neotropical saturniids. Like most of the 23 species of their genus, the larvae are solitary in all instars.

Actual size

The *Arsenura batesii* caterpillar, here in its penultimate instar, has a brown head and brown-black body with two large, lateral, yellow-green blotches and four, long tentacles (lost in the final instar) that are colored yellow white to brown. A further long projection near the posterior end also disappears at the final instar, when the caterpillar largely resembles a stick. The prolegs may be orange.

FAMILY	Saturniidae
DISTRIBUTION	Much of southern Asia and China, Southeast Asia
HABITAT	Tropical and subtropical forests
HOST PLANTS	Deciduous trees and bushes, including Avocado (*Persea americana*), cherry (*Prunus* spp.), lilac (*Syringa* spp.), rhododendron (*Rhododendron* spp.), and willow (*Salix* spp.)
NOTE	One of the world's largest caterpillars
CONSERVATION STATUS	Not evaluated, but common throughout most of its range

ADULT WINGSPAN
9–11 in (228–280 mm)

CATERPILLAR LENGTH
4½–4¹⁵⁄₁₆ in (114–127 mm)

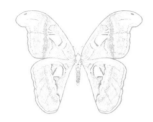

ATTACUS ATLAS
ATLAS MOTH
(LINNAEUS, 1758)

374

The Atlas Moth caterpillar is a translucent greenish blue covered in darker "freckles" and also in non-urticating spines. A row of blackish "hair" runs along the sides of the body. There is a ring of pinkish orange on the hind claspers.

Atlas Moth caterpillars hatch from large, cream-colored eggs with reddish-brown spots. The larvae eat almost anything and sometimes leave one species of host plant for another—a rare trait among caterpillars. All six instars look similar, and with each successive instar the larva covers its body in a white, waxy substance that resembles powder. It is thought that this may help give the appearance of a fungus growing on the caterpillar, making it unappetizing to predators. The powder can become so thick that the actual color of the caterpillar is completely obscured.

The baggy cocoons are usually spun up among the leaves of the food plant and attached by a strong peduncle. They are smaller than might be expected given the size of the emerging adults—among the world's largest moths —and so durable that in Chinese Taipei they are made into coin purses. The brightly colored adult is also known as the Snake's Head Moth for the snakelike markings on the forewing tips.

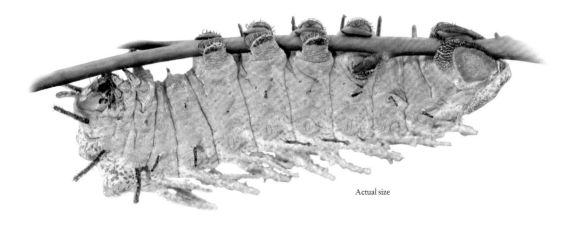

Actual size

FAMILY	Saturniidae
DISTRIBUTION	The Philippines
HABITAT	Tropical and subtropical forests
HOST PLANTS	Many, including lilac (*Syringa* spp.), willow (*Salix* spp.), and plum (*Prunus* spp.)
NOTE	Large caterpillar of one of the largest giant silkmoths
CONSERVATION STATUS	Not evaluated, but limited range

ADULT WINGSPAN
Up to 9⅞ in (250 mm)

CATERPILLAR LENGTH
4⅝ in (120 mm)

ATTACUS CAESAR
CAESAR ATLAS SILKMOTH
MAASEN & WEYMER, 1873

375

The caterpillar of the Caesar Atlas Silkmoth is large and eats many leaves every day, so a small tree cannot support more than one or two without attracting hungry birds or insect parasites. The caterpillar has a defensive ability to spray the enzyme tyrosinase from glands in its integument, which inhibits predators; it is also covered with powdery wax that could give it some protection. If a small tree is completely defoliated, the caterpillar moves to another tree. After the fifth instar, the caterpillar spins an immense papery, silk cocoon.

There are more than a dozen species of Atlas silkmoths in Asia and the Malay Archipelago. Closely related and similar, but smaller, species occur mostly in tropical climates in Asia, Africa, and the Americas. The caterpillars of some species are colorful and ornate, and all spin silk cocoons. Ancestors of these related species probably dispersed before the continents had drifted apart.

The Caesar Atlas Silkmoth caterpillar is blue and greenish blue fading to white dorsally, with many small, round, blue dots. Short horns on the dorsum are white, the spiracles are red with a white border, and a white, waxy dust lightly covers all. The head is blue, and the claspers have a large, blackish area.

Actual size

FAMILY	Saturniidae
DISTRIBUTION	Guiano-Amazonian Basin east of the Andes, from Venezuela south to Bolivia
HABITAT	Tropical forests
HOST PLANTS	Unknown; in captivity has fed on Black Locust (*Robinia pseudoacacia*)
NOTE	Giant silkmoth caterpillar that is protected by poisonous, stinging spines
CONSERVATION STATUS	Not evaluated

ADULT WINGSPAN
2¾–4⅞ in (70–126 mm)

CATERPILLAR LENGTH
3½ in (90 mm)

376

AUTOMERIS CURVILINEA
AUTOMERIS CURVILINEA
SCHAUS, 1906

The beauty of the *Automeris curvilinea* silkmoth caterpillar belies the painful reality of its many poisonous stinging spines. As the caterpillars feed close together, their spines are useful for shielding themselves and their siblings from vertebrate predators. When approached by parasitic flies or wasps, the caterpillars swing their heads back and forth to prevent the parasitoids from depositing eggs on them. Fortunately, they usually avoid contact with humans, as most *Automeris* species stay in the forest canopy. The mature caterpillars spin flimsy cocoons wrapped in leaves.

The *Automeris curvilinea* caterpillar is one of more than 125 colorful *Automeris* species, sometimes called "bull's eye moths" for the prominent false eyes on the hindwings. They belong to a huge subfamily of silkmoths called Hemileucinae found only in the Americas, from Canada to Tierra del Fuego and the Caribbean. All caterpillars in this subfamily are stinging, and those of one genus, *Lonomia*, regularly cause human fatalities.

The *Automeris curvilinea* caterpillar is bright green with five bright yellow, broad dashes on each side. Long spines on its dorsum bristle with slender green and yellow, blue-tipped spines. Rosettes of shorter slender spines radiate from each side. The head is blue, and the feet are red with black tips and covered with white bristles.

Actual size

FAMILY	Saturniidae
DISTRIBUTION	Cerrado area of south central Brazil
HABITAT	Tropical savannahs
HOST PLANTS	*Erythroxylum tortuosum*, *Byrsonima verbascifolia*, and *Bauhinia* spp.
NOTE	Colorful silkmoth caterpillar that has stinging spines
CONSERVATION STATUS	Not evaluated, but common although its habitat is threatened

ADULT WINGSPAN
2⅛–3⅜ in (54–85 mm)

CATERPILLAR LENGTH
2¹⁵⁄₁₆ in (75 mm)

AUTOMERIS GRANULOSA
AUTOMERIS GRANULOSA
CONTE, 1906

377

The yellow caterpillar of *Automeris granulosa* hatches from a yellowish egg and joins its siblings in a tight group. Members of the group take turns to feed, several side by side, at the nibbled edge of a leaf. After molting to the second instar, the caterpillar is brownish yellow with dark spines. Third and fourth instar larvae are white with narrow, dark gray longitudinal lines but, in the sixth instar, become yellow with blue-gray lines and spines. These colors become bright yellow and blue in the sixth and seventh instars.

This caterpillar belongs to the silkmoth subfamily Hemileucinae and like other Hemileucinae members has stinging spines. However, unlike most other members, which spin weak cocoons or none at all, *Automeris granulosa* produces a sturdy, reticulated cocoon. When *A. granulosa* silkmoths were reared in captivity in Brazil, the pupal stage lasted about 26 days, and the time from egg to emerged adult moth was 88 days. The beautiful adult is prized by moth enthusiasts.

The *Automeris granulosa* caterpillar is pale olive green with three, sharply defined, yellow stripes on each side, the wider lateral stripe connecting through the spiracles to the stripe above it, forming a lateral chain of quadrangles. The area between the yellow stripes is blue-green. The head is paler olive, and the feet and the long spines are white.

Actual size

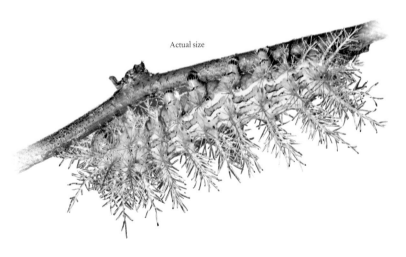

FAMILY	Saturniidae
DISTRIBUTION	Eastern and central United States, west to the Rocky Mountains
HABITAT	Forests
HOST PLANTS	Many, including Black Cherry (*Prunus serotina*) and willow (*Salix* spp.)
NOTE	Caterpillar with spines that can deliver a painful sting
CONSERVATION STATUS	Not evaluated, but common

ADULT WINGSPAN
2–4 in (50–100 mm)

CATERPILLAR LENGTH
2⅜–4 in (60–100 mm)

378

AUTOMERIS IO
IO MOTH
(FABRICIUS, 1775)

The Io Moth caterpillar is green with a brown abdomen and white-and-red (or white-and-orange) longitudinal stripes on its side. There are rows of projections on every segment, which contain glands that secrete poison through numerous branching, stinging spines.

Io Moth caterpillars feed in groups initially; the group size is usually determined by the size of the egg cluster from which they hatch. Females are capable of laying about 300 eggs, but in nature they rarely lay more than 20 at a time. The caterpillars are brown at first, then become green, with orange, yellow, and white stripes that add to their aposematic coloration. They are well defended against predators as they have spines that, if touched, will cause a beelike sting, and may cause a severe reaction. The caterpillar takes two to three months to develop.

Toward the end of their development, the caterpillars become solitary and build a thin, brown cocoon among leaves. The adults are sexually dimorphic, with yellow (sometimes orange or pink) forewings in males and brown in females. When the moth is at rest, the forewings cover the eyespot on the hindwing that distinguishes the species. The range of the Io Moth is more northerly than other member species of the large, mostly Neotropical *Automeris* genus.

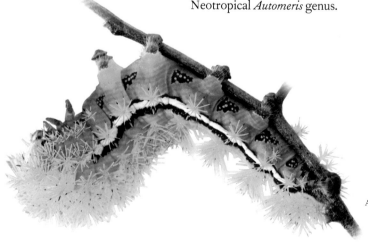

Actual size

FAMILY	Saturniidae
DISTRIBUTION	Guiano-Amazonian region, from Venezuela to Bolivia and Brazil
HABITAT	Tropical forests
HOST PLANTS	Unknown; in captivity has fed on *Erythrina* spp.
NOTE	Decorative and fierce giant silkmoth caterpillar
CONSERVATION STATUS	Not evaluated

ADULT WINGSPAN
3¼–5¾ in (82–147 mm)

CATERPILLAR LENGTH
4¼ in (110 mm)

AUTOMERIS LARRA
AUTOMERIS LARRA
(WALKER, 1855)

The *Automeris larra* silkmoth caterpillar is large and intimidating with a sting that can be quite painful. After hatching from one of many white eggs, the first stage of the caterpillar is white, the second is black with white spines, and the remaining stages are more recognizably green. The caterpillars group tightly together when young and more loosely as they become mature. When they have finished feeding, they spin a thin, papery cocoon wrapped in a leaf.

The adult silkmoth can emerge within six to eight weeks, although one pupa of a similar desert silkmoth species, *Hemileuca burnsi*, is known to have lain dormant for nine years before hatching. There are five species similar to the *Automeris larra* silkmoth distributed throughout much of tropical America from Mexico to Bolivia—all cryptically colored with eyespots on the hindwings and leaflike markings on the forewings. The adults have no mouth for eating and die within a few days after emerging from the cocoon, mating, and egg laying.

The *Automeris larra* caterpillar is a rich orange brown, paler on the dorsum, and covered with tiny, pale orange dots. It has a decorative broad, white lateral band half its body length, long, orange-and-white, bristled spines near the head, several similar, but white, spines on the rear dorsum, and also smaller blue, bristled spines. The head is orange.

Actual size

FAMILY	Saturniidae
DISTRIBUTION	Most of Africa south of the Sahara, Madagascar
HABITAT	Tropical forests and savannahs
HOST PLANTS	Various, including Castor Bean (*Ricinus communis*), Guava (*Psidium guajava*), and mango (*Mangifera* spp.)
NOTE	Caterpillar that is used as a food in southern Africa
CONSERVATION STATUS	Not evaluated, but widespread through much of Africa

ADULT WINGSPAN
4¼–6⅜ in (110–160 mm)

CATERPILLAR LENGTH
3½ in (90 mm)

380

BUNAEA ALCINOE
CABBAGE TREE EMPEROR
(STOLL, 1780)

The strikingly colored Cabbage Tree Emperor caterpillar is big, and in great numbers it can completely defoliate large trees. After finishing feeding in the fifth instar, the caterpillar evacuates its gut and burrows into the soil to form a chamber and pupate. Larger larvae are eaten by gorillas and are also gathered for human food, reportedly at the fifth or sixth instar or even after they have burrowed into the ground. They add excellent nutrients to a diet that is often inadequate, and are also canned and sold in grocery stores.

The beautiful moth is abundant throughout much of Africa, but there is disagreement among lepidopterists as to whether the populations in Kenya and Madagascar, named *Bunaea auslaga*, represent a subspecies or a distinct species. Alcinoe, after whom the Cabbage Tree Emperor was named, was, in Greek mythology, the daughter of King Polybus of Corinth, who was also the adoptive father of Oedipus.

The Cabbage Tree Emperor caterpillar is dark black with orange, ovoid spiracles encircled in deep red. A band of widely spaced white "thorns" adorn each segment and point backward. The head, true legs, and prolegs are black, and the head is particularly large. The anterior horns are black.

Actual size

FAMILY	Saturniidae
DISTRIBUTION	India (Assam), China (Yunnan), eastern Myanmar, and Thailand
HABITAT	Mountain forests
HOST PLANTS	Many, including walnut (*Juglans* spp.), willow (*Salix* spp.), and oak (*Quercus* spp.)
NOTE	Striking, multicolored silkmoth caterpillar
CONSERVATION STATUS	Not evaluated

ADULT WINGSPAN
2⅞–3¾ in (73–95 mm)

CATERPILLAR LENGTH
2⁹⁄₁₆ in (65 mm)

CALIGULA CACHARA
CALIGULA CACHARA
MOORE, 1872

381

The *Caligula cachara* silkmoth caterpillar hatches from a group of irregularly placed eggs, which are partially covered with a protective, brown, lacquer-like cement, secreted by the female silkmoth as she lays. Newly hatched, the caterpillar is mostly blue, but by the end of the first instar it becomes green. By the fifth and final instar it is stunningly marked and colored. Throughout its development the caterpillar remains close to its siblings and, when finished feeding, spins a brown cocoon of open web netting in which the pupa is plainly visible inside.

The *Caligula* genus contains ten species, most of whose eggs overwinter before hatching in the spring. It is now sometimes considered a subgenus of *Saturnia* rather than separated as *Caligula*, or its previous name, *Dictyoploca*. *Caligula* species are distributed mainly in Asia and Japan. The *C. cachara* caterpillar, like most of the others of its genus, possesses very long, pale hair, especially on the dorsum.

Actual size

The *Caligula cachara* caterpillar is mostly blue green on the sides, shading into bluish white dorsally, and electric green below. There is a distinct bright yellow lateral stripe, and a pair of long, orange scoli emerge from the dorsum on the second and the third segments. The legs are yellow with red-and-black bands, and the dorsum has many thin, black, vertical squiggles.

FAMILY	Saturniidae
DISTRIBUTION	Pakistan, north India, China, Burma, and Thailand
HABITAT	Mountain forests at lower elevations
HOST PLANTS	Walnut (*Juglans regia*), oak (*Quercus* spp.), and plum (*Prunus* spp.)
NOTE	Spring silkmoth caterpillar that hatches from an overwintering egg
CONSERVATION STATUS	Not evaluated

ADULT WINGSPAN
4⅞–5¹⁵⁄₁₆ in (125–149 mm)

CATERPILLAR LENGTH
4⅛ in (105 mm)

CALIGULA SIMLA
CALIGULA SIMLA
(WESTWOOD, 1847)

382

The *Caligula simla* caterpillar is large, mostly green with a light blue dorsum and a narrow, yellow stripe below the spiracles, which are dark blue ovals with turquoise borders. The dorsum sprouts dense, long, light blue hair. The feet and head are brownish orange. Shorter stiff, yellow hairs cover the lower sides, legs, and claspers at the rear.

The *Caligula simla* silkmoth follows a seasonal pattern that differs from that of many other species. Instead of breeding in the spring, *C. simla* adults emerge in the fall, mate, and lay eggs that overwinter and hatch in the spring. The tiny caterpillars of the first instar are black, and subsequent instars are colorful combinations of green, blue, red, yellow, black, and white— very different from the mature caterpillars. Later, a silk cocoon of open web netting is spun, and the caterpillar, and soon the pupa, can be easily seen inside.

Other members of the worldwide subfamily (Saturniinae) of silkmoths to which the *Caligula simla* moth belongs, including the Peacock Moth (*Macaria notata*) of Europe and the Luna Moth (*Actias luna*) and Polyphemus Moth (*Antheraea polyphemus*) of North America, share similar shapes and patterns. These usually include a prominent round "eyespot" ringed with various colors on each wing. The caterpillars of some species are mildly stinging, and all species make silk cocoons.

Actual size

FAMILY	Saturniidae
DISTRIBUTION	Eastern United States into Canada
HABITAT	Deciduous woodlands
HOST PLANTS	Deciduous trees and bushes, including apple (*Malus* spp.), Sassafras (*Sassafras albidum*), cherry (*Prunus* spp.), poplar (*Populus* spp.), and Spicebush (*Lindera benzoin*)
NOTE	Caterpillar that is known for making a perfectly camouflaged cocoon
CONSERVATION STATUS	Not evaluated, but one of the more common Saturniidae species

ADULT WINGSPAN
3–3⅜ in (75–85 mm)

CATERPILLAR LENGTH
2–2⅜ in (50–60 mm)

CALLOSAMIA PROMETHEA
PROMETHEA MOTH
(DRURY, 1773)

383

Promethea Moth caterpillars hatch from a cluster of 2 to 12 small, white eggs usually laid in rows on host plants in late spring or early summer. Initially the larvae feed in small groups but become solitary when they reach later instars. This species is single brooded in the north of its range and can have two or more broods in the south. When threatened, the caterpillars regurgitate their food onto themselves to become less palatable to predators. Each of the five instars look different, and in later instars a clear "smiley face" can be seen on the posterior end, right above the claspers.

The cocoons are tightly wrapped in a leaf of the food plant and attached by a strong peduncle. When the leaves drop in the fall, the cocoons remain, looking exactly like a dead leaf. The adult male moth, which flies in the late afternoon, mimics the Pipevine Swallowtail (*Battus philenor*), which is poisonous. The females look like small Cecropia Moths (*Hyalophora cecropia*).

The Promethea Moth caterpillar is very pale with colors ranging from white to a light bluish green. The thoracic segments feature four red protuberances (which may also be orange), and there is a single, yellow protuberance on the eighth abdominal segment. The rest of the body has short, black protuberances.

Actual size

FAMILY	Saturniidae
DISTRIBUTION	Eastern Peru
HABITAT	High-altitude cloud forests
HOST PLANTS	Unknown; in captivity has fed on Laurel Sumac (*Malosma laurina*)
NOTE	Rare, high-altitude silkmoth caterpillar
CONSERVATION STATUS	Not evaluated, but scarce

ADULT WINGSPAN
3–3⅞ in (75–98 mm)

CATERPILLAR LENGTH
2¾ in (70 mm)

CERODIRPHIA HARRISAE

CERODIRPHIA HARRISAE

LEMAIRE, 1975

384

The *Cerodirphia harrisae* silkmoth caterpillar is so rare that the adult female has not yet been officially described. Photographs of its immature stages were obtained by rearing the eggs of the first known female captured in Peru. The newly hatched young were creamy white and congregated in a tight group on the leaf edge to feed, moving by turns to the freshly chewed edge. The caterpillars completed six instars before crawling to the ground to pupate under humid dead leaves wrapped with a few silk threads.

There are 33 species of *Cerodirphia* in Central and South America, with distributions ranging from hot, lowland rain forest to cold, high-altitude cloud forest. Many of the moths are bright pink with a black-and-white banded abdomen. The caterpillars are social and processional (searching for food in single file) and belong to the silkmoth subfamily Hemileucinae, whose members have stinging spines. Many species do not spin silk or make cocoons.

The *Cerodirphia harrisae* caterpillar is pale pinkish cream with areas of scattered filigree. The spiracles are large and white, bordered with caramel. The body is decorated with blue, black-tipped spines studded with smaller yellow or white spines. A pair of long, black spines extends over the yellowish head. The prolegs are green and yellow, and the true legs are black.

Actual size

FAMILY	Saturniidae
DISTRIBUTION	Colombia, Ecuador, Peru, and Brazil
HABITAT	Woodlands, open areas, and rangeland
HOST PLANTS	Wide variety of trees and shrubs, including ash (*Fraxinus* spp.) and beech (*Fagus* spp.)
NOTE	Large silkmoth caterpillar that comes in two color forms
CONSERVATION STATUS	Not evaluated, but locally common

ADULT WINGSPAN
4¼–4⅝ in (110–120 mm)

CATERPILLAR LENGTH
4–4⅝ in (100–120 mm)

CITHERONIA AROA
CITHERONIA AROA
SCHAUS, 1896

Citheronia aroa caterpillars hatch from large, yellowish eggs laid singly or in groups of up to four on the upper surface of host plant leaves seven to ten days earlier. The early instar caterpillars feed mostly by night and rest by day, while later instars also feed openly by day. Mature larvae have two color forms, which probably evolved to blend in with their variable habitats. There are five instars, and development from egg hatch to pupation takes five to six weeks.

When fully fed, the caterpillar becomes a dull turquoise color and descends to the ground, burying itself 5–6 in (130–150 mm) below the surface, where it constructs an earthen chamber. Here it pupates, becoming a dark brown-black chrysalis that usually overwinters. The adults are short-lived, with vestigial mouthparts, meaning no feeding occurs. Females emit a pheromone to attract males for mating and, once mated, spend the rest of their short lives laying eggs.

The *Citheronia aroa* caterpillar is either dark brown black or banded black and white with a red head. The dark form is sometimes marked subspiracularly with a variable, wavy, orange stripe and intersegmental orange bands. The spiracles are outlined in yellowish orange, and the head and true legs are red.

Actual size

FAMILY	Saturniidae
DISTRIBUTION	Eastern United States
HABITAT	Forested areas where host plants are abundant
HOST PLANTS	Various, including ash (*Fraxinus* spp.), Buttonbush (*Cephalanthus occidentalis*), hickory (*Carya* spp.), privet (*Ligustrum* spp.), sweetgum (*Liquidambar* spp.), and walnut (*Juglans* spp.)
NOTE	Caterpillar whose common name reflects its striking appearance
CONSERVATION STATUS	Not evaluated, usually common but now rare in New England

ADULT WINGSPAN
3¾–5¾ in (96–147 mm)

CATERPILLAR LENGTH
6 in (150 mm)

CITHERONIA REGALIS
HICKORY HORNED DEVIL
(FABRICIUS, 1793)

386

The Hickory Horned Devil emerges from the yellowish, semitranslucent eggs of the Regal or Royal Walnut Moth, laid on the upper side of host plant leaves, and can be seen within an egg as hatching time nears. The caterpillar is named for its large size and spiked horns in later instars, which make it look like a small dragon. It uses the horns to scare potential predators, moving its head violently when disturbed. As the horns can be urticating, the caterpillar should be handled with care. The larvae are solitary feeders and grow at a rapid rate, reaching their full length within a month.

When fully grown, the caterpillars climb down to the ground and dig themselves in, creating a protective chamber in which to pupate. They spend the winter there before the adults emerge—moths with a large wingspan and distinctive markings in shades of reddish brown, patched with yellow.

The Hickory Horned Devil, one of the largest North American caterpillars, is colored in a striking translucent green and has orange horns tipped with black on its head. Its thoracic region also features many long horns, with shorter horns running the length of the body. The green coloration takes on a bluish hue before pupation.

Actual size

FAMILY	Saturniidae
DISTRIBUTION	Southeastern Mexico, south to eastern Peru, eastern Bolivia, and Argentina
HABITAT	Tropical forests
HOST PLANTS	Unknown; in captivity has fed on Black Locust (*Robinia pseudoacacia*) and willow (*Salix* spp.)
NOTE	One of only two *Citioica* silkmoth species
CONSERVATION STATUS	Not evaluated

CITIOICA ANTHONILIS
CITIOICA ANTHONILIS
(HERRICH-SCHÄFFER, 1854)

ADULT WINGSPAN
2¹⁄₁₆–4¹⁄₈ in (52–106 mm)

CATERPILLAR LENGTH
2³⁄₈ in (60 mm)

387

The tiny, black *Citioica anthonilis* caterpillar hatches from a group of green eggs deposited on a leaf or stem about one week previously. During the early instars, the larvae remain close together, changing to green in the second instar, and acquiring more color and complex design as they progress. During the fifth and final instar, the caterpillar appears threatening, with many long, spiky, silver tubercles, although these are soft and harmless and cannot sting. After it has finished feeding, the caterpillar burrows into the soil to pupate.

There are only two species of *Citioica*. Adults of both are almost identical—thick bodied and plain brown with two dark lines on the forewing. They belong to the large subfamily Ceratocampinae, found only in the New World, most species feeding on leguminous trees. The caterpillars often host tiny parasitic wasps, which feed on the live caterpillar, killing it, and exit through the integument before spinning miniature cocoons to pupate.

The *Citioica anthonilis* caterpillar is mostly grass green, darker blue green ventrally, with a bold, black, longitudinal stripe passing through the black, oval spiracles, bordered below by a white stripe. The surface is smooth and hairless with small, yellow bumps front and rear. Dorsally, there are four long, pointed, silvery tubercles on most segments.

Actual size

FAMILY	Saturniidae
DISTRIBUTION	Mountains of western United States and northwestern Mexico
HABITAT	Pine forests
HOST PLANTS	Various species of pine (*Pinus* spp.)
NOTE	Periodic edible silkmoth caterpillar
CONSERVATION STATUS	Not evaluated, but sometimes prolific enough to cause significant defoliation

ADULT WINGSPAN
2¾–3⅞ in (70–98 mm)

CATERPILLAR LENGTH
2⁹⁄₁₆ in (65 mm)

COLORADIA PANDORA
PANDORA PINE MOTH
BLAKE, 1863

388

The Pandora Pine Moth caterpillar has a life cycle of two to five years. It hatches and feeds in the fall, becomes lethargic during the winter, then feeds again until late June, when it descends from the tree and burrows into the ground to pupate. It usually remains there at least through the next winter, and sometimes up to five years, before emerging as an adult during the summer to reproduce. Adults are more commonly present in alternating years, when they sometimes cause extensive defoliation of pine trees.

The caterpillar of the Pandora Pine Moth is an important food source for the Paiute people of California. The fully fed caterpillars found on the ground are gathered in great numbers, roasted, washed, then dried and stored for up to two years. After rehydration they are used for soups or finger food. There are about ten species of pine moth in pine forests throughout Mexico and the western United States.

The Pandora Pine Moth caterpillar is light reddish brown with four, white, longitudinal stripes and densely covered with tiny, white dots. Areas between the segments are black. The entire body is sparsely covered with short, fuzzy hairs and stiff dorsal spines surrounded by rosettes of smaller spines that are mildly stinging. The head is reddish brown.

Actual size

FAMILY	Saturniidae
DISTRIBUTION	Eastern central Peru
HABITAT	Mid-elevation cloud forests
HOST PLANTS	Unknown, probably Lauraceae
NOTE	Giant silkmoth species with ancient South American origins
CONSERVATION STATUS	Not evaluated

ADULT WINGSPAN
3½–4⅛ in (90–105 mm)

CATERPILLAR LENGTH
3⅛ in (80 mm)

COPAXA BELLA

COPAXA BELLA

WOLFE, NAUMANN, BROSCH, WENCZEL & NAESSIG, 2005

389

The *Copaxa bella* silkmoth caterpillar is stout and slow moving. After hatching among several rounded, flat, translucent brown eggs with white rims, the caterpillars begin feeding together, but they disperse as they develop. As with most silkmoth caterpillars, each time they molt between instars they fasten their feet to a branch with silk so the old skins will remain as the new caterpillars crawl out. The cocoon of this caterpillar is a stiff, lacquered, silk fishnet with the subsequent pupa easily visible inside.

There are possibly more than 70 species of *Copaxa*, all in Central and South America; the *C. bella* species was discovered only early this century. Their ancestry is a mystery, but they apparently originated on the South American continent, the only genus of their subfamily, the Saturniinae, to do so. Many *Copaxa* caterpillars are hairy, but none are stinging, and although they may occasionally feed on Avocado trees (*Persea americana*), the larvae are not considered pests.

The *Copaxa bella* caterpillar is bright green, lighter dorsally. A diagonal, yellow brushstroke intersects each yellow, oval spiracle, and notable stiff, pink paddles all point forward from broad dorsal tubercles. The entire body is sprinkled with tiny, stumpy, white bristles. The feet are dark brown, and the head is greenish brown. Few long hairs are present.

Actual size

FAMILY	Saturniidae
DISTRIBUTION	Southern Mexico, south to central eastern Peru, northeast Bolivia, and southern Brazil
HABITAT	Tropical forests
HOST PLANTS	Trees of the Sapotaceae family
NOTE	Reclusive caterpillar that feeds on "chewing gum" trees
CONSERVATION STATUS	Not evaluated

ADULT WINGSPAN
3½–4½ in (89–114 mm)

CATERPILLAR LENGTH
3⅛ in (80 mm)

390

COPIOPTERYX SEMIRAMIS

COPIOPTERYX SEMIRAMIS
(CRAMER, 1775)

The *Copiopteryx semiramis* silkmoth caterpillar hatches among several transparent eggs. At first it is yellow with black dorsal bands and black, stiff tubercles, but in subsequent stages it becomes green and mostly smooth. Reared caterpillars are not social and develop through six instars, feeding on *Manilkara chicle*, the sap from which the original chewing gum was made. The mature caterpillar burrows into the ground to pupate, and the adult may emerge in about six weeks or many months, depending on the day length when the caterpillar pupated.

There are five species of *Copiopteryx* in the Arsenurinae subfamily, all with long-tailed adults and similar caterpillars, ranging from southern Mexico to Bolivia. Adults emerge one to three hours after dark, and females call for males by releasing pheromones. It is thought that the long tails help the moth to avoid bat attacks by enlarging its profile. Semiramis, the namesake for this species, was queen regent for her son Adad Nirari III in ninth-century BCE Assyria, and was also the holy spirit of the pagan Babylonian trinity.

The *Copiopteryx semiramis* caterpillar is dark green blending to bluish white on the dorsum. A single, broad, white, diagonal streak on each side is the most obvious mark, and the oval spiracles are red and orange. The feet are dark brown, and the head is green. The caterpillar has twin peaks behind the head.

Actual size

FAMILY	Saturniidae
DISTRIBUTION	New Guinea, northern Australia
HABITAT	Tropical rain forests
HOST PLANTS	Variety of rain forest trees, including Bleeding Heart (*Homalanthus populifolius*), *Dysoxylum* spp., and *Glochidion* spp.
NOTE	Fleshy, blue-green caterpillar that has eyespots to confuse predators
CONSERVATION STATUS	Not evaluated, but not threatened

ADULT WINGSPAN
Up to 10½ in (270 mm)

CATERPILLAR LENGTH
4 in (100 mm)

COSCINOCERA HERCULES
HERCULES MOTH
(MISKIN, 1876)

391

Hercules Moth caterpillars emerge from rusty-colored eggs, laid singly or in small groups on the leaves of host plants. The female moth lays up to 400 eggs that take about two weeks to hatch. The young caterpillars are mostly white and covered in small spines, but become blue in the later stages. The larvae are very large, with a voracious appetite to match. They eat a range of rain forest food plants, especially the leaves of the Bleeding Heart tree.

After three months, the mature caterpillars crawl to the ground to pupate, spinning a brown-colored cocoon wrapped in a leaf for camouflage. The emerging moths, named for their size after the Greek hero Hercules, are the largest in Australia and among the largest in the world. However, they lack any mouthparts so cannot feed and live only for a few days. The males are active at night, flying many miles in search of females.

The Hercules Moth caterpillar is blue green with orangey-red spiracles. The head is also blue green with yellowish-white stripes. The thoracic and abdominal segments bear rubbery, yellow-and-white spines, and the true legs are black. There are two false eyes at the rear end to deter predators.

Actual size

FAMILY	Saturniidae
DISTRIBUTION	Both sides of the Andes, from Venezuela to northern Peru and Bolivia
HABITAT	Mountain forests
HOST PLANTS	Unknown in the wild; in captivity has fed on Laurel Sumac (*Malosma laurina*)
NOTE	Nervous giant silkmoth caterpillar that is seldom still
CONSERVATION STATUS	Not evaluated

ADULT WINGSPAN
3¹⁄₁₆–5 in (78–129 mm)

CATERPILLAR LENGTH
2¾ in (70 mm)

392

DIRPHIA SOMNICULOSA
DIRPHIA SOMNICULOSA
(CRAMER, 1777)

The *Dirphia somniculosa* silkmoth caterpillar is extremely social. The larvae hatch in large numbers from white eggs 60 days after laying and remain close together during most of their development. They are black in all stages and in the laboratory have fed voraciously on Laurel Sumac. They grow to a large size and seem somewhat agitated, moving almost constantly. At the end of the final instar, the caterpillars descend to the ground, and each forms a loose cocoon of silk and debris under leaf litter. Pupation occurs within the cocoon, and it takes several months before the adult moth emerges.

Accidental contact with the *Dirphia somniculosa* caterpillar can result in a painful sting; all caterpillars in the Hemileucinae subfamily of moths to which it belongs have stinging spines. It is a common inhabitant of moderately high-altitude forest throughout most of the Andes and, like most species of giant silkmoth, is not a pest to agriculture.

The *Dirphia somniculosa* caterpillar is black with black feet, black rear claspers, and a black head. Its spiracles are narrow, red ovals with white at each end. The dorsal spines are long and black with many black side spines. White-tipped starbursts of lateral spines provide a color relief from the overall blackness of the caterpillar.

Actual size

FAMILY	Saturniidae
DISTRIBUTION	Only known from two small areas in Guerrero (Mexico) and Zacapa (Guatemala)
HABITAT	Mountainous tropical forests
HOST PLANTS	Unknown in the wild; in captivity has fed on oak (*Quercus* spp.)
NOTE	Giant silkmoth species that has an intriguing geographical range
CONSERVATION STATUS	Not evaluated

ADULT WINGSPAN
2⁹⁄₁₆–3¹¹⁄₁₆ in (59–93 mm)

CATERPILLAR LENGTH
2⅜ in (60 mm)

DIRPHIOPSIS WOLFEI

DIRPHIOPSIS WOLFEI

LEMAIRE, 1992

393

The *Dirphiopsis wolfei* silkmoth caterpillar hatches among tightly grouped, yellow eggs and soon sets off, following pheromone-laced web trails, to join other siblings that are searching for an acceptable feeding spot. It is initially yellow with a black head and black spines. By the fourth instar the caterpillar is reddish brown with white stripes and blue and reddish spines, and in the final fifth and sixth instars it is mostly green. Pupation occurs on the ground under leaf litter and involves very little silk.

Laboratory-reared caterpillars of *Dirphiopsis wolfei* have appeared nervous and hesitant, feeding reluctantly on oak. Of many hatched eggs, few have survived to adulthood. In the wild, the species' unusual range is intriguing because of the great distance between the two known populations—more than 700 miles (1,100 km). *Dirphiopsis wolfei* is one of 18, mostly South American, species within its genus and a member of the huge Hemileucinae subfamily of stinging caterpillars.

The *Dirphiopsis wolfei* caterpillar is green with a broad, reddish-brown spiracular band bordered with white. The spiracles are orange, the head and true legs are reddish, and the feet of the prolegs are black with tiny, white dots. Lateral spines are mostly blue, and dorsal spines are pale reddish. The dorsum and spiracular band are covered with white filigree.

Actual size

FAMILY	Saturniidae
DISTRIBUTION	Guiano-Amazonian region, from Venezuela and Colombia to Bolivia
HABITAT	Tropical forests
HOST PLANTS	Unknown in the wild; in captivity has fed on Laurel Sumac (*Malosma laurina*)
NOTE	Giant silkmoth caterpillar that has an unexpected sting
CONSERVATION STATUS	Not evaluated

ADULT WINGSPAN
3⁷⁄₁₆–5¹⁵⁄₁₆ in (88–149 mm)

CATERPILLAR LENGTH
3¾ in (95 mm)

394

EACLES BARNESI
EACLES BARNESI
SCHAUS, 1905

When the *Eacles barnesi* caterpillar hatches from its transparent egg, it is black with narrow, yellow lines across the dorsum and four long, black horns behind the head. In the second instar it becomes dark pink. In the laboratory the caterpillars feed rapaciously upon Laurel Sumac for about one month before digging into the soil to pupate. Unlike most members of its family, if molested this caterpillar can inflict a painful and itchy sting by erecting the nettle-like, red hairs usually hidden in cavities on the dorsum.

The *Eacles barnesi* moth is similar in size and shape to its cousin, the Imperial Moth (*E. imperialis*) of North and South America. There are 19 species of *Eacles* moths, mostly in South America, and many have colorful caterpillars. They are usually easy to rear and, like most giant silkmoths, are sought by hobbyists and nature students wishing to observe their metamorphosis from egg to adult.

The *Eacles barnesi* caterpillar is dark pink with small, dark bumps and eight bluish horns behind the head. The spiracles are white slits in a blackish sandwich. Reddish stinging hairs on the dorsum are normally hidden, and there is a prominent, thin, dark dorsal stripe. The clasper is blackish with a large, warty, pale patch.

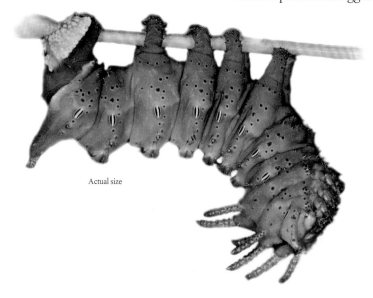

Actual size

FAMILY	Saturniidae
DISTRIBUTION	From Canada south to at least northern Costa Rica, and west to the Rocky Mountains
HABITAT	Deciduous and evergreen forests
HOST PLANTS	American Sweetgum (*Liquidambar styracifluar*), oak (*Quercus* spp.), maple (*Acer* spp.), sassafras (*Sassafras* spp.), and pine (*Pinus* spp.)
NOTE	Caterpillar that comes in two quite different color forms
CONSERVATION STATUS	Not evaluated, but common, though experiencing decline in northeastern United States

ADULT WINGSPAN
3⅛–6⅞ in (80–174 mm)

CATERPILLAR LENGTH
3–4 in (76–100 mm)

EACLES IMPERIALIS
IMPERIAL MOTH
(DRURY, 1773)

395

Imperial Moth eggs are laid in small groups of two to five, and the caterpillars are solitary. Initially orange with black horizontal stripes and long, black filaments, they later turn brown with an orange head, still retaining the filaments. These flexible, though spiny, projections protect the caterpillar from being swallowed by a predator and probably contribute to crypsis, by making the larva look more like a twig. In the last instar the thoracic projections become claw-shaped, spiny hooks, which if swallowed would undoubtedly damage the esophagus of a bird, deterring any future attacks. However, *Eacles imperialis* relies mostly on cryptic coloration to avoid being eaten.

The caterpillar briefly wanders on the ground before burrowing in and pupating in an underground chamber. The adult moths are leaf mimics, cryptic on a background of fallen leaves. Their patterns are highly variable, a feature likely aimed at evading birds, which are skilled at developing search images for a particular "leaf" pattern.

The Imperial Moth caterpillar can be orange with a lighter orange and black head, legs, and spines, and white spiracles—or it can also be a light or dark green with a yellow head. Its polymorphism is probably due to the fact that each color pattern has its advantages and disadvantages in nature, depending on the background on which the caterpillar rests, light conditions, and its most common predator.

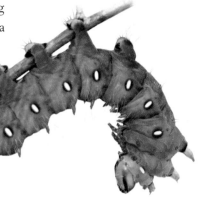

Actual size

FAMILY	Saturniidae
DISTRIBUTION	Southwestern United States, Mexico, and Guatemala
HABITAT	Thorn scrub, canyons, and foothills of desert mountain ranges
HOST PLANTS	Ash (*Fraxinus* spp.), Texas Ranger (*Leucophyllum frutescens*), Ocotillo (*Fouquieria splendens*), and Mexican Jumping Bean (*Sapium biloculare*)
NOTE	Striking caterpillar whose cocoon is used as an ankle rattle
CONSERVATION STATUS	Not evaluated, but generally common

ADULT WINGSPAN
3⅛–4¼ in (80–110 mm)

CATERPILLAR LENGTH
3–3⅛ in (75–80 mm)

396

EUPACKARDIA CALLETA
CALLETA SILKMOTH
(WESTWOOD, 1853)

Calleta Silkmoth caterpillars hatch from large, shiny, white eggs, laid shortly after mating by the adult females, who place them in masses on both surfaces of host plant leaves. First instar caterpillars are black and gregarious but become more solitary as they mature. Feeding takes place at leaf edges, and larvae consume much foliage during development, which usually takes four to five weeks. Pupation occurs within a silk cocoon spun on the host plant very close to the ground.

Adults may develop in as little as eight weeks, or pupae may remain dormant for up to two years. There seem to be two generations annually, with the large, striking, black-winged adult moths appearing in spring and fall. Adults emerge in the evening and mate in daylight the following morning after males locate females by following a pheromone trail. The cocoons of the Calleta Silkmoth are used as ankle rattles by Native Americans during ceremonial dances. *Eupackardia calleta* is the only species in its genus.

The Calleta Silkmoth caterpillar is variably turquoise, green, or pale blue with red and black transverse bands—actually a series of closely spaced red and black protuberances or fleshy tubercles on each segment. The apex of each tubercle is bright blue with small black spines. The true legs are dark, while the prolegs are usually the same color as the body, and the feet are tinged with yellow.

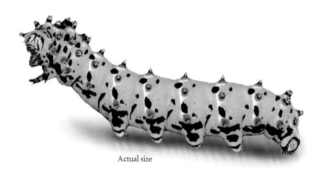

Actual size

FAMILY	Saturniidae
DISTRIBUTION	Angola, Zambia, Malawi, and Mozambique, south to South Africa
HABITAT	Grasslands, low forests, and semideserts
HOST PLANTS	Many trees, especially Mopane (*Colophospermum mopane*)
NOTE	Caterpillar that is a popular South African food
CONSERVATION STATUS	Not evaluated, but widespread, though disappearing in some areas

ADULT WINGSPAN
4⅛–5½ in (105–140 mm)

CATERPILLAR LENGTH
4 in (100 mm)

GONIMBRASIA BELINA

MOPANE WORM

WESTWOOD, 1849

397

The celebrated Mopane, or Mopani, Worm, feeds on the widespread Mopane tree in many habitats of southern Africa. The caterpillars hatch in spring in subequatorial Africa and complete five instars during the summer before burrowing in the ground to pupate for the winter, emerging as moths the following spring. This silkmoth caterpillar is widely eaten, and its commercialization is a multimillion dollar business in southern Africa. It is smoked, dried, or canned in tomato or chili sauce and sold in supermarkets. The adult Mopane Moth and caterpillar also appear on stamps and coins.

There are as many as 39 species of *Gonimbrasia* listed, all African. Adults of *G. belina* are variable and usually colorful, with an orange eyespot on each hindwing. The caterpillars can eat up to 90 percent of the leaves on Mopane trees in an area, but the leaves quickly grow back. Caterpillars once present in great numbers in some areas can no longer be found there due to overharvesting; domestication or reintroduction of the moths followed by better management practices are now being discussed.

The Mopane Worm caterpillar is black with irregular bands of tiny, white ovals grouped closely together, these alternating with bands of yellow on the sides and orange on the dorsum. The short, fleshy scoli are orange, and the head and feet are black. Medium-long white hairs sprawl from the scoli, thoracic shield, legs, and prolegs.

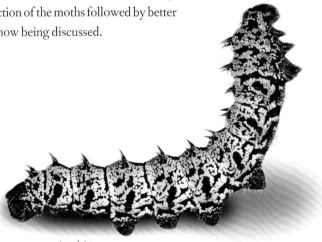

Actual size

FAMILY	Saturniidae
DISTRIBUTION	Southwest Europe
HABITAT	Pine forests, at 2,950–5,900 ft (900–1,800 m) elevation
HOST PLANTS	Pine trees, including *Pinus sylvestris*, *P. laricio*, and *P. uncinata*
NOTE	Colorful caterpillar that is under threat from habitat loss
CONSERVATION STATUS	Not evaluated, but considered vulnerable

ADULT WINGSPAN
2½–3⅜ in (63–85 mm)

CATERPILLAR LENGTH
2¾–3⅛ in (70–80 mm)

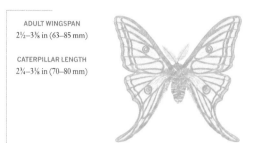

398

GRAELLSIA ISABELLAE
SPANISH MOON MOTH
(GRAËLLS, 1849)

The Spanish Moon Moth caterpillar has an apple-green body and brown head, legs, and prolegs. There is a dark brown stripe down the back, bordered on each side by a broken, white stripe. Laterally, there are alternate, oblique stripes of brown and white. The whole body is covered with tiny, white dots and long, white hairs.

The female Spanish Moon Moth lays up to 150 eggs, either singly or in small groups at the base of young pine needles. The caterpillars hatch within ten days and feed on the pine needles, where they are well disguised. When young, they are gray brown and resemble the twigs on which they rest, while the mature caterpillar often rests with its front segments puffed up to give the appearance of a pinecone. Growth is slow because of the cool temperatures, so the caterpillars take up to eight weeks to reach maturity.

When fully grown, the larvae crawl down from the tree and overwinter in a cocoon on the ground, among pine needles. The cocoon is golden brown, and its silken threads deter predators such as birds. The adult moths fly from spring to early summer. The species has become a rare sight—its decline is the result of adults being collected and, more recently, habitat loss.

Actual size

FAMILY	Saturniidae
DISTRIBUTION	South Africa, from Eastern Cape north to Mozambique
HABITAT	Open forests of Mopane (*Colophospermum mopane*) and *Brachystegia* spp.
HOST PLANTS	Acacia (*Acacia* spp.)
NOTE	Giant silkmoth caterpillar that resembles and mimics small leaves
CONSERVATION STATUS	Not evaluated

ADULT WINGSPAN
4⅝–5 in (120–130 mm)

CATERPILLAR LENGTH
3¼ in (95 mm)

GYNANISA MAIA (MAJA)
SPECKLED EMPEROR
(KLUG, 1836)

399

The Speckled Emperor caterpillar hatches among a small group of semiround eggs marbled brown and white. Initially black with a tan lateral stripe and black head, it becomes green after molting to the second instar. During its five instars, the larva develops long, silver spikes on its dorsum, which serve to break up its green appearance as it feeds on the tiny compound leaflets of the acacia tree, making it difficult to see.

Like the caterpillars of most African giant silkmoths, the Speckled Emperor descends to the ground after its fifth instar to pupate in the subterranean chamber it excavates, and remains there until the following rainy season. The adult Speckled Emperor is similar to the More Speckled Emperor (*Gynanisa nigra*), although *G. nigra* is smaller and darker, with greater speckling and a wider range, throughout much of eastern and central southern Africa. The genus *Gynanisa* includes more than 15 species, all African.

The Speckled Emperor caterpillar is green and densely covered with variously sized white and blue ovals. The dorsum has many toothlike, silver thorns tipped with yellow, and there is a green lateral band with yellow scoli. The spiracles are black ovals, each with a central yellow slit. The head is green, and the feet are black and yellow.

Actual size

FAMILY	Saturniidae
DISTRIBUTION	Areas of southwestern Canada and western United States, south to Baja California
HABITAT	Wide variation, including redwood and pine forests, and riparian areas
HOST PLANTS	Many, including mountain lilac (*Ceanothus* spp.), mountain mahogany (*Cercocarpus* spp.), Bitter Cherry (*Prunus emarginata*), and bitterbrush (*Purshia* spp.)
NOTE	Gregarious, spiny caterpillar that sometimes defoliates large areas of bitterbrush
CONSERVATION STATUS	Not evaluated

ADULT WINGSPAN
2³⁄₁₆–3⁷⁄₁₆ in (56–87 mm)

CATERPILLAR LENGTH
2¾ in (70 mm)

HEMILEUCA EGLANTERINA
ELEGANT SHEEP MOTH
(BOISDUVAL, 1852)

400

The Elegant Sheep Moth caterpillar hatches in the spring from a ring of white eggs deposited on a small twig of its host plant by the adult female during the previous summer or fall. The caterpillar is black in the first instar, but in most populations it develops a pattern of white or yellow longitudinal stripes and clusters of variably colored hairs and spines. The young caterpillars are processional when searching for food and remain tightly grouped, while more mature caterpillars become independent. After about six instars, the caterpillar makes a loose cocoon under leaf litter.

In the United States and Mexico, there are a number of sheep moth (*Hemileuca*) species, both diurnal and nocturnal, belonging to the silkmoth subfamily Hemileucinae, whose caterpillars all have stinging spines. The Elegant Sheep Moth is diurnal, and over its range of distribution it displays a wide variety of color in both caterpillar and adult; in northern and central California, for instance, the moth can be entirely bright pink.

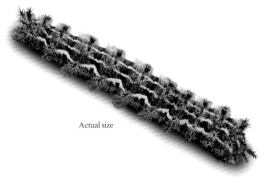

Actual size

The Elegant Sheep Moth caterpillar is usually dull black with three pale lines on each side, the spiracular one zigzagging up and down. It has rosettes of reddish and black spines on each segment and areas of sparse, white hairs dorsally and laterally, with reddish-brown hairs and integument on the true legs and prolegs.

FAMILY	Saturniidae
DISTRIBUTION	Eastern Africa, from KwaZulu-Natal north to Kenya and Uganda
HABITAT	Tropical forests and savannahs
HOST PLANTS	Privet (*Ligustrum* spp.), jasmine (*Jazminium* spp.), and *Rhus* spp.
NOTE	Caterpillars that are polymorphic, taking several different forms
CONSERVATION STATUS	Not evaluated

ADULT WINGSPAN
2–3⅜ in (50–85 mm)

CATERPILLAR LENGTH
2⅜ in (60 mm)

HOLOCERINA SMILAX
VARIABLE PRINCE SILKMOTH
(WESTWOOD, 1849)

401

When the short string of white-and-brown, marbled eggs of the Variable Prince Silkmoth begin to hatch, black caterpillars crawl out. Each sports two bright brown rectangles on its back as it wanders off to join others of its group on a leaf to feed. After the larva molts into its second instar, its color becomes one of several variations acquired among its siblings at that time. When it has completed only four instars, the caterpillar spins a flimsy cocoon of silk and debris above ground, and the adult moth emerges about five weeks later.

This silkmoth caterpillar belongs to the African tribe Ludiini of the subfamily Saturniinae. Some species of this tribe have polymorphic caterpillars, and at least three vastly distinct color patterns have been observed in the Variable Prince Silkmoth final instar. The adult moths are also quite dimorphic, with males and females having differently shaped wings and coloration. The caterpillars of this and other species of the Ludiini tribe possess irritating hairs.

Actual size

The Variable Prince Silkmoth caterpillar may be white with black bands and reddish head, or covered with bluish reticulations bordered with red and a black head, or all black with blue or red spots. Long, curved hairs sprout from stubby tubercles on its dorsum and sides, and white hairs droop over the legs.

FAMILY	Saturniidae
DISTRIBUTION	Southern Canada, most of the United States east of the Rocky Mountains
HABITAT	Forested areas where host plants can be found
HOST PLANTS	Decidous trees and bushes, including cherry (*Prunus* spp.), oak (*Quercus* spp.), maple (*Acer* spp.), Sweetgum (*Liquidambar styraciflua*), and Tree of Heaven (*Ailanthus altissima*)
NOTE	One of the largest North American caterpillars
CONSERVATION STATUS	Not evaluated, but common throughout most of its range

ADULT WINGSPAN
6¹⁄₁₆ in (152 mm)

CATERPILLAR LENGTH
4½ in (114 mm)

402

HYALOPHORA CECROPIA
CECROPIA MOTH
(LINNAEUS, 1758)

The Cecropia Moth caterpillar has a bright green ground color, and protuberances on the thoracic segments are usually red. Along its abdominal segments the protuberances are yellow, and those along the sides of the body are blue. The protuberances may include short, black bristles, but they are not urticating.

Newly hatched from large, cream-colored eggs mottled in reddish brown, Cecropia Moth caterpillars are black and covered in bristles. Although eggs may be laid in small groups, the larvae are solitary feeders and quickly go off on their own. When they enter the second instar they are yellow and still covered in black bristles growing from tubercles that will become blue, orange, or yellow. It is when the caterpillars reach the third instar that they begin to take on the stunning colors for which they are renowned.

Two types of cocoon may be formed during pupation. The first is tightly woven and compact, the second more loose and baggy. Sometimes leaves are incorporated into the cocoon, helping to conceal it from predators during the winter, which is the only time the caterpillars pupate. The large Cecropia Moth, which is ash gray mixed with red, tan, and white, is also called the Robin Moth because in flight it could be mistaken for a bird.

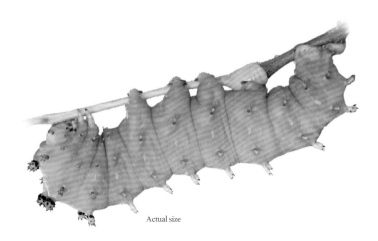

Actual size

FAMILY	Saturniidae
DISTRIBUTION	Southern Mexico to eastern Peru, Bolivia, and Brazil (Pará Province)
HABITAT	Tropical forests or scrub
HOST PLANTS	Unknown in the wild; in captivity has fed on acacia (*Acacia* spp.)
NOTE	Stinging caterpillar that develops through six or seven instars
CONSERVATION STATUS	Not evaluated, but widespread in tropical America

ADULT WINGSPAN
4⅛–5¼ in (105–135 mm)

CATERPILLAR LENGTH
4 in (100 mm)

HYPERCHIRIA NAUSICA
HYPERCHIRIA NAUSICA
(CRAMER, 1779)

403

In its early stages, the *Hyperchiria nausica* caterpillar is a bright yellow member of a tightly knit group and does not acquire the caramel color of a mature caterpillar until the sixth and seventh instars. When roaming from one leaf or branch to another, larvae proceed slowly and deliberately in single file. This caterpillar belongs to the subfamily Hemileucinae, whose members generally require six instars for males and seven instars for the larger females to mature. Most of the tropical species make a silk cocoon wrapped in leaves on or above the ground.

The mature caterpillar resembles a woolly bear caterpillar of Arctiinae, but, unlike the species of that subfamily, this one can sting. The adult moth looks like a dead leaf or animal face with two eyes, a nose, and mouth. *Hyperchiria nausica* is one of between 6 and 20 recognized species of *Hyperchiria*; it was previously classified in *Automeris* because of the eyespots on its hindwings, which are characteristic of members of that genus.

The *Hyperchiria nausica* caterpillar is short, thick-bodied, and blackened yellow in color. It is almost completely covered with protective rosettes of brownish yellow, poisonous spines that are tipped with white. The legs, prolegs, and head are black, and the abdominal feet are pinkish tan.

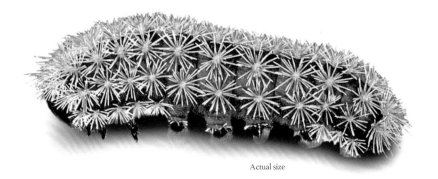

Actual size

FAMILY	Saturniidae
DISTRIBUTION	Eastern Africa, from KwaZulu-Natal north to southern Kenya, coastal areas
HABITAT	Tropical forests and savannahs
HOST PLANTS	Various, including Castor Bean (*Ricinus communis*), Guava (*Psidium guajava*), and mango (*Mangera* spp.)
NOTE	Caterpillar that is used as food in southern Africa
CONSERVATION STATUS	Not evaluated, but declining in some parts of its range

ADULT WINGSPAN
4⅛–5¼ in (105–135 mm)

CATERPILLAR LENGTH
4 in (100 mm)

IMBRASIA WAHLBERGI
WAHLBERG'S EMPEROR SILKMOTH
BOISDUVAL, 1847

The Wahlberg's Emperor Silkmoth begins life as an orange caterpillar during its first instar but, after molting to its second instar, becomes black with orange bands. It feeds in small groups during earlier stages but gains greater independence as it matures. Each instar requires about one week of feeding, and at the end of the fifth instar the caterpillar stops eating, becomes discolored, and discharges its gut content. It then burrows into the ground, where it forms a chamber in which to pupate.

Imbrasia wahlbergi belongs to a tribe, called Bunaeini, of large, colorful African moths that have a round eyespot on each hind wing and produce large, non-stinging caterpillars. The larvae, which are often plentiful, are actively sought for food by several southern African cultures. Overharvesting has greatly reduced some populations in certain areas, but, with proper management, the caterpillars can remain an important food source.

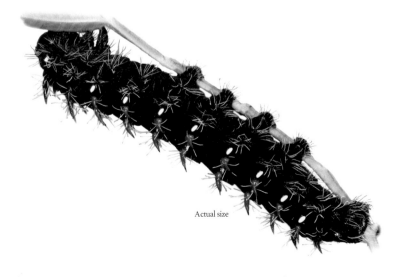

The Wahlberg's Emperor Silkmoth caterpillar is deep black with white, ovoid spiracles. There are a pair of orange, fleshy, backwardly bent sharp "horns" on the dorsum of each segment, with long, white, weak bristles radiating from each horn and regularly spaced over the entire caterpillar. The legs and head are black.

Actual size

FAMILY	Saturniidae
DISTRIBUTION	Central America (Costa Rica, Panama, and Guatemala)
HABITAT	Medium to upper elevation cloud forests
HOST PLANTS	Unknown; in captivity has fed on privet (*Ligustrum* spp.), ash (*Fraxinus* spp.), and Laurel Sumac (*Malosma laurina*)
NOTE	Rare, dramatic-looking caterpillar found in tropical cloud forests
CONSERVATION STATUS	Not evaluated, but likely vulnerable

ADULT WINGSPAN
3½–4 in (90–100 mm)

CATERPILLAR LENGTH
3⅛–3½ in (80–90 mm)

LEUCANELLA HOSMERA
LEUCANELLA HOSMERA
(SCHAUS, 1941)

405

Little is known about the striking *Leucanella hosmera*, which lives at high elevations in Central America. The caterpillars hatch from shiny green, oval eggs, seemingly laid in clusters or batches by female moths, and develop through six instars, with multiple generations likely to occur year-round. They march about in large groups throughout their development, feeding during daylight hours, probably on many different kinds of trees and bushes, although details of host plants are lacking.

The larvae advertise themselves with eye-catching, yellow and white spines, set off against a black background, openly challenging predators to take a mouthful. Those that are tempted discover to their cost something distasteful and likely stinging. Pupation takes place within a tough, brown cocoon, probably attached to the host trees near the ground. The genus *Leucanella* contains nearly 30 species with similar-looking caterpillars, including *L. viridescens*, a voracious, polyphagous eater that can be an economic pest of some crops.

The *Leucanella hosmera* caterpillar is black with striking starbursts of yellow and white spines. There are six starbursts on each segment, the majority yellow, but with some colored white or pink on the anterior three and posterior two segments. The black body is covered with minute, white dots. The head, true legs, and prolegs are black.

Actual size

FAMILY	Saturniidae
DISTRIBUTION	Southern (predominantly southeastern) Brazil, west to northern Bolivia, and south to northeastern Argentina
HABITAT	Forests or scrublands
HOST PLANTS	Many, incuding *Lonicera* spp. and *Solanum* spp.
NOTE	Striking caterpillar with related species throughout South American tropical latitudes
CONSERVATION STATUS	Not evaluated

ADULT WINGSPAN
2¾–3⅞ in (70–98 mm)

CATERPILLAR LENGTH
3 in (75 mm)

LEUCANELLA VIRIDESCENS

LEUCANELLA VIRIDESCENS

(WALKER, 1855)

The caterpillar of *Leucanella viridescens* is a striking sight, as it is usually encountered in tightly knit bunches of black bodies completely covered in dangerous-looking, bright yellow spines, which can deliver painful stings upon contact. When moving from one branch to another, the caterpillars form a single-file procession. After finishing feeding at the end of six instars, they go separate ways in search of a protected cavity or among branches to spin a papery cocoon wrapped in leaves or debris.

There are 28 species in the *Leucanella* genus, most of them colored similarly but with slight variations. They are found in many habitats, from hot lowland scrub to elevations of over 10,000 ft (3,050 m) in the Andes, where the caterpillars are commonly encountered. They occur wherever their preferred host plant grows but are polyphagous; in the laboratory, *L. viridescens* caterpillars are known to eat more than 49 different plant species.

The *Leucanella viridescens* caterpillar is black with white oval spiracles. On each segment there is a band of splayed clusters of overlapping yellow spines, each tipped with a stiff, sharp, stinging bristle. The head, feet, abdominal legs, and claspers are all black, and there are curved gray bristles on all legs.

Actual size

FAMILY	Saturniidae
DISTRIBUTION	Areas of sub-Saharan Africa
HABITAT	Tropical forest patches and savannahs
HOST PLANTS	Many, including Custard Apple (*Anona senegalensis*), *Psidium* spp., and *Sapium* spp.
NOTE	Large caterpillar collected and eaten in parts of Africa
CONSERVATION STATUS	Not evaluated

LOBOBUNAEA PHAEDUSA

BLOTCHED EMPEROR

(DRURY, 1782)

ADULT WINGSPAN
Up to 7⁹⁄₁₆ in (193 mm)

CATERPILLAR LENGTH
4¼ in (110 mm)

407

Hatching from a single egg laid separately on the host tree, the Blotched Emperor silkmoth caterpillar feeds alone. At the end of five instars, it moves to the ground to bury itself for pupation. If conditions are right, the adult moth may emerge within six weeks to produce a second cycle for the year. The larvae used to be abundant, but in many areas of Africa these and other large moth species are now becoming scarce in the wild as a result of overcollecting for food.

Africans, many of whom lack sufficient sources of protein, are actively encouraged to plant more trees and raise caterpillars for eating, which helps protect vegetation and prevent the extinction of certain caterpillar species that are in decline. The Blotched Emperor caterpillars are gathered mostly by children. The larvae's gut content is removed if they have fed on toxic trees, then they are boiled with chili peppers until almost dry and eaten immediately, or stored for up to three months.

The Blotched Emperor caterpillar is pale green, smooth, and plump. It is covered with small, round, darker green dots and has a yellowish lateral line through all segments. The oval spiracles are gold circled by yellow, and the thoracic and abdominal feet are black.

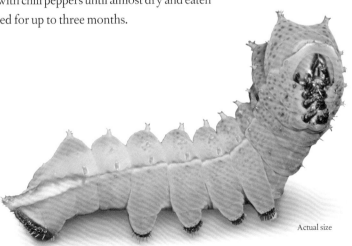

Actual size

FAMILY	Saturniidae
DISTRIBUTION	Southern China, Southeast Asia
HABITAT	Mostly upper mountain forests
HOST PLANTS	Various vines; in captivity it is reared on grapevines (*Vitis* spp.)
NOTE	Caterpillar whose shape and pose blend with vines
CONSERVATION STATUS	Not evaluated

ADULT WINGSPAN
2¾–3½ in (70–90 mm)

CATERPILLAR LENGTH
3 in (75 mm)

408

LOEPA MEGACORE
LOEPA MEGACORE
JORDAN, 1911

The *Loepa megacore* silkmoth caterpillar is entirely black when it hatches. It soon locates its siblings, all of which hatch during the morning of the same day, and joins their group to start feeding at the edge of a leaf. The caterpillar undergoes five instars, shedding the old skin after each instar. In later instars, it appears brownish black with a row of green triangles on each side. At the end of the fifth instar, it makes a narrow cocoon, pointed at both ends, and pupates inside.

There are up to 45 species of *Loepa* spread across Southeast Asia and Malaysia to India, many identified by genetic studies. The caterpillars and adults of most species are very similar in appearance. In spite of the popularity generated among collectors and insect-rearing hobbyists by *Loepa* species, there is little information available regarding the habits of this spectacular and widespread yellow and pink silkmoth genus.

Actual size

The *Loepa megacore* caterpillar is dark grayish brown, covered with black filigree and broad, black patches surrounding the spiracles and extending to the abdominal prolegs. Harmless dorsal rosettes of black spines, and longer black and gray hairs, bestow a shaggy appearance. The head is black, and there is a series of green, triangular lines each side of the body.

FAMILY	Saturniidae
DISTRIBUTION	Guiano-Amazonian basin, from the Atlantic to the eastern Andes, south to central Brazil and northern Bolivia
HABITAT	Forests and scrub, tropical to temperate
HOST PLANTS	Unknown; in captivity has fed on privet (*Ligustrum* spp.) and Laural Sumac (*Malosma laurina*)
NOTE	Stinging caterpillars that en masse can fatally injure humans
CONSERVATION STATUS	Not evaluated

ADULT WINGSPAN
2¾–4½ in (70–114 mm)

CATERPILLAR LENGTH
3 in (75 mm)

LONOMIA ACHELOUS
LONOMIA ACHELOUS
(CRAMER, 1777)

409

The black, newly hatched caterpillar of *Lonomia achelous* eats its eggshell before following the pheromone-laced silk trail of its siblings to join them. In the first instar, it possesses two pairs of long tubercles with forked tips arising from the second and third segments. As it grows, the larva changes shape, color, and the arrangement of its spines. The caterpillars are strictly nocturnal, feeding in the treetop by night and "galloping" fast in procession down the tree trunk to amass on the base each morning. When it has finished feeding, the caterpillar pupates among debris without a silk cocoon.

Stings from *Lonomia* caterpillars cause fatalities every year when people inadvertently come in contact with many larvae at once. However, the sting of a single *L. achelous* caterpillar is not dangerous, and the pain is not as intense as that caused by some other species in the Hemileucinae subfamily. The adult, a species of giant silkmoth, is only moderately sized and resembles a dead leaf.

The *Lonomia achelous* caterpillar is largely black on the dorsum, shading to light brownish gray on the sides with a light brown, double-lobed rectangle on the second, sixth, and seventh segments. Turquoise spines, long on the sides and short on the dorsum, have slender pink branches. There are three thin, white lines on each side, and the abdominal legs are reddish.

Actual size

FAMILY	Saturniidae
DISTRIBUTION	East and southeast African coastal areas, from Somalia south to South Africa
HABITAT	Tropical forest patches and savannahs
HOST PLANTS	Many, including *Bauhinia* spp. and Wild Fig (*Ficus chordate*)
NOTE	Large-headed caterpillar that is edible
CONSERVATION STATUS	Not evaluated

ADULT WINGSPAN
4–4½ in (100–115 mm)

CATERPILLAR LENGTH
3⅜ in (85 mm)

410

MELANOCERA MENIPPE
CHESTNUT EMPEROR
WESTWOOD, 1849

The Chestnut Emperor caterpillar is all black when it hatches from its brown-and-white mottled egg. It soon joins its siblings on a leaf, where they begin feeding. In captivity, the caterpillars have fed on Hong Kong Orchid leaves (*Bauhinia blakeana*) for 73 days, only during daylight, developing through six instars, before evacuating their guts and wandering around to find a place to bury themselves and pupate in the soil. When day length and climatic conditions are right, the beautiful, chestnut-red adult crawls out and spreads its wings.

There are eight species of *Melanocera* throughout Africa. Adults are similar in color and size but with identifying differences. The caterpillars of the Chestnut Emperor, like those of other African wild silkmoths, are used for food by humans. Women climb the trees to collect them, or the larvae are smoked from below, causing them to drop to the ground. It is now prohibited to cut down the trees they feed upon.

The Chestnut Emperor caterpillar is shiny black with a reddish-brown head and markings, which vary over its range of distribution. The skin feels granular and rough to the touch. The dorsum is studded with short, backwardly curved, harmless, fleshy thorns, which assist digging in preparation for pupation. The true legs are black, and the prolegs are brown.

Actual size

FAMILY	Saturniidae
DISTRIBUTION	West Africa, including Guinea, Ivory Coast, Sierra Leone, and Togo
HABITAT	Tropical forest patches and savannah
HOST PLANTS	Unknown; in captivity has fed on Laurel Sumac (*Malosma laurina*)
NOTE	Yellow or black silkmoth caterpillar that is edible
CONSERVATION STATUS	Not evaluated

ADULT WINGSPAN
2–2¾ in (50–70 mm)

CATERPILLAR LENGTH
2⅛ in (55 mm)

MICRAGONE HERILLA

MICRAGONE HERILLA
WESTWOOD, 1849

411

The *Micragone herilla* caterpillar can be seen inside its squarish, translucent yellow egg several days before it hatches. The first instar is pale yellow with a black head and is sparsely covered in stiff, reddish hairs. In captivity, caterpillars have been observed feeding together on leaves of Laural Sumac, developing two totally distinct forms of colors and patterns by the third instar. At the end of the fifth instar, they had all spun a loose cocoon among debris above ground, with the first adult emerging six weeks later.

Micragone herilla adults are sexually dimorphic. As in most species of giant silkmoth, the females are much larger than the males, enabling them to transport their heavy eggs while they search for a preferred host plant on which to place them. The *Micragone* genus contains 31 species that belong to the Micragonini tribe of the Bunaeinae subfamily of Saturniidae, and in Africa most caterpillars of these species are eaten as food.

The *Micragone herilla* caterpillar has at least two color forms. One is mostly yellow, with black bands containing black scoli encircling each segment. The scoli are longer dorsally, all carrying long, curved, black hairs. The other form is mostly black, with red bands and scoli sporting pale yellow hairs.

Actual size

FAMILY	Saturniidae
DISTRIBUTION	Eastern Mexico, south to northern Venezuela, Colombia, and Ecuador west of the Andes
HABITAT	Forests and scrub, tropical to temperate
HOST PLANTS	Leguminous trees; in captivity has fed on *Robinia* spp. and *Acacia* spp.
NOTE	Gregarious caterpillar whose appearance helps distinguish it from similar species
CONSERVATION STATUS	Not evaluated

ADULT WINGSPAN
2⅜–3¹¹⁄₁₆ in (60–94 mm)

CATERPILLAR LENGTH
2½ in (63 mm)

412

MOLIPPA NIBASA
MOLIPPA NIBASA
MAASSEN & WEYDING, 1886

When the tiny caterpillar of *Molippa nibasa* hatches from a brown-and-white egg, it is a greenish flesh color with a black head and translucent spines. After molting to the third instar, it is black with yellow bands of skin sprouting long, branching yellow spines and reddish legs and head. From the outset, the caterpillar is very social, feeding close together with its siblings. After about five weeks, it spins a papery cocoon covered by a leaf or debris in a hidden place.

Molippa nibasa is a member of the Hemileucinae, the subfamily of wild silkmoths with a large number of species—all with stinging caterpillars—that are found only in the Americas. Several of the species related to *M. nibasa* are difficult to tell apart as adults, but their caterpillars are distinct. The name of one of the authors of the species, Weyding, was inadvertently misspelled as Weymer in 1886 by the British entomologist Herbert Druce (1846–1913), an error still widely found today.

The *Molippa nibasa* caterpillar is greenish yellow with scattered small, black, variable blotches and white, oval spiracles. There is a broad, black, subspiracular band with tiny, white dots on each side, partially divided by broad, vertical bars of dark red. A band of long, branched yellow spines protects each segment. The legs and head are red.

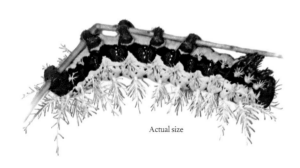

Actual size

FAMILY	Saturniidae
DISTRIBUTION	From Turkey east across much of Asia
HABITAT	Mountains at 3,300–9,850 ft (1,000–3,000 m) elevation, foothills, and orchards
HOST PLANTS	Spiraea (*Spiraea* spp.), pear (*Pyrus* spp.), ash (*Fraxinus* spp.), willow (*Salix* spp.), and cherry (*Prunus* spp.)
NOTE	Tree-feeding, lethargic caterpillar that lives in cool climates
CONSERVATION STATUS	Not evaluated

ADULT WINGSPAN
3⅛–4⅛ in (80–105 mm)

CATERPILLAR LENGTH
3⅛ in (80 mm)

NEORIS HUTTONI
NEORIS HUTTONI
MOORE, 1862

413

Neoris huttoni females live no longer than three days as adults and die within 24 hours of laying their large—¹⁄₁₆ in (2 mm)—oblong, olive-green eggs in neat strings of up to 15 on host plant twigs. The eggs overwinter, hatching in early spring. The caterpillars are lethargic feeders, working from beneath leaves, and are easily disturbed, even by rain. Development takes up to two months, and caterpillars descend to the ground to pupate, with pupation taking place in a creamy-white to reddish-brown cocoon among debris and leaves on the ground.

Most adults emerge from mid to late afternoon in late summer and early fall, with females calling for males that same night. Pairing takes place just after dark and lasts for only a few hours. Both males and females are rapid flyers, attracted to light and also very cold tolerant, with adults readily flying in frosty weather. There are a number of *Neoris huttoni* subspecies, which some authors consider to be distinct species.

The *Neoris huttoni* caterpillar is pinkish black or green grayish in color, and profusely covered with short and long, silver-white hairs. There is a broad, interrupted, yellow-orange stripe mid-laterally, below which the body is darker than above. The spiracles are prominent, orange, and lined with black. The head, true legs, and prolegs are black.

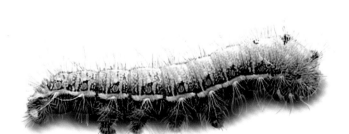

Actual size

FAMILY	Saturniidae
DISTRIBUTION	Central Africa, south of the Sahara
HABITAT	Tropical forests and savannahs
HOST PLANTS	Many, including *Vernicia* spp., *Bauhinia* spp., and *Anacardium* spp.
NOTE	Spiny but edible caterpillar
CONSERVATION STATUS	Not evaluated

ADULT WINGSPAN
3½ in (90 mm)

CATERPILLAR LENGTH
2⁹⁄₁₆ in (65 mm)

NUDAURELIA DIONE
GOLDEN EMPEROR
(FABRICIUS, 1793)

414

The caterpillars of the Golden Emperor silkmoth are tiny and yellow when they hatch from their small group of white eggs early in the morning. During the day they move around searching for the best place to feed, then, just after dark, they line up side by side under the tip of a leaf and begin to eat. In later instars, they separate as their color transforms from yellow to black, with yellow markings. After the fifth instar, they burrow into the ground to pupate.

Nudaurelia dione belongs to a large genus of African moths with 46 species, but the Golden Emperor is one of the most common and widespread among them, occurring mainly within 15 degrees of the equator. The caterpillar is also widely eaten by humans as it is an excellent source of protein. The backward-pointing, hornlike spines, which are designed to assist its eventual burrowing, are also eaten.

The Golden Emperor caterpillar is black with white oval spiracles. A pair of backward-pointing, yellow spines with sparse, radiating white bristles adorn the dorsum of almost every segment. The head, true legs, and prolegs are all black, sparsely studded with white, medium-long bristles. The head is large.

Actual size

FAMILY	Saturniidae
DISTRIBUTION	Across Australia, but mainly near the east coast
HABITAT	Eucalyptus forests
HOST PLANTS	Eucalyptus (*Eucalyptus* spp.) and others, including pine (*Pinus* spp.), apple (*Malus* spp.), and Apricot (*Prunus armeniaca*)
NOTE	Giant silkmoth species with pupation lasting up to ten years
CONSERVATION STATUS	Not evaluated

ADULT WINGSPAN
4¼–5½ in (110–140 mm)

CATERPILLAR LENGTH
3¾ in (95 mm)

OPODIPHTHERA EUCALYPTI
EMPEROR GUM MOTH
(SCOTT, 1864)

415

The Emperor Gum Moth caterpillar hatches in the spring or summer in Australia (October to March) and begins feeding. It is black and white in its first stage and progresses through five larval instars, molting into a larger and differently colored skin each time, ending up mostly green but multicolored in the final stage. The caterpillar spins a very tough and hard cocoon where it pupates. The adult usually emerges the following spring or summer, but depending on rainfall the pupa may remain in the cocoon for up to ten years before emerging.

The genus *Opodipthera* contains from 12 to 20 or more similar species, distributed in Australia, New Guinea, and several nearby islands, and represents most of the saturniid species in Australia. The Emperor Gum Moth caterpillar can be destructive to some wild eucalyptus trees but has not affected commercially grown species. It also feeds on introduced trees such as pine, apple, and apricot, but without significant damage.

The Emperor Gum Moth caterpillar is mostly green, blending into powder blue dorsally, with a white lateral stripe. It has a band of fleshy, bristle-tipped tubercles on each segment, dual-colored either yellow and blue, yellow and purple, red and blue, or red and purple. The spiracles are red, the head is green, and the legs are black and brown.

Actual size

FAMILY	Saturnidae
DISTRIBUTION	Mexico, south through Central America (Guatemala, Honduras, Nicaragua, Costa Rica, and Panama)
HABITAT	Dry and rain forests at mid elevations of 4,000–5,000 ft (1,220–1,525 m)
HOST PLANTS	Oak (*Quercus oleoides*)
NOTE	Silkmoth caterpillar found in the crowns of Central American oaks
CONSERVATION STATUS	Not evaluated

ADULT WINGSPAN
3⅛–4¼ in (80–110 mm)

CATERPILLAR LENGTH
3½–3¾ in (90–95 mm)

416

OTHORENE VERANA
OTHORENE VERANA
SCHAUS, 1900

Othorene verana caterpillars hatch from large, pale green eggs, laid by the female silkmoth about a week earlier. The larvae consume a vast quantity of host plant leaves, first instars eating from the leaf edges while later instars devour entire leaves. The caterpillars are generally solitary and grow rapidly, developing through five instars and reaching maturity in about four weeks. Mature larvae raise the anterior portion of their body when resting. Pupation takes place underground, and there appear to be at least two generations annually in most of the species' range.

Females lay their eggs in the first few hours after sunset. Like the males, they have short lives, as adult *Othorene verana* silkmoths do not feed. There are only four species in the genus *Othorene*, all large silkmoths living in Central America. *Othorene verana* caterpillars appear to live exclusively in the crowns of their host trees in mid-elevation dry forests.

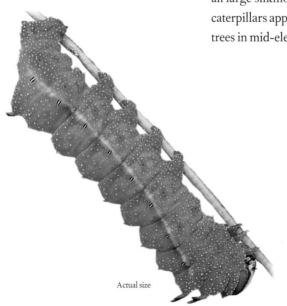

Actual size

The *Othorene verana* caterpillar is green in early instars but usually turns orange in later instars, although some individuals remain green. Small, white spots adorn the body, and there is a lateral, pale subspiracular stripe. The spiracles are slit-like and outlined in black. Four pairs of orange-red, forward-projecting tentacles occur anteriorly.

FAMILY	Saturniidae
DISTRIBUTION	From Mexico south through Central America to Panama
HABITAT	Mostly mid-elevation forests
HOST PLANTS	Unknown in the wild; in captivity has fed on a variety of plants, including species of Leguminosae and Rosaceae
NOTE	Rarely seen montane, stinging silkmoth caterpillar
CONSERVATION STATUS	Not evaluated

ADULT WINGSPAN
2⅜–3 in (60–75 mm)

CATERPILLAR LENGTH
2%₁₆–2¾ in (65–70 mm)

PARADIRPHIA LASIOCAMPINA
PARADIRPHIA LASIOCAMPINA
(R. FELDER & ROGENHOFER, 1874)

417

Paradirphia lasiocampina caterpillars hatch from pale colored eggs laid by the female silkmoth in small batches on host plant leaves. The early instars are gregarious, while later instars live relatively solitary lives. The larvae develop through six instars over a period of between four and six weeks. Before pupation, the caterpillars leave the host plant in search of a pupation site on the ground. A pupation chamber is formed about 4 in (100 mm) below the soil surface, and, unlike most other silkmoth caterpillars, this species does not use silk to line the chamber. The pupa is black and smooth.

Caterpillars of this silkmoth are well defended by their clusters of spines, which can cause mild to severe stinging in some people. There are about 30 described species in the silkmoth genus *Paradirphia*, most of them occurring in Central America and South America.

The *Paradirphia lasiocampina* caterpillar is reddish brown dorsally and white ventrally. Six yellowish, spiny tufts are present on each segment with an elongated tuft posteriorly. True legs and prolegs are crimson red as are the head and two pairs of forward-projecting tentacles located immediately behind the head.

Actual size

FAMILY	Saturniidae
DISTRIBUTION	Mexico, south to Brazil and Bolivia
HABITAT	Dry to humid tropical forests and savannahs
HOST PLANTS	Various trees, including oak (*Quercus* spp.), beech (*Fagus* spp.), hawthorn (*Crataegus* spp) plum (*Prunus* spp.), palm (*Elaeis* spp.), and *Acacia* spp.
NOTE	Stinging caterpillar of moth that ecloses smelling like rotting onions
CONSERVATION STATUS	Not evaluated

ADULT WINGSPAN
3¾–4⅝ in (95–120 mm)

CATERPILLAR LENGTH
2¾–4 in (70–100 mm)

418

PERIPHOBA HIRCIA
PERIPHOBA HIRCIA
(CRAMER, 1775)

Periphoba hircia caterpillars hatch from about 200 ovoid eggs laid by the female moth in straight-line pairs along host plant leaves. Hatching occurs around two to three weeks after ovipositing, depending on temperature. Early instar caterpillars are gregarious and processionary, feeding and resting in groups. By the fourth instar, caterpillars disperse and feed singly at night, resting by day. There are six instars, the larvae taking about 53 days to reach the pupal stage. After wandering for a day, full-fed caterpillars form a cocoon on the ground under debris.

The spines of this caterpillar can cause a painful sting lasting for a number of hours. Newly emerged adult moths give off a pervading and persistent odor of rotting onions. Females use pheromones to call males for mating, and both sexes live for less than a week. Occasionally, *Periphoba hircia* larvae cause economic damage to oil palm plantations.

Actual size

The *Periphoba hircia* caterpillar is blue green dorsally and light green ventrally. Its spiracles are orange, and the prolegs are translucent green. The anal segment and prolegs are outlined in pink. The body is densely covered with small, green spines.

FAMILY	Saturniidae
DISTRIBUTION	Parts of Italy and Austria, across the Balkans to the Caucasus, Turkey, Israel, and Lebanon
HABITAT	Dry, open deciduous woodlands
HOST PLANTS	Mostly oak (*Quercus* spp.), but occasionally poplar (*Populus* spp.) and other deciduous trees
NOTE	Hairy caterpillar that changes from black to green
CONSERVATION STATUS	Not evaluated

ADULT WINGSPAN
2⁷⁄₁₆–3⁷⁄₁₆ in (62–88 mm)

CATERPILLAR LENGTH
2⅜ in (60 mm)

PERISOMENA CAECIGENA

AUTUMN EMPEROR

(KUPIDO, 1825)

419

The female Autumn Emperor moth lays creamy-brown eggs on twigs in rows of up to six, with as many as 100 laid in total. The eggs overwinter, and the caterpillars hatch the following spring when the weather grows warmer. The larvae are gregarious at first, lying together on the upper surface of leaves and feeding on young leaves and the male catkins, but then move apart when they are older to become solitary. They pupate on the ground, where they spin a dark brown, double-walled cocoon among the leaf litter, spending the hot, dry summer as a pupa and emerging in fall as the weather cools.

The large, night-flying Autumn Emperors are, as their name suggests, on the wing in fall, from September to November. The males become active at dusk, while the females fly later in the evening. The moths are short-lived, surviving just a few days. Most mate on the first night, with the females rapidly laying all their eggs soon after.

Actual size

The Autumn Emperor caterpillar is large, green, and covered in short, white hairs. It has a yellow lateral stripe and six small, yellow tubercles on each segment. Each tubercle bears a tuft of long, white hairs.

FAMILY	Saturniidae
DISTRIBUTION	Central Chile
HABITAT	Forests, scrub, and hedgerows
HOST PLANTS	*Maytenus boaria*, *Cryptocarya rubra*, and others
NOTE	Singular, overwintering Hemileucinae caterpillar that has deceptive, large, black "eyes"
CONSERVATION STATUS	Not evaluated

ADULT WINGSPAN
2¹⁵⁄₁₆–3⁷⁄₁₆ in (74–88 mm)

CATERPILLAR LENGTH
3 in (75 mm)

420

POLYTHYSANA APOLLINA
POLYTHYSANA APOLLINA
R. FELDER & ROGENHOFER, 1874

The *Polythysana apollina* caterpillar is unlike those of other genera of the Hemileucinae subfamily of giant silkmoths, in both appearance and habits. It does not have large dorsal spines, is not very urticating, and feeds independently, mostly at night. It is not processionary, and after only five instars it spins a large, fluffy, white, well-insulated cocoon. When prodded, the caterpillar expands the dorsum of its first and second segments to reveal two large, black "eyes," complete with small, white "glints" surrounded by spiny "eyelashes," which have evolved as a defense to scare predators.

The *Polythysana* genus encompasses three species of brightly colored moths with red eyespots, all but one confined to Chile. The adults emerge in March, Chile's early fall, and adult males of several species can be seen simultaneously, undulating here and there over the low forest on warm, sunny days for less than two hours around midday, searching for females by locating pheromone trails. The resulting eggs develop slowly in the coolness of the season.

Actual size

The *Polythysana apollina* caterpillar is pinkish gray with irregular, light flecks. Long, feathery, branching spines curve downward from the sides. The dorsum is smooth with insignificant rosettes of inoffensive spines that are widely spaced. The head is hidden by forward-flattened spines. The first and second segments have dense, short, orange spines.

FAMILY	Saturniidae
DISTRIBUTION	Ecuador, Peru, and Bolivia
HABITAT	Low to medium altitude tropical forests
HOST PLANTS	Unknown; in captivity has fed on *Acacia* spp. and Laurel Sumac (*Malosma laurina*)
NOTE	Nervous caterpillar that likes to "gallop"
CONSERVATION STATUS	Not evaluated

ADULT WINGSPAN
3–4 in (75–100 mm)

CATERPILLAR LENGTH
3⅛ in (80 mm)

PSEUDAUTOMERIS POHLI

PSEUDAUTOMERIS POHLI
LEMAIRE, 1967

421

The *Pseudautomeris pohli* caterpillar, a member of the Hemileucinae subfamily of giant silkmoths, is initially white, acquiring colors and distinctive features as it develops. By the fourth instar its black dorsal spines at front and rear become immensely long, tipped with a white fork, and by the fifth instar the larvae are darker and begin to resemble a mature caterpillar. The caterpillars are processional and feed closely together but are extremely nervous at all times. Simply tilting the branch that they are feeding on can cause all caterpillars of the group to fall to the ground and race in separate directions.

The mature *Pseudautomeris pohli* caterpillar, which can deliver a painful sting if touched, is large, colorful, and constantly in motion. Some individual caterpillars complete only six instars, while others complete seven before spinning a papery cocoon. There are 24 species of *Pseudautomeris*, all in South America, except for one that reaches north to Costa Rica.

Actual size

The *Pseudautomeris pohli* caterpillar is black with widely spaced white dots and red spots, and buff oval spiracles. Each segment is studded with black spines, longer on the dorsum, which radiate blue branches tipped with white. The head is black and white, the true legs are red, and the abdominal feet and prolegs are black with yellow bumps.

FAMILY	Saturniidae
DISTRIBUTION	Most of Africa south of the Sahara and west of Ethiopia and Somalia
HABITAT	Tropical forests and savannahs
HOST PLANTS	Many, including Zebrawood (*Brachystegia speciformis*) and *Bauhinia* spp.
NOTE	Large, cryptic caterpillar that has many natural enemies
CONSERVATION STATUS	Not evaluated, but widespread

ADULT WINGSPAN
3½–4⅝ in (90–120 mm)

CATERPILLAR LENGTH
2¾ in (70 mm)

PSEUDOBUNAEA IRIUS
POPLAR EMPEROR
(FABRICIUS, 1793)

When it hatches, the caterpillar of the Poplar Emperor silkmoth is dark brown with a black head. By the second instar it is olive green, and in the third it is darker olive green and remains so through the fifth and final instar. Its color is lighter on the dorsum and darker on the ventrum because, like many large caterpillars, it hangs upside down. From below, the larva is pale like the sky; from above, it is dark like the earth. When it has finished feeding, the caterpillar descends to burrow into the ground, where it pupates.

Caterpillars have many natural enemies. A study carried out in South Africa revealed that, although bacterial and viral pathogens sometimes kill many Poplar Emperor larvae, parasitoids are more of a threat to the species, especially those that attack the pupae. In fact, the most significant enemies of *Pseudobunaea irius* are hawks, which eat many of the full-grown, surviving caterpillars.

Actual size

The Poplar Emperor caterpillar is green, lighter dorsally and darker ventrally, with orange, oval spiracles, and tiny, dark green dots. There is a distinct, narrow, greenish-white lateral line ending in a white loop around the head. The true legs are black, and there is a brown-and-white bar above the yellow-and-brown claspers.

FAMILY	Saturniidae
DISTRIBUTION	Northeastern South America, along the eastern slope of the Andes, forming a crescent south to Bolivia and southeastern Brazil
HABITAT	Humid forests
HOST PLANTS	Unknown; has eaten many different plants in captivity
NOTE	Well-camouflaged caterpillar that feeds at night
CONSERVATION STATUS	Not evaluated

ADULT WINGSPAN
2⅝ –4¾ in (67–123 mm)

CATERPILLAR LENGTH
3⅛ in (80 mm)

PSEUDODIRPHIA AGIS
PSEUDODIRPHIA AGIS
(CRAMER, 1775)

423

When it hatches from its white egg, the *Pseudodirphia agis* caterpillar is reddish brown with black spines. It is processional and, at all times during its early stages, remains in a dense group, with members working their way toward the nibbled edge of a leaf, taking turns to feed. By the third instar, the caterpillar is mostly brownish black with green spines and varied pink and white patterns on its sides. In the final instar, it becomes flattened in profile. When mature, the caterpillar of *P. agis* burrows into sphagnum moss or similar material and creates a silk-lined chamber for pupation.

Pseudodirphia agis is among 39 currently listed species of its genus, which belongs to the subfamily of Hemileucinae silkmoths. Caterpillars of this genus are urticating, and all feed at night and cluster parallel on a tree branch by day, flattened and well camouflaged. *Pseudodirphia* species are found from southeastern Mexico to Argentina, and many are difficult to tell apart.

Actual size

The *Pseudodirphia agis* caterpillar is flat, colored pale greenish white, and covered with a contrasting dark purplish-gray web forming various sizes and shapes of cells. The larger ones combine to form a large, loose triangle on each segment of each side, pointing toward the rear. The spines are green, with the lateral spines long and the dorsal spines short.

FAMILY	Saturniidae
DISTRIBUTION	Far southeastern Russia, northeastern China, Korean Peninsula, and Japan
HABITAT	Forests
HOST PLANTS	Trees, including oak (*Quercus* spp.), maple (*Acer* spp.), walnut (*Juglans* spp.), and willow (*Salix* spp.)
NOTE	Giant silkmoth caterpillar that squeaks when disturbed
CONSERVATION STATUS	Not evaluated

ADULT WINGSPAN
4¼–5 in (110–130 mm)

CATERPILLAR LENGTH
3⅛ in (80 mm)

RHODINIA FUGAX
RHODINIA FUGAX
(BUTLER, 1877)

424

The *Rhodinia fugax* moth caterpillar hatches in early spring after overwintering as an egg on a twig. In the first instar, it is black on the dorsum and yellow on the sides, but by the third instar the colors are reversed. During this period the dorsum has rows of tubercles studded with curved, black bristles, and the caterpillar groups with its siblings. In the fourth and fifth instars, it becomes solitary and often hangs motionless beneath a branch, its dark green ventrum blending with the darkness below, and the light green dorsum blending with the bright sky above.

The full-grown caterpillars, which are smooth to the touch and squeak when disturbed, spin a cocoon of green silk, shaped like a pitcher, with an open top and a small hole in the narrow base. The moths emerge to mate in the fall, and the females lay their eggs. There are 11 species of *Rhodinia* in Japan.

Actual size

The *Rhodinia fugax* caterpillar is dark green on the ventrum and pale green on the dorsum, with twin, rounded points on the dorsal peak behind the head. A thin, yellow lateral line is clearly visible. The body is smooth but covered with tiny, yellow knobs and a line of blue, bead-like scoli. The spiracles are pale orange, and the head and feet are green.

FAMILY	Saturniidae
DISTRIBUTION	Small area east of Lima, Peru
HABITAT	High desert, scattered shrubs at 6,600–9,850 ft (2,000–3,000 m) elevation
HOST PLANTS	Jatropha (*Jatropha* spp.)
NOTE	Caterpillar that lives at high altitude
CONSERVATION STATUS	Not evaluated

ADULT WINGSPAN
3¾ in (95 mm)

CATERPILLAR LENGTH
3 in (75 mm)

ROTHSCHILDIA AMOENA

ROTHSCHILDIA AMOENA
JORDAN, 1911

425

The *Rothschildia amoena* caterpillar lives in a cold climate high on the Pacific slope of the Andes near the coast. Its range is limited to the range of its host plant, a succulent species of jatropha bush, which grows in a limited, mostly rainless, humid desert area with sparse vegetation. *Rothschildia amoena*, unlike other *Rothschildia* species, has black, heat-absorbing skin, which is helpful for energy and digestion at high altitude. Less mature instars of *R. amoena* have orange tubercles, armed with harmless bristles, on all segments.

The species belongs to a subfamily of silkmoths, the Attacinae, and thus is closely related to the Atlas Moth (*Attacus atlas*) of Asia and the Cecropia Moth (*Hyalophora cecropia*) of North America. The caterpillars of this subfamily do not sting, are usually smooth to the touch, generally have five instars, and spin strong cocoons.

Actual size

The *Rothschildia amoena* caterpillar is black with short, white hairs that create a halo effect around its body. The prominent spiracles are orange, and the boldly banded true legs, prolegs, and head are all orange and black. The profile is very plump.

FAMILY	Saturniidae
DISTRIBUTION	Southeastern Mexico, south to Bolivia, and the Caribbean (Saint Lucia and Martinique)
HABITAT	Tropical forests and scrub
HOST PLANTS	*Cercidium microphillum*; in captivity has fed on plum and wild cherry (*Prunus* spp.)
NOTE	Silkmoth caterpillar that is unusually colorful
CONSERVATION STATUS	Not evaluated

ADULT WINGSPAN
3¾–4⅝ in (95–120 mm)

CATERPILLAR LENGTH
3⅛ in (80 mm)

426

ROTHSCHILDIA ERYCINA
ROTHSCHILDIA ERYCINA
(SHAW, 1796)

When the *Rothschildia erycina* caterpillar hatches from its brownish egg, it is black and yellow and usually among a dozen or more siblings. One by one the larvae make their way to where the first one to hatch has chosen to rest on the underside of an appropriate leaf. As they grow and molt, they change colors and detail in each of their five instars, finally becoming independent and displaying a broad pattern of white, black, green, and bright orange. This is aposematic coloration, warning birds that a caterpillar may be poisonous.

The mature caterpillar spins a strong cocoon, which it carefully hangs from a small branch, and the adult silkmoth emerges about six weeks later. The adult is a much smaller, streamlined version of the Atlas Moth (*Attacus atlas*) of Asia, to which it is related, but is more intricately patterned and more colorful. There are several subspecies of *Rothschildia erycina*, varying widely in size.

The *Rothschildia erycina* caterpillar is white or pale green with a black band on each segment, partly or mostly concealed on the dorsum by a broad, orange band. The tiny scoli, or tubercles, are studded with harmless, short, black spines. The head, true legs, and prolegs are black. The center of the ventrum is pale green.

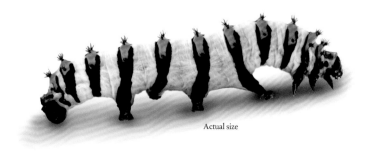

Actual size

FAMILY	Saturniidae
DISTRIBUTION	United States (southeastern Texas), south to southern Peru
HABITAT	Lower altitude tropical forests
HOST PLANTS	Wide variety, including ash (*Fraxinus* spp.), willow (*Salix* spp.), and *Citrus* spp.
NOTE	Giant silkmoth whose cocoons were once used as rattles
CONSERVATION STATUS	Not evaluated, but considered secure

ADULT WINGSPAN
4⅛–4⅝ in (105–120 mm)

CATERPILLAR LENGTH
3½ in (90 mm)

ROTHSCHILDIA LEBEAU

ROTHSCHILDIA LEBEAU
(GUÉRIN-MÉNEVILLE, 1868)

427

The *Rothschildia lebeau* caterpillar hatches from among a group of white eggs and immediately eats the remaining eggshell. Black with yellow spots at this stage, it groups together with its siblings as they feed at the leaf edge. As the caterpillar grows and molts through five instars, its colors change each time, and it moves away from its group. After about five weeks it expels its gut contents and spins a teardrop-shaped, hard, silvery cocoon suspended by a stem of multiple threads fastened to a tree branch.

There are 29 species of *Rothschildia* presently recognized, some with quite colorful caterpillars. They are found only in the Americas and the Caribbean, and distributed from lowland tropical forests to high-altitude habitats in the Andes. Three species have been recorded in the United States. In the past, Native Americans made rattles with small rocks sewn into empty *Rothschildia* cocoons. A window in each wing of *Rothschildia* adults has elicited the name "four eyes" in Mexico.

The *Rothschildia lebeau* caterpillar is green with a broad, white fore edge on most segments of its body, which also has a row of widely spaced, small, yellow tubercles tipped with short bristles on each segment. The head is green, and the true legs are banded in black and yellow. The abdominal legs are black and yellow.

Actual size

FAMILY	Saturniidae
DISTRIBUTION	China and Korea, but naturalized in Japan and other areas across Asia, Europe, North and South America, Africa, and Australia
HABITAT	Lowland forests in China, and in parks and gardens in areas where naturalized
HOST PLANTS	Tree of heaven (*Ailanthus altissima*) and various shrubs
NOTE	Distinctive, silk-producing caterpillar found in many urban areas
CONSERVATION STATUS	Not evaluated

ADULT WINGSPAN
4⅛–5½ in (105–140 mm)

CATERPILLAR LENGTH
2¾–3 in (70–75 mm)

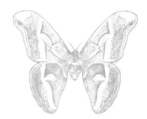

428

SAMIA CYNTHIA
AILANTHUS SILKMOTH
(DRURY, 1773)

The Ailanthus Silkmoth caterpillar—used to produce wild silk—hatches from pale eggs laid in crescent-shaped rows of 10 to 20 on leaves up to 20 days earlier. The young caterpillars are yellow with black-tipped tubercles, becoming pale green with age. They are gregarious at first but move apart in the latter stages, becoming solitary. The caterpillar spins a grayish cocoon attached to the petiole of a leaf on the host plant.

Ailanthus Silkmoths fly in early summer and, in some parts of their range, there may be a second generation in late summer. The silkmoth is native to China and Korea, where its cocoons were first collected for wild silk, but has been introduced across the world in an attempt to establish new silk industries. It is now found in many cities and towns where the ornamental Tree of Heaven is grown. A related species, *Samia ricini*, is fully domesticated for the production of eri silk, the term "eri" being derived from an Assamese word referring to the Castor-oil Plant (*Ricinus communis*) on which the caterpillars feed.

The Ailanthus Silkmoth caterpillar is pale green to white with two rows of distinctive, white tubercles down the back and a lateral row of small, black spots. The head is pale green, and there are ridges of pale blue near the prolegs. The whole body has a powdery appearance.

Actual size

FAMILY	Saturniidae
DISTRIBUTION	Europe, and across northern Asia to Russian Far East and northern China
HABITAT	Moorlands, heathlands, open scrub, field margins, and woodland edges
HOST PLANTS	Various, including birch (*Betula* spp.), willow (*Salix* spp.), Heather (*Calluna vulgaris*), and bramble (*Rubus* spp.)
NOTE	Bright green caterpillar that basks in the sun
CONSERVATION STATUS	Not evaluated

ADULT WINGSPAN
1⁹⁄₁₆–2⅜ in (40–60 mm)

CATERPILLAR LENGTH
Up to 2⅜ in (60 mm)

SATURNIA PAVONIA
EMPEROR MOTH
(LINNAEUS, 1758)

429

Emperor Moth caterpillars hatch from eggs laid in batches 10 to 14 days earlier on the stems of a variety of host plants. The young larvae, which consume part of their eggshell before congregating together, are hairy and black but become green with age. Initially, they are gregarious but disperse when older. When disturbed, the caterpillars release a bitter fluid that deters predators, such as birds, parasitic flies, and ants. They crawl into thick vegetation to pupate.

The pupae overwinter in pear-shaped cocoons, varying in color from white to pale brown, that are attached to a plant stem near the ground. The emerging adult is a spectacular moth, easily identified by its large wingspan and the ornate pattern of eyespots designed to scare predators; when threatened, it flutters its wings to flash the eyespots. The day-flying moths are on the wing from late spring to early summer and do not feed.

The Emperor Moth caterpillar is vivid green in color. Each segment is encircled by a broken black band containing yellow, pink, or orange wartlike spots. Each spot is covered with a tuft of short, black hairs. Small, white hairs cover the rest of the body.

Actual size

FAMILY	Saturniidae
DISTRIBUTION	North America
HABITAT	Woodlands
HOST PLANTS	Honey Locust (*Gleditsia triacanthos*) and Kentucky Coffee Tree (*Gymnociadus dioicus*)
NOTE	Bright green caterpillar that is difficult to spot among leaves
CONSERVATION STATUS	Not evaluated

ADULT WINGSPAN
1⅞–2⅝ in (47–67 mm)

CATERPILLAR LENGTH
2⅛ in (55 mm)

430

SYSSPHINX BICOLOR
HONEY LOCUST
(HARRIS, 1841)

Honey Locust caterpillars hatch from pale green eggs laid in clusters on the underside of leaves of the host plants. The young larvae are gregarious but disperse to become solitary at later instars. They are well camouflaged, with the red-and-white lateral lines helping to break up the body shape and provide countershading. The larvae develop quickly and can be ready to pupate in as little as three weeks. Pupation takes place underground, where the pupae overwinter.

This fast-growing species, named for the caterpillar's principal host plant, the Honey Locust, often has three generations a year. The adult moths fly from April to September, their wing color differing according to the generation from which they eclose, ranging from gray in the first generation to yellow brown in the second, and then dark brown in the final brood. The caterpillar is similar in appearance to that of the Bisected Honey Locust Moth (*Sphingicampa bisecta*), also found on Honey Locust trees.

The Honey Locust caterpillar is lime green with red-and-white lateral lines and speckled with tiny, white dots. There are two pairs of red, thoracic horns and a single, red horn on the posterior segment, as well as several silvered horns on the abdomen. The head is green with a yellow stripe.

Actual size

FAMILY	Saturniidae
DISTRIBUTION	Guiano-Amazonia (northern South America, south and east of the Andes)
HABITAT	Tropical forests and savannahs
HOST PLANTS	Trees of the Bombacaceae family
NOTE	Caterpillar that has two long, straight "horns" at immature stages
CONSERVATION STATUS	Not evaluated

ADULT WINGSPAN
4³⁄₁₆–6¹⁄₁₆ in (107–153 mm)

CATERPILLAR LENGTH
3¾ in (95 mm)

TITAEA LEMOULTI
TITAEA LEMOULTI
(SCHAUS, 1905)

431

The *Titaea lemoulti* caterpillar, when it hatches from its white egg, is banded black and white, with orange head, legs, and tail. It has two long, fork-tipped tubercles behind its head and an erect one on the last segment. All subsequent instars, except the fifth (last) instar, possess two very long "horns" on the dorsum behind the head. In order to avoid detection by birds or other predators, the mature caterpillar is colored in a pattern to somewhat resemble the V shapes formed by the leaves of its host plant, such as *Bombacopsis* species.

The fully grown caterpillar burrows deep into the ground to pupate, as do other members of its subfamily, the Arsenurinae. There are five species of *Titaea* occurring from lowland southern Mexico south to Bolivia and southern Brazil. The adults of most species are gray, brown, or reddish with darker markings, and the males have small, square tails. *Titaea* silkmoths and caterpillars are often common and very large.

The *Titaea lemoulti* caterpillar is mostly bright green and covered with tiny, reddish-brown dots. On each side it has two large, opposing, diagonal patches of dark bluish-gray color outlined in narrow black with broad, light yellow borders. The spiracles are dark with two white dots. The head, feet, and border of the claspers are pale brownish orange.

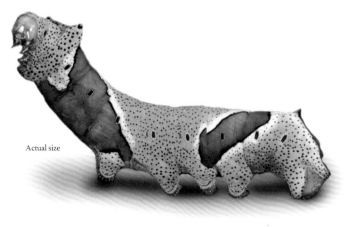

Actual size

FAMILY	Saturniidae
DISTRIBUTION	Western section of Great Escarpment from Cape Town, South Africa, north to southern Namibia
HABITAT	Semidesert
HOST PLANTS	*Eriocephalus* spp.
NOTE	Caterpillar that rolls into a hairy ball when disturbed
CONSERVATION STATUS	Not evaluated

ADULT WINGSPAN
1¾₆–1¾ in (30–45 mm)

CATERPILLAR LENGTH
1¼ in (45 mm)

432

VEGETIA DUCALIS

DUCAL PRINCELING

JORDAN, 1922

Ducal Princeling silkmoth caterpillars hatch from eggs clustered around a twig on the host plant. Unlike most silkmoth species, the new hatchlings do not congregate but continuously wander around, nibbling on leaves at random. By the second instar, the dorsal hairs of the larva become very long, and when it is disturbed it rolls up in a hairy ball. At the end of the fifth instar, the caterpillar quickly spins a cocoon near the base of a bush, decorating it with dead leaves and debris.

Vegetia ducalis normally feeds only on *Eriocephalus* species, although in captivity the larvae are known to accept California Sagebrush (*Artemisia californica*), which has a similar odor, and to thrive on it. The adult moths, which are diurnal, emerge from their cocoons about noon, with the male flying and mating as quickly as ten minutes later. The adults are similar in size, shape, color, and habits to the *Calosaturnia* moths of southern California, where the climate is comparable.

The Ducal Princeling caterpillar is black with two lateral, undulating, longitudinal stripes. Laterally, it is reddish brown, bordered slightly with white. The spiracles are black, bordered with white, and dorsally there are tufts of densely packed, white hairs that resemble long thorns. The head is dark brown, and the legs and prolegs are reddish brown. Long, white hairs obscure much of the body.

Actual size

FAMILY	Sphingidae
DISTRIBUTION	Africa, the Azores, Europe (except the far north), the Middle East, and western Asia; rare summer migrant to Iceland and northern Europe
HABITAT	Cultivated areas and open scrub with solanaceous plants, especially drier and sunnier locations
HOST PLANTS	Numerous Solanaceae, Bignoniaceae, Verbenaceae, and Oleaceae, and several species from other families
NOTE	Horned caterpillar of a species steeped in myth and superstition
CONSERVATION STATUS	Not evaluated, but common in Africa

ADULT WINGSPAN
3½–5 in (90–130 mm)

CATERPILLAR LENGTH
4⅝–5 in (120–130 mm)

ACHERONTIA ATROPOS
DEATH'S HEAD HAWKMOTH
(LINNAEUS, 1758)

433

Death's Head Hawkmoth caterpillars hatch from green or gray-blue eggs, laid singly beneath old leaves of the host plant. Initially pale yellow with a disproportionately long horn, the larva turns dark green as it feeds. In the fourth instar, its coloring becomes more vivid, and, in the final instar, the horn takes on a downcurved shape. The mature caterpillar is large and lethargic but, if threatened, will click its mandibles and even try to bite. It can strip small plants bare and is an occasional local pest. The caterpillars were once a common sight in the potato fields of Europe, but the widespread use of insecticides has reduced their numbers.

Both larvae and the mahogany-brown pupae are cold sensitive, and few survive the winter north of the Mediterranean. The adult moth, which squeaks, is an annual migrant north into central and northern Europe. In the past, it aroused superstitions due to the skull-like marking on the thorax, and the moth featured on a publicity poster for the thriller film *The Silence of the Lambs*; the live moths in the movie, however, were *Acherontia styx*, which is very similar to *A. atropos*.

The Death's Head Hawkmoth caterpillar is green, brown, yellow, or pale creamy yellow. All forms are smooth when fully grown, with seven oblique lateral stripes. These are purple, edged with blue and yellow in the green and yellow forms, but barely visible in the brown form. The head has black cheeks, and the yellow anal horn is warty and downcurved, with a reflexed tip.

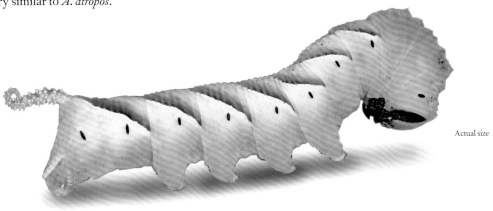

Actual size

FAMILY	Sphingidae
DISTRIBUTION	The tropics, subtropics, and temperate regions of the world (excluding North and South America); rare summer migrant to Iceland and northern Europe
HABITAT	Warm, dry cultivated areas and open scrub or steppe; during migrations almost anywhere except dense forests
HOST PLANTS	Various Convolvulaceae, especially *Convolvulus* spp., *Calystegia* spp., and *Ipomoea* spp.
NOTE	Sluggish caterpillar of a noted migrant with spectacular flying skills
CONSERVATION STATUS	Not evaluated, but common in its range

ADULT WINGSPAN
3¾–5 in (95–130 mm)

CATERPILLAR LENGTH
4–4¼ in (100–110 mm)

AGRIUS CONVOLVULI
CONVOLVULUS HAWKMOTH
(LINNAEUS, 1758)

434

The Convolvulus Hawkmoth caterpillar is initially glaucous, with a straight, black-tipped horn, but gradually acquires a light green coloration through feeding. In the second and third instar, the ground color deepens further, and pale yellow lateral stripes appear. By the fourth instar, different color forms occur, such as brown, green, and, occasionally, yellow. The horn is robust, curved, and smooth.

Convolvulus Hawkmoth caterpillars hatch from spherical eggs deposited singly on the upper side or underside of a host plant leaf. Hiding initially beneath a nearby leaf, the larvae slowly nibble holes through it. By the fourth instar, more is eaten, and growth accelerates. When fully grown, most (but not all) caterpillars hide by day and feed only at night. All are extremely sluggish until later stages, when they may wander rapidly in search of a pupation site. Just before this, they anoint themselves with "saliva," which seems to aid skin darkening, making the caterpillars less visible on the ground.

Although this species penetrates farther north and in greater numbers than the Death's Head Hawkmoth (*Acherontia atropos*), it is more cold sensitive, and fewer of its glossy brown pupae survive northern winters. Even in North Africa and the Middle East, it is not permanently resident. However, in "good" years many full-grown larvae can be found in northern gardens. In the tropics, they can become serious pests of sweet potato.

Actual size

FAMILY	Sphingidae
DISTRIBUTION	Eastern Asia, including Japan and parts of Southeast Asia
HABITAT	Deciduous and coniferous woodlands
HOST PLANTS	Elm (*Ulmus* spp.), poplar (*Populus* spp.), and willow (*Salix* spp.)
NOTE	Well-camouflaged caterpillar that is often found on urban trees
CONSERVATION STATUS	Not evaluated, but generally common

ADULT WINGSPAN
2³⁄₁₆–3¼ in (56–82 mm)

CATERPILLAR LENGTH
2⅜–3⅛ in (60–80 mm)

CALLAMBULYX TATARINOVII
ELM HAWKMOTH
(BREMER & GREY, 1853)

435

The eggs of the Elm Hawkmoth are oval and laid singly by females during spring on the host plants. The caterpillars develop slowly but usually mature by between July and August, protected by the camouflage of their green coloration, which perfectly matches the host plant greenery. There may be a single generation or two generations during the spring to fall period, depending on temperature. Mature caterpillars are often found wandering on the ground looking for a suitable place to build an earthen cell for pupation.

The mahogany-brown pupae overwinter, and adults emerge in late summer, flying, egg laying, and feeding by night. The Elm Hawkmoth comprises a number of subspecies through its geographic range, with some populations distinctly smaller in size. Little is known of the natural enemies of *Callambulyx tatarinovii*, and very few caterpillars collected for rearing have proven to be parasitized. The eggs and mature larvae of *C. tatarinovii* closely resemble those of the Eyed Hawkmoth (*Smerinthus ocellata*).

The Elm Hawkmoth caterpillar is light green, with a distinct, narrow, cream dorsal line and alternate bold and faint, oblique lateral stripes. The bold lateral stripes may be yellow or white, bordered with red. The horn is almost straight and reddish. The entire body is punctuated with many yellow spots.

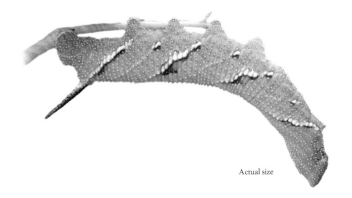

Actual size

FAMILY	Sphingidae
DISTRIBUTION	Africa south of the Sahara, India, China, Japan, Southeast Asia, and Australia
HABITAT	Urban parks and gardens, and lightly wooded areas
HOST PLANTS	Gardenia (*Gardenia augusta*), Coffee (*Coffee arabica*), Wild Pomegranate (*Burchellia bubaline*), and Rhino Coffee (*Kraussia floribunda*)
NOTE	Well-camouflaged, voracious, green caterpillar often found in gardens
CONSERVATION STATUS	Not evaluated, but generally common

ADULT WINGSPAN
1¾–2⅞ in (45–73 mm)

CATERPILLAR LENGTH
2⅛–2⁹⁄₁₆ in (55–65 mm)

436

CEPHONODES HYLAS
COFFEE BEE HAWKMOTH
(LINNAEUS, 1771)

Coffee Bee Hawkmoth caterpillars hatch from eggs laid singly on the underside of terminal host plant leaves, with the dark-colored first and second instars sporting a long, posterior horn. They feed alone and are sluggish but consume a great amount of leaf material. Older larvae, when disturbed, will throw their head back and eject green fluid from their mouth to repel potential predators. Final instar caterpillars wander and become brownish in color, making them less conspicuous as they travel over the ground.

Pupation occurs on, or rarely just below, the surface in a slight cocoon formed of leaves and debris held together with a few strands of silk. Parasitic wasps and flies have emerged from reared larvae, suggesting these natural enemies may be important in population regulation of the species. As its name suggests, the Coffee Bee Hawkmoth mimics bees, flying by day with a distinctive buzzing sound and darting rapidly from flower to flower.

The Coffee Bee Hawkmoth caterpillar is variably colored but frequently green with a prominent lateral, white stripe. The posterior horn is slightly curved, whitish, and speckled with black dots. The head and true legs are green, the prolegs brown or tan. Clusters of white dots appear on the dorsal surface of the first segment, behind the head, and on the posterior segment. A dark form of the caterpillar also occurs.

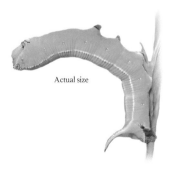

Actual size

FAMILY	Sphingidae
DISTRIBUTION	India, China, Japan, Chinese Taipei, and Korea
HABITAT	Lowland forests, parks, and gardens
HOST PLANTS	Indian Beech (*Millettia pinnata*), Malabar Kino (*Pterocarpus marsupium*), Soybean (*Glycine max*), ox-eye bean (*Mucuna* spp.), and thorn tree (*Acacia* spp.)
NOTE	Caterpillars that are a delicacy in China
CONSERVATION STATUS	Not evaluated, but generally common

ADULT WINGSPAN
3¹¹⁄₁₆–6 in (94–150 mm)

CATERPILLAR LENGTH
3⅜–4 in (85–100 mm)

CLANIS BILINEATA
TWO-LINED VELVET HAWKMOTH
(WALKER, 1866)

437

Two-lined Velvet Hawkmoth caterpillars hatch from smooth and shiny eggs laid singly on the underside of host plant leaves, where they hang, well camouflaged, for much of the time. When resting, mature larvae raise the front part of their body with the head bowed and the true legs held together. Late-stage caterpillars are usually found quite high up, 10–20 ft (3–6 m), toward the ends of branches. Pupation takes place on the ground in a loosely spun cocoon, and there are multiple broods per season in most locations from spring to fall.

No parasitoids have been reported in the caterpillars, although they can suffer from a virus that transforms them into sacs of foul-smelling liquid. Adult moths are nocturnal but usually do not fly until the early hours of the morning. In China, the *Clanis bilineata* caterpillar is an economic pest of Soybean (*Glycine max*) but also considered a delicacy, with a protein content comparable to that of milk or eggs.

The Two-lined Velvet Hawkmoth caterpillar is light to medium green, covered with numerous, tiny, yellow dots. Laterally, there are seven, pale yellow, oblique stripes. The posterior horn is green and relatively reduced compared to other hawkmoth species. The true legs are tan colored, and the head and prolegs are green.

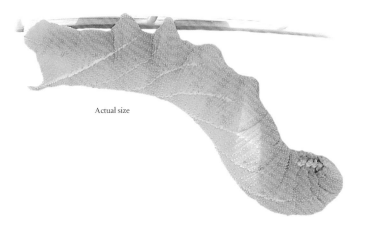

Actual size

FAMILY	Sphingidae
DISTRIBUTION	Africa, Madagascar, entire Mediterranean region, Middle East, Afghanistan, India, southern China, Southeast Asia to the Malay Peninsula, the Philippines, and Hawaii; also central Europe, central Asia, and Japan as a migrant
HABITAT	Dry riverbeds, oases, and scrubby hillsides
HOST PLANTS	Wide range, including Oleander (*Nerium oleander*), grape (*Vitis* spp.), and periwinkle (*Vinca* spp.)
NOTE	Striking caterpillar with eyespots and a horn to deter predators
CONSERVATION STATUS	Not evaluated, but not threatened

ADULT WINGSPAN
3½–4¼ in (90–110 mm)

CATERPILLAR LENGTH
3–3⅜ in (75–85 mm)

DAPHNIS NERII
OLEANDER HAWKMOTH
(LINNAEUS, 1758)

438

Oleander Hawkmoth caterpillars emerge from light green, round eggs, laid singly on young leaves up to 12 days earlier. Initially, the caterpillar is bright yellow green with a long, thin, black horn. It gradually changes to the apple green of the older caterpillar, and then becomes browner in color as it prepares to pupate on the ground among leaf litter. The pupa is light brown with tiny, dark brown spots, in a loosely spun, yellow cocoon.

The species gets its common name from the caterpillar's main food plant, the Oleander. Leaves of the Oleander contain toxins, which are taken up by the larvae to give them protection against predators. The species is one of the most widely distributed hawkmoths. However, although the moths migrate to northern Europe in summer, they cannot survive the cold winters. The adults are active from late spring to early fall, producing four or five generations in a year.

The Oleander Hawkmoth caterpillar is apple green in color, with distinctive, blue-and-white eyespots, ringed in black. The spiracles are black, and the legs are pink. There is a lateral, white stripe with scattered, white spots. The short, orange horn is warty with a black tip.

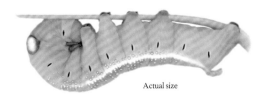

Actual size

FAMILY	Sphingidae
DISTRIBUTION	Temperate Europe across temperate west Asia and southern Siberia to Russian Far East, Japan, and eastern and central China; India (Sikkim, Assam), Nepal, Bhutan, and northern Myanmar; also southern British Columbia (Canada) as an introduction
HABITAT	Ditches and river margins on floodplains, damp forest clearings and margins, even town wastelands, and damp meadows up to 4,920 ft (1,500 m) elevation in the Alps
HOST PLANTS	Herbaceous Onagraceae, especially *Epilobium* spp., and Rubiaceae, often *Galium* spp.
NOTE	Alarming, snakelike caterpillar of attractive pink-and-khaki moth
CONSERVATION STATUS	Not evaluated, but common, especially in warm, damp riverine areas

ADULT WINGSPAN
2⅜–3 in (60–75 mm)

CATERPILLAR LENGTH
2¾–3⅛ in (70–80 mm)

DEILEPHILA ELPENOR
LARGE ELEPHANT HAWKMOTH
(LINNAEUS, 1758)

439

The Large Elephant Hawkmoth caterpillar is initially pale green. In between feeding, by day and night, the young larva rests stretched out beneath a leaf, where it is extremely well camouflaged. Later, larger individuals (now mostly dark brown) feed fully exposed at the top of a plant, preferring the flowers and seed heads to leaves. When not feeding, the caterpillar often hides at the base of the plant, where its dark coloration is of greater advantage. The larva can also swim if it drops from emergent aquatic host plants into water below. It was once an occasional pest of grapevines in southern Europe but is rarely so today.

A striking feature of the species is its defensive behavior. When the caterpillar is alarmed, the head and three thoracic segments are withdrawn into the first and second abdominal segments, which expand greatly, enlarging startling eyespots. Even quite large birds have been known to flee at this sight. There are three other similar Palearctic species—*Deilephila porcellus*, *D. askoldensis*, and *D. rivularis*. All these hawkmoth adults are nonmigratory, nocturnal, and avid flower visitors.

The Large Elephant Hawkmoth caterpillar is, at first, pale green and cylindrical. In the third instar, the first and second abdominal segments enlarge and develop very realistic eyespots that remain brightly colored until pupation. Most larvae also change to the final dark form, but some stay green or occasionally turn blue gray. The horn is short, hooked, and has a white tip.

Actual size

FAMILY	Sphingidae
DISTRIBUTION	Southern and Southeast Asia, from Pakistan east to southern China and Vietnam
HABITAT	Higher elevation areas, parks, and gardens
HOST PLANTS	Privet (*Ligustrum* spp.), honeysuckle (*Lonicera* spp.), and ash (*Fraxinus* spp.)
NOTE	"Sphinx" posturing caterpillar that may defoliate its host plant
CONSERVATION STATUS	Not evaluated

ADULT WINGSPAN
2⅛–3⅜ in (55–86 mm)

CATERPILLAR LENGTH
2¹⁄₁₆–2¾ in (52–70 mm)

440

DOLBINA INEXACTA
COMMON GRIZZLED HAWKMOTH
(WALKER, 1856)

Common Grizzled Hawkmoth caterpillars emerge from spherical, smooth, shiny green eggs laid by the female moth on the underside of host plant leaves. The larvae are not very active, preferring to move little but consume a great deal of vegetation during development, sometimes causing defoliation. The caterpillar spends much of its time resting on the underside of leaves in the typical "sphinx" posture, holding its head above its body. Prior to pupation, it stops feeding, loses its green color, and starts wandering, climbing to the ground in search of a suitable pupation site.

Eventually, the caterpillar burrows about 6 in (150 mm) into the soil to form a chamber, where the pupa overwinters. The hawkmoth adults are nocturnal, feeding and egg laying at night. Like the caterpillars, the *Dolbina inexacta* adults are also well camouflaged when at rest, blending in on tree trunks and other "grizzled" surfaces. One or two generations may occur annually.

Actual size

The Common Grizzled Hawkmoth caterpillar is whitish green with numerous white dots covering the body, more profuse dorsally. There are seven, prominent, oblique white stripes laterally. The whitish-green posterior horn is long and straight. The head is light green with a pair of white stripes either side. The true legs are tan colored, while the prolegs are green.

FAMILY	Sphingidae
DISTRIBUTION	Southern United States, Mexico, and South America
HABITAT	Forests and woods, parks, gardens, and dry hillsides
HOST PLANTS	Poinsettia (*Euphorbia pulcherrima*), other *Euphorbia* spp., Willow Bustic (*Sideroxylon salicifolium*), Saffron Plum (*Bumelia celastrina*), and Satinleaf (*Chrysophyllum oliviforme*)
NOTE	Subtropical, night-feeding caterpillar that is variably colored
CONSERVATION STATUS	Not evaluated, but generally common

ADULT WINGSPAN
3–3⁷⁄₁₆ in (76–88 mm)

CATERPILLAR LENGTH
1–1³⁄₁₆ in (25–30 mm)

ERINNYIS ELLO
ELLO SPHINX MOTH
(LINNAEUS, 1758)

441

Ello Sphinx Moth caterpillars hatch from eggs laid on the leaves, stems, and even spines of the host plants. They are solitary in their, mainly nocturnal, feeding habits, and when inactive can be found along the midvein on the underside of leaves. The caterpillars have a slender posterior horn, which becomes progressively smaller as they mature, and their colors can vary from green to tan to purplish and brown. They are often parasitized by tachinid flies and parasitic wasps.

Pupation occurs in a loose cocoon of silk and debris in a shallow ground indentation. Adults, commonly found in early spring and the fall, emerge after about three weeks and are often seen feeding on periwinkle flowers. Females "call" males for mating by releasing pheromones from glands near the tip of their abdomen. There are one to three generations annually, depending on location and moisture.

The Ello Sphinx Moth caterpillar is highly variable in color, with green, dark, and intermediate forms. One form is dark dorsally with orange-red, black, and white markings laterally in the form of dots and dashes. The spiracles are white. The posterior horn is relatively short and sometimes absent. In some morphs there are eyespots on the anterior end.

Actual size

FAMILY	Sphingidae
DISTRIBUTION	Bolivia, Brazil, Paraguay, and Argentina
HABITAT	Forests
HOST PLANTS	Grapevine (*Vitis* spp.)
NOTE	Large, variably colored caterpillar that pupates underground
CONSERVATION STATUS	Not evaluated, but generally common

ADULT WINGSPAN
3¾–5 in (95–130 mm)

CATERPILLAR LENGTH
3⅜–3¾ in (85–95 mm)

442

EUMORPHA ANALIS
EUMORPHA ANALIS
(ROTHSCHILD & JORDAN, 1903)

Eumorpha analis caterpillars, which specialize in feeding on plants in the grapevine family, come in a green and a dark form, like many hawkmoths, as well as many intermediate forms. Host plant quality is suspected to be a factor in determining the caterpillar color. Development is rapid, and the caterpillars consume a considerable amount of foliage. Pupation occurs a few inches below ground in an earthen cell, and just prior to adult emergence the pupa wriggles to the surface to allow the moth to escape the pupal case unimpeded.

In most warmer, northerly locations there are several generations, and adults eclose and fly throughout the year; but in colder more southerly areas pupae overwinter. Adult moths are nocturnal and feed on many kinds of deep-throated flowers. They are often attracted to lights. Females "call" males at night by releasing pheromones from their abdomen tip, allowing the males to locate them effectively from a mile or more away.

The *Eumorpha analis* caterpillar in its green form is light to medium green and covered profusely with many tiny, white dots. There is no posterior horn. Laterally, there are three white, oval to elongated markings, outlined in black on the final three segments. The head is green, and the spiracles appear pinkish. The true legs and prolegs are green.

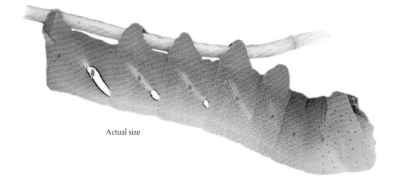

Actual size

FAMILY	Sphingidae
DISTRIBUTION	Southern Canada, United States, Central and South America, and parts of the Caribbean
HABITAT	Rain forests, wet woodlands, and wetlands
HOST PLANTS	Various species in the evening primrose family (Onagraceae), including *Ludwigia* spp.
NOTE	Variably colored caterpillar found in tropical forests and temperate woodlands
CONSERVATION STATUS	Not evaluated

ADULT WINGSPAN
3⅜–3¹³⁄₁₆ in (85–97 mm)

CATERPILLAR LENGTH
2¾–3⅛ in (70–80 mm)

EUMORPHA FASCIATUS
BANDED SPHINX MOTH
(SULZER, 1776)

443

The caterpillars of the Banded Sphinx Moth, also known as the Lesser Vine Sphinx Moth, hatch from large, round eggs laid on the underside of leaves of the host plant. The larvae feed on the leaves, preferring the blade of the leaf either side of the midrib. They move to the ground to pupate in an underground chamber, overwintering there before wriggling to the surface just before eclosion.

The moths appear at dusk and visit flowers to gather nectar. They are seen all year round in the tropics, where there are as many as three generations. To the north, there are two generations, with the moths on the wing from May to July and again from the end of August to October, while in the far north of the species' range there is a single generation flying from August to November. The "sphinx" of the common name probably derives from the caterpillar's habit of rearing up when threatened into a pose like that of an Egyptian sphinx.

The Banded Sphinx Moth caterpillar is variable in appearance, with one form yellow or green with cross stripes of red and black, red head and feet, and lateral subspiracular yellow stripes edged with red, while another is almost completely green. All larvae have black spiracles ringed with white and diagonal white stripes pointing toward the head.

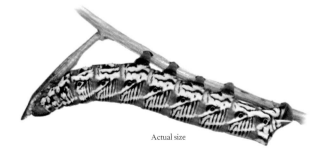

Actual size

FAMILY	Sphingidae
DISTRIBUTION	Nepal, northeastern India, Bangladesh, Myanmar, southern China, Thailand, Vietnam, and Peninsular Malaysia
HABITAT	Forests, parks, and gardens
HOST PLANTS	Various, including Night-scented Lily (*Alocasia odora*), elephant ear (*Colocasia* spp.), angel wings (*Caladium* spp.), zaminkand (*Amorphophallus* spp.), Silver Vine (*Scindapsus pictus*), and Arrowhead Plant (*Syngonium podophyllum*)
NOTE	Caterpillar that pretends to be a snake when disturbed
CONSERVATION STATUS	Not evaluated, but generally common

ADULT WINGSPAN
2¹⁄₁₆–2⁷⁄₁₆ in (53–62 mm)

CATERPILLAR LENGTH
3⅜–3¾ in (85–95 mm)

444

EUPANACRA MYDON
COMMON RIPPLED HAWKMOTH
(WALKER, 1856)

The yellowish-green eggs of the Common Rippled Hawkmoth are laid singly on the leaves of host plants. Prior to hatching, the eggs turn orange, as the first instar, which is orange colored and with a long, posterior horn, becomes visible through the shell. The caterpillars consume a great deal of foliage, so development is rapid, taking little more than a month before pupation, with the pupae formed on or just under the ground. There are multiple generations annually, and adult moths feed nocturnally on a wide variety of flowers.

When disturbed, mature caterpillars are able to retract their head and segments one to four into segment five, which expands that segment, bulging its eyelike markings in a remarkable attempt to intimidate potential predators. With its anterior section "expanded," the caterpillar looks like a snake, the "eyes" appearing to watch from all angles. The eyespots are not visible unless the caterpillar is disturbed.

The Common Rippled Hawkmoth caterpillar is light green with a brown, wavy stripe laterally that extends along the abdomen. The posterior segment is truncated and marked in brown, extending to a short, curved, brown spine. Segments three and four are brighter green and display white dots with brown-and-black markings laterally, forming an "eye" on each side of the caterpillar. The head is light green, and all legs are brown.

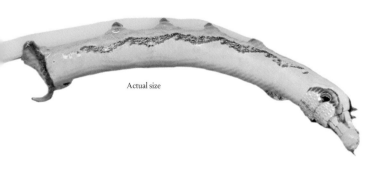

Actual size

FAMILY	Sphingidae
DISTRIBUTION	Eastern Mongolia, Russian Far East, Japan, Korea, and central and eastern China
HABITAT	Temperate woodland margins, tracks and clearings, leafy city parks, and gardens
HOST PLANTS	Honeysuckle (*Lonicera* spp.), as well as snowberry (*Symphoricarpos* spp.) in gardens
NOTE	Sphinxlike caterpillar that is well adapted to gardens and parks
CONSERVATION STATUS	Not evaluated, but very common where honeysuckles are prevalent

HEMARIS AFFINIS
HONEYSUCKLE BEE HAWKMOTH
(BREMER, 1861)

ADULT WINGSPAN
1 $^{11}/_{16}$–2 $^{1}/_{8}$ in (43–54 mm)

CATERPILLAR LENGTH
1 $^{9}/_{16}$–2 in (40–50 mm)

445

The Honeysuckle Bee Hawkmoth caterpillar rests initially along the midrib on the underside of a leaf, occasionally nibbling oval holes. Most larvae feed at night, resting during daylight in a typical sphinxlike pose. Full-grown caterpillars can be found under the terminal twigs of mid-level branches and are common around Beijing, China, on the ornamental *Lonicera maackii* from late August to October. If disturbed, most larvae will drop to the ground. The very dark brown, almost black pupa is formed in a loose brown cocoon among debris on the soil, and overwinters.

This day-flying moth is a bumblebee mimic, flitting noisily like its namesake from flower to flower. In northern China there are two generations per year, with adults on the wing between May and late August; farther south there are three. As with most other members of this 17-species, non-migratory, Holarctic genus, the adults emerge from the pupa with wings fully scaled. After the first flight, loose scales are shed, and clear areas appear.

The Honeysuckle Bee Hawkmoth caterpillar is initially whitish yellow. Fully grown individuals come in two color forms—grass green and blue green—with a pale line from a green head to a purple-blue horn. The dorsal surface is paler, the underside is reddish, and the whole granulose body is covered with pale tubercles.

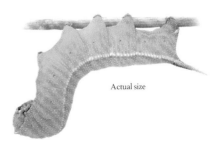

Actual size

FAMILY	Sphingidae
DISTRIBUTION	Europe, tropical Africa, Egypt, India, Asia, and Australia
HABITAT	Wide-ranging, in gardens, parks, and open areas
HOST PLANTS	Many, including grapevine (*Vitis* spp.), *Parthenocissus* spp., *Galium* spp., *Fuchsia* spp., *Epilobium* spp., and *Beta* spp.
NOTE	Snake-mimicking caterpillar of a common migratory hawkmoth
CONSERVATION STATUS	Not evaluated, but usually common

ADULT WINGSPAN
2⅜–3⅛ in (60–80 mm)

CATERPILLAR LENGTH
3⅛–3½ in (80–90 mm)

446

HIPPOTION CELERIO
SILVER-STRIPED HAWKMOTH
(LINNAEUS, 1758)

The Silver-striped Hawkmoth caterpillar is light brown to pinkish tan and covered with small white dots. There are seven dark brown, oblong-shaped markings laterally along the body. The anterior end has two pairs of eyespots and darker brown, broad stripes on the first three segments. The head and all legs are tan colored, and the posterior horn is short, straight, and black.

Silver-striped Hawkmoth females lay variably sized, glossy, bluish-green eggs singly on upper and lower surfaces of host plant leaves, usually near a growing tip. The eggs hatch after five to ten days, depending on temperature, as green first instars with disproportionately long, black horns. The caterpillars are well camouflaged on host plants and spend much time resting on the underside of leaves, with most feeding happening at night. When disturbed, they inflate their anterior eyespot segments to appear snakelike.

Green and brown morphs occur later in larval life, although, unlike many other hawkmoth larvae, the caterpillars do not change color prior to pupation, which takes place on or just below the ground surface in a loosely spun, brown cocoon. In southern parts of its range, the Silver-striped Hawkmoth breeds throughout the year with up to five generations annually, while in northern areas it is a summer migrant. Tachinid fly parasitoids regulate the species by parasitizing caterpillars.

Actual size

FAMILY	Sphingidae
DISTRIBUTION	Peru, Bolivia, Chile, and Argentina
HABITAT	Rangeland, xeric (dry) areas, gardens, and parks
HOST PLANTS	Willowherb (*Epilobium* spp.), four o'clock flower (*Mirabilis* spp.), evening primrose (*Oenothera* spp.), grapevine (*Vitis* spp.), tomato (*Lycopersicon* spp.), purslane (*Portulaca* spp.), and Toothed Spurge (*Euphorbia dentate*)
NOTE	Dramatically black caterpillar, that is sometimes a pest of grapes
CONSERVATION STATUS	Not evaluated, but usually common

HYLES ANNEI
HYLES ANNEI
(GUÉRIN-MÉNEVILLE, 1839)

ADULT WINGSPAN
2⅜–2¾ in (60–70 mm)

CATERPILLAR LENGTH
2⅜–2¾ in (60–70 mm)

Hyles annei adult females lay glossy, green eggs singly on the leaves of their host plants. The caterpillars develop rapidly, taking about a month from egg hatch to pupation, with adults appearing a further three weeks later. Before the adults emerge from the pupal case, pupae wriggle their way to the soil surface to help effect a successful exit. Females release a pheromone from the tip of their abdomen soon after emergence, which attracts males for mating. There are at least two generations a year in northern areas of the range but only one in the south.

Occasionally, *Hyles annei* caterpillars can be an economic pest and cause damage to grapevines, particularly plants that have been newly established in an area with a high moth population. Largely nocturnal, most adults rest by day on stones, low walls, among low vegetation, or even on the ground. They will sometimes fly during the day, however, and visit flowers.

The *Hyles annei* caterpillar is glossy jet black with seven subspiracular, dark red dots. The head, posterior horn, true legs, and prolegs are also dark red, as is a dorsal collar on segment one. There are four distinct, transverse ridges posteriorly on each segment. The spiracles are white.

Actual size

FAMILY	Sphingidae
DISTRIBUTION	Warm, temperate Europe and the Middle East to western China and Mongolia; also areas of United States and Canada as an introduction
HABITAT	Open, dry, sunny locations where *Euphorbia* spp. grow, such as field and woodland edges, coastal sand dunes, and bare mountainsides
HOST PLANTS	Herbaceous species of spurge (*Euphorbia* spp.), occasionally *Rumex* spp. and *Polygonum* spp.
NOTE	Gaudy caterpillar that advertises its toxic nature
CONSERVATION STATUS	Not evaluated, but common in hot, dry areas with plenty of host plants

ADULT WINGSPAN
2¾–3⅜ in (70–85 mm)

CATERPILLAR LENGTH
2¾–3⅜ in (70–85 mm)

HYLES EUPHORBIAE
SPURGE HAWKMOTH
(LINNAEUS, 1758)

448

Young Spurge Hawkmoth caterpillars rest low down on the host plant by day, moving up the stem en masse at dusk to feed. Full-grown larvae feed quite openly, relying on gaudy, aposematic coloration for protection. In all stages, larval feeding powers are prodigious. Vast quantities of leaves and soft stems are consumed between spells of basking, which the caterpillar does more frequently as it grows. If a larva is disturbed, a thick stream of dark green fluid is ejected from its mouth, accompanied by violent lateral body twitching. The green plant slurry is rich in potent, host-derived, toxins and irritants.

The adult of this species is a noted migrant into central Europe and central Asia, where it can be confused with several resident species of *Hyles*, a widespread genus, containing up to 30 similar-looking species and 40 subspecies. *Hyles euphorbiae* has also been introduced into areas of the United States and Canada to control non-native pest species of *Euphorbia* that have taken over grazing lands. The species name is derived from these principal host plants.

The Spurge Hawkmoth caterpillar is off-white with a black head and horn at first. This primary color turns dark olive black, which lightens with feeding. After the first molt, the characteristic bright pattern appears, superimposed on a light greenish to yellow-brown background. With each successive molt, this pattern becomes more startling and gaudy. The dorsolateral line of eyespots may be red, yellow, green, white, or orange.

Actual size

FAMILY	Sphingidae
DISTRIBUTION	South America, including Argentina, Chile, Uruguay, Paraguay, and Brazil
HABITAT	Savannahs and other open areas
HOST PLANTS	Wide range of plants in the families Fabaceae, Nyctaginaceae, Onagraceae, Polygonaceae, Portulacaceae, and Solanaceae
NOTE	South American hawkmoth caterpillar that can be a grape pest
CONSERVATION STATUS	Not evaluated, but generally uncommon

ADULT WINGSPAN
2⁹⁄₁₆–3⅛ in (65–80 mm)

CATERPILLAR LENGTH
3⅜–3½ in (85–90 mm)

HYLES EUPHORBIARUM
HYLES EUPHORBIARUM
(GUÉRIN-MÉNEVILLE & PERCHERON, 1835)

449

Hyles euphorbiarum caterpillars hatch from more than 800 eggs laid singly on the host plants by the female moth. The first instar larvae are green but develop aposematic coloration in later instars—a warning to would-be predators of their distastefulness. The caterpillars feed openly on host plants and develop rapidly. Mature caterpillars pupate in an earthen cell a few inches below ground and form orangish-brown pupae. Before eclosion, the pupae wriggle their way to the soil surface.

The adult moths, which fly throughout the year in most locations within their range, with the greatest numbers seen in March, July, September, and November, are primarily nocturnal, readily attracted to light, and visit many kinds of flowers. In some locations, caterpillars of this species cause economic damage to grapevines. There are about 30 species of hawkmoths throughout the world in the genus *Hyles*, which is thought to have originated and evolved in the Neotropical region.

The *Hyles euphorbiarum* caterpillar is variably colored but usually black with yellow or orange transverse stripes on each segment. On most segments a black-and-white eyespot is located anteriorly, which may be reduced in some populations. The head, legs, and posterior spine are red.

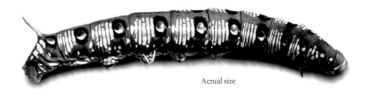

Actual size

FAMILY	Sphingidae
DISTRIBUTION	North America, Europe, central Asia, the Himalayas, and Japan
HABITAT	Meadows, forest edges, parks, and gardens
HOST PLANTS	Bedstraw (*Galium* spp.) and Fireweed (*Epilobium angustifolium*)
NOTE	Red-horned caterpillar whose other coloring may vary
CONSERVATION STATUS	Not evaluated, but not threatened

ADULT WINGSPAN
2⁹⁄₁₆–3½ in (65–90 mm)

CATERPILLAR LENGTH
3–3⅜ in (75–85 mm)

450

HYLES GALLII
BEDSTRAW HAWKMOTH
(ROTTEMBURG, 1775)

The Bedstraw Hawkmoth female lays her small, almost round, blue-green eggs singly on the upper side of leaves and flowers, laying up to five eggs per plant. The newly hatched caterpillar is green with yellow lines and darkens with age. The eyespots and red horn help to scare off predators. The caterpillars feed day and night and, at rest, they can be found lying along the midrib on the underside of a leaf. The older caterpillars crawl down the plant during the day and return at night to feed.

The caterpillar pupates in leaf litter, where it spends the winter. The pupa is light brown and wrapped in a loosely spun, silken net, and the adult emerges in spring. Occasionally, there may be a second generation from adults that emerge early. The species is named for the caterpillar's favorite food, bedstraw plants, although it will feed on other host plants.

The Bedstraw Hawkmoth caterpillar varies in color from olive brown to black. There are a row of startling, yellow eyespots along the dorsal surface, with tiny, yellow eyespots on the sides, and a short, red horn. The underside is pink in all color forms.

Actual size

FAMILY	Sphingidae
DISTRIBUTION	North and South America, from southern Canada to northern Argentina, and the Caribbean
HABITAT	Variety of habitats, including deserts and gardens
HOST PLANTS	Many, including apple (*Malus* spp.), *Amaranthus* spp., beets (*Beta* spp.), *Brassica rapa*, Lettuce (*Lactuca sativa*), and Evening Primrose (*Oenothera biennis*)
NOTE	Polyphagous, abundant caterpillars formerly harvested by Native Americans for food
CONSERVATION STATUS	Not evaluated, but common

ADULT WINGSPAN
2⁷⁄₁₆–3½ in (62–90 mm)

CATERPILLAR LENGTH
3–4 in (76–100 mm)

HYLES LINEATA

WHITE-LINED SPHINX

(FABRICIUS, 1775)

451

White-lined Sphinx caterpillars can be extremely abundant; sometimes thousands are observed crawling on the ground in areas like the Arizona desert. The caterpillars are well adapted to a variety of habitats, being both polyphagous and also able to tolerate a remarkably wide range of temperatures—up to around 113°F (45°C)—changing their orientation on the host plant toward or away from a heat source to maintain a constant internal temperature. Mature larvae pupate in shallow burrows in the soil; adults eclose after two or three weeks and fly both by day and at dusk, hovering over flowers while drinking nectar. The species develops and breeds between February and November, producing two or more generations annually.

Native Americans from the southwest United States consumed the caterpillars as food. In 1884, the entomologist William Greenwood Wright (1830–1912), writing in the magazine *Overland Monthly,* described how Cahuilla people harvested the larvae in early spring, sometimes eating them raw, although most were roasted over hot coals and stored. *Hyles lineata* used to be considered the same species as *H. livornica*—a very similar species from the Old World—but its separate identity is now well established.

The White-lined Sphinx caterpillar is often mostly black in color, with some white and orange spots and broken yellow lateral lines. While it is shaped like other sphinx moth larvae, its coloration can be very variable, often green with two rows of subdorsal eyespots, one on each segment, connected by a black line. The legs are orange, the prolegs are orange or green, the posterior horn is orange and black, and the head is orange or green.

Actual size

FAMILY	Sphingidae
DISTRIBUTION	From northwest Africa, southern Europe, and Asia Minor to northern Pakistan and western Mongolia
HABITAT	Very sunny, dry, stony, mountainous hillsides with scattered clumps of *Euphorbia* and little other vegetation; usually at high altitude in Asia
HOST PLANTS	Herbaceous species of *Euphorbia*, particularly *E. nicaeensis*
NOTE	Huge, gaudy caterpillar of an elusive, local, and scarce species
CONSERVATION STATUS	Not evaluated, but widespread

ADULT WINGSPAN
3⅛–4¼ in (80–110 mm)

CATERPILLAR LENGTH
4–4⅝ in (100–120 mm)

452

HYLES NICAEA
GREATER SPURGE HAWKMOTH
(DE PRUNNER, 1798)

The Greater Spurge Hawkmoth caterpillar initially rests along the midrib on the underside of a leaf, but growth is rapid, and the large, mature larvae usually rest on the stem, often fully exposed. Frenzied bouts of eating are interspersed with long spells of basking. The caterpillars' bright warning coloration helps deter vertebrate predators, but many succumb to parasites. At high altitudes, above 6,600 ft (2000 m), larvae may be jet black to protect them from excessive ultraviolet radiation while also helping them absorb more of the sun's heat as nights and mornings can be extremely cold.

This hawkmoth caterpillar is one of Europe's largest, rivaling that of the Death's Head Hawkmoth (*Acherontia atropos*). The adults are equally huge and prone to wander, often turning up far from known colonies. The specific name, *nicaea*, may be derived from the main host plant, *Euphorbia nicaeensis*, or possibly from the locality where it was first identified—Nice, in southern France. *Hyles* is a widespread genus containing up to 30 similar-looking species and 40 subspecies, most of which occur in Europe, northern Africa, and much of Asia, excluding the south.

The Greater Spurge Hawkmoth caterpillar is initially a rich canary-yellow color, with a blackish horn, but soon becomes apple green with longitudinal rows of black dots. Most full-grown individuals are pale gray with dorsolateral and ventrolateral rows of black-ringed, yellow or red eyespots; the horn is always black. However, many larvae become totally black with small, red eyespots and, sometimes, buff lateral patches. The amount of black pigmentation and size of eyespots are very variable.

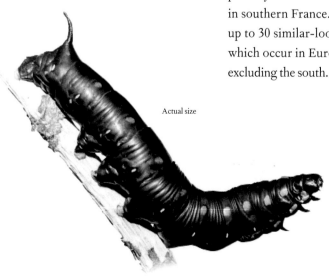

Actual size

FAMILY	Sphingidae
DISTRIBUTION	Northern Europe, Siberia, and eastern Russia into northwest China
HABITAT	Wet woodlands, especially near rivers and lakes
HOST PLANTS	Aspen (*Populus tremula*) and willow (*Salix* spp.)
NOTE	Plump, green caterpillar that is difficult to spot among leaves
CONSERVATION STATUS	Not evaluated, but not threatened

ADULT WINGSPAN
2¹³⁄₁₆–3⅞ in (71–98 mm)

CATERPILLAR LENGTH
2⁹⁄₁₆–3⅛ in (65–80 mm)

LAOTHOE AMURENSIS
ASPEN HAWKMOTH
(STAUDINGER, 1892)

453

Aspen Hawkmoth caterpillars emerge from glossy, green-yellow eggs laid on the underside of leaves of the host plant. The female hawkmoth lays around 100 of the dorsoventrally flattened eggs, which hatch up to 12 days later. The young caterpillars are pale green with faint, yellow markings; oblique stripes appear at later instars. The larvae feed by night on the leaves of their host plants, consuming all of the leaf except the midrib. During the day, they rest under a leaf, and older caterpillars take on a sphinxlike pose, with their head lifted and tucked under the thorax.

The caterpillars crawl down the tree to pupate, burrowing into the soil at the base of a tree or grassy tussock. The pupae overwinter, and the adults eclose and are on the wing in midsummer the following year. The Aspen Hawkmoth is named for the green caterpillar's favorite food plant, the leaves of which also provide perfect camouflage.

The Aspen Hawkmoth caterpillar is bright green and covered in tiny, yellow spots. There are seven distinctive, short, oblique yellow stripes along both sides, the last contiguous with the short, yellow horn. There are two yellow stripes on the head.

Actual size

FAMILY	Sphingidae
DISTRIBUTION	Temperate Europe, the Middle East, east to Lake Baikal, Siberia
HABITAT	Damp, low-lying areas, stream and lake margins, and damp woodland edges
HOST PLANTS	Mainly poplar (*Populus* spp.) and willow (*Salix* spp.); rarely other shrubs and trees
NOTE	Cryptic caterpillar of probably the most common European hawkmoth
CONSERVATION STATUS	Not evaluated, but common and widespread

ADULT WINGSPAN
2¾–4 in (70–100 mm)

CATERPILLAR LENGTH
2⁹⁄₁₆–3⅜ in (65–85 mm)

LAOTHOE POPULI
POPLAR HAWKMOTH
(LINNAEUS, 1758)

454

The Poplar Hawkmoth caterpillar is initially pale green and rough, with small, yellow tubercles and a cream-colored horn. With growth, yellow lateral stripes appear, and the legs and spiracles become pink; the body color, however, usually remains yellowish green with yellow tubercles. Some individuals may be nearly white with cream stripes, or even blue gray. Any of these forms can also be spotted with red.

Initially, young Poplar Hawkmoth larvae rest along the underside of a leaf. After the second molt, however, they adopt a more characteristic upside-down, sphinxlike posture, hanging beneath and blending in with a leaf. The caterpillars are not very active, tending to remain in the same feeding area throughout their life, stripping several shoots bare. Depending on the quality of the host plant, larvae may go through four, five, or even six instars before pupation, overwintering as a pupa.

Most larvae never reach pupation, due to predation and parasitism. The braconid wasp *Microplitis ocellatae* is the most serious parasitoid; its cocoons can sometimes be seen stuck to the skin of caterpillars like little eggs. The genus includes a number of similar-looking species, including *Laothoe austauti* (North Africa), *L. philerema* (central Asia), *L. amurensis* (Palearctic boreal zone), and *L. habeli* (northern China), which most closely resembles *L. populi*. Adult wing color varies from pale buff, through pale browns, reds, and grays to near black.

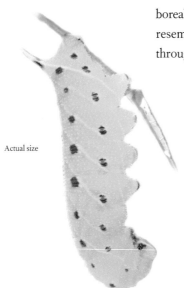

Actual size

FAMILY	Sphingidae
DISTRIBUTION	Areas of South and Southeast Asia, China, Korea, Japan, and Hawaii
HABITAT	Forests, parks, gardens, and abandoned land
HOST PLANTS	Stinkvine (*Paederia foetida*) and *Psychotria rubra*
NOTE	Stout, green caterpillar that feeds on young leaves
CONSERVATION STATUS	Not evaluated, but not threatened

ADULT WINGSPAN
1⅝–2³⁄₁₆ in (42–56 mm)

CATERPILLAR LENGTH
2 in (50 mm)

MACROGLOSSUM PYRRHOSTICTA
MAILE PILAU HORNWORM
BUTLER, 1875

455

Maile Pilau Hornworms hatch from round, white eggs, which the adult female lays singly on the underside of young leaves. The young caterpillars are dark gray and have a shiny appearance. They feed on the youngest leaves and rest on the underside of leaves. The older larvae, which are apple green to bluish green in color, prefer to rest on the twining stems of the host plant. During the hotter summer months, they retreat deeper into the foliage of their host plants.

The caterpillars crawl to the ground to pupate, where they spin a weak silken cocoon in the leaf litter. The adult moths emerge and are on the wing from summer to late fall. The caterpillars are called Maile Pilau Hornworms after the common name of one of their host plants, the Stinkvine or Maile Vine, the leaves of which give off a strong odor when crushed, hence the plant's Latin name, *foetida*.

The Maile Pilau Hornworm caterpillar is pale green with a dark green head. A pale stripe runs from the head to a purple horn, which tapers to an orange point. The body is covered in tiny, white spots, giving a speckled appearance. There are seven oblique, green lateral stripes, and the spiracles are white and red.

Actual size

FAMILY	Sphingidae
DISTRIBUTION	Southern Asia, from eastern India east to southern China, southern Japan, and Southeast Asia
HABITAT	Parks, gardens, mangroves, and wetlands
HOST PLANTS	Skunkvine (*Paederia foetida*) and Noni (*Morinda citrifolia*)
NOTE	Caterpillar that metamorphoses into a day-flying, hummingbird-like hawkmoth
CONSERVATION STATUS	Not evaluated, but usually common

ADULT WINGSPAN
1¹³⁄₁₆–2³⁄₁₆ in (46–56 mm)

CATERPILLAR LENGTH
2¹⁄₁₆–2⁵⁄₁₆ in (53–58 mm)

MACROGLOSSUM SITIENE
CRISP-BANDED HUMMINGBIRD HAWKMOTH
(WALKER, 1856)

456

Crisp-banded Hummingbird Hawkmoth caterpillars hatch from spherical, glossy, green eggs, laid singly on terminal leaves of the host plant. First instars are green and well camouflaged, spending much time resting along the midvein on the undersides of leaves. The caterpillars develop rapidly and have a voracious appetite, feeding primarily on host plant leaves. When nearing pupation, they turn from green to purplish brown and begin to wander, looking for a pupation site, where they weave a few leaves together to create a shelter. Within 48 hours, the caterpillars pupate.

Pupation shelters may be in the lower leaves of a growing plant or among leaf litter on the ground. The pupal period is short, only about 11 to 14 days. Adults are commonly seen flying in daylight hours visiting flowers, such as Golden Dewdrop (*Duranta erecta*) and Lantana (*Lantana camara*) blossoms. They prefer flowers low down on bushes and fly very close to the ground when approaching.

The Crisp-banded Hummingbird Hawkmoth caterpillar is bluish green and covered with tiny, white dots. An indistinct lateral, yellow white line runs the length of the body, leading to the straight posterior horn. The head is green with a yellow stripe on each side. The spiracles, true legs, and prolegs are light brown to tan. A brown morph also occurs.

Actual size

FAMILY	Sphingidae
DISTRIBUTION	Southern Europe and North Africa, east across Asia to the east coast of China and Japan; also in southern England as a resident or migrant
HABITAT	Forest edges, parks, and gardens
HOST PLANTS	Bedstraw (*Galium* spp.) and madder (*Rubia* spp.)
NOTE	Caterpillar that may reach pupation in just 20 days
CONSERVATION STATUS	Not evaluated, but not threatened

ADULT WINGSPAN
1⁹⁄₁₆–1¾ in (40–45 mm)

CATERPILLAR LENGTH
2⅛–2⅜ in (55–60 mm)

MACROGLOSSUM STELLATARUM
HUMMINGBIRD HAWKMOTH
(LINNAEUS, 1758)

457

Hummingbird Hawkmoth caterpillars hatch from up to 200 glossy green eggs, laid singly on separate plants by the female moth six to eight days earlier. The young larvae are initially yellow but become green from the second instar onward. Despite feeding on the top of the food plant, the larvae are difficult to spot thanks to their excellent camouflage. Within a few weeks they move down to the lower stems of the host plant or the ground to pupate, spinning a cocoon among the leaves.

There may be several generations a year. Since it is the adult that overwinters, few of the migrants that fly north survive the cold. Hummingbird Hawkmoths get their common name from their darting flight and the way the moths hover like a hummingbird in front of nectar-rich flowers to feed. They are strong fliers, so they disperse over a large area, well to the north and south of their year-round range.

The Hummingbird Hawkmoth caterpillar is pale green with two lateral, pale stripes. The purple-blue horn is tipped with orange, and the body is covered with small, white spots, which give the caterpillar a speckled appearance. The legs and the tips of the prolegs are orange. Just before pupation, the body turns red brown.

Actual size

FAMILY	Sphingidae
DISTRIBUTION	North and South America, from Canada to Argentina and the Caribbean
HABITAT	Wide variety, including vegetable patches and tobacco fields
HOST PLANTS	Solanaceae, such as tobacco (*Nicotiana* spp.), Tomato (*Solanum lycopersicum*), and *Datura* spp.
NOTE	Caterpillar that is used as a model organism in biological sciences
CONSERVATION STATUS	Not evaluated, but common

ADULT WINGSPAN
4 in (100 mm)

CATERPILLAR LENGTH
2¾ in (70 mm)

458

MANDUCA SEXTA
TOBACCO HORNWORM
(LINNAEUS, 1763)

The Tobacco Hornworm caterpillars normally develop through five larval instars but may molt additional times if the nutritional value of their food is poor. Mature larvae wander, change color, and lose excess liquid before pupation, which occurs in an underground chamber; as with most other sphinx moths, no cocoon is made. The life cycle can be completed within 50 days but is often longer because many pupae will enter diapause during hot summer months. There may be up to four generations annually in Florida, but the average over the species' range is two.

Manduca sexta caterpillars feeding on toxic plants such as tobacco may be better protected from predators than those feeding on less toxic species, such as Tomato (*Solanum lycopersicum*). They are frequently parasitized by the gregarious braconid wasp *Cotesia congregata*, whose larvae feed unseen within the caterpillar, emerging from the body to spin their cocoons, which hang as clusters off their host. Because it is large and easily raised on an artificial diet, the Tobacco Hornworm is often used in developmental and genetic research and has contributed significant insights into the mysteries of metamorphosis.

The Tobacco Hornworm caterpillar is cryptically colored overall green with white diagonal markings edged in black. When not feeding, it assumes a sphinxlike posture, with the head and thorax held upward. Although similar to other *Manduca* species, such as the Tomato Hornworm (*M. quinquemaculata*), the caterpillar can be distinguished by its seven diagonal lines and red or rust-colored posterior horn. It has white true legs, banded in black at the joints, green prolegs, a green head, and pronounced spiracles, which are white or yellow with a black center.

Actual size

FAMILY	Sphingidae
DISTRIBUTION	Russia, Mongolia, China, Chinese Taipei, Hong Kong, Korea, and Japan
HABITAT	Many habitats, including montane forests, cities, and orchards
HOST PLANTS	Fruit trees, including apple (*Malus* spp.), pear (*Pyrus* spp.), plum (*Prunus* spp.), and hawthorn (*Crataegus* spp.)
NOTE	Asian hawkmoth caterpillar that can be a tree fruit pest
CONSERVATION STATUS	Not evaluated, but usually common

ADULT WINGSPAN
2¾–3⅝ in (70–92 mm)

CATERPILLAR LENGTH
3–3¼ in (75–83 mm)

MARUMBA GASCHKEWITSCHII
MARUMBA GASCHKEWITSCHII
(BREMER & GREY, 1853)

459

Marumba gaschkewitschii caterpillars hatch from translucent, jade-green eggs laid singly or in small groups of two to four on host plant leaves. The female usually oviposits on bushes or small trees, and the larvae tend to feed at about 1¾–5 ft (0.5–1.5 m) above the ground. Development is rapid, and the caterpillars are well camouflaged on their host plant. The first few instars rest and feed stretched out along the midrib under a leaf. The larvae go through five to seven instars. Pupation occurs in a silk-free, earthen cell just below the soil surface, and the pupa overwinters.

There are two to three generations annually in most areas, although four or five may occur in warmer areas. The adult moth has been recorded damaging fruit in Korea by piercing it with its proboscis. Caterpillars are also occasionally a pest of apricot trees in Japan. There are five subspecies of *Marumba gaschkewitschii* within its most easterly Asian range.

The *Marumba gaschkewitschii* caterpillar is green with seven oblique, dotted white lines laterally. The entire body is adorned with tiny, white spots. The spiracles are slit-like and outlined in red. The true legs and prolegs are orangish or reddish, and the triangular head is blue green.

Actual size

FAMILY	Sphingidae
DISTRIBUTION	Temperate Europe to temperate west Asia and central Asia
HABITAT	Deciduous woodlands, open river-valley woodlands, and warm, wet mountain scrub or woodlands
HOST PLANTS	Mainly trees of Tiliaceae (linden or lime), Betulaceae, Ulmaceae, and Rosaceae families
NOTE	Caterpillar well adapted to suburban and urban gardens, and parks
CONSERVATION STATUS	Not evaluated, but common, especially in suburban areas

ADULT WINGSPAN
2⅜–3⅛ in (60–80 mm)

CATERPILLAR LENGTH
2⅛–2⁹⁄₁₆ in (55–65 mm)

MIMAS TILIAE
LIME HAWKMOTH
(LINNAEUS, 1758)

460

The Lime Hawkmoth caterpillar generally feeds high in the crown of trees and large shrubs. It tends to rest motionless under a leaf, then creeps up to the leaf edge to feed, exposing no more than its moving head. Parks, cherry orchards, and avenues of urban and suburban trees can support large populations. Upward of 25 overwintering, twiglike, earth-colored pupae can be found at the base of host trees, just beneath surface debris. There are one or two generations a year.

Changing in color to dull greenish pink and grayish brown, most larvae pupate at the base of the host, but some wander off furiously in search of a suitable site, which is when they are seen in parks and gardens. If touched during this stage, the larva will twitch violently from side to side. This is a nonmigratory species, with several relict populations in Turkey, the Caucasus, Iran, and northern Pakistan. The species name, *tiliae*, is derived from the main host plant, species of *Tilia* (linden or lime).

The Lime Hawkmoth caterpillar can be either green or bluish gray, with seven yellow, oblique lateral stripes and yellow body tubercles. A characteristic yellow warty shield covers the anal flap. The horn is blue or purple on top, red and yellow underneath. The caterpillar is noticeably more slender than those of related species, tapering anteriorly to a narrow, triangular head.

Actual size

FAMILY	Sphingidae
DISTRIBUTION	Western areas of North America
HABITAT	Wet woodlands, riparian forest, parks, and gardens
HOST PLANTS	Poplar (*Populus* spp.) and willow (*Salix* spp.)
NOTE	Large, green caterpillar that has distinctive oblique stripes
CONSERVATION STATUS	Not evaluated

ADULT WINGSPAN
5–6½ in (130–165 mm)

CATERPILLAR LENGTH
3½ in (90 mm)

PACHYSPHINX OCCIDENTALIS
WESTERN POPLAR SPHINX
(HY. EDWARDS, 1875)

461

The caterpillars of the Western Poplar Sphinx moth hatch from large, pale green eggs laid in small groups on the leaves of the host plant. The larvae feed on leaves for about five weeks before crawling to the ground, where they dig a shallow burrow in loose soil in which to pupate. The resulting pupa is brown. The species then overwinters in the burrow and ecloses in spring.

Pachysphinx occidentalis is a nocturnal moth, on the wing from June to August in the northern part of the range with a single generation, while in the south of the range the moths are seen from May to September and there are two generations. The adult moths live for about six to ten days and do not feed, instead surviving on fats stored in the body. There are two known subspecies, one found in the south of the United States and the other in Mexico.

The Western Poplar Sphinx caterpillar is light green dotted with white. There are six oblique white stripes and an oblique white band that extends to the short caudal horn. The spiracles are ringed in red, and the legs and the tips of the prolegs are brown.

Actual size

FAMILY	Sphingidae
DISTRIBUTION	Korea, Japan, south to Chinese Taipei, and throughout eastern and central China to Southeast Asia
HABITAT	Forests and woodlands
HOST PLANTS	Paper Mulberry (*Broussonetia papyrifera*), other *Broussonetia* spp., White Mulberry (*Morus alba*), and *Maclura fruticosa*
NOTE	Unique Southeast Asian hawkmoth species that has no close relatives
CONSERVATION STATUS	Not evaluated

ADULT WINGSPAN
2⅜–3½ in (60–90 mm)

CATERPILLAR LENGTH
2–2¾ in (50–70 mm)

PARUM COLLIGATA
PAPER MULBERRY HAWKMOTH
(WALKER, 1856)

462

Paper Mulberry Hawkmoth caterpillars hatch from eggs laid in small masses on the underside of host plant leaves. When first laid by the female hawkmoth, the eggs are whitish but turn yellow prior to hatching in about a week. The caterpillars develop quickly and take about a month to reach pupation, which occurs in a subterranean earthen cell. The adult hawkmoths emerge in as little as 10 to 12 days during warmer times of the year, but the pupae may overwinter in northerly areas.

There are one to two generations a year in northern China, with adults on the wing from May to July. Farther south there may be up to four generations annually with no winter dormancy. Like many hawkmoth larvae, caterpillars of the Paper Mulberry Hawkmoth are often parasitized by small wasps that develop within the caterpillar, emerging to pupate in tiny cocoons festooned around the now-dead larva. *Parum colligata* is the only species of its genus.

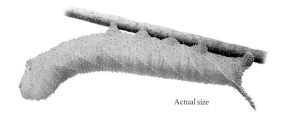

Actual size

The Paper Mulberry Hawkmoth caterpillar is bright green and covered with tiny, raised, white spots giving a granulated appearance. There are seven oblique, lateral, paler stripes, and the head is green, bordered on either side by a white stripe. The legs and prolegs are also green.

FAMILY	Sphingidae
DISTRIBUTION	Northeast India, Nepal, southwest China, northern Thailand, and northern Vietnam
HABITAT	Mountainous forests and woodlands
HOST PLANTS	Unknown in the wild, but likely to be ivy (*Ilex* spp.), which the caterpillar feeds on in captivity
NOTE	Well-camouflaged montane hawkmoth caterpillar
CONSERVATION STATUS	Not evaluated

ADULT WINGSPAN
4–4⅛ in (100–105 mm)

CATERPILLAR LENGTH
3–3⅜ in (75–85 mm)

PENTATEUCHA CURIOSA
HIRSUTE HAWKMOTH
SWINHOE, 1908

463

Hirsute Hawkmoth caterpillars hatch from eggs laid singly on host plants by the female just a few days after mating; the moths rarely live beyond five days. The eggs are relatively small for the size of moth and glossy jade green when first laid, turning shiny bronze within 36 hours and taking about 30 days to hatch. First instars rest on the underside of a leaf along the midrib and feed by eating the leaf tip backward, well camouflaged by their leaf-green color. When not feeding, mature caterpillars cling on to the leaf using the last two sets of prolegs, with the anterior part of the body raised. Growth is rapid, with comparatively little food consumed.

Pupation takes place underground in an earthen cell, with the dark brown pupa appearing to be vulnerable to desiccation. This species occurs in mountainous areas and flies during winter and early spring. The adults eclose in early evening, and mating occurs immediately, the males presumably finding females by locating and following pheromone trails.

The Hirsute Hawkmoth caterpillar is leaf green, blending in perfectly with its surroundings. There is a pale green stripe dorsolaterally, and the posterior horn is bright green. The white spiracles are slit-like and outlined in black. The true legs are orange, and the prolegs are green. The head is green, bordered each side with a white stripe.

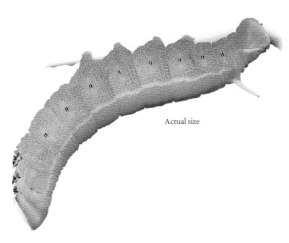

Actual size

FAMILY	Sphingidae
DISTRIBUTION	India, Nepal, Sri Lanka, Thailand, eastern and southern China, Chinese Taipei, Japan, Malaysia, Indonesia, and the Philippines
HABITAT	Various, from open woodlands to metropolitan areas, roadsides, and gardens
HOST PLANTS	Grapevine (*Vitis* spp.), *Cissus* spp., *Leea* spp., *Begonia* spp., *Diffenbachia* spp., and elephant ear (*Caldium* spp., *Colocasia* spp.)
NOTE	Hawkmoth caterpillar that has inflatable eyespots to frighten off predators
CONSERVATION STATUS	Not evaluated

ADULT WINGSPAN
2½–3⅛ in (64–80 mm)

CATERPILLAR LENGTH
2¾–3 in (70–75 mm)

464

PERGESA ACTEUS
GREEN PERGESA HAWKMOTH
(CRAMER, 1779)

Green Pergesa Hawkmoth caterpillars hatch from eggs that are green, shiny, broadly oval, and laid singly on the host plants by the female hawkmoth. The larvae come in two color forms—green or brown—from the fourth instar, but their markings remain the same. Occasionally, a red form is seen. The caterpillars develop rapidly through the instars, taking about a month before they darken in color and wander to the ground to seek a subterranean pupation site. Pupation occurs in an earthen cell and the pupa is pale brown.

Adults fly at night but are particularly active around daybreak and during rainy weather and have been observed drinking from puddles. The caterpillar uses false eyespots as a defensive strategy, increasing their size when threatened. Although well camouflaged and able to ward off potential vertebrate enemies with these false eyes, many Green Pergesa Hawkmoth caterpillars fall victim to parasitic ichneumonid wasps.

Actual size

The Green Pergesa Hawkmoth caterpillar
is pale jade green. On the anterior half of the first abdominal segment there is a pair of false eyespots consisting of a white ellipse with a black base. There is also a row of smaller, green ellipses on the remaining abdominal segments. The posterior horn is much reduced in the final instar, and the true legs are orange.

FAMILY	Sphingidae
DISTRIBUTION	Russian Far East, eastern and central China, Chinese Taipei, Korean peninsula, and Japan
HABITAT	Forest edges, open parklands, and woodlands
HOST PLANTS	Chinese Hickory (*Carya cathayensis*) and Manchurian Walnut (*Juglans mandshurica*)
NOTE	Noisy hawkmoth caterpillar that will hiss and squeak if disturbed
CONSERVATION STATUS	Not evaluated

ADULT WINGSPAN
3¹¹⁄₁₆–5 in (93–130 mm)

CATERPILLAR LENGTH
2⁹⁄₁₆–3⅜ in (65–85 mm)

PHYLLOSPHINGIA DISSIMILIS

BUFF LEAF HAWKMOTH

BREMER, 1861

465

Buff Leaf Hawkmoth caterpillars hatch from eggs laid singly or in small groups by the female hawkmoth; the eggs are pale olive green when freshly laid. Developing through five instars in as little as three weeks, these caterpillars are generally sedentary, feeding on lower branches of the host plant usually 6½–13 ft (2–4 m) above the ground. When disturbed, the larvae thrash out laterally, making a squeaking or hissing sound to dissuade predators. When full grown, the caterpillars darken in color and move to the ground, where they build a silk-free cell among debris, within which they pupate. Unusually, the pupae also hiss or squeak.

The caterpillars are most often found in July and August, with one generation annually in northern areas of its range and possibly two generations in the south. There are currently just two species within the genus *Phyllosphingia*, although others may be identified using molecular techniques.

Actual size

The Buff Leaf Hawkmoth caterpillar is bright green dorsally but blue green ventrally. There are seven oblique, lateral, bumpy stripes, and the green posterior spine is curved downward. The prolegs are green, but the true legs are reddish. The head is green with lateral, white stripes. Some forms of this caterpillar are reddish brown.

FAMILY	Sphingidae
DISTRIBUTION	Southern and southwestern United States to Brazil
HABITAT	Many lowland habitats, including gardens and parklands
HOST PLANTS	Dogbanes (Apocynaceae), including frangipani (*Plumeria* spp.) and Golden Trumpet (*Allamanda cathartica*)
NOTE	Large, conspicuous, aposematic caterpillar capable of defoliating host plant ornamentals
CONSERVATION STATUS	Not evaluated

ADULT WINGSPAN
4¹⁵⁄₁₆–5½ in (127–140 mm)

CATERPILLAR LENGTH
5½–6 in (140–150 mm)

PSEUDOSPHINX TETRIO
TETRIO SPHINX
(LINNAEUS, 1771)

466

Tetrio Sphinx caterpillars hatch from clusters of 50 to 100 large, smooth, pale green eggs laid by the female moth on host plant leaves. The larvae feed ravenously and can devour three large leaves daily. The coloration of the caterpillars is aposematic, warning predators that they are potentially toxic. The caterpillar is able to detoxify poisons it ingests from the sap of its host plants and use them for defense. It is also suggested to be a snake mimic, waving the anterior part of the body around when threatened. When full grown, the larvae pupate in a cell made from soil or debris on the ground. There are several generations annually.

The adult moths are nocturnal and feed from many kinds of flowers, playing an important role in pollinating some species. *Plumeria* species are among the flowers whose fragrance attracts sphinx moths to pollinate them, yet Tetrio Sphinx caterpillars can cause severe damage and may even defoliate the same ornamental species.

The Tetrio Sphinx caterpillar is velvety black with vivid yellow banding and bright red head and legs, speckled with black. The posterior spine is black and arises from a reddish-orange base. Newly molted larvae are gray with light yellow bands, assuming normal coloration after a few hours.

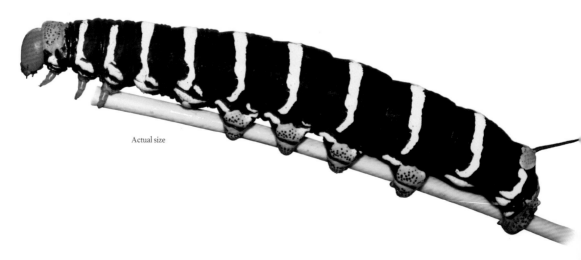

Actual size

FAMILY	Sphingidae
DISTRIBUTION	South Asia, Southeast Asia to northeast China, Korea, and Japan
HABITAT	Forest edges, parks, and gardens
HOST PLANTS	Various shrubs and small trees, including *Ligustrum* spp., *Syringa* spp., and *Paulownia* spp.
NOTE	Caterpillar that has two distinct color forms when mature
CONSERVATION STATUS	Not evaluated, but not threatened

ADULT WINGSPAN
3½–4¾ in (90–122 mm)

CATERPILLAR LENGTH
2¾–4¼ in (70–110 mm)

PSILOGRAMMA INCRETA

PLAIN GRAY HAWKMOTH

(WALKER, 1865)

467

The Plain Gray Hawkmoth lays its eggs singly on the leaves of the host plant. Initially pale green, the eggs become yellow brown as hatching approaches. The emerging caterpillars are yellow with a black horn, becoming greener in color as they develop. They feed on the leaves of their host plants and, at rest, can be found on the underside of large leaves on lower branches. The caterpillars move to the ground, where they burrow into the soil and pupate.

The reddish-brown pupae may overwinter, with adults emerging the following spring. In the northern part of the species' range, there are two generations; the adults fly in late spring to early summer and again in late summer. Farther south there are as many as five generations. The Plain Gray Hawkmoth is primarily an Asian species but has been introduced in Hawaii, and there is a subspecies in Australia.

The Plain Gray Hawkmoth caterpillar is green, mottled with brown or sometimes gray green, and has a medium-length horn. There are a series of seven oblique, white stripes laterally, with a diffuse white patch beneath each one. The spiracles are red.

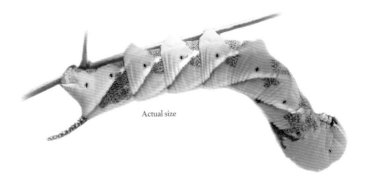

Actual size

FAMILY	Sphingidae
DISTRIBUTION	Northeastern India, Bhutan, Myanmar, northern Thailand, northern Vietnam, and southern and eastern China
HABITAT	Forests and woodlands
HOST PLANTS	Bentham's Rosewood (*Dalbergia bentham*), Lebbeck (*Albizia lebbeck*), and *Lespedeza* spp.
NOTE	Well-camouflaged caterpillar that metamorphoses into a bee-mimicking hawkmoth
CONSERVATION STATUS	Not evaluated

ADULT WINGSPAN
2¹⁄₁₆–2⁵⁄₁₆ in (52–58 mm)

CATERPILLAR LENGTH
1⁹⁄₁₆–2⁹⁄₁₆ in (40–65 mm)

SATASPES XYLOCOPARIS
EASTERN CARPENTER BEE HAWKMOTH
BUTLER, 1875

468

Eastern Carpenter Bee Hawkmoth caterpillars hatch from pale green, smooth, shiny eggs laid singly by the female hawkmoth on the undersides of host plant leaves. The larvae hatch in about four days and feed and develop through five instars. The caterpillars rest in a typical sphinx pose and, when molested, emit brown fluid from the mouth as a defensive reaction. Mature caterpillars darken in color and wander for a day or so before pupating. Pupation occurs within an earthen cell below ground or in a rough cocoon on the soil surface.

The green caterpillars are well camouflaged on their host plants, and to further disguise themselves most individuals have a reddish-brown patch on their bodies that resembles the necrotic area seen on diseased leaves. The adult, a day-flying moth that mimics large bees, particularly carpenter bees in the genus *Xylocopa*, is often seen feeding from flowers in the early morning.

Actual size

The Eastern Carpenter Bee Hawkmoth caterpillar is green with faint oblique, white, lateral stripes, the stripe leading to the tail horn being bolder. Some caterpillars have a large, reddish-brown patch on segments seven and eight. The head is green with two white stripes laterally. The tail horn is green and relatively short.

FAMILY	Sphingidae
DISTRIBUTION	From western Russia to Russian Far East, Mongolia, northeast China, South Korea, and northern Japan
HABITAT	Grassy, mixed, boreal forests, clearings, swamps, and streams
HOST PLANTS	Poplar (*Populus* spp.) and willow (*Salix* spp.)
NOTE	Caterpillar that fools predators by adopting a sphinxlike pose
CONSERVATION STATUS	Not evaluated

SMERINTHUS CAECUS

NORTHERN EYED HAWKMOTH

MÉNÉTRIÉS, 1857

ADULT WINGSPAN
2–2¾ in (50–70 mm)

CATERPILLAR LENGTH
2⅜–2¾ in (60–70 mm)

469

Female Northern Eyed Hawkmoths lay shiny, green eggs singly or in small groups of up to 12 on the underside of host plant leaves. Within seven to eight days, the caterpillars hatch and begin feeding, usually on willow species. They develop between July and September before overwintering as brown-black pupae in earthen cells below the ground. The hawkmoth adults are on the wing in May and June in a single brood, although a second generation may occur in southern parts of the species' range.

The caterpillar's defense is based on camouflage and its adoption of a sphinxlike posture when resting, which breaks up the caterpillar outline, fooling birds foraging for a conventionally worm-shaped caterpillar. However, parasitic flies and wasps likely kill a great many Northern Eyed Hawkmoth caterpillars. The species is closely related to the more widespread Eyed Hawkmoth (*Smerinthus ocellatus*), and both larvae and adults are very similar in appearance.

Actual size

The Northern Eyed Hawkmoth caterpillar is bluish white or yellowish green. Whitish, oblique stripes occur laterally, with the one leading to the tail horn most prominent. The head is green, triangular, and lined on each side with a white stripe. The spiracles and true legs are pinkish, while the prolegs are green.

FAMILY	Sphingidae
DISTRIBUTION	Temperate Europe, northwest Africa, west Asia, central Asia, and western Mongolia
HABITAT	Riverine shingle bars, wet river valleys, damp woodland edges, apple orchards, coastal sandhills, and suburban gardens
HOST PLANTS	Mainly poplar (*Populus* spp.), willow (*Salix* spp.), and apple (*Malus* spp.); also some *Prunus* spp.
NOTE	Caterpillar that parasitoids ravage, killing up to 80 percent annually
CONSERVATION STATUS	Not evaluated, but common, especially in open areas with willows

ADULT WINGSPAN
2¾–3¾ in (70–95 mm)

CATERPILLAR LENGTH
2¾–3½ in (70–90 mm)

SMERINTHUS OCELLATUS

EYED HAWKMOTH

(LINNAEUS, 1758)

470

In open sunny areas, cryptic Eyed Hawkmoth larvae feed quite openly on the tips of new shoots. One large caterpillar can strip several shoots bare; before the use of organic pesticides, the larvae would sometimes devastate apple orchards. The caterpillar is a "sit and hide" species, mimicking leaves when at rest, and taking on the same hue as the leaves around it. Many larvae avoid detection by birds by sitting upside down and using superb body countershading, although most fall victim to the *Microplitis ocellatae* parasitic wasp.

In warmer parts of Europe, there are two or three broods annually, but larvae numbers can fluctuate markedly from year to year, and local populations may die out. If disturbed, the adult Eyed Hawkmoth, which in North Africa has a broader wingspan, up to 4¼ in (110 mm), exposes the large, glaring eyespots on its hind wings suggested by its name. It is one of several similar-looking *Smerinthus* species found across temperate Europe, Asia, and North America.

The Eyed Hawkmoth caterpillar has several color forms, mainly shades of green or gray. Young larvae are whitish green, with a pale pink horn, pale body tubercles, and seven oblique lateral stripes. This color scheme remains the same throughout the larval life, only becoming more contrasting with size, although the horn turns blue. Larger larvae can develop rows of red blotches around the spiracles and elsewhere.

Actual size

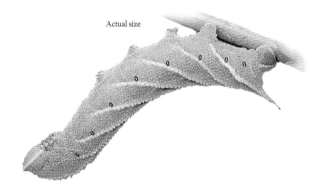

FAMILY	Sphingidae
DISTRIBUTION	North America, southern Canada to central United States
HABITAT	Many, including riparian areas, canyons, savannahs, and shrublands, but particularly associated with forested woodlots
HOST PLANTS	Wild cherry and plum (*Prunus* spp.), lilac (*Syringa* spp.), hackberry (*Celtis* spp.), apple (*Malus* spp.), and serviceberry (*Amelanchier* spp.)
NOTE	Hawkmoth caterpillar that becomes an important pollinator of rare orchids
CONSERVATION STATUS	Not evaluated, but often uncommon

ADULT WINGSPAN
3⅜–4¼ in (85–110 mm)

CATERPILLAR LENGTH
3⅛–3½ in (80–90 mm)

SPHINX DRUPIFERARUM

WILD CHERRY SPHINX

J. E. SMITH, 1797

471

Wild Cherry Sphinx caterpillars hatch from a few hundred eggs laid singly by the female moth on both surfaces of host plant leaves. The eggs hatch about a week after they have been laid. The caterpillars feed rapidly and ravenously on their host plant and complete development within a month. To avoid predation they hide in the day and feed by night. However, many are parasitized by braconid wasps. Mature larvae form pupae in earthen cells about 4 in (100 mm) below the ground and overwinter there. In some cases, the pupal stage extends as long as two years.

There is a single brood annually with the large hawkmoth adults, which are frequently found at flowers such as honeysuckles, flying for about a month in late spring and early summer. In the United States upper Midwest and in Manitoba, Canada, the Wild Cherry Sphinx is an important pollinator of the endangered Western Prairie Fringe Orchid (*Platanthera praeclara*).

The Wild Cherry Sphinx caterpillar is green with seven pairs of white, oblique, lateral stripes, each bordered dorsally with purple. The head is green with a pair of black lines running down laterally. The tail horn is a purple red. The spiracles are orange, and the true legs are yellow.

Actual size

FAMILY	Sphingidae
DISTRIBUTION	Eastern states of Canada and United States
HABITAT	Woodlands, parks, and gardens
HOST PLANTS	Various, including laurel (*Kalmia* spp.), lilac (*Syringa* spp.), ash (*Fraxinus* spp.), and Fringetree (*Chionanthus virginicus*)
NOTE	Bright green caterpillar that has distinct lateral, oblique lines
CONSERVATION STATUS	Not evaluated, but not threatened

ADULT WINGSPAN
3–4⅛ in (75–105 mm)

CATERPILLAR LENGTH
2⁹⁄₁₆ in (65 mm)

SPHINX KALMIAE
LAUREL SPHINX
J. E. SMITH, 1797

The female Laurel Sphinx moth lays her smooth, oval, green-white eggs on leaves of the host plants. The emerging young caterpillars are pale white with a black horn, becoming greener with age. They feed on the underside of leaves, where they are well camouflaged. However, they are vulnerable to attack from parasitoids. The mature caterpillars crawl to the ground, where they burrow into loose soil to pupate. They can overwinter as pupae.

The adult moths are on the wing in summer. There is a single brood in the northern parts of the range, but there may be two or more broods in the south. The genus *Sphinx* is named for the sphinx of mythology—probably for the sphinxlike pose that caterpillars of this genus take when they raise their head off the ground and tuck it into their thorax. The species name, *kalmiae*, may be derived from the laurel host plant but more probably from the Swedish botanist Pehr Kalm (1716–79).

Actual size

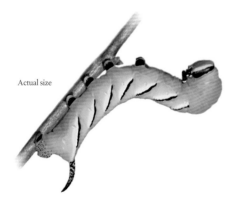

The Laurel Sphinx caterpillar is yellow green to blue green, with seven oblique, lateral, white lines edged with black above and yellow below. The spiracles are orange. The prolegs are green with a yellow band above the black bases. The horn is blue with tiny, black spines.

FAMILY	Sphingidae
DISTRIBUTION	Temperate Europe and northwest Africa to temperate western Asia, central Asia, Siberia, Russian Far East, northern China, and northern Japan
HABITAT	Open scrub and woodland edges, including town suburbs and river valleys, but only north-facing slopes of mountains in the south
HOST PLANTS	Mainly privet (*Ligustrum* spp.), ash (*Fraxinus* spp.), and lilac (*Syringa* spp.); also many *Spiraea* spp. and *Viburnum* spp.
NOTE	Striking caterpillar of a very widely distributed Old World hawkmoth
CONSERVATION STATUS	Not evaluated, but common and widespread

ADULT WINGSPAN
3½–4⅝ in (90–120 mm)

CATERPILLAR LENGTH
3½–4 in (90–100 mm)

SPHINX LIGUSTRI

PRIVET HAWKMOTH

LINNAEUS, 1758

473

Privet Hawkmoth caterpillars hatch from up to 200 eggs laid singly by the female hawkmoth on the underside of leaves. The early instar larvae rest beneath the midrib of a leaf, but when fully grown assume a typical upside-down, sphinxlike pose, clinging to a petiole or stem by their rear legs, with the thoracic segments hunched. Most larvae grasp stripped shoots within 6½ ft (2 m) of the ground. Between four and seven weeks after hatching, the caterpillar turns purplish brown dorsally and moves to the ground to pupate in an earthern cell within soft, loamy soil.

The pupae overwinter, and the adults eclose in June in northern parts of their range but earlier in the south, with a second generation in August. Adults are not very variable, although there are pale forms and sometimes individuals without any pink coloration. Several closely related species inhabit North America and Japan, notably the Wild Cherry Sphinx (*Sphinx drupiferarum*) and *S. constricta*. The species name, *ligustri*, is derived from one of its main hosts, privet.

The Privet Hawkmoth caterpillar is initially light yellow but becomes luminescent green as it feeds and develops its final, oblique, white-and-purple stripes in its third instar. Variation is not great, but some larvae have darker than normal lateral stripes, often complemented by a second, lower purple one. Instead of one posterior, blackish horn, some may have two or more horns in series, each successively smaller. In a rare form, the primary body color of green may be replaced by purple.

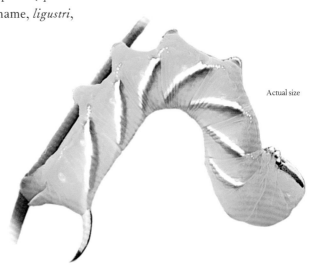

Actual size

FAMILY	Sphingidae
DISTRIBUTION	Most of Europe (except Iberia, Ireland, Scotland, and northern Scandinavia), the Caucasus, southern Turkey, Lebanon, Russia, western Siberia, and northern Kazakhstan
HABITAT	Pine forests, gardens, and parks
HOST PLANTS	Mainly pines, including Scots Pine (*Pinus silvestris*), Swiss Pine (*Pinus cembra*), Norway Spruce (*Picea abies*), and Siberian Spruce (*Picea obovata*); also cedars (*Cedrus* spp.) and European Larch (*Larix decidua*)
NOTE	Well-camouflaged caterpillar that feeds on pines
CONSERVATION STATUS	Not evaluated, but considered secure within its range

ADULT WINGSPAN
2¾–3¾ in (70–96 mm)

CATERPILLAR LENGTH
3–3⅛ in (75–80 mm)

SPHINX PINASTRI

PINE HAWKMOTH

(LINNAEUS, 1758)

474

Pine Hawkmoth caterpillars hatch from around 100 pale yellow eggs laid on pine needles or twigs and immediately consume their eggshell. At first they are a dull yellow with a large head and prominent forked horn but become green in color as they feed, with light yellow stripes at the second instar, creating an effective camouflage. Early instars feed only on the surface of a needle, but later instars consume the entire needle from tip to base. After four to eight weeks, when full grown, the larvae become restless, descend to the ground, and often wander some distance to pupate under moss or fallen needles.

In the north of the species' range, there is one brood annually, and adults fly in June or July. Farther south, there may be two generations, with adults eclosing in May or June and again in August. The larvae are attacked by many parasitoids, especially the parasitic fly *Phryxe erythrostoma*. The Pine Hawkmoth caterpillar and adult are similar in appearance to several other *Sphinx* species, including *S. maurorum* (southwest Europe and northwest Africa), and *S. bhutana*, *S. caligineus*, and *S. yunnana* (eastern Asia).

The Pine Hawkmoth caterpillar, in its final instar, is brown marked with green, or green with brown along the back. It has many thin, dark creases dividing up its smooth, slender body, a black posterior horn, reddish, dark-ringed spiracles, and white and creamy-yellow dashes. The large, glossy head is tan with lighter and darker vertical markings, the prolegs are brown, and the true legs are creamy white.

Actual size

FAMILY	Sphingidae
DISTRIBUTION	Across Canada, from southern British Columbia to Newfoundland, and northeastern United States
HABITAT	Boggy areas, coastal barrens, and deciduous forests
HOST PLANTS	Apple (*Malus* spp.), blueberries and huckleberries (*Vaccinium* spp.), alder (*Alnus* spp.), Carolina Rose (*Rosa carolina*), American Larch (*Larex laricina*), and White Spruce (*Picea glauca*)
NOTE	Canada's commonest hawkmoth caterpillar
CONSERVATION STATUS	Not evaluated, but usually common

ADULT WINGSPAN
2¹¹⁄₁₆–3¾ in (68–95 mm)

CATERPILLAR LENGTH
2¾–3⅛ in (70–80 mm)

SPHINX POECILA

NORTHERN APPLE SPHINX

STEPHENS, 1828

475

Northern Apple Sphinx caterpillars hatch from green eggs that are laid singly by the female hawkmoth on host plant leaves; the eggs hatch within a week of laying. The caterpillars, which are most common between April and September, live and feed solitarily, usually during daylight hours, concealing themselves beneath the leaves. Final instar caterpillars occur in three color forms: green, wine red, and an intermediate form that is a mottled combination of the wine color and green. Pupation occurs in a shallow, subterranean chamber, and the pupa overwinters. There is one brood of this species annually, with adults, often found at lights, flying mostly at night from late May to August.

Sphinx poecila is very closely related to the Apple Sphinx (*S. gordius*), and both were formerly considered a single species. It is very likely that hybridization occurs in overlapping parts of their respective ranges. The Northern Apple Sphinx is considered to be the commonest hawkmoth in Canada.

The Northern Apple Sphinx caterpillar is pale to bright green, with seven oblique, white stripes laterally. The stripes are bordered in black or dark brown. The spiracles are rust colored, and the head is dark green with a pair of light green stripes. The tail horn is green laterally and dark dorsally.

Actual size

FAMILY	Sphingidae
DISTRIBUTION	India, Sri Lanka, Nepal, Myanmar, China, Chinese Taipei, South Korea, Japan, and Indonesia
HABITAT	Open forests, orchards, plantations, gardens, and parks
HOST PLANTS	Grapevine (*Vitis* spp.), arum (*Amorphphallus* spp.), hibiscus (*Hibiscus* spp.), and treebine (*Cissus* spp.)
NOTE	Hawkmoth caterpillar that has inflatable false eyespots
CONSERVATION STATUS	Not evaluated

ADULT WINGSPAN
2¾–4 in (70–100 mm)

CATERPILLAR LENGTH
2¾–3 in (70–75 mm)

THERETRA CLOTHO
COMMON HUNTER HAWKMOTH
(DRURY, 1773)

476

Common Hunter Hawkmoth caterpillars hatch from pale green, smooth eggs laid singly by the female hawkmoth on host plant leaves, mostly on the underside. The larvae feed mostly by night, usually on vines cascading over walls or hanging down from trees. By day they hide within the foliage, against which they are perfectly camouflaged. A further defense mechanism is their pair of false eyespots at the anterior end, which they are able to inflate when threatened. Development through five instars takes three to four weeks. When mature, the caterpillars crawl to the ground and form a flimsy, brown cocoon among debris.

The adults, which are nocturnal or crepuscular, emerge in as little as 14 days and are on the wing from spring to fall, often producing multiple generations. The genus *Theretra* contains about 65 named species that occur in most parts of the world. Common Hunter Hawkmoth caterpillars consume large amounts of vegetation during development and can destroy small vines in garden settings.

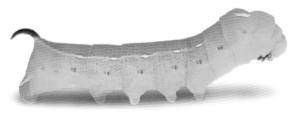

Actual size

The Common Hunter Hawkmoth caterpillar
is bright, almost translucent green tending
to yellowish dorsally. There are two rows of
seven white, round or slit-like eyespots running
down each side. The anterior pair are more
developed and have a green pupil. The spiracles,
true legs, and tail horn are reddish. There is
also a rarer dark form of the caterpillar, where
the body color is a dull brown.

FAMILY	Sphingidae
DISTRIBUTION	Tropical and subtropical South Asia, east Asia, and Southeast Asia to Australia, and north (as a migrant) to Russian Far East
HABITAT	Forest margins, open scrub, orchards, cultivated landscapes, and suburban gardens
HOST PLANTS	*Fuchsia* spp., Sweet Potato (*Ipomoea batatas*), grapevine (*Vitis* spp.), ornamental Busy Lizzie (*Impatiens walleriana*), Taro (*Colocasia esculenta*), and many other species from the families Araceae, Vitaceae, and Onagraceae
NOTE	Gaudy caterpillar that waves its horn as it walks
CONSERVATION STATUS	Not evaluated, but very common and widespread

ADULT WINGSPAN
2⅛–3⅛ in (54–80 mm)

CATERPILLAR LENGTH
2⅜–3⅛ in (60–80 mm)

THERETRA OLDENLANDIAE
TARO HORNWORM
(FABRICIUS, 1775)

477

The Taro Hornworm hatches from pale green eggs laid singly on host plant leaves. It is initially a pale yellowish green but becomes darker by the second instar, and yellow-and-black eyespots appear on most segments. By the fourth instar its basic color is blackish, and it resembles the mature larva. Feeding mainly by day, the caterpillar prefers younger leaves, seedpods, and flower heads, often stripping growing shoots, particularly in the final instar. It is an occasional pest of cultivated grapes and Sweet Potato. The larvae pupate in a loose cocoon among leaf litter; some pupae overwinter.

There are one to two generations a year in northern China, with adults mainly in July and August. However, moths can be found between June and September over most of central and southern China, although it is probable that most are migrants. The adults, but not larvae, are easily confused with several other similar species, most notably *Theretra margarita* and *T. silhetensis*. The caterpillar's common name is derived from Taro, one of its favorite tropical food plants.

The Taro Hornworm is largely dark gray, almost black in color with bands and lines of white speckling and a small, dull head. It has a dorsolateral line of yellow and orange spots on segments two to four, continuing as a stripe of small, gray spots, interspersed with reddish eyespots, up to the base of the horn. Its thin, straight horn is black with a white tip and a yellow ring near the base.

Actual size

FAMILY	Sphingidae
DISTRIBUTION	Sri Lanka, India, Nepal, Bangladesh, Myanmar, Thailand, China, Chinese Taipei, Japan, Vietnam, Malaysia, and Indonesia
HABITAT	Parks, gardens, lightly wooded areas, and waterways
HOST PLANTS	Elephant ear (*Colocasia* spp., *Caladium* spp.), water primrose (*Ludwigia* spp.), rose (*Rosa* spp.), arum (*Arum* spp.), and water caltrop (*Trapa* spp.)
NOTE	Fast-developing caterpillar that could become a biocontrol agent
CONSERVATION STATUS	Not evaluated

ADULT WINGSPAN
2⅜–2¹³⁄₁₆ in (60–72 mm)

CATERPILLAR LENGTH
2⅛–3⅛ in (55–80 mm)

THERETRA SILHETENSIS
BROWN-BANDED HUNTER HAWKMOTH
(WALKER, 1856)

478

Brown-banded Hunter Hawkmoth caterpillars hatch from up to 150 globular, light green, smooth eggs laid singly on the upper and lower surfaces of host plant leaves by the female hawkmoth. Hatching occurs three to ten days after the eggs are laid, according to temperature. The larvae are relatively inactive, spending much of their time resting on the midrib of leaves or on small stems and feeding mainly at night. They are often heavily parasitized by braconid wasps, with up to 160 wasps developing in and emerging from a single caterpillar.

Those that escape parasitization develop through five instars, taking two to five days per instar, and pupate in the ground about two weeks after hatching. The moths emerge from the subterranean pupae after about ten days. This species is also known as the Water Primrose Hawkmoth, and the caterpillars have been considered as a biological control agent for invasive water primrose species in Thailand. As is the case with many hawkmoth caterpillars, there are at least two color forms.

Actual size

The Brown-banded Hunter Hawkmoth caterpillar occurs in green and brown forms. The commoner green form is grass green with seven yellow or green, dark-rimmed false eyespots running down the body laterally. The head and prolegs are green, and the true legs are orange.

FAMILY	Epicopeiidae
DISTRIBUTION	Northeastern India, Bhutan, and southwestern China
HABITAT	Open forests
HOST PLANTS	Unknown
NOTE	Powdery, white caterpillar that feeds high in the canopy
CONSERVATION STATUS	Not evaluated, but seasonally common

NOSSA MOOREI
NOSSA MOOREI
(ELWES, 1890)

ADULT WINGSPAN
2⁹⁄₁₆–3 in (65–75 mm)

CATERPILLAR LENGTH
1⁹⁄₁₆ in (40 mm)

479

Nossa moorei caterpillars feed high in the forest canopy, their presence only revealed by the constant raining of frass at ground level. The species belongs to Epicopeiidae, a small family of Oriental tropical and subtropical moths about which very little is known. Its members are day-flying, butterfly-mimicking moths, some of which are brightly colored and tailed. Typical of this moth family, the caterpillars are covered in white flocculence, which is later incorporated into a waxy cocoon.

The larvae can occur in very large numbers in short windows of time, with the adult moths eclosing spontaneously to maximize reproductive vigor. This species and other members of the genus *Nossa* closely resemble butterflies of the Pieridae family and, like pierids, can be observed soaring high in the forest canopy. In southern China, adult moths are on the wing in early summer, with masses of defoliating larvae reaching maturity by October.

Actual size

The *Nossa moorei* caterpillar is uniformly cylindrical and white with a coarse, powdery, textured skin. There are multiple fine setae throughout, and each segment is topped with a downy tuft of white fluff, more evident on the thoracic and rear segments. The head is black, large, and round. The same black texture is duplicated on the rear segment and claspers. The crochets of the prolegs are particularly prominent.

FAMILY	Geometridae
DISTRIBUTION	Europe, east into Siberia
HABITAT	Lowland woodlands, hedgerows, and gardens
HOST PLANTS	Varied, including *Ribes* spp., *Prunus* spp., and *Crataegus* spp.
NOTE	Distinctive caterpillar that has a black-and-white colored body
CONSERVATION STATUS	Not evaluated, but quite common

ADULT WINGSPAN
1⅜–1⅞ in (35–48 mm)

CATERPILLAR LENGTH
1³⁄₁₆ in (30 mm)

ABRAXAS GROSSULARIATA
MAGPIE
LINNAEUS, 1758

480

Actual size

Magpie caterpillars hatch from eggs laid in late summer on the underside of leaves of the host plant. The larvae quickly seek a place to overwinter—for example, within rolled-up leaves or cracks in walls, and under bark. They emerge in spring to feed on young leaves. Pupation takes place in late spring and early summer, with the caterpillars spinning a flimsy cocoon, either on or near the food plant. The adults eclose and fly in July and August.

The caterpillar is a so-called looper, named for the distinctive method of moving in a looping fashion. Its conspicuous bright colors probably act as a deterrent to predators. Long considered a pest species for its defoliation of soft-fruit bushes and nut trees, its numbers have declined in recent years, possibly due to the use of pesticides. The Magpie adult is a day-flying carpet moth and—with its bright black-and-white wings— is often mistaken for a butterfly.

The Magpie caterpillar has a black head and white body, with a dorsal row of large, black spots along the length of the body, several lateral rows of small, black spots, and a lateral, orange strip. The colorings are variable, with some caterpillars completely black or white.

FAMILY	Geometridae
DISTRIBUTION	Across Europe into central Asia
HABITAT	Deciduous woodland, parks, and gardens
HOST PLANTS	Various deciduous trees, including birch (*Betula* spp.) and oak (*Quercus* spp.)
NOTE	Twiglike looper caterpillar that is perfectly camouflaged on trees
CONSERVATION STATUS	Not evaluated, but not considered to be at risk

ADULT WINGSPAN
1¹⁄₁₆–1³⁄₈ in (27–35 mm)

CATERPILLAR LENGTH
1–1³⁄₁₆ in (25–30 mm)

AGRIOPIS AURANTIARIA
SCARCE UMBER
(HÜBNER, [1799])

481

Actual size

Scarce Umber caterpillars hatch in spring from eggs that have overwintered on the bark of their host plants. The species is common and widespread with the larvae feeding on deciduous leaves from April to early June. The twiglike shape of the caterpillars provides them with perfect camouflage among the leaves and branches of the canopy. When fully fed, they descend from their host trees and pupate on the ground. There is a single generation, with the males on the wing in late fall, from October to November, seeking out females.

During the day, males rest on fences and walls, flying at night, when they are easily attracted to light. Scarce Umber females, however, are virtually wingless, having only nonfunctional, vestigial wings, reduced to small stubs, and spend their short lives crawling up and down the trunks of the host trees, where they lay their eggs. The caterpillars of *Agriopis aurantiaria* are easily confused with the caterpillars of a closely related species, the Dotted Border (*A. marginaria*).

The Scarce Umber caterpillar is twiglike, with a long, slender body that is grayish to yellowish or brownish, with pale and dark brown stripes and dots on the side and sometimes dorsally that give the appearance of tree bark. The head, true legs, and prolegs are orange brown.

FAMILY	Geometridae
DISTRIBUTION	Across Europe, east to the Urals
HABITAT	Woodlands, heaths, parks, and gardens
HOST PLANTS	Deciduous trees, including alder (*Alnus* spp.), beech (*Fagus* spp.), and oak (*Quercus* spp.)
NOTE	Well-camouflaged caterpillar that feeds on deciduous leaves
CONSERVATION STATUS	Not evaluated, but quite common

ADULT WINGSPAN
1¹⁄₁₆–1¼ in (27–32 mm)

CATERPILLAR LENGTH
1³⁄₁₆ in (30 mm)

AGRIOPIS MARGINARIA
DOTTED BORDER
(FABRICIUS, 1777)

482

Dotted Border caterpillars hatch from small, oval, green eggs laid on the tree trunks of their host species. Their defense is their realistically twiglike appearance. They are present from April to June but largely unseen as they remain camouflaged in the tree canopy, where they feed mostly on young leaves. In the final instar, when fully fed, the larvae descend to the ground to pupate and overwinter under the soil as a pupa. There is a single generation, with adults eclosing early the following year, from February to April, although occasionally as early as January and as late as May.

Agriopis marginaria males are night-flying and attracted to light. Females, however, like those of the closely related species Scarce Umber (*A. aurantiaria*), are flightless, with only tiny vestigial wings, and are found resting on tree trunks, where they lay their eggs. The young caterpillars are easily confused with *A. aurantiaria* larvae, which feed on the same host trees. The Dotted Border is common across much of its range, occurring in a wide range of habitats, due to the widespread distribution of the host plants.

Actual size

The Dotted Border caterpillar has a long, slender body typical of the Geometridae. The body is mostly brown with dark, blackish, cross-shaped dorsal and lateral markings and creamy-white patches. The white patches are most pronounced on the middle segments. The head, legs, and prolegs are orange brown.

FAMILY	Geometridae
DISTRIBUTION	Europe to central Asia
HABITAT	Woodlands, scrub, and hedgerows
HOST PLANTS	Blackthorn (*Prunus spinosa*)
NOTE	Well-camouflaged caterpillar that is difficult to spot
CONSERVATION STATUS	Not evaluated, but scarce in much of its range

ADULT WINGSPAN
1¹⁄₁₆ in–1¼ in (27–31 mm)

CATERPILLAR LENGTH
1³⁄₁₆ in (30 mm)

ALEUCIS DISTINCTATA

SLOE CARPET

(HERRICH-SCHÄFFER, [1839])

483

Sloe Carpet caterpillars hatch from eggs laid on the leaves of the host plant and feed on the Blackthorn blossoms. The larvae are active until June and July, when they move to the ground to pupate. The caterpillars spin a cocoon among the leaf litter or in loose soil, and the pupae overwinter. There is a single generation, with the adults emerging in March and on the wing until April or early May, a time that coincides with the flowering of the Blackthorn.

The Sloe Carpet, or Kent Mocha, as it is also known, is scarce across much of its range. It favors places where there are dense thickets of Blackthorn, such as overgrown hedgerows and unmanaged scrub—all areas where there are suckering young plants. Numbers have declined in recent decades due to changing countryside management. Male adults are quite similar in appearance to another Geometridae species, the Early Moth (*Theria primaria*), whose larvae also feed on Blackthorn.

The Sloe Carpet caterpillar is a mottled gray brown in color. There are bands of gray and shades of brown along the length of the body, providing cryptic coloring that creates a resemblance to a length of twig.

Actual size

FAMILY	Geometridae
DISTRIBUTION	From France and United Kingdom north and east across Europe into Russia, and northern China to Japan
HABITAT	Forests and woodlands
HOST PLANTS	Various, including Aspen (*Populus tremula*), Blackthorn (*Prunus spinosa*), honeysuckle (*Lonicera* spp.), and many *Prunus* spp.
NOTE	Well-camouflaged, twiglike caterpillar found on trees and shrubs
CONSERVATION STATUS	Not evaluated, but a common species

ADULT WINGSPAN
1⅜–2 in (35–50 mm)

CATERPILLAR LENGTH
¾ in (20 mm)

ANGERONA PRUNARIA
ORANGE MOTH
(LINNAEUS, 1758)

484

The adult female Orange Moth lays up to 250 eggs in small groups on the underside of leaves of the many host plants. The well-camouflaged caterpillar feeds on those leaves, completes a couple of molts before overwintering, and becomes active again the following spring. The larva pupates on the ground, spinning a loose cocoon among fallen leaves.

The adult moths are on the wing from late May through August. There is a single generation, with the largest numbers of adults seen during the middle of summer, from mid-June to mid-July. The species is relatively common across much of its range, helped by the large number of host plants and their widespread distribution. *Angerona prunaria* is considered a pest, as the caterpillars feed on the leaves of fruit trees of the genus *Prunus*, including cherry and plum, as well as soft fruits, such as currants, gooseberries, and raspberries.

Actual size

The Orange Moth caterpillar has a long and slender body, which is slightly thicker beyond the thorax. The eighth abdominal segment bears a sharp dorsal tubercle, and there are smaller tubercles on other segments. The gray-brown color and slender shape create a twiglike appearance that acts as the perfect camouflage among host plants.

FAMILY	Geometridae
DISTRIBUTION	North Africa, the Middle East, southern Europe, the Balkans, and east to Kazakhstan
HABITAT	Meadows, dry grasslands, steppe, embankments, and olive groves
HOST PLANTS	Various herbaceous species, especially members of Asteraceae, such as carrot (*Daucus* spp.) and ragwort (*Senecio* spp.)
NOTE	Odd-looking caterpillar camouflaged to resemble a spiky fruit
CONSERVATION STATUS	Not evaluated

ADULT WINGSPAN
1³⁄₁₆–1⅝ in (30–42 mm)

CATERPILLAR LENGTH
1⁹⁄₁₆–2 in (40–50 mm)

APOCHIMA FLABELLARIA
MEDITERRANEAN BRINDLED BEAUTY
(HEEGER, 1838)

485

The caterpillars of the Mediterranean Brindled Beauty moth hatch from eggs laid on the stems of the host plant. The larvae are seen from April to June feeding on the young leaves and flowerheads, their cryptic coloration and spines giving the impression of a spiky fruit. When disturbed, they roll into a ball, the spikes deterring predators, such as birds. The mature caterpillars fall to the ground and spin a cocoon under rocks or in the soil, where they pupate and overwinter. The pupa is red brown in color.

The night-flying moths are on the wing from February to April, and there is a single generation. They have a distinctive resting position, with the wings folded, hence their species name, which means "small fan." The forewings are folded and held out in a V shape, while the hindwings are held against the body. In many texts the species is referred to as *Zamacra flabellaria*.

The Mediterranean Brindled Beauty caterpillar has an unusual spiky appearance. The body is green and white with red spiracles ringed in black. Dorsally, there are paired white spines with smaller spines in between and laterally.

Actual size

FAMILY	Geometridae
DISTRIBUTION	Europe and across northern Asia to Korea and Japan
HABITAT	Forests, woodlands, parks, and heathlands
HOST PLANTS	Birch (*Betula* spp.)
NOTE	Caterpillar that is active at night and rarely observed
CONSERVATION STATUS	Not evaluated, but locally endangered

ADULT WINGSPAN
1³⁄₁₆–1⁹⁄₁₆ in (30–40 mm)

CATERPILLAR LENGTH
1 in (25 mm)

486

ARCHIEARIS PARTHENIAS
ORANGE UNDERWING
(LINNAEUS, 1761)

Actual size

Orange Underwing caterpillars hatch from eggs laid in small groups on birch twigs, usually in the angle between a twig and a bud. The larvae hatch at night, which is also the time they feed—initially on catkins before moving on to the leaves. During the day, the caterpillars shelter in a silk web or leaf tent. They move to the ground to pupate in leaf litter or under moss and overwinter there.

Archiearis parthenias adults, produced within a single generation, are active on bright, sunny spring days, when they can be seen flying around the tops of birch trees. The moths are spotted mostly in April and May, although they can appear as early as February. The species is endangered in some parts of its range due to the loss of birch woodland and the expansion of commercial conifer plantations. Many of the colonies are small and fragmented.

The Orange Underwing caterpillar is dark green with several white lines running the length of the body, which is scattered in black dots and covered in sparse hairs. A distinctive, lateral, white stripe runs below the spiracles. The head and legs are paler in color.

FAMILY	Geometridae
DISTRIBUTION	North America, Europe, and across Asia
HABITAT	Deciduous and mixed woodlands
HOST PLANTS	Various trees, including Alder (*Alnus glutinosa*), birch (*Betula* spp.), elm (*Ulmus* spp.), maple (*Acer* spp.), walnut (*Juglans* spp.), and willow (*Salix* spp.)
NOTE	Superbly camouflaged caterpillar that is often overlooked
CONSERVATION STATUS	Not evaluated

ADULT WINGSPAN
1⅜–2⅜ in (35–60 mm)

CATERPILLAR LENGTH
2⅜–2¾ in (60–70 mm)

BISTON BETULARIA

PEPPERED MOTH

LINNAEUS, 1758

487

The female Peppered Moth lays up to 600 yellow eggs, singly, either in bark crevices or on leaves. The caterpillar is a twig mimic; the slender shape and colors of green and brown provide it with excellent camouflage when at rest during the day. The caterpillars move to the ground to pupate, and they overwinter within a cocoon in the soil or in leaf litter. The adults emerge in late spring and early summer. There is usually one generation, with the adults on the wing from April to September.

This is a polymorphic species that has been the subject of research into natural selection, as there are two main adult color forms. The mottled form offers camouflage when the moth rests on bark covered with lichens and mosses, while the melanic or black form equips the moth with camouflage when resting on pollution-darkened bark—an adaptation known as industrialized melanization. However, with decreases in air pollution, the melanic form is now rare.

The Peppered Moth caterpillar looks like a stick. The head is chestnut brown, as are the legs. The body is mostly green, broken up with regular bands of gray brown. The spiracles are also edged in brown. There are small warts and projections to enhance its twiglike appearance.

Actual size

FAMILY	Geometridae
DISTRIBUTION	Southeastern United States, from North Carolina to Texas, Oklahoma, and north Florida
HABITAT	Forests and suburbs
HOST PLANTS	Sweet Gum (*Liquidambar styraciflua*)
NOTE	Caterpillar that has characteristic "horns" on the first thoracic segment
CONSERVATION STATUS	Not evaluated, but uncommon

ADULT WINGSPAN
1⅜ in (35 mm)

CATERPILLAR LENGTH
1⁵⁄₁₆ in (33 mm)

488

CERATONYX SATANARIA
SATAN'S HORNED INCHWORM
GUENÉE, [1858]

The Satan's Horned Inchworm caterpillar has a pair of long, hornlike prothoracic filaments covered with numerous outgrowths and secondary setae. There is also the stump of a filament dorsally on the eighth abdominal segment. The body is gray or rusty brown, with the head, filaments, subdorsal line, and all legs rusty red or beige. The lesser projections (or skin flaps) on the abdominal segments are covered with short, sclerotized spines, which may help the cuticular flaps to maintain their shape, texture, and color.

Little is known of the biology and life cycle of Satan's Horned Inchworm. The genus name *Ceratonyx* is derived from Greek, meaning "punctured horn," and, like the larva's common name, references its strange-looking, hornlike protrusions. In addition to these projections, the larvae also possess cuticular flaps that resemble leaf scars on stems, adding to their ability to camouflage themselves against a diverse array of twigs when at rest. Yet, camouflage is unlikely to be the sole purpose of the "horns," which are also covered in numerous sensory setae, housing sensilla that may help the caterpillar detect approaching predators, alerting it to assume a stiff, twiglike posture.

Ceratonyx satanaria is thought to overwinter as a pupa and produce one brood a year, with adult flight from January to the end of April, peaking in mid-February. Satan's Horned Inchworm is a member of a New World genus that now includes four species, although earlier classifications included 12.

Actual size

FAMILY	Geometridae
DISTRIBUTION	Across Europe to western Siberia
HABITAT	Forests, woods, parks, and gardens
HOST PLANTS	Various deciduous trees and shrubs, including birch (*Betula* spp.), honeysuckle (*Lonicera* spp.), oak (*Quercus* spp.), *Prunus* spp., and willow (*Salix* spp.)
NOTE	Twiglike caterpillar that has a classic looping movement
CONSERVATION STATUS	Not evaluated, but a common species

ADULT WINGSPAN
1⅜–1⅝ in (35–41 mm)

CATERPILLAR LENGTH
1⁹⁄₁₆–1¼ in (40–45 mm)

CROCALLIS ELINGUARIA
SCALLOPED OAK
(LINNAEUS, 1758)

489

Scalloped Oak caterpillars hatch from cube-shaped eggs laid in batches. Having found a suitable leaf, the adult female lays the eggs in a row so that they are touching, lined up along the leaf edge. The eggs overwinter and hatch the following spring; the caterpillars are then seen from April to July, feeding on a variety of deciduous trees and shrubs. They eat mostly leaves but are known to be omnivorous as they will attack and eat smaller caterpillars of the same species. The pupa is red brown.

Crocallis elinguaria moths, all of a single generation, are on the wing at night during the months of July and August and are attracted to light. During the day, they rest on tree trunks and fences. The caterpillar is easily confused with close relatives that live in the same habitat, including the brown form of the Scalloped Hazel (*Odontoponera bidentate*) and the Dusky Scalloped Oak (*Crocallis dardoinaria*).

The Scalloped Oak caterpillar has an elongated, slender body. The colors are variable, ranging from gray to dark brown. The pattern of patches, diamonds, stripes, and longitudinal lines, together with small tubercles, creates the perfect impression of a twig. The head is brown, and there are sparse hairs over the body.

Actual size

FAMILY	Geometridae
DISTRIBUTION	Parts of Southeast Asia and northern Australia
HABITAT	Tropical rain forests
HOST PLANTS	*Carallia brachiata*
NOTE	Caterpillar that, when resting, can resemble a catkin
CONSERVATION STATUS	Not evaluated

ADULT WINGSPAN
3–3⅛ in (75–80 mm)

CATERPILLAR LENGTH
2⅜ in (60 mm)

490

DYSPHANIA FENESTRATA
FOUR O'CLOCK MOTH
SWAINSON, 1833

The Four O'Clock Moth caterpillar takes on an unusual position while resting on the food plant. It has a short thorax, so the legs are tight together, and a long, slender abdomen with two pairs of claspers at the rear, which the caterpillar uses to suspend its body from the leaf and hold it steady, so that it looks like a catkin. The larvae feed primarily on the leaves of *Carallia brachiata*, a small rain forest tree, and in groups may completely defoliate the tree. The mature caterpillar pupates in a fold of a leaf, which it holds in place with silk threads.

The brightly colored, day-flying moth adult gets its name from the time that it tends to appear, later afternoon, hence "Four O'Clock." It is also known as the Peacock Jewel. The caterpillar is equally brightly colored, a clear warning that it is distasteful and should be avoided by predators.

Actual size

The Four O'Clock Moth caterpillar has a yellow-orange head and bright yellow body. There are rows of black spots along the length of the body, which are small on the thorax but larger on the abdomen. The legs and prolegs are yellow red. The body terminates in a large pair of claspers.

FAMILY	Geometridae
DISTRIBUTION	Europe and across northern Asia to Siberia
HABITAT	Woodlands, parks, and gardens
HOST PLANTS	Range of deciduous trees, including birch (*Betula* spp.), oak (*Quercus* spp.), and elm (*Ulmus* spp.)
NOTE	Twiglike caterpillar that is found among leaves of deciduous trees
CONSERVATION STATUS	Not evaluated, but scarce in parts of its range

ADULT WINGSPAN
1⁹⁄₁₆–2 in (40–50 mm)

CATERPILLAR LENGTH
2 in (50 mm)

ENNOMOS AUTUMNARIA
LARGE THORN
(WERNEBERG, 1859)

491

The female Large Thorn lays her eggs on the leaves of deciduous trees in fall, and the eggs overwinter and hatch in spring. The caterpillars remain in the canopy, feeding at night on the leaves of a wide variety of deciduous trees. During the day, the larvae stay motionless on the branches, relying on their camouflage to avoid predation. Their development to the final instar often takes several months. They then pupate in a cocoon spun between leaves. The moths eclose up to six weeks later.

Like all members of the family Geometridae, the caterpillars have a looping movement, giving them the generic name "loopers." In August, the caterpillar drops to the ground, where it pupates in a cocoon in the leaf litter or under moss. The adult emerges a few weeks later. Unusually, the moth is a fall-flying species with the adults on the wing during September and October. The Large Thorn is becoming scarce in some countries, such as the United Kingdom, likely because of urbanization.

Actual size

The Large Thorn caterpillar has a twiglike appearance. The flat, squarish head is brown or gray brown, and the body is brown, in varying shades that resemble tree bark. As well as a long and slender body, there are warts along its length to reinforce the twig camouflage and a pair of large claspers at the end of the abdomen.

FAMILY	Geometridae
DISTRIBUTION	From southern Canada south to Georgia (and possibly to the Florida panhandle) in the eastern United States and to northern California in the west
HABITAT	Deciduous and mixed forests
HOST PLANTS	Alder (*Alnus* spp.), ash (*Fraxinus* spp.), birch (*Betula* spp.), acer (*Acer* spp.), oak (*Quercus* spp.), and poplar (*Populus* spp.)
NOTE	Sticklike caterpillar that blends in with its host plant twigs
CONSERVATION STATUS	Not evaluated, but common

ADULT WINGSPAN
1¹¹⁄₁₆–2⅜ in (43–60 mm)

CATERPILLAR LENGTH
2¾–3⅛ in (70–80 mm)

492

ENNOMOS MAGNARIA
MAPLE SPANWORM
GUENÉE, [1858]

Maple Spanworm caterpillars hatch from overwintering eggs laid in rows on their host plant and can be found between May and August. Both edible and abundant, the larvae have evolved a remarkable ability to camouflage themselves. So as not to create a searchable pattern for predators, the green, brown, or gray coloring of their body varies, and they also blend perfectly with their host plant by assuming a stiff, twiglike position during the day. At maturity, the larvae pupate in a cocoon woven between the leaves of the host plant.

Although listed as feeding on apples, the caterpillars apparently inflict no economic damage. However, like many other inchworms, they have an important ecological significance as food for numerous bird species as well as other insects, such as predatory wasps. The adults fly in a single generation from July to October (depending on the latitude), and, just like the larvae, the moths are masters of camouflage, matching fallen leaves perfectly both in shape and color.

Actual size

The Maple Spanworm caterpillar is a twig mimic with a green, brown, or gray body and tiny, white spotting that simulates the spotting on the bark of the host plant trees. Skin folds and swellings on two or three of the abdominal segments resemble leaf scars. The head is green, flattened, and pointed forward, with a pair of prominent reddish antennae. The metathoracic legs are swollen at their base, resembling another leaf scar.

FAMILY	Geometridae
DISTRIBUTION	Northwest Africa, across Europe, through the Caucasus to northern Iran, and across Russia to northeast China
HABITAT	Woodlands, orchards, grasslands, parks, and gardens
HOST PLANTS	Various deciduous trees and shrubs, including Apple (*Malus pumila*), birch (*Betula* spp.), oak (*Quercus* spp.), *Prunus* spp., and willow (*Salix* spp.)
NOTE	Defoliating caterpillar that is considered a pest species
CONSERVATION STATUS	Not evaluated, but very common

ADULT WINGSPAN
1³⁄₁₆–1⁹⁄₁₆ in (30–40 mm)

CATERPILLAR LENGTH
1³⁄₁₆–1⅜ in (30–35 mm)

ERANNIS DEFOLIARIA
MOTTLED UMBER
(CLERCK, 1759)

Mottled Umber caterpillars hatch from chains of pale brown, oval eggs deposited on the bark of the host plant. Once laid, the eggs overwinter and then hatch the following spring. The larvae are gregarious, staying in a group and spinning leaves together to create a shelter, in which they remain hidden from predators, dropping to the ground if disturbed. Pupation takes place underground in a cocoon.

The female Mottled Umber moth is wingless, has a spiderlike appearance, and remains on the host tree. The nocturnal male moths, all part of a single generation, are on the wing from October to December and are attracted to light. They rest on trees, fences, and walls during the day. The caterpillars damage the flower buds and leaves of the host plant, and infestations of larvae can defoliate entire trees in orchards and woodlands. As a result, the species is considered a pest.

Actual size

The Mottled Umber caterpillar has a slender, elongated shape. It is quite variable in color, with shades of green, orange brown, and dark brown. The head is usually orange brown. The dorsal surface is brown, while the underside is green, with a dark lateral stripe in between. There are widely spaced short hairs.

FAMILY	Geometridae
DISTRIBUTION	Europe and northern Asia to Russian Far East, parts of Canada and northern United States
HABITAT	Moorlands, sandy heaths, and woodlands
HOST PLANTS	Various, including bilberry (*Vaccinium* spp.), birch (*Betula* spp.), Heather (*Calluna vulgaris*), and willow (*Salix* spp.)
NOTE	Slender, twiglike caterpillar that is well camouflaged
CONSERVATION STATUS	Not evaluated, although becoming rare in some regions

ADULT WINGSPAN
1–1⅜ in (25–35 mm)

CATERPILLAR LENGTH
1 in (25 mm)

494

EULITHIS TESTATA
CHEVRON
(LINNAEUS, 1761)

The female Chevron moth lays her creamy-white to pale brown eggs in a line along the edge of a leaf on the host plant. The eggs overwinter, and the caterpillars hatch the following spring, their twiglike appearance providing excellent camouflage as they develop and some protection against predation. Like all members of the Geometridae family, the larvae move with a typical looping action. They pupate on the host plant, spinning a cocoon between the leaves.

The widely distributed nocturnal Chevron is on the wing in summer, from July to August, and is attracted to light. There is usually a single generation, but in some parts of Europe there are two generations. The species name *testata* refers to the base color of the moth's wing, after the Latin *testa*, which means "baked stone," while the common name refers to the distinctive V-shaped, white bar on the forewings. Variations in the base color occur across the moth's range, with different shades of orangey brown and yellow brown.

Actual size

The Chevron caterpillar has a slender, elongated shape typical of its family. The shades of yellow, pale brown, and olive green give the caterpillar the appearance of a twig. There are a number of pale, yellow-brown lines running the length of the body, and the spiracles are brown.

FAMILY	Geometridae
DISTRIBUTION	North America, from Saskatchewan east to Nova Scotia, south to Florida, and west to Texas
HABITAT	Deciduous and mixed woodlands
HOST PLANTS	Ash (*Fraxinus* spp.), basswood (*Tilia* spp.), birch (*Betula* spp.), elm (*Ulmus* spp.), poplar (*Populus* spp.), willow (*Salix* spp.), maple (*Acer* spp.), and other trees
NOTE	Twig-mimicking caterpillar
CONSERVATION STATUS	Not evaluated, but common

EUTRAPELA CLEMATARIA

CURVED-TOOTHED GEOMETER

(J. E. SMITH, 1797)

ADULT WINGSPAN
1½–2³⁄₁₆ in (38–56 mm)

CATERPILLAR LENGTH
2⅜ in (60 mm)

495

Curved-toothed Geometer caterpillars emerge from eggs that are green when laid and turn red just before hatching. The young larvae have a dark brown body, while the older larvae have a greenish, tan, gray, or dark purplish-brown body, helping them to blend in with their host tree. It takes about 40 days for the caterpillar to develop. Adult moths fly from March to August in most of the range but can be found year-round in the south. There are two generations a year.

While not widely known as a pest, *Eutrapela clemataria* has been reported as extremely destructive in small areas of bogs where cranberries are grown, the larvae eating flower buds and blossoms and so adversely affecting production of the fruit. The caterpillars of this species, like many others, suffer from ant predation, although a study examining *E. clemataria* and other moth caterpillars showed that the risk of such predation was less when the range of host plants was increased. The species is the only member of the genus *Eutrapela*.

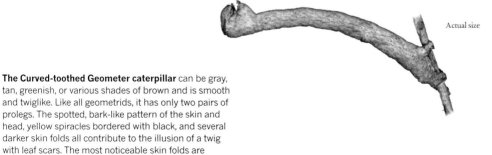

Actual size

The Curved-toothed Geometer caterpillar can be gray, tan, greenish, or various shades of brown and is smooth and twiglike. Like all geometrids, it has only two pairs of prolegs. The spotted, bark-like pattern of the skin and head, yellow spiracles bordered with black, and several darker skin folds all contribute to the illusion of a twig with leaf scars. The most noticeable skin folds are located dorsally, between the first and the second thoracic segments and on the penultimate abdominal segment. The legs are chocolate brown.

FAMILY	Geometridae
DISTRIBUTION	Europe and across Asia to China and Japan
HABITAT	Woodlands, scrub, heathlands, and wet woodlands near streams
HOST PLANTS	Mostly birch (*Betula* spp.), but also Alder (*Alnus glutinosa*), beech (*Fagus* spp.), and hazel (*Corylus avellana*)
NOTE	Green looper caterpillar that is found on a variety of food plants
CONSERVATION STATUS	Not evaluated

ADULT WINGSPAN
2–2⁹⁄₁₆ in (50–65 mm)

CATERPILLAR LENGTH
1³⁄₁₆–1³⁄₈ in (30–35 mm)

GEOMETRA PAPILIONARIA
LARGE EMERALD
LINNAEUS, 1758

Large Emerald caterpillars hatch from eggs laid in late summer. The young larvae are green at first but soon become pale brown with warts, giving them a twiglike appearance. This provides much better camouflage for when they overwinter on dormant trees. The caterpillars are active again in spring, when they become greener in color once more to blend in with the new spring growth of leaves. When fully developed, the larvae move to the ground to pupate in a cocoon among the leaf litter.

The caterpillar is a so-called "looper," moving along branches by drawing its rear end up to the thorax, extending the head and thorax forward until it lies flat, and then repeating the movement. The large, night-flying moth adult is butterfly-like with bright green wings that fade with age. The species occurs in a wide range of habitats, as the caterpillar feeds on many different plants, although its preference is for birch.

The Large Emerald caterpillar is green with a long and slender body. It has a lateral creamy-yellow line, which ends in a terminal segment that is reddish brown in color. There are large posterior claspers, and the head is white with brown markings. Some forms can also be predominantly brown.

Actual size

FAMILY	Geometridae
DISTRIBUTION	Europe, North Africa, and through central Asia to southern Siberia
HABITAT	Scrub, waste ground, verges, parks, and gardens
HOST PLANTS	Mallow (*Malva sylvestris*) and related species, such as *Althaea officinalis* and Bristly Hollyhock (*Alcea setosa*)
NOTE	Caterpillar that raises its body to resemble a twig
CONSERVATION STATUS	Not evaluated, but locally rare

ADULT WINGSPAN
1⁷⁄₁₆–1⁹⁄₁₆ in (36–40 mm)

CATERPILLAR LENGTH
1⁹⁄₁₆ in (40 mm)

LARENTIA CLAVARIA
MALLOW
(HAWORTH, 1809)

497

Eggs laid on the food plants by the night-flying female Mallow moths overwinter, and the larvae hatch the following spring. When not feeding, the caterpillar rests on the upper surface of leaves and holds on firmly with its terminal claspers, raising the rest of the body to give the appearance of a green twig. This makes the larva very difficult to spot and reduces predation. The caterpillars are active from April to July, moving from their host plant to the ground to pupate. The night-flying moths are on the wing from August to November.

Like other members of the Geometridae, the caterpillars are loopers. Their prolegs are found toward the end of the abdomen, giving the caterpillars their characteristic looping movement. There are several subspecies across the range. The dependence of *Larentia clavaria* on a limited range of food plants has contributed to a decline in its numbers, and the moth is now classed as rare in some parts of its range.

Actual size

The Mallow caterpillar is long, slender, and green in color. There are faint bands of pale yellow and green along its length and tiny, white spots. It is covered in short, widely spaced hairs.

FAMILY	Geometridae
DISTRIBUTION	United States, from South Dakota in the Midwest south to Texas, and from Massachusetts to Florida in the east
HABITAT	Pitch pine-scrub oak barrens in the north; woodlands and forests in the south
HOST PLANTS	Apple (*Malus* spp.), *Clethra* spp., *Myrica* spp., cherry (*Prunus* spp.), oak (*Quercus* spp.), *Vaccinium* spp., and likely many other woody plants
NOTE	Caterpillar that is colorful and variable; adult females are wingless
CONSERVATION STATUS	Not evaluated, but relatively common in the south of its range and rare in the north

ADULT WINGSPAN
1⅛–1⅝ in (29–41 mm)

CATERPILLAR LENGTH
1⁹⁄₁₆ in (40 mm)

498

LYCIA YPSILON
WOOLLY GRAY
(FORBES, 1885)

Woolly Gray caterpillars hatch in spring and feed on a variety of woody plants. While most geometrid caterpillars are cryptically colored in subdued gray, brown, or green, this species is brightly colored and patterned, though not aposematic. It seems most likely that the markings serve a dual purpose; their complexity and variability help the caterpillars evade birds, which are skillful at recognizing only specific patterns, and, if the larvae are detected, their bright coloring might also alarm potential predators.

Lycia ypsilon overwinters as a pupa, and, in the north of its range, the males eclose and fly before all the snow has melted. In the south of its range (Florida), it is also one of the earliest moths of the year, flying from January to March in a single generation. The females are wingless and crawl on the ground, making them hardly recognizable as Lepidoptera.

Actual size

The Woolly Gray caterpillar is typical of inchworm moth larvae in shape, with two pairs of prolegs and a long, slender body. It is patterned in yellow, red, maroon, black, and white, with the head and last abdominal segment peppered white. Individuals are variably colored, with a burgundy or gray (light or dark) base color, patterned stripes, and red (or sometimes yellow) spots behind the spiracles.

FAMILY	Geometridae
DISTRIBUTION	Across southern Canada, from Vancouver Island to Nova Scotia, south to much of the United States, excluding California
HABITAT	Deciduous and mixed wood forests and woodlands
HOST PLANTS	Douglas Fir (*Pseudotsuga menziesii*), willow (*Salix* spp.), White Birch (*Betula papyrifera*), hazelnut (*Corylus* spp.), and others; occasionally on crops such as strawberries (*Fragaria* spp.) and carrots (*Daucus* spp.)
NOTE	Caterpillar that has two pairs of long filaments, or tentacles
CONSERVATION STATUS	Not evaluated, but common in riparian habitats in parts of its range

ADULT WINGSPAN
¾–1 in (20–25 mm)

CATERPILLAR LENGTH
1½–2 in (38–51 mm)

NEMATOCAMPA RESISTARIA

HORNED SPANWORM

HERRICH-SCHÄFFER, [1856]

499

The Horned Spanworm, also called the Filament Bearer, is named for its unusual caterpillar, which bears filaments on three of the abdominal segments. The filaments are initially short but increase in length as the caterpillar grows. Their function is not clear, but the sensory hairs on them suggest that they may, for instance, help the larva to detect vibrations produced by an approaching predator. Alerted, the caterpillar would then stop moving; at rest, it is well camouflaged as the filaments provide additional structures that make it look much like a twig.

The moth is also cryptically colored, with colors and a pattern that make it practically invisible against a background of fallen leaves. As a member of the inchworm moth family, Geometridae, which are mostly cryptically colored and palatable, *Nematocampa resistaria* is an essential part of the food chain: birds feed on Geometrids, and especially during colder months, when other insects are less abundant, this moth becomes an important food source.

Actual size

The Horned Spanworm caterpillar has distinctive abdominal projections, or filaments, located dorsally. Otherwise, the caterpillar resembles other inchworms, being a reddish-brown color with a lighter, cream-colored pattern that helps it to blend in with the twigs of the trees on which it feeds.

FAMILY	Geometridae
DISTRIBUTION	United States, southern Canada, Iceland, Europe into western Russia, much of Africa, the Middle East to Iran, and northern India
HABITAT	Woodlands, scrub, parks, and gardens
HOST PLANTS	Various, including *Chrysanthemum* spp., *Polygonum* spp., and dock and sorrel (*Rumex* spp.)
NOTE	Well-camouflaged, twiglike caterpillar that has a wide distribution
CONSERVATION STATUS	Not evaluated

ADULT WINGSPAN
⁹⁄₁₆–⅞ in (15–22 mm)

CATERPILLAR LENGTH
¾–1 in (20–25 mm)

500

ORTHONAMA OBSTIPATA
GEM
(FABRICIUS, 1794)

Actual size

The caterpillars of the Gem moth, also known as the Bent-line Carpet Moth, hatch from yellow, slightly elongated eggs laid singly or in small groups on the underside of leaves of low-growing plants. The larvae, which rest during the day and become active at night, are polyphagous, feeding on a wide variety of food plants while protected by their twiglike camouflage. They pupate in a cocoon in the soil.

The night-flying adult moths are seen from April to November, but in the more tropical parts of its range *Orthonama obstipata* is on the wing all year round. The species is sexually dimorphic, with the females larger and darker than the males and lacking their white lines. Being a strong flier, the moth appears as a migrant over a large area, even flying across sizeable stretches of water. Due to its wide range and variable appearance, the species has been described multiple times by different authors and hence has around 40 synonyms.

The Gem caterpillar has a slender body in shades of brown or green. The markings, including faint dorsal and lateral brown lines running the length of the body and rings of pale brown, give the appearance of a twig. The spiracles are dark brown to black.

FAMILY	Geometridae
DISTRIBUTION	Europe and across Asia to China and Japan
HABITAT	Scrub, waste ground, verges, and gardens
HOST PLANTS	Orache (*Atriplex* spp.) and goosefoot (*Chenopodium* spp.)
NOTE	Odd-looking caterpillar that has perfect camouflage
CONSERVATION STATUS	Not evaluated, but locally rare

ADULT WINGSPAN
1–1³⁄₁₆ in (25–30 mm)

CATERPILLAR LENGTH
¾ in (20 mm)

PELURGA COMITATA

DARK SPINACH

(LINNAEUS, 1758)

501

Dark Spinach caterpillars hatch from clusters of small, creamy-colored eggs laid on the leaves of the host plants. The larvae are active in August and September, feeding on their weedy orache and goosefoot host plants, preferring flowers and seeds to the leaves. The caterpillars move along branches in the looping style typical of the family. The species overwinters underground as a pupa, which is red brown in color. The adult moths are on the wing in July and August, and there is a single generation.

The moths favor weedy ground, such as derelict land, verges, and gardens. However, numbers of this species have fallen steeply in recent decades as the increased use of weed killers on farmland and general loss of derelict land has killed off many of their host plants. For example, in the United Kingdom, the *Pelurga comitata* population reportedly declined by 89 percent between 1968 and 2002, and it is now classed as a priority species.

Actual size

The Dark Spinach caterpillar is unusual in appearance. The upper body is dark brown black and olive green, while the lower side is much paler with a creamy, zigzag line along the side, separating the two areas.

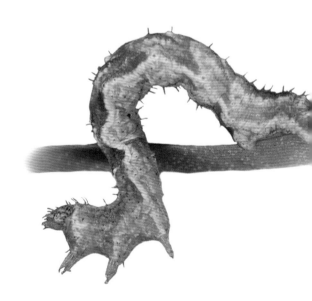

FAMILY	Geometridae
DISTRIBUTION	Europe and across Asia to Japan
HABITAT	Deciduous woodlands
HOST PLANTS	Deciduous trees, including birch (*Betula* spp.), oak (*Quercus* spp.), and willow (*Salix* spp.)
NOTE	Caterpillar whose twiglike appearance provides perfect camouflage
CONSERVATION STATUS	Not evaluated

ADULT WINGSPAN
1⅛–1¼ in (28–32 mm)

CATERPILLAR LENGTH
1³⁄₁₆ in (30 mm)

502

PLAGODIS DOLABRARIA
SCORCHED WING
(LINNAEUS, 1767)

Actual size

The Scorched Wing caterpillar is twiglike and colored in shades of reddish brown and olive green. The upper thorax is darker than the ventral side, and there is a distinctive hump on the abdomen and a dark transverse mark near the posterior end.

Scorched Wing caterpillars hatch from oval, white eggs laid along leaf margins, and they can be seen in woodlands from May to October. To escape predation, the caterpillar relies on camouflage, gripping a twig with its terminal claspers and raising its body up so that it looks like a short length of twig. When fully developed, the caterpillar pupates on the ground in leaf litter, and its chestnut-brown pupa overwinters. There is usually one generation, with the adults on the wing from May to July, but in some locations there may be two generations a year.

The night-flying moth gets its unusual name from the appearance of its wings at rest, the shades of brown resembling a piece of burned paper and providing effective camouflage when it rests on branches during the day. The moths prefer open woodlands with clearings and rides that allow more light to reach the ground and a greater diversity of plant species.

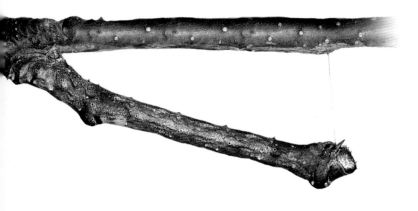

FAMILY	Geometridae
DISTRIBUTION	Europe, except Iberia and Greece, into central Asia and Siberia
HABITAT	Woodlands, hedgerows, parks, and gardens
HOST PLANTS	Various deciduous trees, including birch (*Betula* spp.) and oak (*Quercus* spp.)
NOTE	Well-camouflaged, twiglike caterpillar
CONSERVATION STATUS	Not evaluated

ADULT WINGSPAN
1½–1¾ in (38–44 mm)

CATERPILLAR LENGTH
¾–1 in (20–25 mm)

SELENIA LUNULARIA

LUNAR THORN
(HÜBNER, 1788)

503

Lunar Thorn caterpillars hatch from small, round, red-colored eggs laid on a large range of trees and shrubs on which the larvae feed. They pupate on the plant, attaching to the underside of leaves and branches. The pupa is bright green. There are two generations annually, one in June and another in August and September. The pupa of the second generation overwinters, and the adults emerge in late spring. The moths are on the wing from May to August.

This species gets its common name from the white moon-shaped mark on the wings of the moth and the appearance of the caterpillar, which resembles a thorny twig when at rest. The caterpillar has a looping movement, and when at rest it grasps the twig firmly with its posterior claspers, lifting up the rest of its body to give the appearance of a twig.

Actual size

The Lunar Thorn caterpillar has a long, slender body in shades of olive green, yellow, and brown, banded to look like bark. Numerous tubercles in red and brown enhance the bark-like appearance. The caterpillars vary in color.

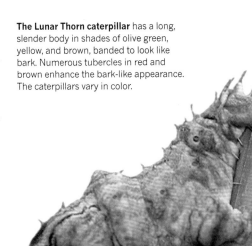

FAMILY	Geometridae
DISTRIBUTION	Widespread in North America, from southern Canada to Georgia, with highest concentrations in the Piedmont plateau region in eastern United States
HABITAT	Flowering fields and gardens
HOST PLANTS	*Aster* spp., black-eyed susan (*Rudbeckia* spp.), snakeroot (*Ageratina* spp.), goldenrod (*Solidago* spp.), and other flowering plants
NOTE	Covert, camouflaged caterpillar
CONSERVATION STATUS	Not evaluated, but not threatened

ADULT WINGSPAN
1¹¹⁄₁₆ in (17 mm)

CATERPILLAR LENGTH
⁹⁄₁₆ in (15 mm)

504

SYNCHLORA AERATA
CAMOUFLAGED LOOPER
(FABRICIUS, 1798)

Actual size

The Camouflaged Looper caterpillar's most identifiable characteristic is its habit of attaching severed flower bits to its body. Like a combat soldier who affixes leaves and branches to his uniform, this tiny looper has mastered the art of "disappearing" into its surroundings. While adorned in its flowered regalia, the caterpillar has been observed swaying back and forth, as if to simulate part of the flower fluttering in the breeze. The disguising cloak is shed after a molt, then quickly replenished with a new array of plant material. The larvae feed on flower heads.

Mature caterpillars are present from April to October, with two generations in the northern range and up to four broods possible in the south. A middle instar larva overwinters. In many texts, *Synchlora aerata* is given the common name of Wavy-lined Emerald in reference to the appearance of the green adult moth, which is nocturnal and attracted to light. Several *Synchlora* subspecies occur south of Pennsylvania.

The Camouflaged Looper caterpillar has a base color that is primarily brown and black, with white wavy lines along the abdomen. The small head is brown and mottled. The physical appearance is variable, highly influenced by the host plant and the collection of flower material the larva gathers and attaches to its body.

FAMILY	Geometridae
DISTRIBUTION	Northeast Himalayas, southern China, Chinese Taipei, and Southeast Asia to Borneo
HABITAT	Montane forests
HOST PLANTS	Oak (*Quercus* spp.)
NOTE	Caterpillar that precisely mimics fresh host plant growth
CONSERVATION STATUS	Not evaluated, but not uncommon

ADULT WINGSPAN
2–2¾ in (50–70 mm)

CATERPILLAR LENGTH
1⁹⁄₁₆ in (40 mm)

TANAORHINUS VIRIDILUTEATUS

TANAORHINUS VIRIDILUTEATUS
(WALKER, 1861)

505

The *Tanaorhinus viridiluteatus* caterpillar is typical of geometrid looper or inchworm moths. It has well-developed anal claspers but only one pair of prolegs and a limbless gap between the true legs and prolegs, which means it moves in a looping fashion. At rest, however, the larva will usually secure itself to the host plant using the rear limbs only and levitate the rest of the body in an arch. The caterpillars feed at the extremes of branchlets, often completely consuming entire leaves and resting inconspicuously at the base of the naked petioles and stipules. Pupation occurs within a light, silken cocoon in a folded host plant leaf.

Both caterpillar and adult moth are masterfully camouflaged. The abdominal segments of the caterpillar possess hornlike growths resembling leaf stipules and new leaf buds. These lengthen as the caterpillar matures. Body markings also become woodier in appearance with maturity to reflect the graduation to more substantial and developed foliage. The moth is leaf green, with subtle markings resembling anomalies on a leaf's surface.

The *Tanaorhinus viridiluteatus* caterpillar has pairs of bladelike, hook-tipped, and finely furred extensions on its first five and final abdominal segments, the largest on segments three to five. The rest of the body is gnarled and roughly textured with regular patterns of brown and white on a green base. There are lateral, knot-like markings on the penultimate abdominal segment.

Actual size

FAMILY	Geometridae
DISTRIBUTION	Across northern Europe into western Siberia
HABITAT	Woodlands, woodland margins, grasslands, bogs, and uplands
HOST PLANTS	Low-growing plants such as bedstraw (*Galium* spp.), knotgrass (*Polygonium* spp.), and *Vaccinium* spp.
NOTE	Camouflaged, twiglike caterpillar that has a looping movement
CONSERVATION STATUS	Not evaluated

ADULT WINGSPAN
¾–1¹⁄₃₂ in (20–26 mm)

CATERPILLAR LENGTH
1–1³⁄₁₆ in (25–30 mm)

506

XANTHORHOE SPADICEARIA
RED TWIN-SPOT CARPET
(DENIS & SCHIFFERMÜLLER, 1775)

The caterpillars of the Red Twin-spot Carpet moth hatch from creamy-yellow, oval eggs laid on the underside of leaves either singly or in small clusters. The larvae, which mature quite quickly, feed on leaves and then crawl to the ground to pupate, where they spin a silky cocoon among the leaf litter. The pupa is red brown in color.

The moths may be seen during the day but are most active at dusk. There are usually two generations a year, with the moths on the wing from early May to June and July to August. The pupae of the second generation overwinter. In the more northerly parts of the range, there is a single generation flying from June to July. The adult moth, which has two dark spots close to the outer edge of each forewing, is very similar in appearance to the closely related Dark-barred Twin-spot Carpet (*Xanthorhoe ferrugata*).

Actual size

The Red Twin-spot Carpet caterpillar is long and slender with a twiglike appearance. It is dark brown with several pale lines running the length of the body. Dorsally, there is a series of orange-and-black diamond shapes surrounded by white dots. Short hairs cover the body.

FAMILY	Notodontidae
DISTRIBUTION	Europe and across Asia to Japan
HABITAT	Wet, lowland forests
HOST PLANTS	Birch (*Betula* spp.), poplar (*Populus* spp.), and willow (*Salix* spp.)
NOTE	Caterpillar that rears up and looks larger when threatened
CONSERVATION STATUS	Not evaluated, but locally rare

ADULT WINGSPAN
1–1⅜ in (25–35 mm)

CATERPILLAR LENGTH
1⁹⁄₁₆ in (40 mm)

CERURA ERMINEA
LESSER PUSS MOTH
ESPER, 1783

507

Lesser Puss Moth caterpillars hatch from reddish-brown, flattened eggs laid on the underside of leaves of the food plant. Usually, the larvae remain high in the tree canopy as they develop. When disturbed, the caterpillar acts defensively by rearing up to look larger. It overwinters as a pupa, protected within a tough cocoon made from wood shavings. The adults are on the wing from May to July, with the caterpillars seen from June to August.

The *Cerura erminea* caterpillar is very similar in appearance to its close relative the Puss Moth (*C. vinula*) but lacks the red collar and false eyes of the Puss Moth larva and also has a vertical white mark halfway along the abdomen. Once more common, the Lesser Puss Moth is suffering from the loss of lowland forests rich in native black poplar trees. As a result, the species is often classed as scarce or rare within its range.

Actual size

The Lesser Puss Moth caterpillar is green, with a dark brown dorsal band extending along the length of the body into two long tails. The band forms a V shape laterally and is edged in white. The spiracles are ringed in brown.

FAMILY	Notodontidae
DISTRIBUTION	Southern Canada, from eastern Alberta east to Quebec, south through United States to Florida, and west to Texas
HABITAT	Deciduous woodlands near rivers and lakes
HOST PLANTS	Cherry (*Prunus* spp.), poplar (*Populus* spp.), and willow (*Salix* spp.)
NOTE	Caterpillar defended by face-like pattern, flagella, and formic acid
CONSERVATION STATUS	Not evaluated

ADULT WINGSPAN
1–1%₆ in (25–40 mm)

CATERPILLAR LENGTH
2–2⅜ in (50–60 mm)

508

CERURA SCITISCRIPTA
BLACK-ETCHED PROMINENT
WALKER, 1865

Black-etched Prominent caterpillars, like other related prominents, hatch from eggs laid on the underside of host plant leaves and may initially feed together. When resting, the larvae are quite cryptic, but, like other members of their genus, it is their appearance when threatened that has gained them notoriety. As they raise and retract their head to face an opponent, the thoracic region swells, exposing and emphasizing the red coloration and two black eyespots, so that the caterpillar resembles a face with an open mouth. Simultaneously, it raises its "tail"—black flagella equipped with bright orange-red extensions—and may also squirt formic acid.

While such defenses will not deter a predating Blackbird (*Turdus merula*) or Praying Mantis (*Mantis religiosa*), they seem effective against some parasitoids such as *Cotesia* wasps, which are possibly repelled by scents stored in the flagella. Nevertheless, these larvae are not completely immune to parasitism. Mature caterpillars pupate on a leaf or stem in a cocoon of silk and plant tissue, and the adults fly from March to October in one or two generations depending on location. The genus *Cerura* comprises 20 species of moths that all have striking larvae.

Actual size

The Black-etched Prominent caterpillar is green with a white subspiracular stripe. The head is beige with black, lateral, vertical stripes. The last abdominal segment has extendable, black flagella about half of the caterpillar's length, with orange-red tips everted when the caterpillar is disturbed. The thorax bears two white stripes dorsally, terminating in two prominent, black spots in front. The first thoracic segment is colored pink dorsally.

FAMILY	Notodontidae
DISTRIBUTION	Europe, east to central Asia; also eastern China
HABITAT	Damp woodlands and scrub
HOST PLANTS	Poplar (*Populus* spp.) and willow (*Salix* spp.)
NOTE	Caterpillar that has menacing eyespots and a red collar
CONSERVATION STATUS	Not evaluated

ADULT WINGSPAN
2⁹⁄₁₆–3 in (58–75 mm)

CATERPILLAR LENGTH
3⅛ in (80 mm)

CERURA VINULA
PUSS MOTH
(LINNAEUS, 1758)

509

Puss Moth caterpillars hatch from red-colored eggs laid in small groups on the upper side of leaves on the food plant. The young larvae are black but gradually develop a green-and-black coloring—a disruptive pattern that breaks up their shape, helping to conceal them from predators. The caterpillar overwinters in a tough cocoon formed from wood shavings, which is attached to a tree trunk or post. The adult moths are on the wing from April to July.

The Puss Moth caterpillar is best known for its conspicuous defensive behavior. When disturbed, it takes on a menacing appearance, rearing up and pulling its head into its thorax, causing it to swell and reveal a bright red collar and two false eyes, while the tails curl forward and the red filaments appear. To further deter a predator, the caterpillar can spray acid from glands behind its head.

The Puss Moth caterpillar is bright green in color with a dorsal, dark brown-black band, outlined in white. The abdomen ends in two tails, each terminating in an extendable, red, whiplike filament. The red collar and false eyes are not visible unless the caterpillar is disturbed.

Actual size

FAMILY	Notodontidae
DISTRIBUTION	Europe and across Asia to China, Korea, and Japan
HABITAT	Wet woodlands; only shingle banks in United Kingdom
HOST PLANTS	Poplar (*Populus* spp.) and willow (*Salix* spp.)
NOTE	Hairy caterpillar that has rows of black and orange spots
CONSERVATION STATUS	Not evaluated, but scarce or rare in parts of its range

ADULT WINGSPAN
1⁷⁄₁₆ in (37 mm)

CATERPILLAR LENGTH
1³⁄₁₆ in (30 mm)

CLOSTERA ANACHORETA
SCARCE CHOCOLATE-TIP
(DENIS & SCHIFFERMÜLLER, 1775)

510

Female Scarce Chocolate-tip moths lay their slightly flattened eggs on the underside of leaves of the host plant. The young caterpillars emerge within a few days and feed on the host plant leaves. When fully grown, they pupate in a loose silken cocoon within a rolled-up leaf. The second generation of adult moths emerges three to four weeks later. The caterpillars of the second generation are active until September and then pupate. These pupae overwinter, and the adults emerge in spring. There are usually two generations a year, with adults flying in April to May and again in July and August.

This species is rare in the United Kingdom and at risk across some its range due to the loss of suitable habitats and because it has a narrow range of host plants. In the past, poplars were common in the landscape of Europe but are now planted less frequently. The moth is similar to the more widespread Chocolate-tip (*Clostera curtula*) but distinguished by a white line on its chocolate-colored wingtips.

Actual size

The Scarce Chocolate-tip caterpillar has a black head and hairy dark body with four, fine white lines running the length of the body. There is a row of lateral black spots with orange spots below. There are tufts of white and creamy-brown hairs and a raised red spot flanked by two white spots on the abdomen.

FAMILY	Notodontidae
DISTRIBUTION	North America, south from Quebec to Florida, west to Manitoba and Texas
HABITAT	Fields, woodlands, and roadside edges
HOST PLANTS	Legumes (Fabaceae), including bush clover (*Lespedeza* spp.) and locust tree (*Gleditsia* spp.)
NOTE	Brightly colored caterpillar that has a distinctive black spot
CONSERVATION STATUS	Not evaluated, but rarely found in northern parts of its range

DASYLOPHIA ANGUINA
BLACK-SPOTTED PROMINENT
(J. E. SMITH, 1797)

ADULT WINGSPAN
1³⁄₁₆–1⅝ in (30–41mm)

CATERPILLAR LENGTH
1⁹⁄₁₆ in (40 mm)

511

The Black-spotted Prominent caterpillar is widely sought but seldom seen. In the eastern United States, it is most commonly encountered at high elevations by searching roadside locust trees. Inactive larvae are often found resting on the stems of young trees. When threatened, the caterpillar assumes the classic "prominent pose"—arching its back and posterior in an effort to appear more intimidating to would-be predators. A distinguishing black spot on the dorsum is its most defining characteristic. Its bright coloration suggests possible toxicity to predators. When fully developed, the larvae pupate in soil or leaf litter, and the pupae overwinter.

The brownish adult moths, which are somewhat mundane in comparison to their colorful larvae, fly from April to September. A single brood of caterpillars typically peaks in late summer, but they can be found from May to November. The *Dasylophia anguina* caterpillar is similar in size and design to the more common Red-humped Oakworm (*Symmerista canicosta*), a ubiquitous oak feeder that shares its range.

The Black-spotted Prominent caterpillar has a distinctive red head and a large, black mark on the dorsum. Yellow, lavender, and orange stripes traverse the length of its body. A series of thin, black stripes bisect the lavender portions. A pair of fake eyes and short "antennae" are evident on the posterior.

Actual size

FAMILY	Notodontidae
DISTRIBUTION	North America, from Quebec to Florida, west to Arkansas
HABITAT	Woodlands, barrens, and field edges
HOST PLANTS	Oak (*Quercus* spp.), chestnut (*Castanea* spp.), witch hazel (*Hamamelis* spp.), and blueberry (*Vaccinium* spp.)
NOTE	Communal caterpillar that lives in dry woodlands
CONSERVATION STATUS	Not evaluated, but not considered threatened

ADULT WINGSPAN
1½–2⅛ in (38–55 mm)

CATERPILLAR LENGTH
1¾ in (45 mm)

DATANA CONTRACTA
CONTRACTED DATANA
WALKER, 1855

512

During its initial larval stage, the Contracted Datana caterpillar is yellow in color. Throughout a period of rapid growth, the overall appearance changes at successive instars, eventually culminating in its most recognizable, striped mature form. Like the larvae of all *Datana* species, this caterpillar is gregarious, typically observed feeding and resting in large groups. Young caterpillars skeletonize leaves, while mature caterpillars consume entire leaves, leaving the largest veins intact. Defoliation typically occurs one branch at a time.

Nearing transformation, single caterpillars migrate away from the colony, eventually dropping from the tree to pupate beneath the soil. Adult moths usually emerge in July to mate and deposit single-layered clusters of eggs on the underside of leaves. *Datana contracta* caterpillars share many characteristics with *D. ministra* and other *Datana* species, both in appearance and behavior. Radcliffe's Dagger Moth (*Acronicta radcliffei*) is often confused with *D. contracta*. It looks strikingly similar but lacks the characteristic, arched "*Datana* pose" when alarmed.

The Contracted Datana caterpillar is mostly black with creamy-yellow stripes. The head is black, with a red-orange prothoracic shield clearly visible on later instars. (Young larvae have black shields.) The entire body is clothed in long, dense, wispy setae, which are whitish in color.

Actual size

FAMILY	Notodontidae
DISTRIBUTION	North America, from Nova Scotia to north Florida, west to Kentucky
HABITAT	Deciduous woodlands
HOST PLANTS	Blueberry (*Vaccinium* spp.), linden (*Tilia* spp.), and witch hazel (*Hamamelis* spp.)
NOTE	Gregariously feeding, aposematic caterpillar, that is striped black and yellow
CONSERVATION STATUS	Not evaluated, but common

ADULT WINGSPAN
1%₁₆–2³⁄₁₆ in (40–56 mm)

CATERPILLAR LENGTH
1¼–2 in (45–50 mm)

DATANA DREXELII
DREXEL'S DATANA
HY. EDWARDS, 1884

513

Drexel's Datana caterpillars hatch from eggs laid in clusters on the leaves of their host plant. The young larvae are gregarious, feeding together and skeletonizing leaves when newly hatched. In later instars, the leaf is consumed in its entirety except for the midrib. When one branch is stripped, the caterpillars move en masse to another. In the final instar, the larvae descend to the ground to pupate, and the pupa overwinters. There is usually one generation a year in the north and two in the south of the species' range.

When disturbed, the larvae assume a characteristic posture with their rear and front raised. Later instars are colored aposematically, a clear signal to predators of unpalatability, amplified by their group feeding. In the final instar, *Datana* caterpillars also have a ventral gland that secretes various chemicals, which, unlike those in some other notodontid species, are not sprayed as a defense but may play some role in communication. While *Datana* larvae can be distinctive, the adult moths of this genus are notoriously difficult to tell apart.

The Drexel's Datana caterpillar is shiny black, with a head as tall as it is wide, almost square in shape though tapering slightly toward the top. The body is black with eight yellow stripes and with the dorsal anterior half of the prothoracic segment yellow orange. The true legs and prolegs are citron yellow or orange tipped with black. The caterpillar is covered with long, thin, white setae.

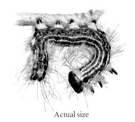

Actual size

FAMILY	Notodontidae
DISTRIBUTION	Southeast United States
HABITAT	Bogs and swamps
HOST PLANTS	Azalea (*Rhododendron* spp.) and Bog Rosemary (*Andromeda polifolia*)
NOTE	Colorful caterpillar that has lateral yellow stripes
CONSERVATION STATUS	Not evaluated

ADULT WINGSPAN
1¹⁵⁄₁₆–2 in (40–50 mm)

CATERPILLAR LENGTH
2 in (50 mm)

514

DATANA MAJOR
AZALEA CATERPILLAR
GROTE & ROBINSON, 1866

The Azalea Caterpillar hatches from a cluster of up to 100 small, white eggs laid by the female moth on the underside of leaves. The first instars are gregarious and feed together. They have huge appetites and quickly skeletonize leaves. The older caterpillars eat whole leaves and can defoliate an entire plant. When disturbed, the larvae raise their front and rear ends and sometimes drop below the leaf, hanging on a silken thread. Initially, the caterpillars are yellow with seven, red, longitudinal stripes but gain color with each molt.

After its final molt, the caterpillar crawls to the ground, where it pupates in the soil and overwinters. The moth adults are on the wing from June to August, with the caterpillars active from July to October. There is usually one generation a year, but there may be a partial second generation in the southern part of the range.

The Azalea Caterpillar is black with eight broken yellow stripes that run along the length of its body and a red posterior segment. The head, legs, and prolegs are glossy red. There are tufts of fine white hairs over the thorax and abdomen.

Actual size

FAMILY	Notodontidae
DISTRIBUTION	United States and southern Canada, east of the Rocky Mountains, south to Florida; also California
HABITAT	Woodlands, parks, barrens, and orchards
HOST PLANTS	Oak (*Quercus* spp.), apple (*Malus* spp.), birch (*Betula* spp.), basswood (*Tilia* spp.), willow (*Salix* spp.), and other woody trees
NOTE	Commonly encountered, showy caterpillar
CONSERVATION STATUS	Not evaluated, but rated "globally secure" by the National Center for Genome Resources, although declining in northeast United States

ADULT WINGSPAN
1⁹⁄₁₆–2¹⁄₁₆ mm (40–53 mm)

CATERPILLAR LENGTH
1¼–2 in (45–50 mm)

DATANA MINISTRA
YELLOWNECKED CATERPILLAR
(DRURY, 1773)

515

If encountered, the Yellownecked Caterpillar can exhibit some impressive showmanship. When threatened, it raises its head and tail portions, curling them backward dramatically over the dorsum. This static posture of defense, maintained long after the threat has ended, is a clearly recognizable characteristic of the *Datana* genus. The caterpillars are gregarious throughout the larval cycle, and are usually seen feeding and resting in tight groups. Young larvae skeletonize leaves. Mature caterpillars consume entire leaves, leaving only the thickest veins intact.

Adult moths fly in June and July, the female laying masses of up to 100 white eggs, deposited on the underside of leaves. A single generation of caterpillars is typical, with a second brood more likely in the south of the range. Mature caterpillars burrow into the soil to overwinter as pupae. In the northeast of the species' range, a conspicuous decline in population has been attributed to the influx of European parasitoids, which were imported to quell Gypsy Moth (*Lymantria dispar*) infestation.

The Yellownecked Caterpillar is predominantly black in color, with four lateral, yellow lines traversing both sides. Numerous fine, white setae encompass the entire body. The head is black. A bright yellow to orange plate behind the head, sometimes obscured by setae, differentiates the Yellownecked Caterpillar from similarly patterned *Datana* species.

Actual size

FAMILY	Notodontidae
DISTRIBUTION	Spain and southern France into central Europe, southern Russia, and Turkey
HABITAT	Dry woodlands and scrub, riparian forests, dry grasslands, and rocky slopes near wooded areas
HOST PLANTS	Elm (*Ulmus* spp.)
NOTE	Well-camouflaged caterpillar that often frequents hot, dry places
CONSERVATION STATUS	Not evaluated, but endangered in parts of its range

ADULT WINGSPAN
1⅜–1⁹⁄₁₆ in (35–40 mm)

CATERPILLAR LENGTH
1⁹⁄₁₆ in (40 mm)

DICRANURA ULMI
ELM MOTH
([DENNIS & SCHIFFERMÜLLER], 1775)

516

Elm Moth caterpillars emerge from white eggs that are laid singly on the upper surface of elm leaves. The larvae are seen on the leaves from May through to July, the young ones resting along the midrib on the underside of leaves, while the older ones rest along twigs, their cryptic coloration providing excellent camouflage. The mature caterpillars move to the ground, where they pupate in a cocoon just under the surface of the soil. The species overwinters as a pupa and ecloses in spring.

The night-flying moths are on the wing from March to May, and there is a single generation a year. *Dicranura ulmi* has disappeared from much of its former range and is currently at risk through the loss of its habitat, particularly hot, dry grassland, as a result of changes in agricultural management and industrial and tourist developments. Its numbers in southern Europe are more robust.

The Elm Moth caterpillar varies in color from shades of green to brown. The body is covered in small, yellow spots, with a dark dorsal line and yellow lateral lines. Two red-brown tubercles lie behind the brown head, with two dorsal brown tubercles on the abdomen. Two antenna-like filaments extend from the abdomen to give the appearance of a false head.

Actual size

FAMILY	Notodontidae
DISTRIBUTION	From western Europe and North Africa to the Urals, Asia Minor, and the Caucasus
HABITAT	Forests, wooded hedgerows, parklands, and gardens
HOST PLANTS	Oak (*Quercus* spp.)
NOTE	One of the less conspicuous caterpillars of its family
CONSERVATION STATUS	Not evaluated

ADULT WINGSPAN
1⁷⁄₁₆–1¹³⁄₁₆ in (36–46 mm)

CATERPILLAR LENGTH
1⅜–1⁹⁄₁₆ (35–40 mm)

DRYMONIA RUFICORNIS

LUNAR MARBLED BROWN

(HUFNAGEL, 1766)

517

Lunar Marbled Brown caterpillars hatch from rounded, light blue eggs laid in small groups on twigs or leaves from late March to early June. The larva is solitary and feeds until July, living openly without making a shelter, resting on the underside of a leaf, often high in the tree. When fully fed, it descends the tree and builds a cocoon on the ground close by, in which the stout, smooth, and shiny black pupa is formed and then overwinters.

The *Drymonia ruficornis* caterpillar is well camouflaged on the underside of a leaf. Several related species, also on oak, have quite similar caterpillars but with a slightly different combination of features, usually with red in the stripe along the sides and more broken lines. The Marbled Brown (*D. dodonaea*), for instance, has a pair of broken lines close together along the middle of the back.

Actual size

The Lunar Marbled Brown caterpillar is green when small and has raised black spots with short, black hairs. When larger, it develops two well-separated, bright yellow stripes on the back and sides. In the final instar, the stripes become finer with a white component and the body is blue green with a smooth, white bloom, especially on the back.

FAMILY	Notodontidae
DISTRIBUTION	Northeast India, southern China, Chinese Taipei, and mainland Southeast Asia
HABITAT	Lower-altitude montane forests
HOST PLANTS	Fagaceae, including chinkapin (*Castanopsis* spp.) and oak (*Quercus* spp.)
NOTE	Caterpillar that has elaborate reticulate markings and disruptive patterns
CONSERVATION STATUS	Not evaluated, but not uncommon

ADULT WINGSPAN
1¾–2⅛ in (45–55 mm)

CATERPILLAR LENGTH
1¾ in (45 mm)

518

FENTONIA BAIBARANA
FENTONIA BAIBARANA
MATSUMURA, 1929

The *Fentonia baibarana* caterpillar is unornamented and smooth with well-developed anal claspers, in contrast to other notodontid prominent moth larvae, which can be grotesquely shaped, with an absurd posture, and may lack the claspers. The *F. baibarana* larvae pupate in the soil in a loose, silken cocoon. The species has two or three generations during the spring and summer months, from April to the end of October, with the final generation of the year overwintering as pupae to re-emerge around late March the following year.

Although the caterpillars are intricately patterned with bold geometric markings and bright colors, they remain remarkably difficult to see among foliage. They represent a classic example of disruptive patterning and coloration, whereby their body markings create false edges and boundaries, thus disguising their outline and shape. In contrast, the adult moth is drab and cryptic, although well camouflaged against tree bark or leaf litter.

Actual size

The *Fentonia baibarana* caterpillar has reddish-brown, abdominal, reticulate, striate markings, like a network of veins, originating from the dorsal midline on a paler fawn base. The head is disproportionately large with similar striate markings. From the mid-abdominal segments caudally, there are a series of bright yellow spots on a dark brown dorsal stripe. The thoracic segments are green on their lateral aspect.

FAMILY	Notodontidae
DISTRIBUTION	Western Europe and North Africa east to the Urals and Asia Minor, extending into southwest Russia and Kazakstan; also Mongolia and Xinjiang province in northern China
HABITAT	Forests, hedgerows, parklands, and gardens
HOST PLANTS	Poplar (*Populus* spp.) and willow (*Salix* spp.)
NOTE	Caterpillar that has unusual hind appendages
CONSERVATION STATUS	Not evaluated

ADULT WINGSPAN
1¾–1⅞ in (44–48 mm)

CATERPILLAR LENGTH
1⁵⁄₁₆–1½ in (34–38 mm)

FURCULA BIFIDA

POPLAR KITTEN
(BRAHM, 1787)

519

The Poplar Kitten caterpillar hatches from a black, hemispherical egg laid on the upper side of a leaf during the summer and feeds and rests among the leaves of its host plant. When fully developed, it pupates within a well-camouflaged, hard, strong cocoon incorporating bark and wood and chewed slightly into a branch, trunk, or fence post. The pupa overwinters.

Adults and caterpillars of the genus *Furcula* (kitten moths) resemble those of the closely related but much larger moths of the genus *Cerura* (puss moths) and share very similar habits, forms, and life histories. The feline derivation arises in two ways. The pair of small protuberances behind the head of the younger caterpillar give it a catlike appearance from behind, and the adults are very furry and rest with the particularly hirsute front legs stretched out, in a feline manner. Differences between the numerous kitten moths in all their life stages are in many cases only slight.

Actual size

The Poplar Kitten caterpillar is green and dark brown, with two small protuberances behind the head. When mature, it has an irregular brown, saddlelike mark along the back. The hind prolegs are modified into a long, thin, two-pronged structure. To ward off predators, both ends are raised, and each produces a reddish, tentacle-like flagellum.

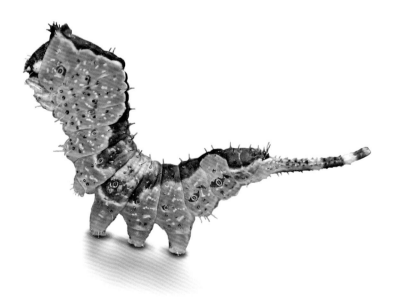

FAMILY	Notodontidae
DISTRIBUTION	Europe and into Asia
HABITAT	Wet woodlands, scrub, and heathlands
HOST PLANTS	Deciduous trees, including birch (*Betula* spp.), poplar (*Populus* spp.), sallow (*Salix caprea*, *Salix cinerea*), and other willows (*Salix* spp.)
NOTE	Bright green caterpillar that has a long tail spur
CONSERVATION STATUS	Not evaluated

ADULT WINGSPAN
1¹⁄₁₆–1⅜ in (27–35 mm)

CATERPILLAR LENGTH
1⅜ in (35 mm)

520

FURCULA FURCULA
SALLOW KITTEN
(CLERK, 1759)

Sallow Kitten caterpillars hatch from eggs laid in small groups on the surface of leaves, then feed on a range of trees and shrubs. When fully developed, the caterpillars make a tough cocoon by gnawing off bits of bark, which they intertwine with silk. The cocoon is securely fixed to a tree trunk or branch. The pupa overwinters, and the moth adults eclose the following spring. There is only one generation a year in the northern parts of the species' range but usually a second generation in the south.

Furcula furcula moths are on the wing from April until late August. The common name is derived from the appearance of the adult, which has a furry head and front legs, like a kitten. Though they tend to be smaller, Sallow Kitten larvae can be confused with those of the Puss Moth (*Cerura vinula*). Both the adult and caterpillar are also very similar in appearance to the closely related Poplar Kitten (*Furcula bifida*).

The Sallow Kitten caterpillar is bright green with a prominent, forked tail spur. There is a brown stripe running from the head along the top of the body, which extends laterally to create a V shape when viewed from the side, as it does in the similar Puss Moth caterpillar. There are numerous small lateral spots ringed in white.

Actual size

FAMILY	Notodontidae
DISTRIBUTION	From Afghanistan southeast through northern India and Nepal to southern China and Chinese Taipei, and Southeast Asia to Borneo
HABITAT	Low- and medium-altitude montane forests
HOST PLANTS	Various, including members of Fagaceae
NOTE	Caterpillar that has a cryptic body posture and multiple horns
CONSERVATION STATUS	Not evaluated, but its life history is largely unknown

ADULT WINGSPAN
2–2⁷⁄₁₆ in (50–62 mm)

CATERPILLAR LENGTH
1⁹⁄₁₆–1¾ in (40–45 mm)

HARPYIA MICROSTICTA

HARPYIA MICROSTICTA
(SWINHOE, 1892)

521

Harpyia microsticta caterpillars are gargoyle-like in appearance with a horned head capsule, thoracic and abdominal horns and spines, and an expanded, angular tail segment. The anal claspers are absent, in contrast to the larvae of other genera of prominent moths, where the claspers can be present or modified into taillike structures. Later instar caterpillars feed on one half of the leaf only, progressively advancing along the stripped midrib to the tip. When alarmed, they raise their tail end and flatten their head against the substrate, making them look like a ragged leaf edge with exposed brown vein ends.

The adult moths are easily recognizable by their salt-and-pepper speckled markings and distinct black triangle on the forewing leading edge. Three recognized subspecies (formerly separate species) occur across the expansive range—*Harpyia microsticta microsticta*, *H. microsticta baibarana*, and *H. microsticta dicyma*. The genus *Harpyia*, including members with similar adult and larval morphology, is represented in Europe, northern Africa, and in Asia as far north as southeast Russia.

The *Harpyia microsticta* caterpillar is green with a saddle mid-body in shades of brown. There are brown, twin-pronged horns along the dorsal midline, the largest being closest to the head and curving toward the rear. The claspers are absent, leaving four pairs of prolegs, and the rear segment is a swollen wedge shape topped with a forward-pointing bifid horn. The large, brown, rectangular head capsule has squat, blunt horns.

Actual size

FAMILY	Notodontidae
DISTRIBUTION	Across Europe, east to the Urals and Turkey
HABITAT	Forests, woodlands, and parks
HOST PLANTS	Beech (*Fagus* spp.) and oak (*Quercus* spp.)
NOTE	Strangely shaped caterpillar camouflaged as part of a leaf
CONSERVATION STATUS	Not evaluated, but is regionally endangered

ADULT WINGSPAN
1⁹⁄₁₆–2¹⁄₁₆ in (40–52 mm)

CATERPILLAR LENGTH
1⅜–1⁹⁄₁₆ in (35–40 mm)

HARPYIA MILHAUSERI
TAWNY PROMINENT
(FABRICIUS, 1775)

522

The caterpillars of the Tawny Prominent moth hatch from distinctive eggs laid on the underside of leaves in small clusters. Each egg, with its rings of brown, looks a little like an eyeball. The larvae feed on leaves, nibbling along the margins, and the mature caterpillar spins a cocoon of silk mixed with pieces of chewed wood to create a strong structure, which is fixed into a crack in the bark or occasionally on the ground. The species overwinters as a pupa and ecloses in spring.

The moths are nocturnal and on the wing from May to June, and there is a single generation, with an occasional partial second generation in late summer. The caterpillar has a strikingly disruptive shape and coloration, giving it the appearance of a damaged leaf, which provides excellent camouflage when at rest on a twig. The species is in decline in Europe due to the loss of oak woodland.

Actual size

The Tawny Prominent caterpillar has a shiny, green body with a yellow dorsal line and many cream-yellow dots. There are raised, thornlike protuberances on the first to fifth abdominal segments and at the rear end. A brown head and brown legs and prolegs, plus irregular patches of brown on the body, complete the illusion of a damaged leaf.

FAMILY	Notodontidae
DISTRIBUTION	North America, from southeastern Canada to Florida, west to Texas
HABITAT	Forests, woodlands, and roadside edges
HOST PLANTS	Woody trees, including basswood (*Tilia* spp.), oak (*Quercus* spp.), beech (*Fagus* spp.), cherry (*Prunus* spp.), and witch hazel (*Hamamelis* spp.)
NOTE	Notodontid prominent caterpillar commonly encountered in eastern United States forests
CONSERVATION STATUS	Not evaluated, but common in its range

ADULT WINGSPAN
1½–2³⁄₁₆ in (38–56 mm)

CATERPILLAR LENGTH
1¾ in (45 mm)

HETEROCAMPA BIUNDATA
WAVY-LINED HETEROCAMPA
WALKER, 1855

523

Like many "prominent" caterpillars, which are named for the protruding tuft of hair on the forewing of adults, the Wavy-lined Heterocampa prefers to feed and rest on leaf edges, where an effective camouflage helps it to hide in plain sight. Early instar larvae sport a pair of fleshy "antlers" behind the head. These defining appendages become fully manifested during the third instar. During the fourth and fifth (final) instars, the prothoracic growths are minimalized or nonexistent, as the caterpillar's appearance becomes more bulbous and reddish prior to pupation. The prepupal caterpillar overwinters in soil or leaf litter, and the adult moth flies from April to August.

Heterocampa biundata shares its range with two similar species, *H. obliqua*, a specific oak feeder, and *H. guttivitta*, which is usually less vibrantly patterned. Two generations of mature larvae occur from May to November. Wavy-lined and other *Heterocampa* caterpillars are often found infested with the larvae of parasitic chalcid wasps, which deposit their eggs on the caterpillar's skin, leaving the offspring to consume the caterpillar from within.

Actual size

The Wavy-lined Heterocampa caterpillar is variable in pattern, predominantly light green with a recognizable X-shaped saddle marking in the center of the dorsum. Brown or white splotches are often present along the sides of the body. The head color ranges from light brown to dull reddish pink purple. Young individuals are more vibrantly antlered and typically darker than more mature larvae.

FAMILY	Notodontidae
DISTRIBUTION	North America, from southern and southeastern Canada to Florida, west to Texas
HABITAT	Fields, woodlands, and roadside edges
HOST PLANTS	Elm (*Ulmus* spp.)
NOTE	Caterpillar that mimics a leaf
CONSERVATION STATUS	Not evaluated, but common throughout its range

ADULT WINGSPAN
1³⁄₁₆–1⁹⁄₁₆ in (30–40 mm)

CATERPILLAR LENGTH
1⁹⁄₁₆ in (40 mm)

NERICE BIDENTATA
DOUBLE-TOOTHED PROMINENT
WALKER, 1855

524

Double-toothed Prominent caterpillars hatch from eggs laid on their host plant from June onward, and are present into early November. The larvae are communal, so when one is found, a search of the surrounding elm foliage will likely yield many others. They are more often discovered on young sapling elms than on larger, more mature trees. Nature has provided the caterpillar with an effective deterrent against predation. Like many larvae that use disguise as defense, it is a master of mimicry, carving out a section of leaf, then positioning its resting body within the cavity to create a convincing new leaf edge.

The pupa overwinters underground, and adult moths fly from April to September. Two generations are typical throughout the range. Resembling a miniature stegosaurus dinosaur, the Double-toothed Prominent caterpillar is easily recognized by its jagged, double-toothed dorsum. This attribute clearly defines the species and differentiates it from similarly colored notodontid prominent larvae located within its range.

The Double-toothed Prominent caterpillar is bright to olive green with lighter green to pale white on the upper abdomen of mature specimens. A reddish stripe, often bordered with cream or yellow, appears laterally along each side of the thorax. A coarse keel of jagged "teeth" spans the dorsum. The claspers of this species are disproportionally small for a caterpillar of its size.

Actual size

FAMILY	Notodontidae
DISTRIBUTION	Europe, western Asia
HABITAT	Wet woodlands and coppiced woodlands
HOST PLANTS	Various trees and bushes but mostly poplar (*Populus* spp.) and willow (*Salix* spp.)
NOTE	Caterpillar whose distinctive appearance gives the species its name
CONSERVATION STATUS	Not evaluated, but regionally endangered

ADULT WINGSPAN
1¾–2⅛ in (45–55 mm)

CATERPILLAR LENGTH
1³⁄₁₆ in (30 mm)

NOTODONTA TRITOPHUS

THREE-HUMPED PROMINENT

(DENNIS & SCHIFFERMÜLLER, 1775)

525

The Three-humped Prominent caterpillar has, as its name indicates, three rear-pointing humps on its back. Its strange appearance is further exaggerated when it raises its posterior end while feeding. After hatching from eggs laid on their host plant, the larvae are active from June to September, when they can be found feeding in the canopy of trees such as poplar and willow. The species overwinters as a pupa, and the adult moths are on the wing from April to August. There is usually a single generation, and in some places there may be two generations.

This species, clearly named for its caterpillar, thrives in light woodland, where sunlight can reach the woodland floor. However, this type of habitat is in decline and under threat in some regions as a result of clearance and through neglect, as woodlands become less managed. *Notodonta tritophus* is sometimes classified as *N. phoebe*.

Actual size

The Three-humped Prominent caterpillar is brown or olive green, with three backward-facing humps on the abdomen and a further forward-pointing hump at the end of the abdomen. The spiracles are ringed in white, and there are many tiny, white spots scattered over the body. The head is brown with many small, black spots.

FAMILY	Notodontidae
DISTRIBUTION	Europe and North Africa to central Asia, east to China
HABITAT	Wet woodlands, coppiced woodlands, river valleys, and gardens
HOST PLANTS	Poplar (*Populus* spp.) and willow (*Salix* spp.)
NOTE	Camouflaged caterpillar that raises its tail segment when feeding
CONSERVATION STATUS	Not evaluated

ADULT WINGSPAN
1⁹⁄₁₆–1¾ in (40–45 mm)

CATERPILLAR LENGTH
1³⁄₁₆ in (30 mm)

NOTODONTA ZICZAC
PEBBLE PROMINENT
(LINNAEUS, 1758)

526

Pebble Prominent caterpillars hatch from round, white eggs laid singly on the leaves of poplar and willow in late spring. The caterpillars are active from June to early October. The young larvae feed on one side of the leaves, creating fenestrations, but the older caterpillars feed along the leaf margins. The caterpillars crawl to the ground, where they pupate in a loose cocoon underground. The pupa is red brown in color. The adult moths are on the wing from April to September, and there are often two generations.

The caterpillar relies on camouflage to avoid predation. The first instar is green, but on molting it becomes brown and more twiglike. The characteristic humps that give rise to the species name appear in the later stages. At rest, the caterpillar hangs from the underside of a stem, raising the end of the abdomen and the head. Its genus name, *Notodonta*, means "back-tooth," referring to the toothlike margin of the forewings.

Actual size

The Pebble Prominent caterpillar is light orange to grayish brown in color. There are two backward-facing humps and a hump at the end of the abdomen, where there is a ringed "eyespot" pattern. A white lateral stripe runs the length of the body at the level of the spiracles. The spiracles are ringed in white. The head is mottled with white markings and spots.

FAMILY	Notodontidae
DISTRIBUTION	Eastern and western United States, southern Canada
HABITAT	Woodlands, forests, and roadside edges
HOST PLANTS	Oak (*Quercus* spp.), poplar (*Populus* spp.), willow (*Salix* spp.), and other deciduous trees
NOTE	Foliage-mimicking caterpillar that feeds on leaf edges
CONSERVATION STATUS	Not evaluated, but widespread and not threatened

ADULT WINGSPAN
1³⁄₁₆–1¼ in (30–45 mm)

CATERPILLAR LENGTH
1⁹⁄₁₆ in (40 mm)

OLIGOCENTRIA SEMIRUFESCENS
RED-WASHED PROMINENT
(WALKER, 1865)

527

The Red-washed Prominent is often referred to as the "rhinoceros" caterpillar for its pronounced "horn," which projects forward, sometimes extending beyond the head when at rest. Early instars feed by skeletonizing leaves from the top or bottom. The mature caterpillar, like similar prominent species, feeds along the leaf edges. It carves out a section of leaf material, then positions its body into the decimated cavity to create an illusion of dying leaf material. The coloration of larvae is variable to aid their ability to effectively conceal themselves by blending into adjacent foliage.

Two generations of mature caterpillars occur from June throughout the summer and early fall. Prepupal larvae overwinter in soil or leaf litter, and adult moths fly from May to September. The larva's striking design is similar to two *Schizura* caterpillar species (*S. unicornis* and *S. leptinoides*), but it lacks defining *Schizura* characteristics, such as dark patches behind the head and green thoracic saddles.

The Red-washed Prominent caterpillar ranges in coloration from brown or yellow to pink, with intermittent dark spots extending from the subdorsal to subventral regions. Intricate, wormlike patterns are often present in mature specimens. The head color typically matches the body color, with vertical brown lines extending from the vertex to the antennae.

Actual size

FAMILY	Notodontidae
DISTRIBUTION	Western Europe to the Urals, Turkey, and the Caucasus; also western and southern Siberia (east to Lake Baikal), northwestern Mongolia, and northwest China
HABITAT	Many habitats, including forests, wooded or scrubby wetland margins, heathland, hedgerows, parks, and gardens
HOST PLANTS	Willow (*Salix* spp.) and poplar (*Populus* spp.)
NOTE	Common and rather distinctive green, sleek-looking caterpillar
CONSERVATION STATUS	Not evaluated, but very common

ADULT WINGSPAN
1¹¹⁄₁₆–2⅛ in (43–55 mm)

CATERPILLAR LENGTH
1½–1⁹⁄₁₆ in (38–40 mm)

PTEROSTOMA PALPINA
PALE PROMINENT
(CLERCK, 1759)

The Pale Prominent caterpillar hatches from a hemispherical, light green egg; the female moth lays her eggs in small groups on the underside of a leaf. The larvae can be found from June to September in two broods. They live and feed openly among the foliage of the tree or bush, where they are well camouflaged. When fully developed, the caterpillars descend to the ground, where the dark brown, shiny pupa is formed in a cocoon on or slightly under the surface, near the base of the tree, covered with soil and debris.

This is a very common species, with distinctive markings. Although green with white stripes is a popular livery among moth caterpillars, few if any species living on willow and poplar (at least in the Western Palearctic) possess the exact same combination of features as the Pale Prominent. The brownish adults fly in two broods, from April to September.

The Pale Prominent caterpillar is green when young with tiny, black spots; in later stages it is blue green with a white bloom on the back or bright green. There are four white lines along the back, a yellow or yellowish-white line along the sides, and fine, dark edging above. The body tapers at both ends, and the head is set at a low angle.

Actual size

FAMILY	Notodontidae
DISTRIBUTION	Europe and across Asia to Russian Far East, China, and Japan
HABITAT	Forests, hedgerows, scrublands, parks, and gardens
HOST PLANTS	Many trees and shrubs, including oak (*Quercus* spp.), Hornbeam (*Carpinus betulus*), lime (*Tilia* spp.), hazel (*Corylus* spp.), hawthorn (*Crataegus* spp.), and rose (*Rosa* spp.)
NOTE	Caterpillar that throws its head backward when alarmed
CONSERVATION STATUS	Not evaluated

ADULT WINGSPAN
1⁷⁄₁₆–2 in (37–50 mm)

CATERPILLAR LENGTH
1⁵⁄₁₆–1⁷⁄₁₆ in (34–36 mm)

PTILODON CAPUCINA
COXCOMB PROMINENT
(LINNAEUS, 1758)

529

Coxcomb Prominent caterpillars hatch from whitish, hemispherical eggs laid in small groups on the leaves of their host plant. They can be found feeding, usually singly, in two generations from June to October, on tall trees as well as small bushes. When fully developed, the larvae descend to the ground to form the pupa on or under the surface in a cocoon that incorporates soil and debris. This stage overwinters.

Notodontid caterpillars are highly varied in shape. When threatened, some raise the front and hind ends as a warning, and some have elaborate appendages that they raise and wave around. In this situation, the Coxcomb Prominent throws its head right back toward the red warts on the raised hind end, with the true legs held together, making it look spiky and less like a caterpillar. Other caterpillars, such as the Sprawler (*Asteroscopus sphinx*), a noctuid, can similarly throw back the head but without raising the hind end.

Actual size

The Coxcomb Prominent caterpillar is green with raised black spots when young, its head black or green with a pair of black spots. Later it becomes bright green, blue green with a white bloom on the back, or pinkish with a green tinge. The mature caterpillar has red legs, a yellow, red-edged stripe along the sides, and two red, wartlike projections on the hind protuberance.

FAMILY	Notodontidae
DISTRIBUTION	From western Europe (the Pyrenees and southern England) through central Europe, north to the Baltic Sea, and east to the Black Sea
HABITAT	Forests, hedgerows, and scrublands, especially on calcareous soils
HOST PLANTS	Mainly Field Maple (*Acer campestre*) and probably other *Acer* spp.
NOTE	Caterpillar that is unique among European maple-feeding species
CONSERVATION STATUS	Not evaluated

ADULT WINGSPAN
1³⁄₁₆–1¾ in (30–45 mm)

CATERPILLAR LENGTH
1³⁄₁₆–1⁵⁄₁₆ in (30–34 mm)

530

PTILODON CUCULLINA
MAPLE PROMINENT
([DENIS & SCHIFFERMÜLLER], 1775])

Maple Prominent caterpillars hatch from domed, bluish-white eggs laid on leaves in small groups and in two broods, from May to August. The larvae feed and rest openly among the leaves, usually singly. The smooth, brown pupa is formed in a flimsy cocoon among leaf litter on the ground, and this is the overwintering stage.

When half-grown, the Maple Prominent caterpillar looks much like the Coxcomb Prominent (*Ptilodon capucina*), also found on *Acer* species. It differs, however, in that its head has two dark, roughly vertical stripes and less deeply divided warts on the hind protuberance. When mature, the caterpillar more closely resembles *Notodonta* species, especially the Pebble Prominent (*N. ziczac*), although the latter does not feed on *Acer*, lacks the Maple Prominent's double-tipped wart on its hind protuberance, and has larger dorsal humps. The very similar *Ptilodon saerdabensis*, found in Turkey, the Causasus, and Iran, was formerly considered to be a subspecies of *P. cucullina*.

Actual size

The Maple Prominent caterpillar is whitish green or whitish brown, with darker green or brown along the back, forming a broad band in the frontal third of the body, reduced otherwise to three thin lines or dashes. It has a series of dorsal humps, the largest on the fifth segment. The protuberance near the hind end has a pink, double-tipped, wartlike projection.

FAMILY	Notodontidae
DISTRIBUTION	From the Pyrenees across the Alps to the Balkans
HABITAT	Scree and sheltered mountain valleys
HOST PLANTS	Poplar (*Populus* spp.) and willow (*Salix* spp.)
NOTE	Brown, hairy caterpillar that is found in alpine areas
CONSERVATION STATUS	Not evaluated

ADULT WINGSPAN
$^{11}/_{16}$–¾ in (18–20 mm)

CATERPILLAR LENGTH
$^{9}/_{16}$–¾ in (15–20 mm)

RHEGMATOPHILA ALPINA
ALPINE PROMINENT
(BELLIER DE LA CHAVIGNERIE, 1881)

531

The caterpillars of the Alpine Prominent moth hatch from eggs laid on leaves of poplar and willow. The dark brown colors of the caterpillar provide good camouflage when it rests along twigs and stems. The brown pupa overwinters and ecloses the following spring.

The small, gray-brown moth is on the wing from May to September. There are usually two generations, although in the northerly parts of the range there may be just one. The moths are active at night, and during the day they rest on bark, where their gray-patterned wings blend perfectly with the background colors of the trees. The name Alpine Prominent comes from the projecting tuft of hair on the trailing edge of the forewing and the distribution of the species in alpine regions. The latter is also reflected in the scientific name, *alpina*. The genus *Rhegmatophila* is small with just three species.

The Alpine Prominent caterpillar has a brown body that tapers a little. There are bands of slightly different shades of brown, with whorls of tubercles that bear tufts of pale brown hairs. There are two black ventral spots on the thorax, and the head is brown with dark brown markings.

Actual size

FAMILY	Notodontidae
DISTRIBUTION	North America, from southeastern Canada to Florida, west to Texas
HABITAT	Fields, parks, and woodland forests
HOST PLANTS	Basswood (*Tilia* spp.), oak (*Quercus* spp.), beech (*Fagus* spp.), blackberry (*Rubus* spp.), and morning glory (*Ipomoeae* spp.)
NOTE	Caterpillar that imitates a leaf
CONSERVATION STATUS	Not evaluated, but considered secure within its range

ADULT WINGSPAN
1⁷⁄₁₆–1⅞ in (36–47 mm)

CATERPILLAR LENGTH
1⁹⁄₁₆ in (40 mm)

SCHIZURA IPOMOEAE
MORNING GLORY PROMINENT
DOUBLEDAY, 1841

532

The Morning Glory Prominent caterpillar is a master of disguise. Like other *Schizura* species, the larva carves out a section of leaf material, then positions its body into the decimated cavity. This behavior allows the caterpillar to hide in plain sight, appearing to hungry birds and other predators as curled leaf material. A single brood of larvae is common in the north. Second generations have been documented in the south, with mature caterpillars existing from June throughout the summer and fall months. *Schizura ipomoeae* caterpillars are often found parasitized by braconid wasp larvae.

This caterpillar's species name is misleading as the larvae are more often found feeding on trees and shrubs other than morning glory. The Morning Glory Prominent caterpillar can be easily mistaken for the similarly shaped Unicorn Caterpillar (*Schizura unicornis*). Close examination of distinguishing head and dorsal markings is the best way to differentiate between the two. Adult moths are nocturnal, and fly from April to September.

Actual size

The Morning Glory Prominent caterpillar is light brown with wormlike patterns along the abdomen and a contrasting green saddle on the thoracic region. The first and fifth abdominal segments are horned and distinctly humped. White chevron dorsal markings and a striped head are its most defining characteristics.

FAMILY	Notodontidae
DISTRIBUTION	North America
HABITAT	Deciduous woodlands and forests, parks, and gardens
HOST PLANTS	Various trees and shrubs, including Alder (*Alnus glutinosa*), birch (*Betula* spp.), hawthorn (*Crataegus* spp.), poplar (*Populus* spp.), and willow (*Salix* spp.)
NOTE	Cleverly camouflaged caterpillar that squirts acid if disturbed
CONSERVATION STATUS	Not evaluated

ADULT WINGSPAN
1⁵⁄₁₆–1⅜ in (24–35 mm)

CATERPILLAR LENGTH
1³⁄₁₆ in (30 mm)

SCHIZURA UNICORNIS
UNICORN CATERPILLAR
(J. E. SMITH, 1797)

533

Female Unicorn moths lay small, round eggs on the underside of leaves of the host plant. They hatch into odd-looking caterpillars that are active from May to October. The larvae then drop to the ground, where they spin a protective cocoon in the leaf litter. This species overwinters as a caterpillar and pupates within its cocoon in spring. The adult moths are on the wing from February to September in the more southerly zones of their range and May to August in the north. There is a single generation each year.

The Unicorn gets its name from the appearance of the caterpillar. It has a large horn on the first abdominal segment, and is also equipped with a cervical gland from which it squirts formic acid, if disturbed. The horn, together with the disruptive coloration created by the green band across the thorax, provide excellent camouflage. The adult moth also relies on camouflage, wrapping its wings around its body to form a tube that looks like a piece of broken twig.

The Unicorn Caterpillar has a mottled brown body with two bright-green thoracic segments and two faint, dark lateral lines. There is a conspicuous, hornlike dorsal hump on the first abdominal segment and a much smaller one on the eighth abdominal segment.

Actual size

FAMILY	Notodontidae
DISTRIBUTION	Across central and southern Europe, parts of northern Europe into southern Russia, Turkey, Iran, and Iraq
HABITAT	Oak woodlands, grasslands, and scrub areas near oak
HOST PLANTS	Oak (*Quercus* spp.)
NOTE	Twiglike caterpillar that feeds on oak leaves
CONSERVATION STATUS	Not evaluated, but endangered in parts of its range

ADULT WINGSPAN
1⅜–1⁹⁄₁₆ in (35–40 mm)

CATERPILLAR LENGTH
2 in (50 mm)

534

SPATALIA ARGENTINA
ARGENTINE MOTH
(DENIS & SCHIFFERMÜLLER, 1775)

Argentine Moth caterpillars hatch from eggs laid on the older leaves of oak trees. Typically, the larvae start feeding at the tip of the leaf blade and work down either side of the midrib. Their twiglike appearance provides excellent camouflage among the foliage. The caterpillars move to the ground to pupate, spinning a loose cocoon, and the dark brown pupa overwinters, with the adult moths eclosing in late spring.

The night-flying Argentine Moth, which, confusingly, is not found anywhere in South America, is also known as the Silver Stain-toothed Spinner. The moths are on the wing from April to August, with two generations, although there may only be one in the north of the range. *Spatalia argentina* is under threat from the loss of oak woodlands, especially those of an open structure with young growth. Also, the moth is killed when foresters spray trees to destroy defoliating moths such as the Gypsy Moth (*Lymantria dispar*), which also feeds on oak.

The Argentine Moth caterpillar has a mottled, pale brown body with an orange-brown head and legs. Faint white lines run the length of the body. Two dark brown knobs behind the thorax, a brown-edged transverse ridge on segment ten, and another knob on segment eleven create the appearance of a twig with buds.

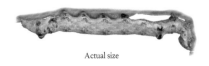

Actual size

FAMILY	Notodontidae
DISTRIBUTION	Europe, east across Asia (excluding the south) to Japan
HABITAT	Woodlands, especially beech (*Fagus* spp.), scrub, hedgerows, and orchards
HOST PLANTS	Various trees and shrubs, including apple (*Malus* spp.), beech (*Fagus* spp.), birch (*Betula* spp.), Hazel (*Corylus avellana*), lime (*Tilia* spp.), and willow (*Salix* spp.)
NOTE	Caterpillar whose bizarre outline is the perfect disguise
CONSERVATION STATUS	Not evaluated

ADULT WINGSPAN
2–2¾ in (50–70 mm)

CATERPILLAR LENGTH
2¾ in (70 mm)

STAUROPUS FAGI

LOBSTER MOTH

(LINNAEUS, 1758)

535

Lobster Moth caterpillars hatch from small, round eggs laid singly on leaves. The young larvae, which in their initial instar feed only on their eggshell, are mimics of ants or small spiders, and, if disturbed, they wiggle like an injured ant. They are also aggressive and will attack any approaching small insect. With each molt, the caterpillars take on a more extreme appearance, with long legs and an enlarged abdomen. The mature larvae move to the ground, where they pupate in a cocoon in the leaf litter and overwinter. The adult moths are on the wing in May and June.

The species gets its common name from the strange crustacean-like appearance of the red-brown caterpillar, providing a perfect disguise against its host plant. When disturbed, the caterpillar raises its head and "tail" over its body and extends its long legs forward. It can also squirt formic acid from glands below the thorax to deter predators, such as small birds or ichneumon wasps that may try to parasitize it.

Actual size

The Lobster Moth caterpillar is red brown in color, with a large head and extra-long, ant-like legs. It has a series of raised dorsal humps on segments four to seven and a much enlarged posterior segment, ending in thin claspers that extend out behind the abdomen.

FAMILY	Notodontidae
DISTRIBUTION	Europe into Asia
HABITAT	Woodlands, parks, and gardens
HOST PLANTS	Mostly oak (*Quercus* spp.); also others, including beech (*Fagus* spp.), birch (*Betula* spp.), Hazel (*Corylus avellana*), and sweet chestnut (*Castanea* spp.)
NOTE	Caterpillar that has thousands of detachable hairs with a hazardous irritant
CONSERVATION STATUS	Not evaluated

ADULT WINGSPAN
1–1⅜ in (25–35 mm)

CATERPILLAR LENGTH
1³⁄₁₆ in (30 mm)

THAUMETOPOEA PROCESSIONEA
OAK PROCESSIONARY
(LINNAEUS, 1758)

536

Female Oak Processionary moths lay their eggs in rows on twigs and branches. The eggs overwinter, and the caterpillars emerge in April. The larvae are gregarious and spin communal silk nests under branches and on trunks, leaving trails of white silk. They also defoliate much of their host tree, making it more vulnerable to disease. As they develop through each instar, the caterpillars molt and pupate within the nest, giving it a brown appearance. The adult moths are on the wing from May to September.

These caterpillars are named after their habit of walking in nose-to-tail processions along branches and on the ground. Their hairs contain thaumetopoein, an intensely irritating compound, which can cause rashes and allergic reactions in people who come into contact with them. Airborne hairs and even hairs on the ground retain their irritant effect. Where infestations are reported, public authorities take immediate action by spraying trees and, in rare cases, burning infested areas.

The Oak Processionary caterpillar is gray brown with a dark head, a single dark dorsal stripe, and a white lateral line. Clumps of long, white hairs arise from red tubercles. In addition, the body is covered in thousands of shorter hairs that contain an irritant.

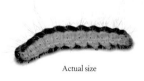

Actual size

FAMILY	Erebidae
DISTRIBUTION	The Andes of Colombia and Ecuador
HABITAT	Humid montane forests and forest borders
HOST PLANTS	*Erato polymnioides* and *Miconia* spp.
NOTE	Caterpillar with setae that break off and cause skin irritation
CONSERVATION STATUS	Not evaluated, but not considered threatened

ADULT WINGSPAN
2⁹/₁₆–3¹/₁₆ in (65–77 mm)

CATERPILLAR LENGTH
2⅜–2¾ in (60–70 mm)

AMASTUS AMBROSIA

AMASTUS AMBROSIA
(DRUCE, 1890)

537

The robust *Amastus ambrosia* caterpillar is found as a solitary individual on its host plant, and it is not known how many eggs the adult lays in a single batch. The caterpillar is covered in brown, hairlike setae and, prior to pupation, creates a cocoon of these hairs around its pupa, presumably providing good protection from predators. At least one bird, the Fawn-breasted Tanager (*Pipraeidea melanonota*), however, has been observed to consume the spiky larvae. The caterpillar's bright crimson spiracles on an otherwise dark-colored body might be a form of warning coloration, although they are mostly obscured from view by setae.

When disturbed, larger individuals tend to thrash about wildly, while smaller ones drop quickly from the host plant to the ground below. The caterpillar's setae can break off when handled and cause considerable skin irritation. Until recently, the subspecies pictured here (*Amastus ambrosia thermidora*, Hampson, 1920) was considered a full species.

The *Amastus ambrosia* caterpillar has a reddish-brown head, and its body's ground color is dull gray brown with mostly indistinct, cream-colored and black markings. It bears moderately dense, short, stiff setae that are reddish brown at their base, dark centrally on the shaft, and with red-brown tips. Its true legs are black, and its prolegs are similar in color to the body but with darker pads.

Actual size

FAMILY	Erebidae
DISTRIBUTION	Around the Mediterranean, east to central Asia and areas of northwest China
HABITAT	Dry scrub and grasslands
HOST PLANTS	Various members of Fabaceae, especially *Genista* spp. and *Spartium* spp.
NOTE	Caterpillar whose bright colors indicate it is distasteful
CONSERVATION STATUS	Not evaluated, but common

ADULT WINGSPAN
2¹⁵⁄₁₆–3¼ in (74–82 mm)

CATERPILLAR LENGTH
1⁹⁄₁₆ in (40 mm)

538

APOPESTES SPECTRUM
APOPESTES SPECTRUM
(ESPER, 1787)

Apopestes spectrum caterpillars hatch from eggs laid on various members of the Fabaceae plant family, in particular Spanish Broom (*Spartium junceum*). Their elaborate colors make the caterpillars conspicuous and are a clear warning to would-be predators to stay away. The larvae are active from late spring to midsummer, when they pupate. The moths eclose in July and then go into hibernation, overwintering and emerging the following spring as temperatures rise. There is one generation a year, with the moths on the wing in spring.

This species was previously classified as *Noctua spectrum*. It has been the subject of extensive studies in China, where increasing numbers of the larvae and adults are resulting in the damage of nationally important wall murals in caves in the northwest of the country. The clinging action of the prolegs as the caterpillars move across the surface of the murals is found to damage the pigments, as do the acidic droppings of the adult moths.

The *Apopestes spectrum* caterpillar is yellow, white, and black. The head is black and white, while the body is black with creamy-white dorsal lines and a broad lateral band of yellow with black spots. There are further broken lines in white and yellow, and spots edged in white.

Actual size

FAMILY	Erebidae
DISTRIBUTION	Southern Canada, United States (from New York west to Washington State, Oregon, and California); also Europe and across Asia, excluding the far north and south
HABITAT	Gardens, parks, river valleys, and meadows
HOST PLANTS	Wide variety of herbaceous plants, including nettle (*Urtica* spp.), Bracken (*Pteridium aquilinum*), dock (*Rumex* spp.), blackberry (*Rubus* spp.), and honeysuckle (*Lonicera* spp.)
NOTE	Spectacularly hirsute caterpillar popularly known as a "woolly bear"
CONSERVATION STATUS	Not evaluated, but declining in many countries

ADULT WINGSPAN
1¾–2⁹⁄₁₆ in (45–65 mm)

CATERPILLAR LENGTH
2⅛–2⅜ in (55–60 mm)

ARCTIA CAJA
GARDEN TIGER MOTH
(LINNAEUS, 1758)

539

Garden Tiger Moth caterpillars hatch from bluish-yellow eggs laid by the female moth in batches of up to 50 some 10 days earlier on the underside of host plant leaves during July and August. The caterpillars immediately disperse to live solitary lives. After a couple of instars, the partially grown larvae hibernate and then resume feeding and growth in spring. Development is usually completed by the end of May, when prepupal caterpillars are often seen wandering, on the look out for pupation sites. Pupation takes place in a sparsely spun cocoon made of silk and caterpillar hairs.

The adult moth emerges in July and is spectacularly colored in brown, white, red, and blue, its striking appearance advertising distastefulness to predators. Despite being covered in long hairs, the caterpillar is frequently parasitized by wasps and flies. The Garden Tiger Moth was formerly common in the United Kingdom but has steadily declined by an estimated 30 to 40 percent over the past 20 years.

The Garden Tiger Moth caterpillar is densely clothed in long, black hairs and some shorter, white hairs dorsally. It is black dorsally and brownish orange below, with four white, lateral spots on each segment. The head is black but usually hidden within the bordering hairs.

Actual size

FAMILY	Erebidae
DISTRIBUTION	Europe and across northern and central Asia to Japan
HABITAT	Deciduous and mixed forests
HOST PLANTS	Various trees, including beech (*Fagus* spp.) and birch (*Betula* spp.)
NOTE	Caterpillar that uses sudden movement and falling to avoid predation
CONSERVATION STATUS	Not evaluated

ADULT WINGSPAN
1⅜–1¼ in (35–45 mm)

CATERPILLAR LENGTH
2 in (50 mm)

540

ARCTORNIS L-NIGRUM
BLACK V
(MÜLLER, 1764)

Black V caterpillars hatch from eggs laid by the female moth in small groups on the surface of leaves of the host trees, and feed there. When not feeding, the larvae rest on the leaves and, if disturbed, they flick to one side and sometimes fall deliberately in an effort to avoid predation. The second or third instar caterpillars move to the ground, where they spend the winter in the leaf litter, sometimes crawling into rolled-up leaves.

The larvae emerge in spring and resume feeding to complete their growth. They then pupate in a cocoon, which they spin among leaves. There is usually one generation, but occasionally a partial second generation appears in late summer and early fall. The moths, which fly from May to July, have a prominent V-shaped mark on their forewings—hence their common name. The species is widely distributed across Asia, where there are several subspecies.

The Black V caterpillar has a black head and chestnut-brown body, covered with long hairs in brown, black, and creamy white. The hairs on the anterior and posterior segments are extra long. There are pale white dorsal stripes.

Actual size

FAMILY	Erebidae
DISTRIBUTION	Southern China, Chinese Taipei, Southeast Asia, the Philippines, New Guinea, and northeastern Australia
HABITAT	Lowland to montane forests and cultivated land
HOST PLANTS	Many, including Mango (*Mangifera indica*), fig (*Ficus* spp.), Oil Palm (*Elaeis guineensis*), Sal (*Shorea robusta*), *Eucalyptus* spp., and *Brassica* spp.
NOTE	Vibrantly colored, hairy caterpillar that has a startling defense strategy
CONSERVATION STATUS	Not evaluated, but very common

ADULT WINGSPAN
1⁹⁄₁₆–2⅜ in (40–60 mm)

CATERPILLAR LENGTH
2 in (50 mm)

CALLITEARA HORSFIELDII
HORSFIELD'S TUSSOCK MOTH
(SAUNDERS, 1851)

541

Despite their striking, yellow hirsuteness, *Calliteara horsfieldii* caterpillars can remain remarkably inconspicuous on the shaded underside of leaves. If disturbed, however, they arch their thorax, exposing a contrasting pitch-black, intersegmental membrane. This may be sufficiently startling to deter a predator or even give the impression of a vertebrate's opening eye. The species completes seven larval instars over a 50 to 60 day period and then pupates within inverted hollows created between silk-woven host plant leaves. A silken pad is first laid, and then a see-through but sturdy twin-layered cocoon incorporating the lengthy hairs of the caterpillar is constructed. The pupal period lasts 9 to 14 days.

The extreme polyphagy of this species of tussock moth larva, currently encompassing around 50 host plant genera across 30 families and including species of economic significance, reflects an impressive capacity to adapt to its environment and add to its geographic distribution. The adult male moths are dimorphic, as are the genders; the females are paler and larger. Typical of the non-feeding Lymantriine tribe of moths, the adults live for only four to eight days.

The Horsfield's Tussock Moth caterpillar is pearly white beneath its almost fluorescent yellow coat of lengthy secondary setae. There are four dense tussocks on the dorsum of the first four abdominal segments and a longer brush of hairs on the rear. Between the first and second abdominal segments (equating to the first two dorsal tussocks) is the usually concealed or partially concealed black, intersegmental membrane.

Actual size

FAMILY	Erebidae
DISTRIBUTION	Europe and across Asia to Japan
HABITAT	Deciduous woodlands, scrub, and parklands
HOST PLANTS	Various trees and shrubs, including birch (*Betula* spp.), hawthorn (*Crataegus* spp.), and willow (*Salix* spp.), as well as hop (*Humulus* spp.)
NOTE	Hairy, potentially defoliating caterpillar that was once dubbed "hop-dog"
CONSERVATION STATUS	Not evaluated

ADULT WINGSPAN
1¹⁵⁄₁₆–2³⁄₈ in (40–60 mm)

CATERPILLAR LENGTH
1¹⁵⁄₁₆–2 in (40–50 mm)

542

CALLITEARA PUDIBUNDA
PALE TUSSOCK
(LINNAEUS, 1758)

Pale Tussock caterpillars hatch from 300 to 400 eggs laid in batches on the underside of leaves. Initially, the larvae are gregarious but become solitary in the latter stages. The brightly colored hairs are a warning to deter potential predators, as they can cause irritation and detach easily, filling the mouth of an animal or the bill of a bird. The caterpillars crawl to the ground to pupate, spinning their cocoon among the leaf litter, and overwinter in the cocoon. The moths are on the wing from April to June with just one generation a year.

The Pale Tussock caterpillar is green yellow with a dorsal row of black marks. The head and body are covered in tufts of creamy-white hairs. There are four conspicuous dorsal tufts of yellow hairs and, at the end of the abdomen, extra-long tufts of red-brown hairs form a tail spur.

The hairy caterpillars were once a major pest of hops and were named "hop-dogs" by the hop workers. Nowadays, the fields of hops are sprayed, and the species is far less common. However, population explosions occur periodically in forests and woodlands, causing defoliation of trees. Generally the trees suffer no long-term damage as the defoliation occurs late in the growing season.

Actual size

FAMILY	Erebidae
DISTRIBUTION	North America, from southeastern Canada to northern Georgia, west to Texas
HABITAT	Mountainous woodlands and mesic forests
HOST PLANTS	Meadow-rue (*Thalictrum* spp.)
NOTE	Caterpillar that mimics sawfly larvae
CONSERVATION STATUS	Not evaluated

ADULT WINGSPAN
1⁹⁄₁₆–1⁹⁄₁₆ in (33–40 mm)

CATERPILLAR LENGTH
1⅜ in (35 mm)

CALYPTRA CANADENSIS
CANADIAN OWLET
(BETHUNE, 1865)

543

The Canadian Owlet caterpillar feeds specifically on *Thalictrum* plants. The larvae are most commonly spotted in damp woodlands at higher elevations, in terrain where their host plant grows. A single generation is typical within its normal range, with two generations possible farther west. In the eastern United States, mature caterpillars can be found from early July throughout the summer. The pupae overwinter. The adult moths are fruit piercers (subfamily Calpinae) and are attracted to light. They fly from June to September. Some moths reportedly stray to the south and west of the species' general range.

In habit and coloration, the Canadian Owlet caterpillar has a propensity to imitate sawfly larvae (Hymenoptera), both in habit and coloration. Despite having a full set of usable prolegs, the caterpillar propels itself forward in a looping fashion. When alarmed, it tucks its head under its body—another sawfly-like trait. However, the caterpillar can be easily differentiated from a similarly patterned sawfly larva by counting the prolegs. It has four, while sawflies always have six or more.

The Canadian Owlet caterpillar is yellow and white with intermittent black markings along its length. Its yellow head has black spots on either side. The four prolegs range in color from yellow to black, depending on age and instar. The true legs are red or orange. Early instars are pale yellow with a series of dark spots.

Actual size

FAMILY	Erebidae
DISTRIBUTION	North Africa, Mediterranean Europe, Turkey, southern Russia, and central Asia
HABITAT	Warm oak forests
HOST PLANTS	Oak (*Quercus* spp.)
NOTE	Twiglike caterpillar that is well hidden on oak trees
CONSERVATION STATUS	Not evaluated, but classed as rare in some parts of its range

ADULT WINGSPAN
2%₁₆–3⅝ in (65–92 mm)

CATERPILLAR LENGTH
2–2¾ in (50–70 mm)

CATOCALA DILECTA
CATOCALA DILECTA
(HÜBNER, 1808)

544

The caterpillars of the *Catocala dilecta* moth hatch in late spring from eggs that overwinter in the bark crevices of the host trees where they were laid. The hatching coincides with the opening of the leaves, and the larvae are then active from April to June. Their bark-like cryptic coloration provides the perfect camouflage, so they are difficult to spot when resting on twigs and branches. The caterpillars pupate in the tree canopy, spinning their wispy, yellow cocoon between leaves.

The adult moths are on the wing from late June through September, with a single generation each year. Reliance on a single host plant means numbers of this species are declining, a situation exacerbated by the clearance of oak forest, especially in the Mediterranean countries, and the use of pesticides against forest pests, which usually wipes out all moth and butterfly species and not just the target species.

Actual size

The *Catocala dilecta* caterpillar has a light brown body with many small, dark brown dots. Rows of reddish-brown tubercles run the length of the abdomen, each bearing a short, dark hair. There is a fringe of bristles on both sides of the thorax and abdomen. The head is a mottled brown, with hairs that are of a similar color to the body.

FAMILY	Erebidae
DISTRIBUTION	From eastern Spain east through southern and central Europe, southern Russia, and the Causasus to China, Korea, and Japan
HABITAT	Damp woodlands and scrub in river valleys
HOST PLANTS	Willow (*Salix* spp.) and, less often, poplar (*Populus* spp.)
NOTE	Smooth caterpillar that lives mainly on smooth-barked trees
CONSERVATION STATUS	Not evaluated

ADULT WINGSPAN
2⁹⁄₁₆–3⅛ in (65–80 mm)

CATERPILLAR LENGTH
2⅛–3⅛ in (55–65 mm)

CATOCALA ELECTA
ROSY UNDERWING
(VIEWEG, 1790)

545

The Rosy Underwing caterpillar hatches from an egg that is grayish with purple blotches and laid in a bark crevice on a trunk or branch. The egg overwinters, and the caterpillar hatches in April or May and is fully grown in June or July. Its habits are similar to those of closely related species such as the Red Underwing (*Catocala nupta*). The larva feeds at night, resting by day on a twig, branch, or trunk. The pupa is formed on the food plant, in a cocoon, either between spun leaves or in a crevice.

Like many *Catocala* caterpillars feeding on willow and poplar, the Rosy Underwing is rather smooth, giving good camouflage on the bark of many of those trees. It is distinguished from the Red Underwing by the large, yellow (rather than brown or gray) wart on the sixth segment and the brighter, more orange-brown, paired warts along its back. The adults, which have pink and black hindwings, fly from July to September in one generation.

Actual size

The Rosy Underwing caterpillar is gray brown and covered in tiny, irregularly shaped, dark spots, sometimes forming irregular stripes. It has a fringe of short hairs low down along each side and pairs of small, orange-brown warts along the back. Its eighth segment has a raised, yellowish wart, and there is a yellow-brown or dark gray band and two large, brownish or yellowish warts on the eleventh segment.

FAMILY	Erebidae
DISTRIBUTION	Western Europe (including southern England, where it has recently recolonized) to the Middle East, and across Asia to Russian Far East and Japan
HABITAT	Usually woodlands and well-wooded areas
HOST PLANTS	Mainly Aspen (*Populus tremula*) and other poplars (*Populus* spp.); also ash (*Fraxinus* spp.), oak (*Quercus* spp.), willow (*Salix* spp.), European Beech (*Fagus sylvatica*), and other trees
NOTE	Cryptic caterpillar that appears to move some distance before feeding
CONSERVATION STATUS	Not evaluated, but common within most of its range

ADULT WINGSPAN
3½–4⅜ in (90–112 mm)

CATERPILLAR LENGTH
2⁹⁄₁₆–3 in (65–75 mm)

CATOCALA FRAXINI
CLIFDEN NONPAREIL
(LINNAEUS, 1758)

546

The Clifden Nonpareil caterpillar hatches from an egg that is grayish and domed with vertical ridges. The egg overwinters on a branch or trunk, and the caterpillar hatches in May or June. Observed in captivity, at first the larva crawls constantly for many hours before settling to feed. Later, it feeds at night and rests along a branch, like other *Catocala* caterpillars, stretched out with the legs often splayed sideways. The caterpillar is fully grown in July and forms a pupa on the food plant, in a cocoon, often between leaves drawn together.

The behavior of the young caterpillar suggests that it normally lives high in the tree and must move a long way before feeding. When larger, it is sluggish when handled. The caterpillar is rarely seen in the wild; its shape, color, and fringe of lateral hairs make it highly cryptic. The adults, with blue-and-black hindwings, fly from August to October in a single generation. They regularly migrate outside the breeding range. The species' alternative common name is Blue Underwing.

The Clifden Nonpareil caterpillar is rather smooth and flattened, with a fringe of short hairs low down along each side. It is colored blue gray or gray brown and covered in tiny, gray and blackish spots that form a dark band on the eighth and ninth segments. It has a low hump on the eighth segment and on the eleventh segment, which also has a narrow, angled, black band.

Actual size

FAMILY	Erebidae
DISTRIBUTION	United States and southern Canada
HABITAT	Oak forests and suburbs where oaks are present
HOST PLANTS	Oak (*Quercus* spp.), including black, red, and white oaks
NOTE	Cryptic caterpillar that blends with lichens on tree bark
CONSERVATION STATUS	Not evaluated, but common

ADULT WINGSPAN
2⅞₆–3¼ in (65–82 mm)

CATERPILLAR LENGTH
2⅜–2¾ in (60–70 mm)

CATOCALA ILIA
ILIA UNDERWING
CRAMER, 1780

547

Ilia Underwing caterpillars hatch in the spring from eggs that have overwintered. The species is most common in eastern North America, where studies suggest that the larvae prefer to feed on the Bear or Scrub Oak (*Quercus ilicifolia*). In Connecticut, they hatch in the last few days of April and feed on fresh growth, which is the only food suitable for the young caterpillars. *Catocala* larvae are all cryptic, their colors blending with the bark on which they rest, although *C. ilia* occasionally mimics brighter lichens. Their camouflage defense and diet of large, nontoxic trees indicate that the larvae are palatable to birds and other predators. Following pupation, adults eclose and fly from June to September.

Worldwide there are more than 150 different *Catocala* species, providing a uniform group collectively referred to as "underwings" for their mostly bright-colored hindwings, which are hidden at rest beneath cryptic forewings. In the western United States, the subspecies *C. ilia zoe* is quite distinctive and possibly represents a different species from the eastern populations. *Catocala ilia* is most easily confused with *C. aholibah*, another oak-feeding species.

Actual size

The Ilia Underwing caterpillar is mottled green and black or gray and black with small dorsal bumps, resulting in a cryptic color pattern that matches the lichens on oak branches. The larva is flattened ventrally, with fringes of hairs that enhance its camouflage against the substrate. When the larva is flipped over, its strikingly different ventral surface can be seen—pink purple in color with a wide, black, transverse stripe on each segment.

FAMILY	Erebidae
DISTRIBUTION	Western Europe to Asia Minor, across Asia to northern China, Russian Far East, and Japan
HABITAT	Woodlands, wetland margins, hedgerows, and gardens
HOST PLANTS	Willow (*Salix* spp.) and poplar (*Populus* spp.)
NOTE	Typical *Catocala* caterpillar, feeding by day and hiding by night
CONSERVATION STATUS	Not evaluated, but common within most of its range

ADULT WINGSPAN
2¾–3½ in (70–90 mm)

CATERPILLAR LENGTH
2⅜–2¾ in (60–70 mm)

548

CATOCALA NUPTA
RED UNDERWING
(LINNAEUS, 1767)

The Red Underwing caterpillar hatches in April or May from a gray-brown, ribbed egg that overwinters where it was laid in a crevice on the trunk of the host plant. In captivity, the larva is initially very restless. In the wild, it ascends the tree to feed at night, hiding in a crevice on the trunk or branch by day. It is fully fed in July. The pupa is formed on the food plant, in a cocoon either between spun together leaves or in a crevice.

Catocala nupta is not dissimilar to the Clifden Nonpareil (*C. fraxini*), although that species lacks the pairs of brown warts and the dark hairs of this caterpillar. The caterpillars of a number of other *Catocala* species feed on willows and poplars and are similar in appearance and life history to the Red Underwing, only differing from one another in the detail of warts, humps, pattern, and coloring. Adult Red Underwings fly from August to October in a single generation.

Actual size

The Red Underwing caterpillar is gray brown or gray, with a subdued pattern of fine, irregular lines and spots. It has two rows of small, brown, or reddish warts along the back and a fringe of short hairs low down along each side. It also has a sparse covering of stiff, dark hairs and raised humps on the darker eighth segment (with larger warts) and on the eleventh.

FAMILY	Erebidae
DISTRIBUTION	Western Europe, east to the Urals, North Africa, and Asia Minor
HABITAT	Warm, usually dry woodlands with abundant oak
HOST PLANTS	Oak (*Quercus* spp.), including Pedunculate Oak (*Q. robur*), Sessile Oak (*Q. petraea*), and Downy Oak (*Q. pubescens*)
NOTE	Less smooth *Catocala* caterpillar, reflecting its knobbly food plants
CONSERVATION STATUS	Not evaluated

ADULT WINGSPAN
2⅜–3⅛ in (60–80 mm)

CATERPILLAR LENGTH
2⅛–2%₁₆ in (55–65 mm)

CATOCALA SPONSA
DARK CRIMSON UNDERWING
(LINNAEUS, 1767)

549

The Dark Crimson Underwing caterpillar hatches in April from a slightly flattened, spherical, brown-and-yellow egg that has overwintered. The larva is fully grown in June. It feeds at night, when it is most easily found, and conceals itself by day on a branch or trunk, often in a crevice. The pupa, which like that of other *Catocala* species, is covered in a bluish-white bloom, is formed on the food plant, in a cocoon constructed either between spun leaves or in a crevice or hollow on a limb.

Compared to *Catocala* species that feed on willow and poplar, the Dark Crimson Underwing caterpillar is less smooth in appearance, with larger warts and humps, reflecting the more irregular, knobbly appearance of oak twigs and branches. The caterpillar of *C. dilecta*, also found on oak in warm forests, is very similar but lacks the pale patches of *C. sponsa*. Adult Dark Crimson Underwings fly from late July to mid-October in one generation.

Actual size

The Dark Crimson Underwing caterpillar is brown, grayish, or blackish with several whitish or yellowish patches or bands. It has pairs of brown or blackish warts along the back and others along the sides, and a rather large hump and wart on the eighth segment and another hump with two warts on the eleventh. It is sparsely covered with stiff, dark hairs.

FAMILY	Erebidae
DISTRIBUTION	The Himalayas, northeast India, southern China, Chinese Taipei, Borneo, Sulawesi, and southern islands of Japan
HABITAT	Low- and medium-altitude montane forests
HOST PLANTS	Lichen, moss, and algae
NOTE	Hairy caterpillar that grazes microflora from tree trunks and branches
CONSERVATION STATUS	Not evaluated, but can be very common

ADULT WINGSPAN
2⅛–2⅜ in (55–60 mm)

CATERPILLAR LENGTH
1⅜–1⁹⁄₁₆ in (35–40 mm)

550

CHRYSAEGLIA MAGNIFICA
CHRYSAEGLIA MAGNIFICA
(WALKER, 1862)

Chrysaeglia magnifica caterpillars feed on lichens (a symbiosis of fungi and algae) and their associated mosses and algae. As a result, the larvae are not routinely found on foliage except perhaps to molt or pupate. They graze on tree trunks, rocks, and even man-made structures, where such microflora abound, usually in damp, dark habitats. Their diet makes them distasteful to potential predators and they confer that defense to the adult moths. The nature and composition of the caterpillar diet, which is not highly nutritious, also means that the larval stage is prolonged. The caterpillars relocate to feed or pupate, escaping threats by descending on silken lines.

Pupation occurs beneath a silken sheet on the surface of a leaf. The adult moths are metallic blue and orange and fly nocturnally from July to September. *Chrysaeglia magnifica* and other members of their Lithosiini tribe can collectively be called lichen moths, and *C. magnifica* caterpillars also have the extremely long body hairs that are typical of their tribe.

The *Chrysaeglia magnifica* caterpillar is long and slender with fine, black-and-white, pinstripe longitudinal markings. Each segment bears multiple tubercles from which lengthy white setae emerge, which are longest and most numerous laterally and to the front and rear. The underbelly and prolegs are red or pink. The anterior thoracic tubercles and the rear segment have a blue hue.

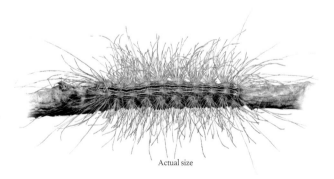

Actual size

FAMILY	Erebidae
DISTRIBUTION	Eastern Ecuador
HABITAT	Edges of humid and semi-humid montane forests
HOST PLANTS	Unknown genus in the plant family Boraginaceae
NOTE	Caterpillar whose coloration and behavior suggest it is unpalatable
CONSERVATION STATUS	Not evaluated, but not considered threatened

CROCOMELA ERECTISTRIA

CROCOMELA ERECTISTRIA
(WARREN, 1904)

ADULT WINGSPAN
1¹¹⁄₁₆–1¹⁵⁄₁₆ in (43–49 mm)

CATERPILLAR LENGTH
1⅜–1¾ in (35–45 mm)

The *Crocomela erectistria* caterpillar is fairly slender but appears larger due to its setae-covered body. It is unequivocally aposematic, calling attention to itself with bright yellow and metallic blue markings. The caterpillars of this species feed in groups of 10 to 15 individuals, resting on the upper surface of leaves—behavior that adds to their warning advertisement. When disturbed, they lift their front and rear portions from the surface of the leaf and thrash them about vigorously.

Within its fairly restricted known range, *Crocomela erectistria* is relatively rare, with only one group of caterpillars ever having been collected and reared (in northeastern Ecuador). Because the caterpillars have been found feeding in large groups, it is assumed that eggs are laid in large groups also; other aspects of the life cycle of this species, however, remain undiscovered. The brightly colored adults, seen more frequently than the caterpillars, are weak, daytime fliers, sometimes observed feeding at moist sand or roadside puddles.

Actual size

The *Crocomela erectistria* caterpillar has a shiny black head and a black body crossed by thick, bright yellow, transverse stripes and with sporadic metallic blue flecking. It is covered with sparse, short to mid-length, fairly stiff, black setae, and the anterior and posterior segments bear several very long, soft, slightly plumose setae, some with white tips.

FAMILY	Erebidae
DISTRIBUTION	Eastern Ecuador
HABITAT	Cloud forest edges, second growth, and regenerating landslides
HOST PLANTS	*Chusquea scandens* and *Baccharis latifolia*
NOTE	Rarely seen caterpillar that is vulnerable to parasitization
CONSERVATION STATUS	Not evaluated, but not considered threatened

ADULT WINGSPAN
1¼–1½ in (32–38 mm)

CATERPILLAR LENGTH
1⅜–1⁹⁄₁₆ in (35–40 mm)

552

DESMOTRICHA IMITATA
DESMOTRICHA IMITATA
(DRUCE, 1883)

Desmotricha imitata caterpillars are solitary in nature, and that, together with their tendency to rest on the lower side of leaves, make them difficult to detect on the dense foliage of their host plants. Beating methods are used most frequently to knock and collect the larvae from the plant. It is not known if the caterpillars are palatable to predators, but they are known to be parasitized by several species of wasps from the families Braconidae and Ichneumonidae. Parasitoid flies in the family Tachinidae also attack the larvae.

The red, black, and blue coloration of adults makes *Desmotricha imitata* an adept wasp mimic as it flies about during the day. Nevertheless, adults are known to be seized from the air by at least one forest bird species, the Smoke-colored Pewees (*Contopus fumigatus*). Given that the geographic range of its host plants is fairly substantial, *D. imitata* may also have a range that extends beyond eastern Ecuador.

Actual size

The *Desmotricha imitata* caterpillar is stout and roughly square in cross section. It has a uniformly bright orange head and a tricolored body that is bright white dorsally, strongly marked with black, and dull yellow white ventrally. The larva is sparsely covered with long to very long, soft, slightly plumose setae, the shorter ones entirely white and the longest ones gray apically.

FAMILY	Erebidae
DISTRIBUTION	The Andes of Colombia, south to Bolivia
HABITAT	Humid cloud forest interiors, especially along streams
HOST PLANTS	Many, including *Schefflera dielssi*, *Oreopanax* spp., *Alloplectus tetragonoides*, *Chusquea scandens*, *Dendrophorbium* spp., and *Critoniopsis occidentalis*
NOTE	Caterpillar that twitches its body spasmodically when disturbed
CONSERVATION STATUS	Not evaluated, but not considered threatened

ADULT WINGSPAN
2⅝–2¹⁵⁄₁₆ in (66–74 mm)

CATERPILLAR LENGTH
2⅜–2¾ in (60–70 mm)

DYSSCHEMA PALMERI
DYSSCHEMA PALMERI
(DRUCE, 1910)

553

Despite the fairly bright coloration of *Dysschema palmeri* caterpillars, they are infrequently encountered. The larvae feed solitarily, usually resting on the lower side of their host plant leaves, well out of sight from potential predators. When disturbed, they spasmodically twitch their head and rear sections, waving their long setae about over their body. So far as is known, the setae do not cause any sort of skin irritation.

The adults of *Dysschema palmeri* resemble, and probably mimic, several of the larger species of clear-winged butterflies (Ithomiinae) within their range. They do not generally fly about during the day, instead resting below leaves with their wings folded back over their bodies in a roughly triangular shape, taking flight only when disturbed. It is not clear when this species' eggs are laid, but possibly around dusk and dawn, when visibility is poor and the adults' resemblance to toxic butterflies may still be an advantage.

The *Dysschema palmeri* caterpillar has a uniformly shiny black head and a velvety black body with dark purple markings around the intersegmental areas, thin, broken lines of bright white dashes, and a mid-dorsal bright crimson stripe. It is sparsely covered with fairly stiff, black setae, each segment also bearing a few long, slightly plumose, soft, white setae.

Actual size

FAMILY	Erebidae
DISTRIBUTION	North America, from southern Canada and Maine, south to Florida, and west to Texas
HABITAT	Fields and roadside edges
HOST PLANTS	Milkweed (*Asclepias* spp.) and dogbane (*Apocynum* spp.)
NOTE	Gregarious caterpillar commonly encountered throughout most of its range
CONSERVATION STATUS	Not evaluated, but globally secure, although possibly rare at the peripheries of its range

ADULT WINGSPAN
1¼–1¹¹⁄₁₆ in (32–43 mm)

CATERPILLAR LENGTH
1⅜ in (35 mm)

EUCHAETES EGLE

MILKWEED TUSSOCK

(DRURY, 1773)

The Milkweed Tussock caterpillar, which hatches from a mass of eggs laid on the underside of leaves, is visually unique. While early instars are slightly hairy and gray, later instars, at first glance, look more like a dust mop or pile of discarded yarn than a life form. The larvae are social feeders until the end of the third instar. Early instar larvae feed by skeletonizing leaves, and older caterpillars have developed the trait of pre-severing leaf veins to reduce the sap flow of toxic glycosides. A large colony of these caterpillars can quickly defoliate an entire milkweed (or dogbane) plant and, unlike Monarch caterpillars (*Danaus plexippus*) and other milkweed feeders, prefer mature plants over younger, more succulent plants.

As a defense when threatened, the caterpillars freeze in place, then slightly curl their bodies and drop from the host plant into the leaf litter below. In addition, both caterpillars and adult Milkweed Tussock moths are also chemically protected by the milkweed toxins the larvae ingest. Two generations per year are typical throughout all but the far northern fringe of their range. The species is also known as the Milkweed Tiger Moth.

The Milkweed Tussock caterpillar is densely tufted. Its blackish-gray abdomen sports multiple white and black lashes, predominantly along the thoracic and anterior regions. Orange-brown tufts of setae on the dorsum are curled upward and meet along the centerline. The head is black.

Actual size

FAMILY	Erebidae
DISTRIBUTION	Across Europe into Russia, the Middle East as far as Turkmenistan and Iran
HABITAT	Evergreen riparian forests, scrub, hedgerows, wastelands, parks, and gardens
HOST PLANTS	Various, including borage (*Borago* spp.), dandelion (*Taraxacum* spp.), Hemp (*Cannabis sativa*), agrimony (*Eupatorium* spp.), and nettle (*Urtica* spp.)
NOTE	Hairy caterpillar that feeds on a range of food plants
CONSERVATION STATUS	Not evaluated, but a protected species in the European Union

ADULT WINGSPAN
2¹⁄₁₆–2⁹⁄₁₆ in (52–65 mm)

CATERPILLAR LENGTH
1⁹⁄₁₆–2 in (40–50 mm)

EUPLAGIA QUADRIPUNCTARIA
JERSEY TIGER
(PODA, 1761)

555

The female Jersey Tiger moth lays her smooth, round eggs on the underside of leaves. When the young caterpillars hatch, they overwinter on their host plant and become active again in spring. The mature larvae crawl to the ground to pupate in a cocoon among the leaf litter, and the resulting pupae are reddish brown in color.

The Jersey Tiger, as strikingly hued and patterned as its big cat namesake, flies during the day and night. The moths are on the wing from July to September, and there is a single generation. A migratory species, *Euplagia quadripunctaria* flies long distances during the summer months and, due to warming conditions, is expanding its range north. The subspecies *E. quadripunctaria rhodensensis* can be found in large numbers in the Valley of Butterflies on the Greek island of Rhodes. Here, encouraged by a perfect microclimate, clouds of brightly colored moths take to the air in a spectacular display.

The Jersey Tiger caterpillar is black with a wide, yellow-orange dorsal stripe and lateral row of cream spots. A ring of orange-brown tufts of hair is found on each segment. The head is black.

Actual size

FAMILY	Erebidae
DISTRIBUTION	Areas of North America, Europe, the Middle East, and across Asia to Siberia and Japan
HABITAT	Woodlands, scrub, parks, and gardens
HOST PLANTS	Various deciduous trees, including alder (*Alnus* spp.), ash (*Fraxinus* spp.), birch (*Betula* spp.), chestnut (*Castanea* spp.), oak (*Quercus* spp.), and willow (*Salix* spp.)
NOTE	Caterpillar with long hairs that may irritate predators
CONSERVATION STATUS	Not evaluated

ADULT WINGSPAN
1⅜–1¼ in (35–45 mm)

CATERPILLAR LENGTH
1³⁄₁₆ in (30 mm)

556

EUPROCTIS SIMILIS
YELLOW-TAIL
(FUESSLY, 1775)

Female Yellow-tail moths lay their eggs in batches on leaves of the host plant. The young caterpillars hatch and are gregarious at first, staying together in large groups, but as they mature they disperse and become solitary. Unusually, this species overwinters on the food plant. The caterpillars become active again in spring, feeding on young leaves, completing their development and pupating by June. The pupa is brown black, enclosed in a creamy-white cocoon.

The common name of the species is derived from the defensive posture of the adult. When disturbed, the moth—also known as the Gold-tail Moth or Swan Moth—lies on its side and raises the tip of its yellow abdomen so that it sticks out beyond the back of the wings. The moths are night-flying and on the wing during July and August. Both larvae and moths have irritating hairs to deter predators. Contact with the hairs can trigger a rash or even an allergic reaction.

The Yellow-tail caterpillar is black with a broad, red-orange dorsal stripe and lateral rows of white spots, and a further red-orange stripe below the spiracles. There is a distinctive hump with an orange stripe on the first segment, and black tubercles appear on segments four and eleven. The body is covered with tufts of long, black hairs. The head is black.

Actual size

FAMILY	Erebidae
DISTRIBUTION	Southern Mexico, south through Central America and western South America to Bolivia
HABITAT	Mid-elevation forests and forest edges, including severely degraded habitat
HOST PLANTS	Many, including *Nectandra* spp., *Erythrina edulis*, *Desmodium* spp., *Rubus* spp., *Wercklea ferox*, and *Chusquea scandens*
NOTE	Distinctive, densely tufted caterpillar
CONSERVATION STATUS	Not evaluated, but not considered threatened

ADULT WINGSPAN
1⁹⁄₁₆–1¾ in (39–45 mm)

CATERPILLAR LENGTH
1¾–2⅛ in (45–55 mm)

HALYSIDOTA ATRA
HALYSIDOTA ATRA
(DRUCE, 1884)

557

The singular appearance of the *Halysidota atra* caterpillar establishes its identity, although few other species of this genus are known. Its characteristic forward and rearward oriented tufts are moveable and wave about prominently when the caterpillar is walking. At rest, the tufts serve to completely conceal the head. Younger caterpillars tend to feed gregariously, but individuals go their separate ways some time during the final instar. Groups of caterpillars waving their contrasting yellow tufts and white tufts can appear quite menacing.

Early and middle instars, which have a bright pink ground color and sparser secondary setae, look completely different from the final instars. Older larvae are parasitized by *Distatrix* wasps (Braconidae) and several unknown species of tachinid flies. Both the braconids and tachinid parasitoids develop internally, emerging to pupate outside of the host soon after the infected larva begins to show visual signs of being parasitized.

The *Halysidota atra* caterpillar is short and robust-looking due to its dense coating of dark brown setae and bulbous, shiny, brown head. The second and third thoracic segments bear pairs of dorsal and subdorsal tufts composed of soft, very long, pale orange, whitish, or yellowish setae, those on the second thoracic segment oriented forward and those on the third thoracic segment swept rearward.

Actual size

FAMILY	Erebidae
DISTRIBUTION	North America, from southern and southeastern Canada to Florida, and west to Texas
HABITAT	Woodlands and forests
HOST PLANTS	Many woody shrubs and trees, including alder (*Alnus* spp.), ash (*Fraxinus* spp.), birch (*Betula* spp.), elm (*Ulmus* spp.), oak (*Quercus* spp.), and willow (*Salix* spp.)
NOTE	Conspicuous, hirsute caterpillar
CONSERVATION STATUS	Not evaluated, but usually common

ADULT WINGSPAN
1⁹⁄₁₆–1¾ in (40–45 mm)

CATERPILLAR LENGTH
1⅜ in (35 mm)

HALYSIDOTA TESSELLARIS
BANDED TUSSOCK MOTH
(J. E. SMITH, 1797)

Banded Tussock Moth caterpillars hatch from eggs laid in masses on the underside of host plant leaves. There are two generations each summer, and although not gregarious the larvae are conspicuous, preferring to rest openly on the upper side of leaves. This suggests they are not palatable to predators, either because of their tufted bodies or the chemical defenses they have acquired from host plants. Pupation occurs in a gray, silken cocoon laced with many of the caterpillar's hairs. The pupae overwinter.

As with many moth species, the caterpillars of the Banded Tussock Moth are more attractive than the adults. However, handling is not recommended as the profuse setae of the larvae can cause irritation to some people. The adult moths fly by night and are frequently attracted to lights. This species often co-occurs with the similar Sycamore Tussock Moth (*Halysidota harrisii*) and Florida Tussock Moth (*Halysidota cinctipes*).

The Banded Tussock Moth caterpillar is variably yellow brown to gray black, with conspicuous black tufts and white tufts extending from the anterior and posterior ends of the body. Dark setae usually form a middorsal line along the body. The head is black.

Actual size

FAMILY	Erebidae
DISTRIBUTION	Northeastern Ecuador to southeastern Peru
HABITAT	Forest and montane stream edges
HOST PLANTS	*Barnadesia parviflora, Adenostemma harlingii, Browallia speciosa,* and *Solanum* spp.
NOTE	Caterpillar with urticating hairs that can cause an itchy rash
CONSERVATION STATUS	Not evaluated, but not considered threatened, although not common

ADULT WINGSPAN
1½–1¾ in (38–44 mm)

CATERPILLAR LENGTH
2⅛–2¾ in (55–70 mm)

HYPERCOMPE OBSCURA
HYPERCOMPE OBSCURA
(SCHAUS, 1901)

559

Hypercompe obscura caterpillars are solitary feeders and, although they live on a wide variety of host plants, not common. The larvae can be found on almost any portion of the host plant, but they frequently rest along the midvein on the underside of the leaf. When disturbed, they tend to drop down from the plant and curl up into a ball, flourishing their long, urticating setae.

When at rest, these caterpillars are positioned with their tufts of setae held in close proximity, giving them the appearance of being completely covered in hairs. When feeding or moving, however, they expose the bare intersegmental areas, presumably making themselves more vulnerable to attack by parasitoids than while at rest. When the larvae are handled, their setae tend to break off and cause skin irritation—especially between the fingers. The adult is one of many boldly marked members of the genus *Hypercompe*, a trait that makes "tiger moth" an apt name for this and related genera.

The *Hypercompe obscura* caterpillar has a uniformly shiny, dark brown to black head and a robust, dark brown body. A wash of pale gray gives it a frosted appearance. The mid-length, reddish-brown, stiff setae are arranged into distinct rings along the length of the body, and the posterior segments also bear a few sparse, long, soft, pale setae emanating from subdorsal and lateral verrucae.

Actual size

FAMILY	Erebidae
DISTRIBUTION	From southern Ontario, Canada, south through United States to Florida, and west to the Midwest, Oklahoma, and Texas
HABITAT	Meadows and forest edges as well as disturbed areas
HOST PLANTS	Wide range of frequently toxic plants, from dandelion (*Taraxacum* spp.) to willow (*Salix* spp.)
NOTE	Woolly-bear caterpillar frequently found crawling on the ground
CONSERVATION STATUS	Not evaluated, but common

ADULT WINGSPAN
3 in (76 mm)

CATERPILLAR LENGTH
3⅛ in (80 mm)

560

HYPERCOMPE SCRIBONIA
GIANT LEOPARD MOTH
(STOLL, 1790)

The woolly-bear caterpillars of the Giant Leopard Moth eat large volumes of leaves as they mature, and are often seen crawling in search of a suitable food source. Their characteristic black spines, while probably a good defense against birds, are harmless to humans. Unlike many other caterpillars, the larvae easily switch between host plant species, and sometimes search out specific plants that are toxic in order to self-medicate against parasitoids. Adults, too, are chemically defended. Before pupation, the caterpillar makes a loose black cocoon.

The Giant Leopard Moth is part of a large subfamily (Arctiinae) of tiger moths with more than 10,000 species, which, until recently, was considered its own family. Tiger moths are characteristically colorful, which indicates their distastefulness. *Hypercompe scribonia* is no exception, being white with dark, sometimes shiny blue spots, and occasionally orange spots on the shiny blue abdomen. When disturbed, it falls to the ground and rolls its abdomen, exposing its aposematic colors.

The Giant Leopard Moth caterpillar, while initially mostly orange with wide, black stripes, becomes black with thin, red stripes toward later instars, hidden by black, dense, thick spines. When it rolls into a ball, sensing danger, it exposes red, circular stripes between its segments—an aposematic warning to predators. The red stripes can also be seen when the caterpillar is crawling, but not when it is resting on its host plant.

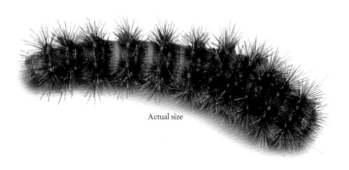

Actual size

FAMILY	Erebidae
DISTRIBUTION	North and Central America, southern Scandinavia, eastern Europe, western Russia, Mongolia, and northern China to Japan
HABITAT	Forests, wet woodlands, parks, and orchards
HOST PLANTS	Variety of deciduous and ornamental trees, and also shrubs, including hickory (*Carya* spp.), walnut (*Juglans* spp.), apple (*Malus* spp.), and maple (*Acer* spp.)
NOTE	Widely distributed caterpillar that damages forests and orchards
CONSERVATION STATUS	Not evaluated

ADULT WINGSPAN
⁹⁄₁₆–¹¹⁄₁₆ in (15–17 mm)

CATERPILLAR LENGTH
1³⁄₁₆–1³⁄₈ in (30–35 mm)

HYPHANTRIA CUNEA
FALL WEBWORM
(DRURY, 1773)

561

The female Fall Webworm moth lays her eggs in clusters of several hundred on the underside of leaves of the food plant. The caterpillars are social, living together in huge communal webs constructed over the tips of branches of the host plants. When feeding, they stay within the protection of the web, which they steadily expand. The species overwinters in a brown cocoon made of bits of twig and silk. The moths are on the wing from midsummer to early fall. There is one generation in the northern part of their range and two in the south.

Once found only in North America, the species has been introduced around the world, and is now an invasive pest in Europe and Asia. It can break out in epidemic numbers, causing economic damage to forests and orchards. The young caterpillars eat the upper surface of leaves, while the older ones eat the whole leaf, often defoliating entire trees.

The Fall Webworm caterpillar has 12 pairs of small tubercles bearing tufts of long hairs. It is otherwise highly variable in color with a yellow to green body, a dorsal black stripe, and a lateral yellow stripe. The head is either black or red.

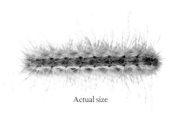

Actual size

FAMILY	Erebidae
DISTRIBUTION	The Andes of Venezuela, Colombia, and Ecuador
HABITAT	Mid-elevation cloud forests and forest edges
HOST PLANTS	Bombacaceae (one unidentified species is currently the only confirmed host plant)
NOTE	Rare, black, fuzzy caterpillar that resembles many in its family
CONSERVATION STATUS	Not evaluated, but not considered threatened

ADULT WINGSPAN
1⅞–2¹/₁₆ in (48–52 mm)

CATERPILLAR LENGTH
2⅜–2¾ in (60–70 mm)

562

IDALUS VENETA

IDALUS VENETA
(DOGNIN, 1901)

Despite their somewhat aposematic, red-on-black coloration, *Idalus veneta* caterpillars are rather docile when handled, usually dropping from the host plant. Although their stiff setae tend to break off, they are not particularly irritating to the skin. It is likely, however, that the setae are still effective defenses against vertebrate predators who wish to avoid damage to their mouths and mucosal membranes.

Only several caterpillars of *Idalus veneta* have been found and reared, and the life cycle has never been completely described. The reason for the rarity of the caterpillars is unknown and is particularly puzzling considering how common adults can be at lights. The moth adults can be seasonally absent, however, and it is possible that they show altitudinal migrations and only reproduce during certain times of year. The species' alternate spelling, "*venata*," is believed to be a printer's error as Paul Dognin, the French entomologist who first described it, subsequently referred to it as "*veneta*."

Actual size

The *Idalus veneta* caterpillar has a large, shiny, black head with a pale pink patch around the epicranial suture. Its body is a deep, velvety black, with pale rose-colored intersegmental areas and prolegs and small, red marks ventrolaterally. The black setae are stiff and slightly plumose (appearing serrate along the shaft) and largely obscure most of the other body markings.

FAMILY	Erebidae
DISTRIBUTION	From England, Wales, and western Europe, including southern Scandinavia, to Russian Far East, China, and Japan
HABITAT	Woodlands, mature scrub, and hedgerows
HOST PLANTS	Lichens growing on trees and bushes, including *Physcia stellaris* and *Xanthoria parietina*
NOTE	Highly cryptic, lichen-feeding caterpillar unlike other hook-tips
CONSERVATION STATUS	Not evaluated

ADULT WINGSPAN
1–1⁷⁄₁₆ in (25–36 mm)

CATERPILLAR LENGTH
⁷⁄₈–1 in (22–25 mm)

LASPEYRIA FLEXULA
BEAUTIFUL HOOK-TIP
([DENIS & SCHIFFERMÜLLER], 1775)

563

The Beautiful Hook-tip caterpillar hatches from a domed, grayish egg laid singly or in small groups, either on a twig or on the food plant. It hides among lichens by day, only coming out to feed at night. The larvae are found particularly among lush growths of lichen in shady places rather than in drier, more open habitat. Caterpillars of the fall generation hibernate while still small. The pupa is blackish brown with greenish bands between the segments.

The moth adults fly in one to three generations, depending on climate, from May to September and somewhat resemble the hook-tips of the family Drepanidae. However, the *Laspeyria flexula* caterpillar is quite unlike Drepanidae hook-tip larvae, having normal prolegs instead of spinelike hind extensions. It is also somewhat unusual in feeding on primitive lichens rather than plants and is highly cryptic on its hosts, with a lichen-like pattern and fringe of small, pale, membranous, growths along the sides.

The Beautiful Hook-tip caterpillar is quite slender and varies in color. Some individuals are light gray green with a striking, irregular, and partly diamond-shaped pattern along the back and complex, darker green mottling. Others are plainer, darker green, or dull, greenish gray, with a more subtle pattern. The front two pairs of prolegs are highly reduced in size, so the caterpillar walks with a looping action.

Actual size

FAMILY	Erebidae
DISTRIBUTION	Western and eastern areas of North America, Europe, North Africa, western and central Asia, Russian Far East, and Japan
HABITAT	Woodland edges and hedgerows, parks, and gardens
HOST PLANTS	Deciduous trees, including willow (*Salix* spp.) and poplar (*Populus* spp.)
NOTE	Widely distributed and destructive caterpillar
CONSERVATION STATUS	Not evaluated, but widespread and common

ADULT WINGSPAN
1⁷⁄₁₆–2 in (37–50 mm)

CATERPILLAR LENGTH
1³⁄₈–1¼ in (35–45 mm)

LEUCOMA SALICIS
WHITE SATIN MOTH
(LINNAEUS, 1758)

The female White Satin Moth lays eggs in clusters on tree trunks and covers them in a white froth. The caterpillars hatch and feed for a few weeks before overwintering in a silken web. The larvae then emerge the following spring, start feeding again, and by midsummer are ready to pupate. They then spin a loose cocoon, which they attach to almost any surface. As well as defoliating trees, the caterpillars wander widely in their search for food or a place to spin their loose cocoon, often within rolled leaves. The pupa is glossy black with tufts of hair.

The caterpillar is a major forest pest as it defoliates vast swathes of deciduous forest, especially in North America, where, since its arrival in the 1920s, it has spread across the continent. European species of parasitic wasps have been introduced to help control the species. The white, glossy-winged *Leucoma salicis* moth flies in summer.

The White Satin Moth caterpillar is brown black with a row of distinctive oblong-shaped, white to yellow blotches along the back and a yellow lateral line. It also has dorsal and lateral tufts of long, reddish-brown hairs.

Actual size

FAMILY	Erebidae
DISTRIBUTION	Colombia, Ecuador, Bolivia, and likely Peru
HABITAT	Humid montane forest borders, pastures, and river edges
HOST PLANTS	Many, most commonly *Boehmeria bullata*, *B. caudate*, *B. pavoni*, and *B. ulmifolia*; also species of several other families, including Poaceae and Fabaceae
NOTE	Uniquely shaped caterpillar that is highly polyphagous
CONSERVATION STATUS	Not evaluated, but not considered threatened

ADULT WINGSPAN
1¾–2¹⁄₁₆ in (44–52 mm)

CATERPILLAR LENGTH
2¹⁄₁₆–2⁹⁄₁₆ in (52–58 mm)

LOPHOCAMPA ATRICEPS
LOPHOCAMPA ATRICEPS
(HAMPSON, 1901)

565

Lophocampa atriceps caterpillars feed in groups until the third instar, moving to separate leaves or plants to feed solitarily in later instars. Although the highly polyphagous caterpillar has been reared a number of times and has an extensive geographical distribution, its complete life cycle has not been described, and little is known of its natural enemies. Like other species in its cloud forest habitat, however, it likely has many parasitoid and other enemies. When disturbed, the larvae raise their posterior-most abdominal segments and wiggle them back and forth like a dog's tail, perhaps brandishing this tuft of hairs to potential predators.

In shape, the somewhat singular caterpillar is elongate, tapering sharply behind the head, and slightly swollen around the first abdominal segments, giving it a partially humpbacked appearance. Two erect, flat-topped tufts of setae, one forward and one rearward, further add to the larva's strange silhouette. The hairs of these tufts are mildly irritating to the skin, but the remaining soft, white setae appear to be non-urticating.

The *Lophocampa atriceps* caterpillar has a shiny, black head and velvety, black body, the latter covered with fine, bright, white flecking and speckling, as well as with sparse, long, soft, finely plumose white setae. The dorsum of the third thoracic segment to the seventh and eighth abdominal segments bears dense clusters of erect, black setae.

Actual size

FAMILY	Erebidae
DISTRIBUTION	Eastern North America
HABITAT	Forests and parks
HOST PLANTS	Various, in particular hickory (*Carya* spp.) and walnut (*Juglans* spp.)
NOTE	Caterpillar that is protected by a covering of irritating hairs
CONSERVATION STATUS	Not evaluated, but common within its range

ADULT WINGSPAN
1⁷⁄₁₆–2⅛ in (37–55 mm)

CATERPILLAR LENGTH
1⁹⁄₁₆–1¾ in (40–45 mm)

566

LOPHOCAMPA CARYAE
HICKORY TUSSOCK MOTH
HARRIS, 1841

Hickory Tussock Moth caterpillars hatch from eggs laid in clusters on leaves of the food plant. The young larvae are gregarious, staying together in large groups, skeletonizing leaves, although as they get older they disperse and lead solitary lives. The caterpillars crawl to the ground to pupate in late summer and fall, spinning a loose cocoon in the leaf litter, where they overwinter. The adult moths are on the wing during early summer, with a single generation a year.

Unlike the hairs of some of its relatives, the hairs of this caterpillar—particularly the longer black lashes—contain an irritant to deter predators. Even the cocoon is protected, as the caterpillar incorporates its hairs as it spins the silk. Most people get an itchy skin rash if they handle the caterpillars without gloves. The microscopic barbs of the hairs attach to the skin of fingers and can also be rubbed accidentally into the eye.

The Hickory Tussock Moth caterpillar is covered in tufts of long, white hairs, with a row of black tufts along the dorsal line. There are four long, black pencils of hair, two near the head and two at the posterior. The head is black, and there are lateral lines of black spots.

Actual size

FAMILY	Erebidae
DISTRIBUTION	Canada, eastern and western United States
HABITAT	Deciduous and mixed forest
HOST PLANTS	Various, but mostly poplar (*Populus* spp.) and willow (*Salix* spp.)
NOTE	Brightly colored caterpillar whose hairs can cause an allergic reaction
CONSERVATION STATUS	Not evaluated

ADULT WINGSPAN
1⅜–1¾ in (35–45 mm)

CATERPILLAR LENGTH
2 in (50 mm)

LOPHOCAMPA MACULATA

SPOTTED TUSSOCK MOTH

HARRIS, 1841

567

Spotted Tussock Moth caterpillars hatch from eggs laid on the leaves of the host plant and become active in summer and early fall, feeding on leaves of a range of deciduous food plants. The mature larvae move to the ground to pupate in loose cocoons formed from silk and hairs. They overwinter, and the night-flying adults emerge in early summer. There is a single generation a year.

The eye-catching colors of the *Lophocampa maculata* caterpillar, which are variable in different parts of its range, are a warning to predators that the hairs are of an irritating nature. The irritant can also cause skin rashes and even allergic reactions in humans who are tempted to pick up the larvae, which should not be handled without gloves. The Spotted Tussock common name refers to the caterpillar's tufts of hair. The species is also known as the Mottled Tiger or Spotted Halisidota.

The Spotted Tussock Moth caterpillar is densely covered with hairs. The head and end of the abdomen are black, while the rest of the body is yellow to orange red with a dorsal row of black tufts. Additionally there are tufts of long, white hairs at the head and posterior ends.

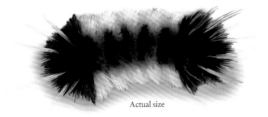

Actual size

FAMILY	Erebidae
DISTRIBUTION	Southern Canada and eastern United States, with occasional outbreaks in the west; also Europe, North Africa, and Asia
HABITAT	Temperate forests
HOST PLANTS	More than 500 coniferous and broadleaved tree species, including oak (*Quercus* spp.), willow (*Salix* spp.), and birch (*Betula* spp.)
NOTE	Widely distributed pest caterpillar that can inflict significant economic damage
CONSERVATION STATUS	Not evaluated, but widespread and pestiferous

ADULT WINGSPAN
1¼–2⁷⁄₁₆ in (32–62 mm)

CATERPILLAR LENGTH
2–2¾ in (50–70 mm)

LYMANTRIA DISPAR
GYPSY MOTH
(LINNAEUS, 1758)

568

The Gypsy Moth flies in late summer, with each female laying as many as 1,000 eggs on tree trunks, branches, and even vehicles. The eggs overwinter and hatch in early May. The young caterpillars climb to the canopy of host trees, where they stay, feeding gregariously during the day. Early instars nibble holes in leaves, while later instars consume entire leaves. The mature fifth or sixth instar caterpillars crawl down the trunk by day and rest under bark, returning to the canopy at night to feed. They pupate in flimsy silken cocoons on the same tree, under bark or in crevices. The moths eclose after about two weeks.

Two distinct strains of the species, Asian and European, have been identified, both virtually identical in appearance and both capable of causing equally extensive damage. The pattern of outbreak alternates between one or two years of light infestation with little visible damage, followed by up to four years when the trees experience moderate to severe defoliation, leading to a population collapse.

The Gypsy Moth caterpillar is covered in urticating hairs. The body is colorful, with two rows of five blue and two rows of six red spots, each spot bearing a tuft of yellow-brown hairs. The head is yellow and black. The legs are red.

Actual size

FAMILY	Erebidae
DISTRIBUTION	Eastern Ecuador, most of eastern Peru
HABITAT	Montane cloud forest borders and nearby second-growth habitats
HOST PLANTS	*Erato polymnioides* and *Miconia* spp.
NOTE	Caterpillar patterned in orange, red, and brown, like the adult
CONSERVATION STATUS	Not evaluated, but not considered threatened

ADULT WINGSPAN
1⅝–1⅞ in (42–48 mm)

CATERPILLAR LENGTH
2⅛–2¾ in (55–70 mm)

MELESE PERUVIANA

MELESE PERUVIANA

(ROTHSCHILD, 1909)

Melese peruviana caterpillars are found feeding as solitary individuals, resting on the top of their host plant leaves. When disturbed, they generally lift their thorax from the leaf surface, wiggle it about a little halfheartedly, and then drop from the plant to crawl quickly out of sight into the leaf litter. In their final instar, they are large, attractive caterpillars, with fewer setae than many members of the Erebidae. The long, quite tightly packed tufts of setae that project forward over the head appear almost like the antennae of an adult insect and are perhaps used to sense air currents produced by the approach of a potential predator.

The adults of *Melese peruviana* are, despite their relatively bright red-and-yellow patterning, well camouflaged when resting among the dead foliage where they generally spend the day. They are commonly attracted to lights at certain times of the year and nearly absent during others. The species was first described from Peru and later found in eastern Ecuador; its range likely includes at least southeastern Colombia.

The *Melese peruviana* caterpillar is slender, with a uniformly orangeish head and a complexly patterned body. The ground color is slightly beige, with olive, orange, or black washing to some areas and bright red spots, of various sizes, dorsally. There are only sparse, dull orange setae, the most noticeable arranged in two, forward-projecting tufts on the pronotum.

Actual size

FAMILY	Erebidae
DISTRIBUTION	From eastern Canada and eastern United States to the Caribbean, south to central Argentina
HABITAT	Fields, grasslands, gardens, and open areas
HOST PLANTS	Various grasses, including crops such as Oat (*Avena sativa*), Pearl Millet (*Pennisetum glaucum*), Sugarcane (*Saccharum officinarum*), Sorghum (*Sorghum bicolor*), Wheat (*Triticum aestivum*), and Maize (*Zea mays*)
NOTE	Striped caterpillar that feeds on grasses
CONSERVATION STATUS	Not evaluated, but common

ADULT WINGSPAN
1⁵⁄₁₆–1¹¹⁄₁₆ in (33–43 mm)

CATERPILLAR LENGTH
2–2⅜ in (50–60 mm)

570

MOCIS LATIPES
STRIPED GRASS LOOPER
(GUENÉE, 1852)

Striped Grass Looper caterpillars hatch from eggs laid on grass blades; a single female will lay more than 250. Young larvae usually just scrape the top surface of the leaf, while the later instars feed on entire leaves. To avoid predators and parasitoids, they feed at night and spend the day in a shelter at the base of the leaf. When disturbed, the caterpillars drop to the ground, where they blend perfectly with the dry stems and leaves of their host plants. Development over seven instars takes about 27 days at temperatures of around 68°F (20°C). The larvae pupate inside a grass shelter.

Mocis latipes adults fly from May through December in northern locales but can breed continuously in the tropics. The caterpillar can be a damaging pest of pasture grasses, sorghum, maize, and rice, with occasional outbreaks on sugarcane in tropical areas such as the Caribbean. It is controlled by parasitoids such as sarcophagid flies, and braconid, chalcid, and ichneumonid wasps. *Mocis latipes* represents a large genus of about 40 species, many of which are similar-looking.

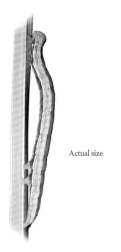

Actual size

The Striped Grass Looper caterpillar has only three pairs of abdominal prolegs, missing them on abdominal segments three and four. By the third instar, it acquires its characteristic striated coloration of thin, horizontal, brown, white, and beige stripes that extend through the head. The white subspiracular stripe is a little more prominent than the rest; below it, the body is darker. There may be a number of black markings along the subdorsal line on abdominal segments one and two.

FAMILY	Erebidae
DISTRIBUTION	The Andes of Colombia, Ecuador, and Peru
HABITAT	Primary and secondary cloud forests, especially along streams and other areas of disturbance
HOST PLANTS	Various, including *Columnea ericae*, *Chusquea scandens*, *Rubus* spp., *Miconia* spp., and *Tibouchina lepiota*
NOTE	Caterpillar that is difficult to see despite its contrasting coloration
CONSERVATION STATUS	Not evaluated, but not considered threatened

ADULT WINGSPAN
1½–1¾ in (38–44 mm)

CATERPILLAR LENGTH
1⅜–1⁹⁄₁₆ in (35–40 mm)

NEONERITA HAEMASTICTA

NEONERITA HAEMASTICTA

DOGNIN, 1906

571

The boldly patterned *Neonerita haemasticta* caterpillar is not often found; most of those larvae encountered to date have been discovered only by beating potential host plants. Despite its somewhat striking and contrasting color pattern, the caterpillar can be quite cryptic, generally resembling a fungus-infected portion of the host plant. Its long, delicate, plumose setae are non-urticating and likely help the larva to detect air disturbed by an approaching enemy. It is unknown if the shorter setae within the dorsal tufts are urticating, but this seems quite possible.

Although currently only known to occur from Colombia to Peru, this highly polyphagous species may also eventually be found in the Andes of Venezuela and Bolivia. *Neonerita haemasticta* has been reared numerous times in northeastern Ecuador and has not yet been found to be a host for any parasitoid flies or wasps. The biology and behavior of adult moths are virtually unknown.

The *Neonerita haemasticta* caterpillar is generally black, with intricate yellow patterning and two rows of short, densely packed, bright white tufts of setae along its dorsum. The setae of the first and seventh tufts are approximately twice as long as the rest. Otherwise, the body bears sparse, soft, long, dark setae, many of which are plumose and tipped with white.

Actual size

FAMILY	Erebidae
DISTRIBUTION	North America, Europe, and across northern Asia
HABITAT	Woodlands, hedgerows, parks, and gardens
HOST PLANTS	Various tree species, including oak (*Quercus* spp.), willow (*Salix* spp.), and poplar (*Populus* spp.)
NOTE	Bizarre-looking, tufty caterpillar with irritating hairs
CONSERVATION STATUS	Not evaluated

ADULT WINGSPAN
1–1⅜ in (25–35 mm)

CATERPILLAR LENGTH
1⅜– 1⁹⁄₁₆ in (35–40 mm)

572

ORGYIA ANTIQUA
RUSTY TUSSOCK MOTH
(LINNAEUS, 1758)

Unusually, the female Rusty Tussock Moth is flightless, so she lays her large clutch of 200 to 300 eggs on her empty cocoon. The eggs overwinter, and the caterpillars emerge in spring. The caterpillars are active between May and September, feeding on a variety of deciduous species. The hairy young larvae disperse by ballooning on threads of silk. They pupate in crevices in bark or on fences, spinning a black, hairy cocoon. The adult moths eclose from July to October. There are usually one or two generations, but in some places there are three a year.

The Rusty Tussock Moth, also named the Vapourer, has a wide distribution across the northern hemisphere. The caterpillar is considered a pest, as large aggregations on trees can lead to defoliation and damage to parkland and orchard trees. If the caterpillar is handled without gloves, its hairs, which detach easily, can cause skin irritation.

Actual size

The Rusty Tussock Moth caterpillar is gray black and covered in tufts of yellow-brown hairs, which arise from rings of red tubercles. There are four extra-long dorsal tufts of creamy-white to yellow hairs, two long, black tufts either side of the head, and a terminal tuft of long, black hairs. The legs are red, while the prolegs are orange red.

FAMILY	Erebidae
DISTRIBUTION	Spain
HABITAT	Heathlands, grasslands, and alpine slopes up to 6,600 ft (2,000 m) elevation
HOST PLANTS	*Cytisus* spp. and *Genista* spp.
NOTE	Colorful caterpillar that is covered in irritating hairs
CONSERVATION STATUS	Not evaluated

ORGYIA AUROLIMBATA

ORGYIA AUROLIMBATA

(GUENÉE, 1835)

ADULT WINGSPAN
1 in (25 mm)

CATERPILLAR LENGTH
1³⁄₁₆ in (30 mm)

573

The female *Orgyia aurolimbata* moth is wingless and hairy and has very limited mobility, so rarely leaves the cocoon. The caterpillars hatch from eggs she lays usually either in or on her cocoon and are seen between April and early August, often basking in the sun near the top of the host plant. When threatened, they quickly drop into the vegetation. The larvae are covered in urticating hairs to deter predators, and, when mature, they spin a thick silken cocoon and cover it with the urticating hairs for added protection. The cocoon is attached to stems of the host plant.

The day-flying male moths are on the wing from June to September, and there is a single generation. They are attracted to the wingless females by pheromones. The males' life span is just four or five days, but the females live for up to 13 days, starting to lay eggs, whether fertilized or not, after three to five days.

Actual size

The *Orgyia aurolimbata* caterpillar is covered in irritating hairs. Its body is gray brown with a black dorsal strip bounded by orange, plus lateral yellow, black, and brown stripes. There are four pairs of orange dorsal tufts, a black tuft at the end of the abdomen, and two long, black tufts that extend each side of the head.

FAMILY	Erebidae
DISTRIBUTION	Colombia and Ecuador
HABITAT	Lower temperate and upper subtropical forest borders
HOST PLANTS	*Mikania micrantha*
NOTE	Caterpillar of a rare and poorly studied species
CONSERVATION STATUS	Not evaluated, but not considered threatened

ADULT WINGSPAN
1³⁄₁₆–1½ in (30–38 mm)

CATERPILLAR LENGTH
2–2⅛ in (50–55 mm)

PHAIO CEPHALENA

PHAIO CEPHALENA
(DRUCE, 1883)

574

The distinctive *Phaio cephalena* caterpillar is very rarely encountered, although the adults can be fairly common, at least in northeastern Ecuador. Indeed, to date, the caterpillars of *P. cephalena* have only been found and reared twice, and in both instances they were discovered as solitary individuals, suggesting that the adult female lays her eggs one at a time. The reasons for the scarcity of larval records are unknown, but, as some species of their host plant genus commonly grow as high-climbing vines, the larvae may, in fact, be much more common in the poorly explored canopy habitats within their geographic range.

The moth adults are remarkably good mimics of wasps, likely relying on most vertebrates' aversion to handling wasps as a form of protection. Despite being frequently seen in the day, the moths appear also to fly at night and are frequent arrivals to porch lights and blacklight traps. The genus *Phaio*, commonly misspelt as *Phaeo*, includes 13 species, found in Central America, Cuba, and South America.

The *Phaio cephalena* caterpillar is fairly stout but tapers sharply at the thorax. Its head is uniformly yellow, matching the general ground color of the body, which is sparsely spotted or washed with shades of brown and black. The thoracic segments are nearly bare, while the middle abdominal segments have sparse, short, soft, pale setae. The most notable setae are packed into long tufts, pairs of which occur on both the first and seventh abdominal segments.

Actual size

FAMILY	Erebidae
DISTRIBUTION	Eastern Australia, Papua New Guinea, and New Caledonia
HABITAT	Subtropical forests below 1,970 ft (600 m) elevation
HOST PLANTS	Menispermaceae, including Carronia Vine (*Carronia multisepalea*) and *Pycnarrhena australiana*
NOTE	Spectacular caterpillar that has false eyes and a defensive posture
CONSERVATION STATUS	Not evaluated, but the southern subspecies is endangered

ADULT WINGSPAN
5–6¾ in (130–170 mm)

CATERPILLAR LENGTH
4⅝ in (120 mm)

PHYLLODES IMPERIALIS
IMPERIAL FRUIT-SUCKING MOTH
DRUCE, 1888

575

The female Imperial Fruit-sucking Moths lay their eggs singly on vines of the Menispermaceae family, choosing young leaves on low-growing plants in heavily shaded areas. Once hatched, the caterpillars feed on leaves and, at rest, lie flat against the stems, their dead-leaflike camouflage making them difficult to spot among the foliage. The mature larvae crawl to the ground and pupate in a loose cocoon in the leaf litter. The resulting pupae have a bronze color.

The *Phyllodes imperialis* caterpillar is often referred to as the "big-headed" caterpillar after its remarkable defensive behavior. When threatened, it arches its body and bends its head down to reveal a pair of pale black, blue, and yellow eyespots and rows of white, teethlike markings to startle any would-be predator. There are several subspecies, including the endangered Southern Pink Underwing moth, which is named for the bright pink patches on its hindwing.

The Imperial Fruit-sucking Moth caterpillar is olive green to gray brown, with several thin, pale, wiggly dorsal lines and oblique shading that resemble veins of a leaf. Two large, blue-black eyespots and white markings are found on the first abdominal segment, with another brown-and-red mark behind. The last abdominal segments are elongated with a ventral black mark outlined in white.

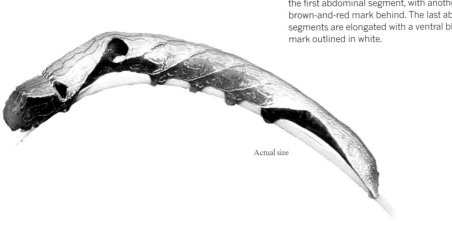

Actual size

FAMILY	Erebidae
DISTRIBUTION	Venezuela, Ecuador, Peru, and likely also Colombia
HABITAT	Cloud forests and forest borders
HOST PLANTS	Many, including *Saurauia bullosa* Wawra, *Anthurium* spp., *Baccaris* spp., *Brunnelia* spp., *Erythrina edulis*, and *Tibouchina lepidota*
NOTE	Common caterpillar that has distinctive, long tufts of orange setae
CONSERVATION STATUS	Not evaluated, but not considered threatened

ADULT WINGSPAN
2¾–3⅛ in (70–80 mm)

CATERPILLAR LENGTH
2⁹⁄₁₆–3 in (65–75 mm)

576

PRAEAMASTUS ALBIPUNCTA
PRAEAMASTUS ALBIPUNCTA
(HAMPSON, 1901)

The *Praeamastus albipuncta* caterpillar, always a great find for a novice caterpillar hunter, is remarkable in appearance for its subtle but attractive coloration and contrasting distinctive orange tufts of setae. In addition, unlike many other erebid caterpillars, the larvae of this species only occasionally attempt to escape by dropping from the host plant and are, instead, easily herded into a collecting bag. Or, gripping tightly to their host plant leaf, they may be plucked and collected along with the leaf.

The caterpillar is highly polyphagus and is the known host for a variety of species of parasitoid flies (Tachinidae) and wasps (Braconidae). In some cases, up to 17 braconids may develop internally, emerging in the final instar to pupate away from the caterpillar. Adult moths are frequent arrivals around lights during the night, but little is known of their behavior or natural history.

The *Praeamastus albipuncta* caterpillar has a dull brown head with a very fine, reticulated pattern. The body ground color is pale olive green, with several large, pale yellow patches on the thorax and ventrolaterally. The first to seventh abdominal segments have several dense tufts of dull orange setae, laterally and dorsally. The tenth abdominal segment bears the longest tuft, projecting rearward and tipped with black.

Actual size

FAMILY	Erebidae
DISTRIBUTION	Southern Canada, United States, and northern Mexico
HABITAT	Forests, temperate rain forests, grasslands, and meadows
HOST PLANTS	Various, including grasses, dandelion (*Taraxacum* spp.), nettle (*Urtica* spp.), plantain (*Plantago* spp.), and dock (*Rumex* spp.)
NOTE	Caterpillar that is equipped with "antifreeze" to survive cold winters
CONSERVATION STATUS	Not evaluated

ADULT WINGSPAN
1¾–2⅛ in (45–55 mm)

CATERPILLAR LENGTH
1⁹⁄₁₆–2 in (40–50 mm)

PYRRHARCTIA ISABELLA

BANDED WOOLLY BEAR

(J. E. SMITH, 1797)

577

Banded Woolly Bear caterpillars hatch from eggs laid in large clusters on bark in the fall and then overwinter. In the northern parts of their range, the larvae can survive freezing temperatures in winter due to the presence of "antifreeze" chemicals in their cells. In spring, they become active and start feeding. The adult moths are on the wing in summer.

The *Pyrrharctia isabella* caterpillar is known as a woolly bear because of its hairy appearance. The hairs are impressive but are not irritants, although some people may contract a rash when handling them. The caterpillar's defense is to curl up when disturbed. Folklore has it that the width of the orange band in the middle section of the caterpillar is an indicator of the severity of the coming winter. In fact, it is an indicator of age; the bigger and therefore older caterpillars have a narrower orange band. The species is also known as the Isabella Tiger Moth.

Actual size

The Banded Woolly Bear caterpillar is orange red in the center, with a black head, thorax, and posterior segments. It is covered in dense tufts of hairs.

FAMILY	Erebidae
DISTRIBUTION	Eastern Ecuador, but also likely into northern Peru and possibly southern Colombia
HABITAT	Forest edges, light gaps, and revegetating landslides
HOST PLANTS	Many, especially the evergreen bamboo *Chusquea scandens*; also *Piper baeʒanum* and *P. augustum* Rudge, *Miriocarpa* spp., *Boehmeria bullata*, and *Casearia* spp.
NOTE	Caterpillar that plucks and weaves its setae into a cocoon
CONSERVATION STATUS	Not evaluated, but not considered threatened

ADULT WINGSPAN
¾–⅞ in (19–22 mm)

CATERPILLAR LENGTH
¾–1 in (20–25 mm)

578

SAURITA MOSCA
(DOGNIN, 1897)

The highly polyphagous *Saurita mosca* caterpillars are found singly on their many and varied hosts but most commonly on bamboo, which is ubiquitous within their montane habitats. Through most of their life cycle, the larvae are relatively devoid of setae but grow dense tufts in the final instar. Prior to pupation, they pluck these tufts of setae and spin them into a thin, dome-shaped cocoon, within which they pupate. This barrier of setae and silk is presumably an effective defense against predators such as ants.

The caterpillars are heavily parasitized by braconid and ichneumonid wasps, as well as by tachinid flies. Most of these parasitoids emerge from late instars and pupate away from the host, except for some of the ichneumonids, which pupate inside the *Saurita mosca* larva, and one tachinid species that emerges from the pupa. The adult moths, which fly by both day and night, are convincing wasp mimics, although little is known of their general behavior.

The *Saurita mosca* caterpillar is dull, dark gray, appearing greenish due to the leafy contents of the gut showing through. It bears indistinct white highlights laterally. The head is shiny black laterally but pale whitish around the ecdysial suture, dividing the black into two hemispheres. The lateral tufts of short, dull orange setae on the first and seventh abdominal segments are so densely packed that they appear to be solid projections.

Actual size

FAMILY	Erebidae
DISTRIBUTION	North America, from southern Canada south to Mexico
HABITAT	Woodlands, scrub, wastelands, parks, and gardens
HOST PLANTS	More than 100 low-growing plants, trees, and shrubs, including Pinto Bean (*Phaseolus vulgaris*), Soybean (*Glycine max*), and Corn (*Zea mays*)
NOTE	Caterpillar covered in hairs that can irritate
CONSERVATION STATUS	Not evaluated

ADULT WINGSPAN
1¼–2¹⁄₁₆ in (32–53 mm)

CATERPILLAR LENGTH
2⅜ in (60 mm)

SPILOSOMA VIRGINICA
YELLOW WOOLLY BEAR
(FABRICIUS, 1798)

579

Yellow Woolly Bear caterpillars hatch from yellow eggs laid in large clusters of up to 100 on the underside of leaves. The young caterpillars are gregarious but disperse as they get older. The larvae are active from May to November. There are two generations over the summer months, with the pupae of the second generation overwintering on the ground in the leaf litter. The cocoon consists of silk threads and brown hairs, so it is well disguised among the dead leaves.

The common name relates to the caterpillar's covering of dense tufts of long, soft, orange-brown, irritating hairs, which can cause a skin rash if the larvae are handled without gloves. This species has a uniform color unlike the related Banded Woolly Bear (*Pyrrharctia isabella*), which, as its name suggests, has bands of black and orange. The *Spilosoma virginica* adult is better known as the Virginia Tiger Moth.

The Yellow Woolly Bear caterpillar is covered in tufts of hairs, which vary in length. They are usually orange brown with some variation from yellow to red brown, but, whichever the color, it is uniform throughout. Beneath the hairs, the body is yellow with a dark lateral line and spiracles ringed in white.

Actual size

FAMILY	Erebidae
DISTRIBUTION	Poorly known, with records only from western Colombia and eastern Ecuador
HABITAT	Cloud forests and secondary growth of montane forests at around 7,200 ft (2,200 m) elevation
HOST PLANTS	*Gunnera* spp., *Miconia* spp., and *Tibouchina lepidota*
NOTE	Caterpillar that in most instars feeds in large groups
CONSERVATION STATUS	Not evaluated, but not considered threatened

ADULT WINGSPAN
1⅝–1⅞ in (41–47 mm)

CATERPILLAR LENGTH
2–2⁹⁄₁₆ in (50–65 mm)

580

SYMPHLEBIA PALMERI
SYMPHLEBIA PALMERI
(ROTHSCHILD, 1910)

Symphlebia palmeri caterpillars feed and rest in groups during most of their life, apparently separating only late in the final instar or just prior to pupation. Pupation has not been observed in nature but seems likely to occur away from the host plant, as caterpillars held in bags in the laboratory often exhibit a day or more of wandering behavior prior to pupation. Their setae, unlike those of many other erebid species, appear to be only mildly urticating. When the caterpillar is in motion, or stretched out to feed, the otherwise uniform coating of setae across the back will appear broken into segmental rings or individual tufts, giving it an orange-on-black, potentially aposematic look.

The adults are common visitors to lights at night and, despite their striking coloration, are remarkably cryptic when at rest, with wings folded over their back, making them appear much like a dead, damaged, or moldy leaf. The patchy nature of the known range of the species suggests that it may go overlooked in many areas or that more than one species may be involved.

Actual size

The *Symphlebia palmeri* caterpillar is robust, with a shiny, black head and a velvety, dark purple to black body. The dorsum bears a few indistinct white lines crossing the body near intersegmental areas, and the spiracles are bright white. The dorsum is also covered with many dense tufts of short, soft, deep orange setae, creating a nearly uniform covering.

FAMILY	Erebidae
DISTRIBUTION	Southeastern United States, from the coastal regions of Georgia into Florida, the Caribbean, Central America, and northern South America
HABITAT	Cultivated landscapes, pinelands, and coastal areas
HOST PLANTS	Oleander (*Nerium oleander*), Devil's-potato (*Echites umbellata*), Rubber Vine (*Cryptostegia grandiflora*), and Desert Rose (*Adenium obesum*)
NOTE	Caterpillar that is gregarious in early instars and during pupation
CONSERVATION STATUS	Not evaluated, but common

ADULT WINGSPAN
1 $^{11}/_{16}$ in (43 mm)

CATERPILLAR LENGTH
1 $^{9}/_{16}$ in (40 mm)

SYNTOMEIDA EPILAIS
POLKA-DOT WASP MOTH
(WALKER, 1854)

581

Polka-dot Wasp Moth caterpillars hatch from pale, spherical eggs laid in clusters of 12 to 75, usually on the underside of Oleander leaves. Their native host plants are Devil's-potato or Rubber Vine, which are now relatively local and rare; Oleander, introduced from the Mediterranean, has become the major food source, and the larvae are now considered a pest on this ornamental plant. Early instars feed gregariously and can completely defoliate their host plants. Later stages feed solitarily, but pupation can occur in groups; the pupae are encased in thin cocoons into which hairs from the caterpillars are woven.

The larvae may be protected from bird predation by their setae, aposematic coloring, and toxic chemicals. Predatory stink bugs, parasitic tachinid flies, wasps, and fire ants are their main enemies. The Polka-dot Wasp Moth breeds year-round in south Florida and the Caribbean but is killed by cold spells in the northern part of its range, which it recolonizes the following spring. The species belongs to a fascinating wasp-mimicking group within the erebid family, which includes many moths that are diurnal, brightly colored, and sometimes practically impossible to tell apart from real wasps.

The Polka-dot Wasp Moth caterpillar is orange with clumps of black hairs arising from black tubercles. The hairs on the first thoracic and last abdominal segments are longer than those in the middle of the body. The legs and prolegs are black, and the head is orange.

Actual size

FAMILY	Erebidae
DISTRIBUTION	Western North America, Europe, Middle East, and area of central Asia, east to northern China; also New Zealand
HABITAT	Scrub, grasslands, and wastelands
HOST PLANTS	Ragwort (*Senecio jacobaea*) and Groundsel (*Senecio vulgaris*)
NOTE	Striking caterpillar that gains its toxicity from its host plant
CONSERVATION STATUS	Not evaluated, but quite common

ADULT WINGSPAN
1⅜–1¼ in (35–45 mm)

CATERPILLAR LENGTH
1³⁄₁₆ in (30 mm)

TYRIA JACOBAEAE
CINNABAR
(LINNAEUS, 1758)

582

Cinnabar moth caterpillars hatch from eggs laid in clutches of up to 40 on the underside of leaves. The caterpillars are yellow at first but soon gain their characteristic stripes. The larvae are gregarious and quickly defoliate food plants, hence their value as a control agent for Ragwort, which is classed as a noxious weed. For this reason, the species has been introduced to New Zealand and Tasmania. When fully developed, the caterpillars pupate on the ground, and the pupae overwinter in a sparse cocoon within the leaf litter. The red, day-flying adult moths are on the wing from May to July.

The caterpillar is one of the world's most poisonous larvae, gaining its toxicity from the host plant, Ragwort, which is rich in alkaloids. The larva is unaffected by the poison it accumulates but is noxious to any predators. The poison passes from caterpillar to adult. The Cinnabar moth is named for the adult's bright color; cinnabar, a scarlet to brick-red mineral, was once used as an artist's pigment.

The Cinnabar caterpillar is brightly colored with bands of orange and black. There are sparsely scattered long, white hairs and shorter, dark hairs. The head, legs, and prolegs are black.

Actual size

FAMILY	Erebidae
DISTRIBUTION	North and South America, from Nova Scotia south to Argentina, including the Caribbean
HABITAT	Meadows and forest edges, in close association with host plants
HOST PLANTS	Rattlebox plants in the genus *Crotalaria* (Fabaceae family)
NOTE	Brightly striped caterpillar whose colors advertise its unpleasant taste
CONSERVATION STATUS	Not evaluated, but common

ADULT WINGSPAN
1³⁄₁₆–1¾ in (30–45 mm)

CATERPILLAR LENGTH
1³⁄₁₆–1⁹⁄₁₆ in (30–40 mm)

UTETHEISA ORNATRIX
ORNATE BELLA MOTH
(LINNAEUS, 1758)

583

Initially, Ornate Bella Moth caterpillars feed in groups on the underside of leaves but are solitary during later instars as they seek out and penetrate the seedpods of rattlebox plants in order to consume nutritious, alkaloid-rich seeds. From these they derive not only nutrients but also toxic chemicals called pyrrolizidine alkaloids, which are passed on to the adult and help protect against predators at all stages of development. Growing rapidly, the larvae reach the final instar in three weeks and spin spiderweb-like silk, with which to enclose the pupae in a loose layer.

For its intriguing handling of the toxic alkaloids, which are also transformed by males into pheromones, the species has served as a model for research into chemical ecology. It is closely related to about 40 members of the genus *Utetheisa*, which form several distinct subgenera and are found mostly in the tropics. The adult is a colorful, day-flying moth.

The Ornate Bella Moth caterpillar is aposematically colored in orange-and-black stripes, with a red head—a pattern memorable to potential predators. The width of the orange stripes is variable, with some individuals more orange-colored, some speckled with white, while others are almost entirely black. The primary setae are black, with longer white setae at the front and back.

Actual size

FAMILY	Erebidae
DISTRIBUTION	North America, from Ontario to Nova Scotia, south to Florida, and west to Texas
HABITAT	Barrens, pocosins, woodlands, and forests
HOST PLANTS	White and evergreen oak (*Quercus* spp.), blueberry (*Vaccinium* spp.), and spruce (*Picea* spp.)
NOTE	Caterpillar that, if disturbed, hurls itself from the host plant
CONSERVATION STATUS	Not evaluated, but uncommon

ADULT WINGSPAN
1⅜–1¹¹⁄₁₆ in (35–43 mm)

CATERPILLAR LENGTH
2 in (50 mm)

584

ZALE AERUGINOSA
GREEN-DUSTED ZALE MOTH
(GUENÉE, 1852)

Green-dusted Zale Moth caterpillars are very active in early instars and may wander some distance before feeding, usually on young leaves. In later instars, older leaves and spruce needles are consumed. The nontoxic plant diet of these caterpillars makes them prey for birds and other natural enemies. While they are cryptically colored and rest pressed against the host plant stem, when disturbed they may "jump" off their perch, which offers some additional protection against predators and parasitoids. The pupae overwinter in leaf litter. There are two generations a year in the south of the range but a single generation in the north. Adults fly from January to October.

Caterpillars of the genus *Zale* resemble those of underwing moths in the genus *Catocala*. However, *Zale* adults are less spectacular than *Catocala* underwing moths and more subdued in color. The species' name, *aeruginosa*, derives from the Latin for "copper rust," reflecting the bluish-green colors sprinkled on the moth's black forewings. The most recent revision of the genus *Zale* lists 39 species in North America north of Mexico, most of which feed on large trees.

The Green-dusted Zale moth caterpillar has a long, slender body, tapering anteriorly and posteriorly. The front pairs of prolegs are small, and the last prolegs extend behind the body. The caterpillar's washed-out colors of brown, beige, whitish, and cream blend to resemble tree bark. There is a vague, dark brown, spiracular stripe. The ventral surface is whitish, and the head is white with a pattern of fine, brown dots and stripes. The eighth abdominal segment is squared and bears skin flaps that look like leaf scars. A few primary setae are thin, short, and hardly visible, except for their white bases.

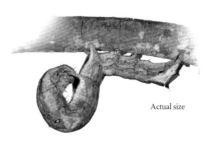

Actual size

FAMILY	Euteliidae
DISTRIBUTION	Northern Thailand, southeast China, Chinese Taipei, and Japan
HABITAT	Forests and gardens
HOST PLANTS	*Spondias* spp.
NOTE	Candy-striped caterpillar that has a distinctive swollen thorax
CONSERVATION STATUS	Not evaluated, and relatively unknown, although not uncommon regionally

ADULT WINGSPAN
1³⁄₁₆–1³⁄₈ in (30–35 mm)

CATERPILLAR LENGTH
1³⁄₈ in (35 mm)

PHALGA CLARIRENA

PHALGA CLARIRENA
(SUGI, 1982)

585

Encountering the *Phalga clarirena* caterpillar for the first time, people often describe it as having a tumor or as looking parasitized because of its swollen front end. The larva's basic body shape is, however, characteristic of several genera of euteliid moths from Asia. Nothing about the appearance or behavior of the immature stages of *P. clarirena* suggests it uses crypsis as a defense, but its yellow and orange colors are classically aposematic in nature as predation deterrents. Several larvae can between them defoliate host plant branches. The caterpillars turn a deep blue prior to pupation and head down to the soil to metamorphose.

There are two to three generations during the summer months, and, in the species' less tropical range, overwintering occurs as the adult moth. Unlike the caterpillar in their markings and behavior, the adults are masters of camouflage and will hang from vegetation and spiderwebs, looking very much like windblown, dry leaves.

Actual size

The *Phalga clarirena* caterpillar has a characteristic euteliid body shape and color scheme, with a bulbous thorax boldly marked with black tiger stripes on a pale blue background. The blue continues down its topside to the rear, where there is a deep blue spot on the anal segment. It has bright yellow flanks. The head is a vivid orange but usually tucked away out of sight when the larva is not feeding.

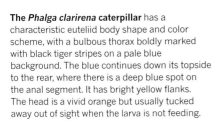

FAMILY	Nolidae
DISTRIBUTION	Across Europe into western Asia and northern Iran
HABITAT	Woodlands and parks
HOST PLANTS	Oak (*Quercus* spp.)
NOTE	Bright green caterpillar that hides among leaves
CONSERVATION STATUS	Not assessed, but locally common

ADULT WINGSPAN
1⁹⁄₁₆–1¹⁵⁄₁₆ in (40–50 mm)

CATERPILLAR LENGTH
⁹⁄₁₆ in (15 mm)

586

BENA PRASINANA
SCARCE SILVER-LINES
(LINNAEUS, 1758)

Actual size

Scarce Silver-lines caterpillars hatch from eggs laid on the leaves of the host plant. The young larvae are appropriately red brown in color, helping to conceal them as they overwinter on twigs, becoming active again in spring. The older caterpillars are green and found on the underside of leaves in May and June. The mature larvae pupate on twigs and the underside of leaves within a silken cocoon.

The adult night-flying moths, all produced in a single generation, are seen on the wing from late May to August, when they are attracted to lights. The species is rarely seen in open spaces, and populations tend to be largest in ancient oak woodland and other woodlands where mature oak trees are found. Numbers of Scarce Silver-lines moths have declined due to the loss of oak woodland across Europe, fueled by a preference for planting faster-growing trees. The genus *Bena* comprises eight species and belongs to the tuft moths of the family Nolidae.

The Scarce Silver-lines caterpillar is plump in shape, tapering to the end of the abdomen. It is bright green, with scattered white dots and sparse, fine hairs. There are two pale yellow dorsal lines and a series of pale oblique lines on the sides. The head and legs are much paler in color.

FAMILY	Nolidae
DISTRIBUTION	The Himalayas, South and Southeast Asia, southern China, and Chinese Taipei
HABITAT	Lowland and montane forests
HOST PLANTS	Myrtaceae, including *Eugenia* spp., Rose Myrtle (*Rhodomyrtus tomentosa*), *Campomanesia xanthocarpa*, *Cleistocalyx operculatus*, and Java Plum (*Syzygium cumini*)
NOTE	Caterpillar that mimics unpalatable berries
CONSERVATION STATUS	Not evaluated, but common

ADULT WINGSPAN
1%16–1¼ in (40–45 mm)

CATERPILLAR LENGTH
1 in (25 mm)

CAREA VARIPES

CAREA VARIPES
WALKER, 1856

587

Carea varipes caterpillars, like many species of the Careini tribe of Oriental nolid moths, have a grossly swollen thorax, a feature that is present at all instar stages but exaggerated in the final instar. This is thought to be designed to mimic unpalatable berries as a defense against predation, notably by birds. The thorax can appear full and glossy or deflated and wrinkled, perhaps indicating larval nutritional status. When disturbed, the caterpillars assume a raised tail and head posture and may regurgitate their gut contents as a droplet of green liquid.

Unlike other related larvae, which occur in large, defoliating masses, *Carea varipes* caterpillars occur singly and at widely separate locations, usually resting on the top side of host leaves. The larvae pupate in rolled host plant leaves or sandwiched between leaves. The adult moths eclose after nine days and are on the wing throughout the spring and summer months.

Actual size

The *Carea varipes* caterpillar is notable for its enlarged, spherical thoracic segments. The caterpillar is colored in varied shades of glossy green with sparsely spaced primary setae. A mottled white band runs from the thorax to a broad, paler green, conical horn at the rear. There is a similar white patch between the horn and the claspers. The prolegs and their crochets are prominent. The head is red.

FAMILY	Nolidae
DISTRIBUTION	Southern Europe, central Europe, southern Russia, southern Scandanavia, and Turkey
HABITAT	Warm and wet riparian forests
HOST PLANTS	White Poplar (*Populus alba*)
NOTE	Pale gray caterpillar that is covered in fine hairs
CONSERVATION STATUS	Not evaluated

ADULT WINGSPAN
¾–1 in (20–25 mm)

CATERPILLAR LENGTH
⁹⁄₁₆–¾ in (15–20 mm)

EARIAS VERNANA
SILVER POPLAR SPINNER
(FABRICIUS, 1787)

The caterpillars of the Silver Poplar Spinner moth hatch from eggs that are laid on the leaves of the host plant. The larvae feed on leaves at the end of branches, where they live within a leaf shelter secured by silk. The mature caterpillar spins a pale brown cocoon on a stem or twig in which to pupate and overwinter.

The night-flying moths are on the wing from April to August, and there are usually two generations a year, from April to June and July to August. In the south of the range, there may even be a third generation, while in the more northerly regions a single generation is more usual. The species is less common in the north, where it is increasingly rare due to the loss of riparian forests. *Earias vernana* is often mistaken for *E. clorana*, which is found in similar habitats.

Actual size

The Silver Poplar Spinner caterpillar has a plump, sluglike body, which is covered in long, pale hairs. It is overall a pale gray color with darker gray patches and dark spiracles. There is a faint dorsal line.

FAMILY	Nolidae
DISTRIBUTION	Europe, North Africa, the Middle East, and across northern Asia to Russian Far East; also Canada (British Columbia)
HABITAT	Woodlands, hedgerows, and gardens
HOST PLANTS	Rosaceae family, including *Cotoneaster* spp., *Malus* spp., *Prunus* spp., and *Sorbus* spp.
NOTE	Small, hairy caterpillar that is found in woodlands and hedgerows
CONSERVATION STATUS	Not evaluated, but not considered to be at risk

ADULT WINGSPAN
⁹⁄₁₆–¾ in (15–20 mm)

CATERPILLAR LENGTH
¾ in (20 mm)

NOLA CUCULLATELLA
SHORT-CLOAKED MOTH
(LINNAEUS, 1758)

589

The caterpillars of the Short-cloaked Moth hatch from ribbed, round eggs laid on leaves of the host plant. The young larvae feed briefly after hatching, before crawling into small cracks in the bark, where they spend the winter. They become active again in late spring. The cocoon, which is found attached to twigs, is made from silk, hairs, and pieces of wood, giving it a pale brown coloration.

The moths—nocturnal and attracted to light—are on the wing in June and July, and there is a single generation. *Nola cucullatella* is widespread across Europe and northern Asia, and recently it has been discovered near the port of Vancouver in British Columbia, Canada, where it most likely arrived in cargo. The species gets its common name from the dark, basal coloring on the wings, which, when the wings are folded at rest, creates the appearance of a short cloak. The species name comes from the Latin word *cuculla*, meaning "hood."

Actual size

The Short-cloaked Moth caterpillar is small and hairy, chestnut brown in color, and with a dorsal white stripe, which is broken on the fourth segment. There is a ring of tubercles on each segment, bearing tufts of long, white hairs. The head is dark brown.

FAMILY	Nolidae
DISTRIBUTION	Across Europe into western Russia; also Japan
HABITAT	Woodlands, scrub, parks, and gardens
HOST PLANTS	Various deciduous trees, including alder (*Alnus* spp.), birch (*Betula* spp.), and oak (*Quercus* spp.)
NOTE	Plump, green caterpillar that pupates in a boat-shaped cocoon
CONSERVATION STATUS	Not evaluated

ADULT WINGSPAN
1³⁄₁₆–1¾ in (30–45 mm)

CATERPILLAR LENGTH
1⅜ in (35 mm)

590

PSEUDOIPS PRASINANUS
GREEN SILVER-LINES
(LINNAEUS, 1758)

The caterpillars of the Green Silver-lines moth emerge from reddish-brown, flattened eggs laid in batches of up to 250 on the underside of leaves. The larvae rest on the leaf underside during the day and become active at night. They crawl down the tree to pupate in a crevice in the bark or in the leaf litter, spinning a brown, boat-shaped cocoon. The pupa overwinters and ecloses the following spring.

The adult moths are bright green in color and are on the wing in June and July. There is usually a single generation, but occasionally a second generation appears in late summer. The moths rest on tree trunks during the day, fly at night, and are attracted to light. The common name derives from the distinctive silvery, diagonal lines on the green forewings. The equally green caterpillars can be seen from mid-July to as late as October.

Actual size

The Green Silver-lines caterpillar is short and plump. It is green, with a dorsal band of yellow-green cross shapes and dots, bounded on both sides by a yellow line. There are similar lateral markings. The end of the abdomen has a forked clasper edged in red. The head is dark green, bounded by a yellow band.

FAMILY	Noctuidae
DISTRIBUTION	Europe into central Asia, North Africa
HABITAT	Woodlands, gardens, and parks
HOST PLANTS	Trees, such as Sycamore (*Acer pseudoplatanus*), Horse Chestnut (*Aesculus hippocastanum*), lime (*Tilia* spp.), poplar (*Populus* spp.), and oak (*Quercus* spp.)
NOTE	Easily recognized caterpillar resembling a tuft from a yellow carpet
CONSERVATION STATUS	Not evaluated, but regionally endangered

ADULT WINGSPAN
1⅜–1¾ in (35–45 mm)

CATERPILLAR LENGTH
1⁹⁄₁₆ in (40 mm)

ACRONICTA ACERIS
SYCAMORE
LINNAEUS, 1758

591

Caterpillars of the Sycamore moth hatch from checkered, black-and-white eggs. The young larvae are covered in long hairs, which are initially ginger but become more yellow with age. The dense tufts of hairs—a distinctive feature—can cause mild irritation, deterring would-be predators. Although brightly colored, the caterpillar is not poisonous. When disturbed, the caterpillar curls into a distinctive U shape.

The pupae overwinter under bark, especially the bark of old poplar trees. There is a single generation each year, with the adult moths on the wing from late spring to midsummer. The Sycamore is under threat from the loss of its woodland habitat. In particular, the felling of old hybrid poplar trees that were once common along roads and on farmland has affected numbers. The genus *Acronicta* contains around 150 species, most of which have conspicuous, brightly colored, hairy caterpillars.

The Sycamore caterpillar is covered in dense tufts of long, yellow hairs with four pairs of red-orange tufts. There is a distinctive dorsal line of white spots encircled in black.

Actual size

FAMILY	Noctuidae
DISTRIBUTION	North America, east of the Rockies
HABITAT	Deciduous woodlands and forests
HOST PLANTS	Woody plants, including maple (*Acer* spp.), alder (*Alnus* spp.), birch (*Betula* spp.), elm (*Ulmus* spp.), willow (*Salix* spp.), and ash (*Fraxinus* spp.)
NOTE	Caterpillar that can cause a severe rash when roughly handled
CONSERVATION STATUS	Not evaluated, but usually common

ADULT WINGSPAN
2–2⁹⁄₁₆ in (50–65 mm)

CATERPILLAR LENGTH
2–2⅛ in (50–55 mm)

592

ACRONICTA AMERICANA
AMERICAN DAGGER
HARRIS, 1841

American Dagger caterpillars hatch from green eggs laid singly by the female moth and live solitary lives feeding on the leaves of their host trees. Early instar caterpillars graze leaves, skeletonizing irregular patches of lower leaf tissue, while later instars eat entire leaves. The larvae rest with their head curled to one side. Prepupal caterpillars wander and excavate a pupation crypt in soft wood or bark. A tough cocoon is constructed within the excavation, in which pupation occurs. There is a single generation of caterpillars in the north of the range, where the pupae are most likely to overwinter, but two or three in the south. Although harmless to the touch, roughly handled American Dagger caterpillars can produce a severe rash.

The larvae appear between July and October, while the adult moths fly between April and September. The caterpillars are frequently attacked by parasitic wasps, which deposit numerous eggs in each larva. After consuming the caterpillar's insides, the wasp larvae pupate in the cadaver, emerging later as adult wasps.

The American Dagger caterpillar is large and densely covered with long, white or pale yellow setae from which emerge black, diverging tufts of black hair on the first and third abdominal segments. A single long, black hair tuft is present on the eighth abdominal segment.

Actual size

FAMILY	Noctuidae
DISTRIBUTION	Europe, excluding United Kingdom, into Asia, east to western Siberia
HABITAT	Scrubby grasslands, clearings in woodlands, and dry forest fringes
HOST PLANTS	Various, including Heather (*Calluna vulgaris*), *Rubus* spp., oak (*Quercus* spp.), and *Vaccinium* spp.
NOTE	Black caterpillar that has tufts of brightly colored hairs
CONSERVATION STATUS	Not evaluated, but locally rare

ADULT WINGSPAN
1⁷⁄₁₆–1⅝ in (36–42 mm)

CATERPILLAR LENGTH
1⁹⁄₁₆ in (40 mm)

ACRONICTA AURICOMA
SCARCE DAGGER
(DENNIS & SCHIFFERMÜLLER, 1775)

593

The Scarce Dagger is a distinctive caterpillar that is seen in a wide range of wooded habitats over spring and summer, although populations are declining in many areas as a result of increased urbanization. There are two generations a year, with the adult moths on the wing from April to June and then again in July to August. The pupae of the second generation overwinter, either on the host plant or in the leaf litter below it.

Like most caterpillars within the genus *Acronicta*, the Scarce Dagger has brightly colored hairs that deter predators. In some parts of its range, the caterpillar is considered a pest as it feeds on the leaves of economically valuable plants such as bilberries. The adult, a former resident but now rare immigrant in the United Kingdom, is one of a number of so-called "dagger moths" within *Acronicta*, named for the black, dagger-shaped markings on the upper side of the moth's forewing.

The Scarce Dagger caterpillar has a black head and dark brown to black body. Each segment has a ring of prominent tubercles, all of which bear tufts of long hairs; some are black, while others are orange in color.

Actual size

FAMILY	Noctuidae
DISTRIBUTION	From Morocco and northern Spain through central and eastern Europe (excluding the Mediterranean basin), Asia Minor, and the Caucasus; southern Scandinavia east across Russia and southern Siberia to China, Korea, Japan, and Russian Far East
HABITAT	Damp woodlands, marshes, bogs, and other damp habitats
HOST PLANTS	Alder (*Alnus* spp.) and birch (*Betula* spp.); also Rowan (*Sorbus aucuparia*)
NOTE	Caterpillar of damp places that has a broad, yellow stripe
CONSERVATION STATUS	Not evaluated, but local in western Europe due to habitat loss from drainage

ADULT WINGSPAN
1¹¹⁄₁₆ in (43 mm)

CATERPILLAR LENGTH
1⁹⁄₁₆–1⁵⁄₈ in (40–42 mm)

ACRONICTA CUSPIS
LARGE DAGGER
(HÜBNER, [1813])

594

Large Dagger caterpillars hatch from eggs laid singly on the underside of host plant leaves. The larvae feed openly on the leaves as they develop. In their final instar, when their distinctive yellow dorsal stripe fades to white, the caterpillars descend to pupate in a cocoon formed within rotten wood or under bark. There are one or two generations annually from June to October. The daggerlike markings on the forewings of adults give this and other *Acronicta* species their common name.

At early instars, the Large Dagger and the Grey Dagger (*Acronicta psi*), which shares a similar geographic range, are quite difficult to tell apart. When more mature, both have a broad, yellow stripe along the back, but the Large Dagger is easily distinguished by the much smaller protuberance on the fourth segment and the distinct, long pencil of black, white-tipped hairs extending from it. The Large Dagger is also more restricted by food plant and habitat than related species.

Actual size

The Large Dagger caterpillar is gray and moderately hairy, with a broad, yellow dorsal stripe that turns white before pupation. Laterally, it has orange-red marks and blocks of irregular, black spots. There is also a broad, white stripe low down. On the fourth segment, a low, blackish hump arises, with a long pencil of distinctly black, longer, white-tipped hairs. Near the tail end is a gray hump.

FAMILY	Noctuidae
DISTRIBUTION	Europe, excluding the far north of Scandinavia; Asia Minor, the Caucasus, Syria, northern Iran, and western Russia
HABITAT	Woodlands, hedgerows, and poplars growing in open country
HOST PLANTS	Poplar (*Populus* spp.), including Aspen (*P. tremula*) and Black Poplar (*P. nigra*); occasionally willow (*Salix* spp.)
NOTE	Species named for the large head of its caterpillar
CONSERVATION STATUS	Not evaluated

ADULT WINGSPAN
1½–1¾ in (38–44 mm)

CATERPILLAR LENGTH
1¼–1⅜ in (32–35 mm)

ACRONICTA MEGACEPHALA
POPLAR GREY
([DENIS & SCHIFFERMÜLLER], 1775)

595

The Poplar Grey caterpillar hatches from a flattened, rounded, partly translucent egg laid singly on the leaves. It feeds openly on the leaves, resting on or under a leaf, with the front end curled around almost facing the tail end in a characteristic posture that resembles a bird dropping. The caterpillars can be found in one or two generations from June to September. The pupa is formed in a strong cocoon on the tree under loose bark, in a crevice or rotten wood, or in the ground, and this stage overwinters.

With its large head and long, shaggy fringe of whitish hairs, the Poplar Grey caterpillar is unlikely to be confused with any other larva feeding on poplars and willows. The specific name of the Poplar Grey, *megacephala*, is derived from the ancient Greek words *megas*, meaning "large" or "mighty," and *cephala*, meaning "head."

Actual size

The Poplar Grey caterpillar has a slightly flattened body and is brown, gray, or greenish, with a large, boldly striped head. Tufts of long, whitish hairs extend out all around the sides and from the head. The back has sparse, shorter, blackish hairs, dark wavy bands, small, brown warts, a fine peppering of white dots, and a large, creamy-white, mask-like blotch on the tenth segment.

FAMILY	Noctuidae
DISTRIBUTION	Europe, including southern and eastern Scandinavia, east across Russia and central Asia to Magadan, Mongolia, China, and Korea
HABITAT	Woodlands, scrub, and often parklands and gardens
HOST PLANTS	Wide variety of broadleaved trees and bushes, including willow (*Salix* spp.), lime (*Tilia* spp.), oak (*Quercus* spp.), hawthorn (*Crataegus* spp.), and rose (*Rosa* spp.)
NOTE	Hairy caterpillar that has a yellow dorsal stripe and hump
CONSERVATION STATUS	Not evaluated

ADULT WINGSPAN
1⁵⁄₁₆–1¾ in (33–45 mm)

CATERPILLAR LENGTH
1⁷⁄₁₆–1⁹⁄₁₆ in (36–40 mm)

ACRONICTA PSI
GREY DAGGER
(LINNAEUS, 1758)

596

The Grey Dagger caterpillar hatches from a domed, whitish egg laid singly on a leaf. It feeds and rests openly on the food plant, in one or two generations from June to October. When the larva is fully fed, the yellow stripe along its back turns white and it is then often seen descending the tree to seek out a cavity under loose bark, rotten wood, or the ground, in which it forms a glossy brown pupa in a tough cocoon. This stage overwinters, with adults eclosing the following summer.

Few noctuid caterpillars are noticeably hairy, but the Acronictinae (dagger moths, named for the forewing markings) are an exception. The adult Grey Dagger and Dark Dagger (*Acronicta tridens*) are almost identical, but the caterpillars are easy to tell apart. The Dark Dagger has a much shorter protuberance on the fourth segment, and the dorsal stripe is orange-red and yellow (or white and yellow) with a fine dark central line; it also has some white spots.

Actual size

The Grey Dagger caterpillar has sparse, long hairs on the back and shorter, denser ones on the sides. It is gray in color, with a broad, clear, yellow stripe on the back. Laterally, there are vertical, orange-red marks, blocks of irregular, black spots, and a broad, white or sometimes yellowish stripe low down. It has a long, fleshy, blackish protuberance on the fourth segment and a gray hump near the tail end.

FAMILY	Noctuidae
DISTRIBUTION	From eastern Spain, western France, the British Isles, and southern Scandinavia, east to western Turkey and western Russia
HABITAT	Wide variety, including woodlands, moorlands, marshes, hedgerows, rough ground, and gardens
HOST PLANTS	Many herbaceous plants, including Common Sorrel (*Rumex acetosa*), Meadowsweet (*Filipendula ulmaria*), and grasses (*Poaceae*); also broadleaved trees, including willow (*Salix* spp.) and hawthorn (*Crataegus* spp.)
NOTE	Caterpillar that lies dormant in its cocoon before pupating
CONSERVATION STATUS	Not evaluated, but widespread

ADULT WINGSPAN
1³⁄₁₆–1⁹⁄₁₆ in (30–40 mm)

CATERPILLAR LENGTH
1⅜–1⁹⁄₁₆ in (35–40 mm)

AGROCHOLA LITURA
BROWN-SPOT PINION
(LINNAEUS, 1761)

597

Brown-spot Pinion caterpillars hatch from brownish eggs. These are laid in small batches, probably on dead vegetation, twigs, or bark, and they overwinter before the larvae hatch. At first the caterpillar consumes the foliage of herbaceous plants and grasses and, when larger, may ascend to feed on bushes and trees. A single, annual generation develops from April to June. When fully fed, the caterpillar forms a cocoon in the ground, in which it lies dormant for several weeks before pupating. The moths fly from late August to November.

The Brown-spot Pinion caterpillar feeds mainly at night when larger and so is not often seen. It closely resembles the equally widespread Beaded Chestnut (*Agrochola lychnidis*), with similar color variation. In its brown form, the Brown-spot Pinion can be distinguished from the Beaded Chestnut by the blackish spot above and behind each spiracle (absent in the green form). The similar, brown-colored Flounced Chestnut (*A. helvola*) caterpillar has a bolder white stripe along the sides.

The Brown-spot Pinion caterpillar is smooth and light green, yellowish green, brown, or brick red in color, peppered with a subtle pattern of fine white (in the light green form) or brownish mottling. There are three fine, pale lines along the back and a scattering of round white dots. A fairly broad, white, yellow, or white-and-yellow stripe extends along the sides, narrowly edged above with dark brown or green.

Actual size

FAMILY	Noctuidae
DISTRIBUTION	Western Europe, including Iberia and most of the British Isles, and east through central and southern Europe to the Balkans and Crimea
HABITAT	Wide variety, including grasslands, heathlands, woodland farmlands, and gardens
HOST PLANTS	Various herbaceous plants, including dock (*Rumex* spp.), Knot-grass (*Polygonum aviculare*), lettuce (*Lactuca* spp.), and grasses (*Poaceae*)
NOTE	Small, brown cutworm caterpillar, but not a pest species
CONSERVATION STATUS	Not evaluated

ADULT WINGSPAN
1³⁄₁₆–1¼ in (30–32 mm)

CATERPILLAR LENGTH
1³⁄₁₆–1⁵⁄₁₆ in (30–33 mm)

598

AGROTIS PUTA
SHUTTLE-SHAPED DART
(HÜBNER, [1803])

Shuttle-shaped Dart caterpillars hatch from brownish, globular eggs laid in batches on the food plant. They feed at night close to the ground and hide by day, often slightly beneath the surface. There are two or three generations, which may overlap so that, depending on climate, caterpillars are found from May throughout the summer and fall, usually overwintering fully grown in the ground, where the pupa is formed in March or April. The moths fly from April to June, July to August, and, in some places, in the fall, although in lesser numbers.

The *Agrotis puta* caterpillar is typical of larvae of this genus, whose members are also known as cutworms because they eat through the bases of stems, mostly near the ground, and through roots. Some species, although not *A. puta*, are agricultural pests in different parts of the world. *Agrotis* larvae tend to look very similar to one another, and those of the closely related *A. catalaunensis* are almost indistinguishable from the Shuttle-shaped Dart, as are the adults. This has led to some uncertainty over the exact distribution of *A. puta*.

Actual size

The Shuttle-shaped Dart caterpillar is slightly squat and a dirty, mottled brown color. Along the back, there is a rather faint, thin, pale central stripe and two pale, broad, wavy, diffuse stripes, sometimes vaguely forming open V marks. Dark, thinly scattered, short hairs arise from small, black plates in pairs along the back. The body is paler laterally and underneath, with black spiracles.

FAMILY	Noctuidae
DISTRIBUTION	Europe, from east of the Pyrenees and northern Italy, the British Isles, and southern Scandinavia, east to the southern Urals, the Caspian Sea, and western Russia
HABITAT	Woodlands, moorlands, heathlands, scrub, hedgerows, parklands, and gardens
HOST PLANTS	Hawthorn (*Crataegus* spp.), Blackthorn (*Prunus spinosa*), plum (*Prunus* spp.), apple (*Malus* spp.), and, in northern British Isles, Rowan (*Sorbus aucuparia*)
NOTE	Twiggy caterpillar found on plants in the rose family
CONSERVATION STATUS	Not evaluated

ADULT WINGSPAN
1⅜–2 in (35–50 mm)

CATERPILLAR LENGTH
1¾–2 in (45–50 mm)

ALLOPHYES OXYACANTHAE
GREEN-BRINDLED CRESCENT
(LINNAEUS, 1758)

599

The Green-brindled Crescent caterpillar hatches from a ribbed, whitish egg laid on a twig or branch; the eggs overwinter. The caterpillar feeds from April to June, resting along a twig, well camouflaged by its twiglike appearance. When fully fed, it forms a glossy brown pupa in a strong cocoon in the ground. The adults emerge in the fall, in one annual generation. *Allophyes oxyacanthae* is one of several species that look very similar to one another in all life stages. The others, *A. cretica*, *A. asiatica*, *A. alfaroi*, and *A. corsica*, are restricted to warm, dry, scrubby habitats around the Mediterranean, whereas *A. oxyacanthae* occurs more widely and in cooler, damper climates.

The caterpillar of the closely related Double-spot Brocade (*Meganephria bimaculosa*) is similar but much browner. The Green-brindled Crescent caterpillar also bears a general resemblance to certain *Catocala* species (Erebidae), although they have a side-fringe of hairs, mostly grow larger, and rarely feed on plants of the Rosaceae family.

The Green-brindled Crescent caterpillar is rather cylindrical, slightly warty, and twiglike, with a large, mottled brown head. The body is light to dark gray brown, with irregular, fine, subdued brown stripes, and is often heavily variegated with greenish white in habitats with abundant lichen growth. There is an orange-brown or whitish, open, dark-edged V on the fourth segment and a double-pointed hump at the tail end.

Actual size

FAMILY	Noctuidae
DISTRIBUTION	United States, southern Canada, North Africa, Europe, the Middle East, northern India, Russia, Mongolia, northern China, Korea, and Japan
HABITAT	Deciduous woodlands, parks, and gardens
HOST PLANTS	Deciduous trees and shrubs, especially oak (*Quercus* spp.)
NOTE	Green caterpillar that has a pyramid-shaped dorsal hump
CONSERVATION STATUS	Not evaluated, but a relatively common species

ADULT WINGSPAN
1⁹⁄₁₆–2¹⁄₁₆ in (40–52 mm)

CATERPILLAR LENGTH
1⅜–1⅝ in (35–42 mm)

600

AMPHIPYRA PYRAMIDEA
COPPER UNDERWING
(LINNAEUS, 1758)

The Copper Underwing female moth lays her eggs, singly or in small groups, on the bark of host trees or shrubs. The eggs overwinter and hatch in spring, the caterpillars appearing as early as April and as late as June. The larvae feed on leaves of the host plant and then pupate inside a leaf shelter, which they build by rolling a leaf and securing it with threads of silk.

There is a single generation, with the night-flying adults on the wing from August to October. The species name of the Copper Underwing comes from the pyramid-shaped hump at the end of the caterpillar's abdomen, and this feature is reflected in some of the moth's alternative names, such as the Pyramidal Green Fruitworm and the Humped Green Fruitworm. Although called an "underwing," *Amphipyra pyramidea* is not a true underwing moth of the *Catocala* genus as the adult moth lacks banded or all-black hindwings.

The Copper Underwing caterpillar has a distinctive dorsal hump with a yellow point on the eighth abdominal segment. The body is apple green with sparse white dots. There is a lateral yellow-and-white stripe, which is not visible on two of the segments. The spiracles are ringed in black.

Actual size

FAMILY	Noctuidae
DISTRIBUTION	Europe, excluding northernmost Scandinavia; Asia Minor, the Middle East, western central Asia east to China (Xinjiang), and southern Siberia to Lake Baikal; Sakhalin Island (Russian Far East)
HABITAT	Grasslands, from gardens to arable fields, wastelands, woodland rides, coastal dunes, and high moorlands; mainly at higher altitudes in warmer climates
HOST PLANTS	Grasses, including Cock's-foot (*Dactylis glomerata*), fescue (*Festuca* spp.), reed-grass (*Calamagrostis* spp.), and hair-grass (*Deschampsia* spp.)
NOTE	Caterpillar that lives in a chamber in grass tussocks
CONSERVATION STATUS	Not evaluated, but widespread and common

ADULT WINGSPAN
1¾–2⅛ in (45–55 mm)

CATERPILLAR LENGTH
1⁹⁄₁₆–2⅛ in (40–45 mm)

APAMEA MONOGLYPHA
DARK ARCHES
(HUFNAGEL, 1766)

601

The Dark Arches caterpillar hatches from an oval, whitish egg laid in small groups on a grass leaf sheath or seed head that provides its first meal. The larva soon descends the plant to live in a silken chamber among the roots and lower stems, which it feeds on until it is full grown. There is one main generation; the caterpillars overwinter, feeding in mild weather, then leave the chamber in April or May to form a pupa in the ground. The main adult flight is from June to August, with small numbers of a second generation in the fall.

Although it is an abundant species throughout its range and in many habitats, the Dark Arches caterpillar, unless it is disturbed from its home, is seldom seen until it leaves the chamber to pupate, which usually happens at night. Its lifestyle is identical to that of the closely related and almost indistinguishable Light Arches (*Apamea lithoxylaea*) and Reddish Light Arches (*A. sublustris*) caterpillars.

The Dark Arches caterpillar is gray brown, stout, smooth, and shiny. On most segments, there are four hardened, blackish plates in a trapezoidal arrangement on the back and another group on the side. On the first three segments, some plates are elongated and transverse. The head is blackish or brown, and the first and last segments have a large, black plate.

Actual size

FAMILY	Noctuidae
DISTRIBUTION	India, Southeast Asia, Japan, Fiji, New Guinea, and Norfolk Island (Australia)
HABITAT	Woodlands, forests, and urban areas
HOST PLANTS	China Ramie (*Boehmeria nipononivea*) and *Boehmeria australis*
NOTE	Aposematic caterpillar that produces a fruit-piercing moth
CONSERVATION STATUS	Not evaluated, but locally common

ADULT WINGSPAN
3⅜–3½ in (85–90 mm)

CATERPILLAR LENGTH
2⅜–2⁹⁄₁₆ in (60–65 mm)

602

ARCTE COERULEA
RAMIE MOTH
GUENÉE, 1852

Ramie Moth caterpillars hatch and then feed primarily on canopy shoots of the host plant; development on these shoots is better than on understory shoots. The larvae are able to withstand the chemical defenses of the host plant and likely store the poisons they ingest, making them unpalatable to vertebrate predators. This is advertised by the aposematic coloration of the caterpillars. They develop through six instars and pupate in a slight, silken cocoon between leaves drawn together with silk.

Two to three generations of caterpillars of this species appear during spring to fall, with large outbreaks sometimes occurring from August to September, causing severe damage to the ramie host plants, which produce a fiber crop used for fabric production. The caterpillars are frequently gregarious and change their coloration according to population density. Under crowded conditions, the black transverse stripes on the thorax and abdomen increase in size in later instars. Solitary caterpillars have black heads, whereas the head is brown under crowded conditions.

The Ramie Moth caterpillar is primarily creamy white with transverse, black banding dorsally. A black stripe runs down each side, and spiracles are outlined in black and highlighted by red blotches. The prolegs and head are black.

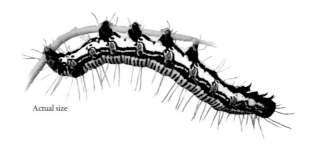

Actual size

FAMILY	Noctuidae
DISTRIBUTION	Mediterranean Europe, throughout Africa and the Middle East, South and Southeast Asia, Japan, and northern Australia
HABITAT	Coastal sand dunes
HOST PLANTS	Lilies, including *Amaryllis* spp., *Crinum* spp., *Narcissus* spp., and *Pancratium maritimum*
NOTE	Black-and-white caterpillar that feeds on lily plants
CONSERVATION STATUS	Not evaluated, but endangered in Europe

ADULT WINGSPAN
1⁹⁄₁₆ in (40 mm)

CATERPILLAR LENGTH
1⁹⁄₁₆–2 in (40–50 mm)

BRITHYS CRINI

LILY BORER

FABRICIUS, 1775

The caterpillars of the Lily Borer moth, also known as Kew Arches, hatch from pale yellow eggs laid in small clusters on the leaf of a host plant. Initially, the young caterpillars mine into the fleshy leaves where they feed, but later they emerge to feed externally, moving down the leaves toward the bulb. The bright colors of the larvae warn predators that they are unpalatable, the poisons coming from the leaves of the host plants. To pupate, the caterpillars move to the ground, just under the surface in sandy soil.

The adult moth is on the wing from March through September, and there are two, possibly three, generations each year, with the caterpillars of the final brood overwintering. In Europe, the loss of coastal dunes has led to a steep decline in the species, while in other parts of the world *Brithys crini* is considered a pest as the caterpillar damages ornamental lilies found in parks and gardens.

The Lily Borer caterpillar has a plump, black body with rings of white spots. The head, prolegs, and posterior segment of the abdomen are red brown, while the legs are black. Sparse black hairs protrude from the body.

Actual size

FAMILY	Noctuidae
DISTRIBUTION	Europe and across central Asia, through Siberia to Korea; also eastern and western parts of North America, where introduced
HABITAT	Shingle banks, waste ground, and verges
HOST PLANTS	Common Toadflax (*Linaria vulgaris*) and Dalmatian Toadflax (*Linaria dalmatica*)
NOTE	Colorful caterpillar that is sometimes used as a biological weedkiller
CONSERVATION STATUS	Not evaluated, but locally rare

ADULT WINGSPAN
1–1¼ in (25–32 mm)

CATERPILLAR LENGTH
1⁹⁄₁₆ in (40 mm)

604

CALOPHASIA LUNULA
TOADFLAX BROCADE
[HUFNAGEL, 1766]

Toadflax Brocade caterpillars hatch from up to 80 eggs laid by the female moth on the food plant. The larvae have a huge appetite and grow quickly, first eating all the young leaves and flower buds, and then stems and older leaves. There are usually two generations a year and they may overlap in summer, the moths flying in May to August, and July and August. The reddish-brown pupae of the second generation overwinter, either in the soil, or on or within the lower stems of the host plant.

The distribution of this species is limited by the occurrence of its food plant. It feeds mostly on Common Toadflax but has been seen on naturalized varieties and closely related species. In some parts of the world where its host plant is considered a noxious weed, the moth has been used for biological control, for example in parts of southern Europe, and also in North America, where the species was introduced during the 1960s.

The Toadflax Brocade caterpillar is yellow and black with irregular markings. Laterally, it is gray white with yellow, black, and gray lines, and black and white spots. The body is covered in short hairs.

Actual size

FAMILY	Noctuidae
DISTRIBUTION	From northern Spain through most of Europe (excluding extreme south), north to Iceland and Scandinavia, and across Russia and Siberia to Russian Far East
HABITAT	Cool, open habitats, such as high moorlands and mature coastal dunes; lesser numbers in warmer, open, scrubby habitats
HOST PLANTS	Wide range, including Heather (*Calluna vulgaris*), Broom (*Cytisus scoparius*), Bracken (*Pteridium aquilinum*), Bramble (*Rubus fruticosus*), willow (*Salix* spp.), Sea Buckthorn (*Hippophae rhamnoides*), and Colt's-foot (*Tussilago farfara*)
NOTE	Pretty, striped caterpillar that is easily found in the daytime
CONSERVATION STATUS	Not evaluated

ADULT WINGSPAN
1¼–1⅝ in (32–42 mm)

CATERPILLAR LENGTH
1⁹⁄₁₆–1¾ in (40–45 mm)

CERAMICA PISI
BROOM MOTH
(LINNAEUS, 1758)

605

Broom Moth caterpillars hatch from yellowish-gray eggs laid in batches on the food plant. The larvae are inconspicuous when small but when larger are often seen resting or feeding openly by day, although they are more active at night. At the final instar, they form a pupa in a fragile cocoon in the ground, and the pupa overwinters. The caterpillars can be found from July into the fall; just one generation is produced annually. The adults fly from June to August.

The Broom Moth caterpillar can occur in almost any scrubby, relatively undisturbed habitat. It is tolerant of cold and is most abundant in northern areas, particularly on acid moorland, whereas it may be uncommon in lowland gardens. With its set of four bright yellow stripes, the *Ceramica pisi* larva is quite distinctive and can be confused with few other caterpillars. Middle instars of the Sword-grass caterpillar (*Xylena exsoleta*) are also green with yellow stripes, but their stripes are narrower, and, unlike *C. pisi*, their back and sides have the same color.

The Broom Moth caterpillar is smooth and slightly tapered at both ends. It is green or brown, always with four bold, bright yellow stripes, edged with white and finely edged with black, two on each side. The back is either plain dark brown or green, and the sides are usually paler and finely mottled with light green or brown.

Actual size

FAMILY	Noctuidae
DISTRIBUTION	Western and central Europe, Iceland, Scandinavia, east across Russia and Siberia to Sakhalin Island and Magadan; introduced in Newfoundland (northeastern Canada)
HABITAT	Open grasslands, particularly in acid uplands and in cool climates; also in warmer lowlands on well-drained soils
HOST PLANTS	Grasses, particularly hard-bladed species such as Mat-grass (*Nardus stricta*), Purple Moor-grass (*Molinia caerulea*), and Sheep's-fescue (*Festuca ovina*); also sedge (*Cyperaceae*)
NOTE	Usually inconspicuous caterpillar that can sometimes occur in "armies"
CONSERVATION STATUS	Not evaluated

ADULT WINGSPAN
1⁵⁄₁₆–1⁹⁄₁₆ in (24–39 mm)

CATERPILLAR LENGTH
1³⁄₁₆–1⅜ in (30–35 mm)

CERAPTERYX GRAMINIS
ANTLER MOTH
(LINNAEUS, 1758)

606

The Antler Moth caterpillar hatches in March or April and develops, feeding mainly at night, until June. At later instars, when larger, it hides deep in the grass by day. Although the larvae are difficult to detect, small birds such as the Whinchat (*Saxicola rubetra*) spot and often take them. When fully fed, the caterpillar forms a pupa in a chamber in the ground among grass roots. There is one generation annually, and the distinctively marked, brown-and-cream adults fly from July to September by day and by night. The tough, rounded, grayish eggs are laid as the female flies low over grassland, and this stage overwinters.

Caterpillar numbers sometimes build up to high levels, and the larvae may defoliate large areas of hillside, feeding by day and night in "armies." Later instars of the closely related Feathered Gothic (*Tholera decimalis*) and Hedge Rustic (*T. cespitis*) look very similar and have very similar habits.

Actual size

The Antler Moth caterpillar is gray brown, smooth, and rather shiny and tapers toward the tail end. It has three well-separated, whitish or very light brown stripes along the back, a more diffuse, pale stripe laterally, and a broad, pale stripe low down on the sides. The head is brown and the spiracles are black.

FAMILY	Noctuidae
DISTRIBUTION	Canada and United States
HABITAT	Deciduous forests
HOST PLANTS	Mainly beech (*Fagus* spp.), but also other trees, including birch (*Betula* spp.), maple (*Acer* spp.), and oak (*Quercus* spp.)
NOTE	Hairy caterpillar with distinctive yellow markings on its black head
CONSERVATION STATUS	Not evaluated

ADULT WINGSPAN
1½–1⅞ in (38–48 mm)

CATERPILLAR LENGTH
1³⁄₁₆–1⁹⁄₁₆ in (30–40 mm)

CHARADRA DERIDENS
LAUGHER
(GUENÉE, 1852)

607

The Laugher caterpillar is characterized by a large, boldly marked head possibly designed to startle birds that attack its leaf shelters. The larvae feed on the leaves of a variety of broad-leaved trees, in particular beech, and can cope with tough, older leaves that are often avoided by other species of caterpillar. They gain protection by spinning a silken nest between a couple of leaves, in which they rest, or sometimes they will use the shelters of other caterpillars.

The caterpillar spins a light cocoon on the food plant, in which it overwinters and pupates the following spring. There are two generations a year. In the northern part of the range, the moths are on the wing from May to August, but farther south they have a much longer season. Also called the Marbled Tuffet, this species is a type of owlet moth. It gets the name Laugher from the pattern of markings on the moth's forewing, which resembles a laughing face.

The Laugher caterpillar has a creamy-white to yellow body, which is covered in tufts of long, gray-white hairs. At first, the caterpillar has a yellow head, but by the final instar it is shiny black with a small, yellow triangle flanked by two yellow crescents.

Actual size

FAMILY	Noctuidae
DISTRIBUTION	Australia, including Tasmania
HABITAT	Farmlands, parks, and gardens
HOST PLANTS	Beans (Fabaceae), canola (*Brassica* spp.), Chard (*Beta vulgaris*), sunflower (*Helianthus* spp.), Tomato (*Solanum lycopersicum*), and tobacco (*Nicotiana* spp.)
NOTE	Green, semi-looping caterpillar that is found on many crops
CONSERVATION STATUS	Not evaluated

ADULT WINGSPAN
1³⁄₁₆ in (30 mm)

CATERPILLAR LENGTH
1⁹⁄₁₆ in (40 mm)

608

CHRYSODEIXIS ARGENTIFERA
TOBACCO LOOPER
(GUENÉE, 1852)

Tobacco Looper caterpillars hatch from round, white eggs laid separately on the underside of leaves. Described as a semi-looper, the larva has two pairs of prolegs rather than four and so has a partly looping movement, similar to that of the true loopers of the Geometridae. Each caterpillar spins a white cocoon on the underside of a leaf, which it camouflages with bits of leaves and even droppings. The adult moth emerges after a pupation of around three weeks.

This species is an agricultural pest that attacks many crops. Young caterpillars feed on one side of the leaf, creating distinctive "feeding windows," but as they molt and get larger, they chew holes in the leaves. Mature caterpillars feed from the leaf margin, sometimes defoliating whole plants. The caterpillars also damage crops such as tomatoes by chewing the unripe fruits, and they bore into the pods of beans and peas to reach the seeds inside.

Actual size

The Tobacco Looper caterpillar is green with a dark, green dorsal line bordered by two white stripes. Laterally, there are several fine, white lines and a row of black spots. The body tapers toward the head.

FAMILY	Noctuidae
DISTRIBUTION	Southern Europe (from Spain to the Balkans), north to United Kingdom and Denmark, east to the Caspian Sea, the Middle East, Macaronesia, Africa, and Madagascar
HABITAT	Farmlands, gardens, and other open places with herbaceous plants
HOST PLANTS	Cruciferous vegetables (*Brassica* spp.) and other Fabaceae, Common Nettle (*Urtica dioica*), bean (*Phaseolus* spp.), pelargonium (*Pelargonium* spp.), Tomato (*Solanum lycopersicum*), and chrysanthemum (*Chrysanthemum* spp.)
NOTE	Widespread, semi-looper caterpillar that is a tropical crop pest
CONSERVATION STATUS	Not evaluated, but common or abundant throughout most of the range

ADULT WINGSPAN
1¼–1¾ in (32–44 mm)

CATERPILLAR LENGTH
1⁵⁄₁₆–1½ in (34–38 mm)

CHRYSODEIXIS CHALCITES
GOLDEN TWIN-SPOT
(ESPER, 1789)

609

The Golden Twin-spot caterpillar hatches from a greenish-white, domed egg laid singly on a leaf. When very small, the larva is green with scattered black spots, each with a short, dark hair. At first, it grazes the underside of the leaf, creating translucent windows, and if disturbed drops down on a silk thread. When larger, the caterpillar eats all or most of the leaf and excavates fruit and unripe seedpods, but without burrowing inside. It forms a black-and-green pupa in a white cocoon, often spun under a leaf. The species can breed continuously with up to nine generations annually.

Like other members of its subfamily Plusiinae, the Golden Twin-spot caterpillar walks rather like an inchworm or looper caterpillar (family Geometridae). The body shape, tapering from back to front, is also a characteristic of the group. The caterpillar is a major crop pest in Africa, the Middle East, and southern Europe. Adults migrate farther north, sometimes imported with produce, and occur as pests in glasshouses.

Actual size

The Golden Twin-spot caterpillar is green or yellowish green with six irregular, whitish stripes along the back and a broader white stripe and small, black spots along each side. There are three pairs of prolegs, and the body tapers toward the head.

FAMILY	Noctuidae
DISTRIBUTION	Southern and eastern Europe into southern Russia, Turkey, and the Middle East
HABITAT	Hot, dry, and often rocky slopes, from sea level up to 6,600 ft (2,000 m) elevation
HOST PLANTS	Figwort (*Scrophularia* spp.)
NOTE	Brightly colored caterpillar that turns into a drab adult
CONSERVATION STATUS	Not evaluated

ADULT WINGSPAN
1⁹⁄₁₆ in (40 mm)

CATERPILLAR LENGTH
1⅜–2 in (35–50 mm)

610

CUCULLIA BLATTARIAE
CUCULLIA BLATTARIAE
(ESPER, 1790)

Cucullia blattariae caterpillars hatch from small, round eggs with a ridged surface, positioned singly, or in small groups, on the host plant close to flower buds. The larvae feed on the buds and leaves. The mature caterpillars move to the ground to pupate, spinning a loose cocoon in the leaf litter.

The night-flying adult moths are on the wing from late March to May—maybe early August at higher elevations—and there is a single generation. Some authorities refer to the species as *Shargacucullia blattariae* (Esper, 1790), *Shargacucullia* being a subgenus. Despite having a distinctive caterpillar and a European distribution, very little is known about this species, which may in part be due to the drab nature of the adult moth and confusion with similar-looking caterpillars. For example, the more reported Water Betony (*Shargacucullia scrophulariae*) has a comparable distribution and its caterpillar also feeds on members of the figwort family.

The *Cucullia blattariae* caterpillar is brightly colored. The body is white with a dorsal line of large, black crosses and many smaller, black spots along the sides. There are two pale yellow lateral stripes. The head and true legs are brown. Widely spaced black hairs occur across the whole body.

Actual size

FAMILY	Noctuidae
DISTRIBUTION	Mountain ranges in Europe, including the Pyrenees, Alps, Apennines, Carpathians, and Caucasus Mountains
HABITAT	Dry and sunny rocky slopes, scree, and outcrops up to 5,900 ft (1,800 m) elevation
HOST PLANTS	Bellflower (*Campanula* spp.)
NOTE	Alpine caterpillar that is found on rocky mountain slopes
CONSERVATION STATUS	Not evaluated, but classed as endangered in parts of its range

ADULT WINGSPAN
1½–1⅝ in (38–42 mm)

CATERPILLAR LENGTH
1⅜–2 in (35–50 mm)

CUCULLIA CAMPANULAE

CUCULLIA CAMPANULAE

FREYER, 1831

611

Cucullia campanulae caterpillars hatch from eggs laid on the leaves of the host plant, which is the source of the species' scientific name. The larvae are active over the summer months, from June to September, feeding on clumps of bellflowers growing among the rocks. At higher altitudes, mostly in the Alps, they are seen only in August. Parasitism of the caterpillars by parasitic wasps is common, and healthy, unparasitized individuals tend to be found hiding on the ground or under plants during the day, only emerging at night to feed. The caterpillars overwinter under rocks, which is where pupation takes place.

This alpine moth is on the wing from late May to July, and there is a single generation. The species has been endangered by the loss of its habitat due to reforestation, changes in farming management from traditional to more intensive cultivation, overgrazing, and new tourist developments in alpine areas.

The *Cucullia campanulae* caterpillar is white with many black spots and a lateral, broken, pale yellow stripe just below the spiracles. There are widely spaced short hairs across the body. The legs are black.

Actual size

FAMILY	Noctuidae
DISTRIBUTION	Across Europe into Russia, Turkey, and the Caucasus; also the Altai Mountains of central Asia
HABITAT	Shady habitats, such as forests and woodland margins, glades, and tracks up to 6,600 ft (2,000 m) elevation
HOST PLANTS	Compositae, including Lettuce (*Lactuca sativa*) and *Sonchus* spp.
NOTE	Eye-catching caterpillar that likes shady places
CONSERVATION STATUS	Not evaluated

ADULT WINGSPAN
1⁹⁄₁₆–2 in (40–50 mm)

CATERPILLAR LENGTH
2 in (50 mm)

612

CUCULLIA LACTUCAE
LETTUCE SHARK
(DENIS & SCHIFFERMÜLLER, 1775)

The caterpillars of the Lettuce Shark moth hatch from conical-shaped, ridged eggs laid on the underside of leaves of the host plant. The larvae feed on buds, flowers, and young fruits and will move to the leaves if nothing else remains. Their bright colors warn would-be predators of their unpleasant taste. The mature caterpillars move to the ground to pupate in a cocoon just below the surface of the soil. The pupa is orange brown and overwinters.

The night-flying adult moths are on the wing from May to July but are rarely attracted to light. During the day, they rest on tree trunks. In the northern parts of the species' range there is a single generation, but to the south there may be a second generation flying in August and September. The adults are very similar in appearance to the Shark Moth (*Cucullia umbratica*) and the Chamomile Shark (*C. chamomillae*).

The Lettuce Shark caterpillar has a white background with a central dorsal stripe of either yellow or orange between two rows of large, black spots. A lateral stripe of, again, either yellow or orange carries smaller black spots. The head, true legs, and prolegs are black.

Actual size

FAMILY	Noctuidae
DISTRIBUTION	North America, from southeastern Canada to Florida, west to Texas
HABITAT	Woodlands and forest edges
HOST PLANTS	Porcelainberry (*Ampelopsis* spp.), Virginia Creeper (*Parthenocissus quinquefolia*), and grapevine (*Vitis* spp.)
NOTE	Caterpillar that rests and feeds exposed, usually on young foliage
CONSERVATION STATUS	Not evaluated, but relatively uncommon

ADULT WINGSPAN
1⅜–1¹³⁄₁₆ in (35–46 mm)

CATERPILLAR LENGTH
1⁹⁄₁₆ in (40 mm)

EUDRYAS GRATA
BEAUTIFUL WOOD NYMPH
(FABRICIUS, 1793)

613

The Beautiful Wood Nymph caterpillar feeds and rests in plain sight, suggesting probable toxicity to predators. One generation is typical in its northern range, with two or more possible farther south. Larvae occasionally rest on wood structures such as fence posts or park benches, lending credence to the common name. At pupation, the caterpillar digs into bark or stems, parts of which it uses to construct its cocoon. The pupa overwinters, and the moth ecloses in early summer.

The caterpillar is relatively uncommon throughout its range and usually happened upon while searching for something else. At first glance, *Eudryas grata* larvae are almost indistinguishable from those of the closely related Pearly Wood Nymph (*E. unio*), although there is a subtle difference in the prolegs: *E. grata* has a single black basal spot, while *E. unio* has two. Adults of both species also look strikingly similar—both are convincing bird-dropping mimics that rest in plain sight.

The Beautiful Wood Nymph caterpillar has alternating rings of orange, black, and white, with intermittent black spots extending across the abdomen and into the leg regions. The head and prothoracic shield are orange with black spots. Few visible setae are present.

Actual size

FAMILY	Noctuidae
DISTRIBUTION	Parts of southeast Canada, United States, the Caribbean, and through Central America; parts of Brazil, Argentina, and Paraguay; Spain, the Canary Islands, and Madeira
HABITAT	Forests, coastal rain forests, and conifer forests
HOST PLANTS	Wood sorrel (*Oxalis* spp.)
NOTE	Odd-looking, slug-shaped caterpillar in shades of brown
CONSERVATION STATUS	Not evaluated, but rare in much of its range

ADULT WINGSPAN
¾–1 in (20–25 mm)

CATERPILLAR LENGTH
1³⁄₁₆ in (30 mm)

GALGULA PARTITA
WEDGLING MOTH
GUENÉE, 1852

614

The caterpillars of the Wedgling Moth hatch from conical, ridged eggs laid on the underside of leaves, either singly or in small clusters. The larvae feed only on wood sorrel, an unusual food plant as the leaves have high levels of the relatively unpalatable salt oxalate, which means few animals feed on them. In fact, this species is the only one to feed on wood sorrel in North America.

The nocturnal Wedgling Moth is on the wing at different times of year, depending on the location. In the south of their range, the moths are seen from March to November, while in the north they fly from May to September. There are several generations a year. As *Galgula partita* feeds only on a single plant species, any loss of that plant's habitat leads to a decline in moth numbers due to the lack of alternative host plants, hence the Wedgling's increasing rarity in some regions.

The Wedgling Moth caterpillar has an unusual, enlarged body section that creates a tapered, sluglike shape. The dorsal surface is dark brown, and the sides are a mottled, pale brown, edged in white. The enlarged section bears rings of white dots, and there are several pale dorsal lines. The head is brown with white stripes.

Actual size

FAMILY	Noctuidae
DISTRIBUTION	Europe, from Iberia, southern England, and southern Scandinavia across Russia and southern Siberia to Sakhalin Island; Morocco and Algeria, the Middle East, and across western and central Asia to northern China and Japan
HABITAT	Warm, open, dry habitats, including gardens
HOST PLANTS	Carnation (*Dianthus caryophyllus*), Sweet-william (*Dianthus barbatus*), and pinks (*Dianthus* spp.); also campion (*Silene* spp.), including Bladder Campion (*Silene vulgaris*)
NOTE	Caterpillar that hatches among flower heads, hiding elsewhere when larger
CONSERVATION STATUS	Not evaluated

ADULT WINGSPAN
⅞–1³⁄₁₆ in (22–30 mm)

CATERPILLAR LENGTH
1³⁄₁₆–1⁵⁄₁₆ in (30–34 mm)

HADENA COMPTA
VARIED CORONET
([DENIS & SCHIFFERMÜLLER], 1775)

615

The Varied Coronet caterpillar hatches from a pale brown egg laid singly inside a flower of the host plant. It feeds on the unripe seeds, living at first inside the seedpods. When too large to be concealed within them, it hides by day on the ground and climbs up to feed at night. The pupa is formed in a cocoon in the ground, and it overwinters in this stage. There is one generation of caterpillars in the north of its range and two in the south, from June until early September.

Although the caterpillar often occurs on cultivated *Dianthus* plants, it does not appear to be considered a pest, perhaps because it feeds largely in the flower heads after the flowers have faded. The *Hadena compta* larva is similar to several others in the genus *Hadena* and it often shares the host plant with the Lychnis moth (*H. bicruris*). While both have a dorsal, brown stripe in the final instar, in the Lychnis this is usually formed into larger, well-defined chevrons.

The Varied Coronet caterpillar is pale gray brown or tawny, finely and irregularly mottled with darker brown. Along the back, there is a broad, slightly blurred, dark brown stripe, sometimes formed into a series of blotches or small chevrons. Each segment has a pair of dark dots. The sides are mottled and much paler below the dark-ringed spiracles.

Actual size

FAMILY	Noctuidae
DISTRIBUTION	Europe, from southern United Kingdom and Denmark to western Russia, Asia Minor, and the Caucasus; North Africa across to the Middle East, Iraq, and Iran, and Afghanistan to the Himalayas
HABITAT	Warm, dry, disturbed open ground, including cultivated land, quarries, verges, and embankments
HOST PLANTS	Prickly Lettuce (*Lactuca serriola*), Great Lettuce (*Lactuca virosa*), Wall Lettuce (*Lactuca muralis*), and cultivated lettuce (*Lactuca sativa*)
NOTE	Caterpillar that eats only flowers and lettuce seed heads
CONSERVATION STATUS	Not evaluated

ADULT WINGSPAN
1–1⅜ in (25–35 mm)

CATERPILLAR LENGTH
1³⁄₁₆–1⅜ in (30–35 mm)

HECATERA DYSODEA
SMALL RANUNCULUS
([DENIS & SCHIFFERMÜLLER], 1775)

616

The Small Ranunculus caterpillar hatches from a brownish egg laid singly or in small groups on a flower bud. The caterpillar feeds on the flowers and unripe seed heads. When small, it rests openly on the inflorescence but, when larger, may hide lower down during the day. The caterpillar is found from June and July to early fall, in one or two generations depending on climate. The pupa is formed in a chamber in the ground and overwinters, with the moths flying from June to August.

Due to their pale color, which provides excellent camouflage, Small Ranunculus caterpillars are difficult to detect when they rest outstretched along the stems of the food plant. They can be numerous on an individual plant and will feed on stems and lower leaves if the normal food is exhausted. Eggs are laid on cultivated lettuces only if the plants have been allowed to flower, but the caterpillars could be a pest where lettuces are being grown as a seed crop.

The Small Ranunculus caterpillar is pale green, brownish green, or pale brown above and green on the sides, often with a whitish stripe. There may be a double, fine, dark pencil line along the back, other dark lines, and pairs of dark dots on each segment, but often the markings other than the black spiracles are faint.

Actual size

FAMILY	Noctuidae
DISTRIBUTION	Northern areas of Africa, the Middle East, and much of southern Asia; migrant farther north, reaching the British Isles, southern Scandinavia, southern Urals, and northern China
HABITAT	Dry, generally warm, open habitats, including semidesert and cultivated land, including gardens; also open habitats as a migrant
HOST PLANTS	Many, including Asteraceae, such as Pot Marigold (*Calendula officinalis*); Geraniaceae, such as pelargonium (*Pelargonium* spp.); and Solanaceae such as Deadly Nightshade (*Atropa belladonna*); also crops such as Chickpea (*Cicer arietinum*) and Soybean (*Glycine max*)
NOTE	Caterpillar that can be a crop pest in subtropical climates
CONSERVATION STATUS	Not evaluated

ADULT WINGSPAN
1¼–1⅝ in (32–42 mm)

CATERPILLAR LENGTH
1¼–1½ in (32–38 mm)

HELIOTHIS PELTIGERA

BORDERED STRAW
([DENIS & SCHIFFERMÜLLER], 1775)

617

The Bordered Straw caterpillar hatches from a ribbed, white-and-purple egg laid on the food plant. When young, it often lives in a slight web near a shoot tip but feeds openly when larger. It forms a pupa in a fragile cocoon in the ground. In warm climates, this species may breed continuously. Farther north it has one to three generations, with migrants arriving from the south from May onward. It does not survive temperate winters.

In subtropical areas, the Bordered Straw caterpillar is often reported as a pest on a wide variety of crops, although this may sometimes be due to misidentification. In more northern, temperate areas such as the British Isles, it is not recorded as a crop pest and is more likely to be found on garden plants such as *Calendula* and *Pelargonium*. Some green forms of the extremely variable Scarce Bordered Straw (*Helicoverpa armigera*) caterpillar closely resemble the Bordered Straw.

The Bordered Straw caterpillar has sparse, rather short but quite noticeable whitish hairs on all segments, arising from small, pale, or dark warts. It may be light or dark green, banded green and pinkish, or dark purplish gray tinged with green. Most forms have distinct lighter and darker stripes, including a bright whitish or yellowish line along the sides.

Actual size

FAMILY	Noctuidae
DISTRIBUTION	Across much of United States south of the Great Lakes; Mexico, Central America, and the Caribbean
HABITAT	Meadows and forest edges, as well as disturbed habitat
HOST PLANTS	Fruit of Giant Ground Cherry (*Physalis peruviana*), Chinese Lantern (*Physalis alkekengi*), and other *Physalis* spp.
NOTE	Caterpillar with a unique diet that protects it from parasitoids
CONSERVATION STATUS	Not evaluated, but common

ADULT WINGSPAN
1¹⁄₁₆–1¼ in (27–31 mm)

CATERPILLAR LENGTH
1½ in (38 mm)

HELIOTHIS SUBFLEXA
SUBFLEXUS STRAW MOTH
(GUENÉE, 1852)

618

The Subflexus Straw Moth caterpillar often feeds in seclusion within lantern-shaped *Physalis* fruits, a food resource exclusive to this species, which has developed a unique ability to mature on a diet lacking linolenic acid. In other caterpillars, linolenic acid is necessary for producing a chemical, volicitin, in their saliva. However, the chemical also triggers the production of volatiles by the plants on which the caterpillars feed, which are then detected by parasitoids, enabling them to locate the larvae. The *Heliothis subflexa* caterpillar avoids such parasitoid attacks because of its linolenic-acid-free diet.

The moth and its caterpillar are very similar to a close relative, the Tobacco Budworm (*Heliothis virescens*). However, *H. virescens* is a significant pest on many different crops, from soybeans to cotton, while *H. subflexa* is innocuous. The two exemplify how a seemingly minor adaptation can radically change the economic significance of an insect species. *Heliothis subflexa* is barely known, while the diet of *H. virescens* costs millions of dollars a year.

The Subflexus Straw Moth caterpillar is cryptically colored green with longitudinal dark stripes, dark spiracles, and black, sclerotized dorsal verrucae. It has only very few short hairs.

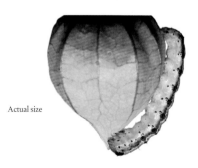

Actual size

FAMILY	Noctuidae
DISTRIBUTION	Mostly in southern United States, dispersing north to Canada annually; widely in the Caribbean; sporadically in Central America and South America
HABITAT	Disturbed agricultural habitats
HOST PLANTS	Many, including tobacco (*Nicotiana* spp.), beggarweed (*Desmodium* spp.), Japanese Honeysuckle (*Lonicera japonica*); lupine (*Lupinus* spp.), sunflower (*Helianthus* spp.), and Velvetleaf (*Abutilon theophrasti*)
NOTE	Caterpillar that is a pest of many crops
CONSERVATION STATUS	Not evaluated, but common

ADULT WINGSPAN
1⅛–1⅜ in (28–35 mm)

CATERPILLAR LENGTH
1–1⁷⁄₁₆ in (25–36 mm)

HELIOTHIS VIRESCENS

TOBACCO BUDWORM

(FABRICIUS, 1777)

619

Tobacco Budworm caterpillars hatch from spherical eggs laid on blossoms, fruit, and fresh growth, developing rapidly through all instars in as few as 17 days. The larvae are ferocious eaters and can even cannibalize each other. Young larvae bore into buds and blossoms, but, when the latter are consumed, caterpillars will move onto leaves. There are up to seven instars, although most commonly five or six. Pupation occurs in the soil, but the pupa survives the winter generally only in southern United States, or in greenhouses and other sheltered locations. Eclosing adults may disperse north to New England and even southern Canada by late summer.

The species derives its common name from its tobacco host plant, but it is also an important pest of many other species, from Alfalfa (*Medicago sativa*) to Soybean (*Glycine max*). Natural enemies include spiders, birds, and wasp parasitoids in the genera *Trichogramma*, *Cardiochiles*, *Cotesia*, and *Microplitis*. These can be used to control the larvae and have been the object of numerous research projects related to integrated pest management, providing an environmentally friendly alternative to pesticides.

The Tobacco Budworm caterpillar is variable in ground color from pale yellowish green to dark brown or shades in between. The cryptic coloration is assisted by many narrow, white stripes dorsally and one pronounced, broad, subspiracular band. Other banding may be narrow or incomplete. The head color varies from orange to brown or green, the legs are dark brown, and the prolegs and ventral surface are dark to translucent green. Numerous black, thornlike spines originate from sclerotized black bases and are relatively short.

Actual size

FAMILY	Noctuidae
DISTRIBUTION	Europe, from Iberia and southern England north to the southern and eastern coasts of Scandinavia, and east across southern Siberia to Lake Baikal; Canary Islands; North Africa; the Middle East through Iran, Turkmenistan, Tajikistan, and eastern Kazakhstan
HABITAT	Dry, open habitats, particularly on calcareous soils, and including coastal shingle, rocky places, scrub, farmlands, grasslands, and gardens
HOST PLANTS	Many, including Caryophyllaceae, such as campion (*Silene* spp.); Fabaceae, such as rest-harrow (*Ononis* spp.) and Pitch Trefoil (*Psoralea bituminosa*); Asteraceae, such as Smooth Hawk's-beard (*Crepis capillaris*); Malvaceae, such as cotton (*Gossypium* spp.); and Solanaceae, such as Tomato (*Solanum lycopersicum*)
NOTE	Caterpillar feeding on flowers and seeds, sometimes a crop pest
CONSERVATION STATUS	Not evaluated

ADULT WINGSPAN
1¼–1⁷⁄₁₆ in (32–37 mm)

CATERPILLAR LENGTH
1³⁄₁₆–1⁷⁄₁₆ in (30–35 mm)

620

HELIOTHIS VIRIPLACA
MARBLED CLOVER
(HUFNAGEL, 1766)

The Marbled Clover caterpillar hatches from a domed, ribbed, whitish egg laid singly on a flower of the food plant. It feeds on the flowers and ripening seedpods and forms a pupa in a slight cocoon in the ground, which is the overwintering stage. There are one or two generations annually, the second being sometimes partial, in July and August and in the fall.

The caterpillars are recorded as pests of crops such as bean and cotton, but often a closely related species is, in fact, responsible. Larvae in the genus *Heliothis* are highly variable, and differences between some closely related species are slight, so that the markings of a single individual can be misleading. The caterpillar of the Shoulder-striped Clover (*H. maritima*), which is confined to western Europe and highly localized on heathland, coastal dunes, and saltmarshes, is similar to this species but tends to show greater contrast in the stripes, as does that of *H. adaucta*, a more eastern species.

Actual size

The Marbled Clover caterpillar is slender, tapering at both ends. It is green or pale brown marked with darker brown and has irregular, fine, pale markings on the back, a pale, central, dark stripe, and a white or yellow stripe to each side. There are one or two broad, yellow or white stripes lower along the sides.

FAMILY	Noctuidae
DISTRIBUTION	Southern Canada and United States, except southern and central areas
HABITAT	Wet meadows, woodlands, riparian areas, and hop fields
HOST PLANTS	Hops (*Humulus* spp.), nettles (*Urtica* spp.), and woodnettle (*Laportea* spp.)
NOTE	Looper caterpillar that can be a serious pest of hops
CONSERVATION STATUS	Not evaluated, but usually common

ADULT WINGSPAN
1–1¼ in (25–32 mm)

CATERPILLAR LENGTH
1–1³⁄₁₆ in (25–30 mm)

HYPENA HUMULI
HOP LOOPER
HARRIS, 1841

621

Hop Looper caterpillars hatch from eggs laid singly in spring on host plants, such as hops, some three days earlier. Prior to ovipositing, the adult females overwinter in caves or old buildings. After hatching, the caterpillars develop rapidly, taking just 14 days at 78.8°F (26°C) to reach maturity. The larvae move with a characteristic looping motion and feed mostly by night. Pupation occurs in a slight, silken nest between leaves or on the soil surface, and the adult moths emerge after nine to ten days. There are two to three generations annually.

Hop Loopers sometimes occur in large numbers in hopyards in the northwestern United States and can defoliate hop vines. However, up to 70 percent of Hop Looper caterpillar populations may be parasitized by the many species of parasitic wasps and flies that attack them. Adult moths can sometimes disappear from an area, and it is suspected that they migrate. As adults are attracted to a combination of acetic acid and 3-methyl-l-butanol, traps containing this mixture are proving an effective monitoring tool for this species.

The Hop Looper caterpillar is pale green with thin, white, subdorsal and spiracular stripes. The setae are fine and pale golden orange to brown, and the spiracles are small. The head is shiny green, usually with scattered minute black spots, which also occur on the body.

Actual size

FAMILY	Noctuidae
DISTRIBUTION	North Africa, Europe (except northern Scandinavia), east across southern Russia; the Middle East to Afghanistan, northern India, and China to the Pacific coast
HABITAT	Many, but particularly gardens and other cultivated land, rough ground with lush herbaceous vegetation, and saltmarsh edges
HOST PLANTS	Many herbaceous and woody plants, but often Lamb's-quarters (*Chenopodium album*) and other Chenopodiaceae; also willowherb (*Epilobium* spp.), tobacco (*Nicotiana* spp.), Tomato (*Solanum lycopersicum*), tamarisk (*Tamarix* spp.), and elm (*Ulmus* spp.)
NOTE	Leaf-feeding caterpillar that also eats tomatoes
CONSERVATION STATUS	Not assessed, but very common

ADULT WINGSPAN
1⅜–1¾ in (35–45 mm)

CATERPILLAR LENGTH
1⁹⁄₁₆–1¾ in (40–45 mm)

LACANOBIA OLERACEA
BRIGHT-LINE BROWN-EYE
(LINNAEUS, 1758)

622

The Bright-line Brown-eye caterpillar hatches from a yellowish-green egg and, throughout its development, feeds at night, usually on the foliage of its food plant, hiding low down by day. The pupa is formed in a fragile cell in the ground, and this stage overwinters. There are one or two generations per year depending on climate. The adults, also known as Tomato Moths, first appear in May, and caterpillars can be found from June until late fall.

The caterpillar is most abundant on rich soils where its favored host plants proliferate. It is well known to gardeners and horticulturalists as a pest on tomatoes—as well as eating the leaves it destroys the fruit by burrowing into it. The closely related Dog's-tooth caterpillar (*Lacanobia suasa*) is very similar but its two outer dorsal stripes are formed into oblique, rather than curved, dashes and may be extended to the middle line to form a W on each segment. Green forms may lack this, making the two species indistinguishable from each other.

The Bright-line Brown-eye caterpillar is smooth and cylindrical and can be light green, darker green, brown, or pinkish brown in color. It is densely peppered with tiny, white spots and more thinly scattered black, white-ringed spots. On the back, it sometimes has three diffuse dark stripes, the outer two broken into slightly curved dashes. There is a bright yellow stripe along each side, often edged above by a diffuse dark stripe.

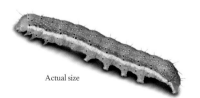

Actual size

FAMILY	Noctuidae
DISTRIBUTION	From northern Portugal, central Spain, western France, the British Isles, and southern Scandinavia across Europe to Asia Minor, the Caucasus, and western Siberia
HABITAT	Woodlands, wooded hedgerows, other open country with large oak trees, parklands, and mature gardens
HOST PLANTS	Mainly oak (*Quercus* spp.); also recorded on poplar (*Populus* spp.), willow (*Salix* spp.), plum (*Prunus* spp.), and elm (*Ulmus* spp.)
NOTE	Light bluey-green caterpillar with white hairs and white warts
CONSERVATION STATUS	Not evaluated

ADULT WINGSPAN
1⁵⁄₁₆–1⅝ in (34–42 mm)

CATERPILLAR LENGTH
1½–1⅝ in (38–42 mm)

LITHOPHANE ORNITOPUS
GREY SHOULDER-KNOT
(HUFNAGEL, 1766)

623

The Grey Shoulder-knot caterpillar hatches from a ribbed, whitish egg laid on a twig or branch; eggs are laid either singly or in small groups. The larva feeds singly on the leaves of its host plant, where it also rests, as it develops through all instars from April to June. The caterpillar then descends the tree and constructs a tough, silk cocoon in the ground and lies dormant within it for several weeks before forming a pupa. The adults emerge from September and are active until November, overwintering and mating in the spring. There is just one generation a year.

The larvae of a large number of noctuid moth species feed on oak and other trees in the spring, and are also often colored green or blue green, with white patterning and lines. However, unlike the Grey Shoulder-knot caterpillar, most have a prominent white stripe along the sides or along the center of the back, or along both, and have short, inconspicuous hairs. The absence of that stripe and the presence of longer, whitish hairs in the *Lithophane ornitopus* caterpillar make it distinctive.

The Grey Shoulder-knot caterpillar is quite stout and cylindrical, with a large head, and is light bluey green in color. When small, it is relatively plain, but it later develops a fine covering of small, white spots and irregular markings. It has three fine, slightly irregular, broken, white lines along the back and scattered, raised, white warts that give rise to fairly short, fine, white hairs.

Actual size

FAMILY	Noctuidae
DISTRIBUTION	Europe, excluding northern Scandinavia; Libya; Asia Minor; the Middle East; across southern Asia through Iran, Afghanistan, and northern India to southeast China; across Russia and Siberia to Japan and Russian Far East
HABITAT	Cultivated land, but also in many other, mainly open habitats
HOST PLANTS	Many wild and cultivated species, especially Cabbage (*Brassica oleracea*) and other *Brassica* crops, clematis (*Clematis* spp.), Apple (*Malus pumila*), and plants of Solanaceae, Fabaceae, and Asteraceae
NOTE	Highly polyphagous caterpillar that is notorious for ruining cabbage crops
CONSERVATION STATUS	Not evaluated, but widespread and common

ADULT WINGSPAN
1⅜–2 in (35–50 mm)

CATERPILLAR LENGTH
1¾–2 in (45–50 mm)

MAMESTRA BRASSICAE
CABBAGE MOTH
(LINNAEUS, 1758)

624

The Cabbage Moth caterpillar hatches from a brownish egg laid in large batches of 20 to 100 on the underside of a leaf. It feeds, mainly at night, usually on the leaves, and hides by day, often, when large, at ground level around the plant base. The caterpillar burrows 2–4 in (50–100 mm) into the ground to form its pupa in a thin cocoon. There are several generations a year, and the larvae are most abundant in late summer and fall. The night-flying adults are present from May to October.

The caterpillar is a pest in many countries on a range of herbaceous crops, but it is most frequent and damaging on cabbages, burrowing deep into the hearts, rendering them inedible. Superficially, it resembles other noctuids, and well-marked brown forms with dark bars along the back, emboldened near the tail end, are not dissimilar to some members of the Noctuinae subfamily. Green forms also resemble other species. However, in both brown and green forms, the build, large head, obscurely mottled pattern, and lifestyle of the Cabbage Moth caterpillar help to identify it.

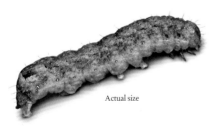

Actual size

The Cabbage Moth caterpillar is stout, with a large head. It is quite cylindrical, but the tail end is slightly humped. It is dirty gray brown, green, or (rarely) blackish above and much paler below the white, black-ringed spiracles. The whole surface is finely mottled so that the overall pattern is generally rather blurred. Some individuals have pinkish or yellowish rings.

FAMILY	Noctuidae
DISTRIBUTION	From northern Spain across most of Europe (excluding northern Scandinavia), Asia Minor, parts of the Middle East, and east across Asia to Russian Far East, Japan, and China
HABITAT	Shaded and more open habitats—woodlands, scrub, hedgerows, and gardens
HOST PLANTS	Many herbaceous and woody plants, including dock (*Rumex* spp.), Common Nettle (*Urtica dioica*), plantain (*Plantago* spp.), elder (*Sambucus* spp.), Wild Barberry (*Berberis vulgaris*), and dogwood (*Cornus* spp.)
NOTE	Distinctive caterpillar that has chevrons and a hump
CONSERVATION STATUS	Not evaluated

ADULT WINGSPAN
1⅜–2 in (35–50 mm)

CATERPILLAR LENGTH
1⅝–1¾ in (42–45 mm)

MELANCHRA PERSICARIAE

DOT MOTH

(LINNAEUS, 1761)

625

Dot Moth caterpillars hatch from pinkish-brown eggs laid singly or in batches on the food plant. The larvae feed on the foliage, generally resting low down on the food plant when larger, but may be active by day as well as at night. They develop slowly from July into the fall and then form a pupa in a cocoon in the ground, which overwinters. There is one generation a year. The adults, whose dark brown forewings feature a prominent white mark that gives the species its common name, fly in June and July.

The caterpillar's distinctive markings make it fairly unmistakable. However, because it often rests in a hunched position, as do caterpillars of the Plusiinae subfamily (which are less smooth), it could be confused with the Spectacle (*Abrostola tripartita*) and Dark Spectacle (*A. triplasia*). Both also have dorsal chevrons and a tail hump but are smaller, with a smaller head and stronger white markings. Middle instar Dot Moth caterpillars, which are humped but lack darkened chevrons, might be mistaken for the Burnished Brass (*Diachrysia chrysitis*), although that caterpillar has three instead of four pairs of prolegs.

Actual size

The Dot Moth caterpillar is green, light brownish green, or dark brownish green marbled with pink. It is smooth, with a noticeable hump at the tail end, and is thickset when mature, with a large head. Dorsally, it has a thin, white stripe down the middle and a series of dark chevrons, emboldened on the third and fourth segments and on the hump.

FAMILY	Noctuidae
DISTRIBUTION	Europe (excluding northern parts of Scandinavia), Asia Minor, southwest Russia, the Caucasus, and Turkmenistan
HABITAT	Woodlands, hedgerows, open areas with scrub, and gardens
HOST PLANTS	Herbaceous plants, including dock (*Rumex* spp.), Primrose (*Primula vulgaris*), Common Nettle (*Urtica dioica*), and Foxglove (*Digitalis purpurea*); woody plants and bushes, including Bramble (*Rubus fruticosus*) and willow (*Salix* spp.)
NOTE	Plump, pale brown caterpillar that has conspicuous black side spots
CONSERVATION STATUS	Not evaluated

ADULT WINGSPAN
1¾–2 in (45–50 mm)

CATERPILLAR LENGTH
2–2⅛ in (50–55 mm)

626

NOCTUA FIMBRIATA
BROAD-BORDERED YELLOW UNDERWING
(SCHREBER, 1759)

Broad-bordered Yellow Underwing caterpillars hatch from greenish eggs, laid in batches on the host plant in early fall. The caterpillar feeds at night, throughout the fall, then overwinters near the ground and completes its growth in April or May, again feeding at night and hiding near the ground during the day. The pupa is formed in the ground. There is one generation a year, with the adults emerging in July and estivating for several weeks before breeding.

In the caterpillar's final instar, the light brown color and bold, black lateral spots make the Broad-bordered Yellow Underwing conspicuous after dark, when it is easily found by using a flashlight as it comes up to feed. Earlier instars, without the black spots, are less distinctive. *Noctua tirrenica*, only described as a distinct species in 1983 and found in southern Europe, North Africa, Turkey, and the Caucasus, has very similar habits and a very similar caterpillar. Its adults resemble pale forms of *N. fimbriata*.

Actual size

The Broad-bordered Yellow Underwing caterpillar is plump, pale tawny brown, faintly mottled with darker brown, and paler below the white, black-outlined spiracles. Along the back are a very fine, whitish, central line and pale, dark-edged markings, squared off at the tail end. In the final instar, there is a conspicuous, rounded, black spot next to most of the spiracles.

FAMILY	Noctuidae
DISTRIBUTION	Canada and United States, Europe, North Africa, and western Asia into northern India
HABITAT	Grasslands, verges, parks, and gardens
HOST PLANTS	Wide variety of herbaceous plants and crops, including strawberry (*Fragaria* spp.), freesia (*Freesia* spp.), *allium* spp., Tomato (*Solanum lycopersicum*), and potato (*Solanum* spp.)
NOTE	Caterpillar that is widely resented as a garden pest
CONSERVATION STATUS	Not evaluated, but common

NOCTUA PRONUBA

LARGE YELLOW UNDERWING
(LINNAEUS, 1758)

ADULT WINGSPAN
2–2⅜ in (50–60 mm)

CATERPILLAR LENGTH
1³⁄₁₆–1⁹⁄₁₆ in (30–40 mm)

627

Female Large Yellow Underwing moths lay large clusters of up to 1,000 eggs, on the underside of host plant leaves. The eggs are slow to hatch, taking up to five weeks. The young larvae overwinter and feed on milder winter days but become fully active in the spring. The caterpillar, also known as a cutworm, is a notorious garden pest. It emerges at night to feed on low-growing plants, severing the shoots at ground level. By day, it seeks shelter in clumps of plants where it may continue to feed and curls up into the typical cutworm C shape when disturbed.

The larvae complete their growth and pupate in a small chamber underground. The adult moths eclose from the chestnut-brown pupae and are on the wing during summer to early fall. They are strong fliers and, being migratory, can travel long distances. Accidentally introduced into Canada in 1979, the species has now spread west across North America.

The Large Yellow Underwing caterpillar is brown or olive green. Each segment bears a dorsal, black-and-cream dash, either side of a dark-brown, median line. The head is brown with two thick, black lines.

Actual size

FAMILY	Noctuidae
DISTRIBUTION	Central Europe into southwest Russia, Georgia, and Turkey
HABITAT	Dry, open steppe, grasslands, and undisturbed areas around cultivated fields
HOST PLANTS	Spurge (*Euphorbia* spp.)
NOTE	Caterpillar from the steppes of Europe that effectively defoliates spurge
CONSERVATION STATUS	Not evaluated, but common

ADULT WINGSPAN
1⁵⁄₁₆–1¹⁄₁₆ in (24–27 mm)

CATERPILLAR LENGTH
1–1³⁄₁₆ in (25–30 mm)

OXICESTA GEOGRAPHICA
GEOGRAPHICAL MOTH
(FABRICIUS, 1787)

The Geographical Moth caterpillars hatch from eggs laid in large batches of around 300 on the underside of leaves of the host plant. The larvae are gregarious at first, living together in groups of up to 30. They spin a silken web over the host plant, feeding first on the flower buds and young leaves before moving on to the older leaves, capable of defoliating the whole plant. The fifth instar is solitary and pupates in a pale yellow cocoon that it spins on a stem. The species overwinters as a pupa.

The moths appear in April and May, and there is usually a single generation. Trials have been conducted into the use of *Oxicesta geographica* as a biological control agent for leafy spurge (*Euphorbia esula*), an aggressive perennial weed found across North America, where it is a major problem on grasslands and has proved resistant to control through herbicides.

Actual size

The Geographical Moth caterpillar is dark brown, with two white, lateral stripes and a series of transverse, orange bars. The spiracles are ringed in white. There are tufts of white and brown hairs, and the head is dark brown.

FAMILY	Noctuidae
DISTRIBUTION	Northern and southern Europe, east across Asia to Korea and Japan
HABITAT	Coniferous forests, especially those of a cool, humid type
HOST PLANTS	Various coniferous trees, especially spruce (*Picea* spp.); also pine (*Pinus* spp.) and European Larch (*Larix decidua*)
NOTE	Hairy caterpillar found in the canopy of conifer trees
CONSERVATION STATUS	Not evaluated, but locally rare

ADULT WINGSPAN
1⁹⁄₁₆–2 in (40–50 mm)

CATERPILLAR LENGTH
1¾–2⅛ in (45–55 mm)

PANTHEA COENOBITA
PINE ARCHES
(ESPER, 1785)

629

Pine Arches caterpillars hatch from yellow eggs laid in clusters on the needles of coniferous trees in late summer. The larvae stay in the canopy, where they feed on needles as they develop. Their coloring and form provide excellent camouflage, so they can rest openly, lying along narrow branches. The hairs of the caterpillars also contain irritants, giving the caterpillars some protection from predation. The mature larvae move to the ground, incorporating hairs into their cocoon for extra protection and also for camouflage when they pupate.

The chestnut-brown pupa overwinters, and the adult then emerges the following spring. The moths, which have a distinctive black-and-white, zigzag patterning, are on the wing from early May to August, and there is one generation a year. The range of this moth species has been expanding in recent decades due to the increased establishment of conifer plantations, especially plantations of spruce.

The Pine Arches caterpillar is reddish brown in color and covered in tufts of hairs. There is a series of dorsal, creamy-white marks and several amber-colored lateral lines. Distinctive yellow tufts of hair arise on segments two, three, four, and eleven. The head is brown with a central, red-brown flash.

Actual size

FAMILY	Noctuidae
DISTRIBUTION	Across Europe into Russia as far as the Urals; Turkey, Armenia, Syria, and parts of North Africa
HABITAT	Woodlands, scrub, grasslands, parks, and gardens
HOST PLANTS	Wide range of low-growing plants, particularly White Dead Nettle (*Lamium album*)
NOTE	Plump, green caterpillar found in almost any European habitat
CONSERVATION STATUS	Not evaluated, but common

ADULT WINGSPAN
1¾–2 in (45–50 mm)

CATERPILLAR LENGTH
1¾ in (45 mm)

PHLOGOPHORA METICULOSA
ANGLE SHADES
(LINNAEUS, 1758)

630

The caterpillars of the Angle Shades moth hatch from pale brown, slighty flattened, and ridged eggs laid in clusters on the leaves of a wide variety of host plants. The larvae feed mostly on leaves, and those of the second generation overwinter and become active again in spring. The mature caterpillars drop to the ground, where they spin a flimsy cocoon, either among the leaf litter or in the soil. The pupa is chestnut brown in color.

The nocturnal Angle Shades moths, which are attracted to light, are seen from May to October, and there are two or more generations each year. During the day, they rest on fences, trees, and shrubs, their disruptive coloration and wing shape giving the appearance of a withered leaf, which provides excellent camouflage. This is a migrant species that can fly long distances. It is often seen in large numbers along coastlines.

The Angle Shades caterpillar is yellow green with many tiny dots that create a mottled appearance. The body has a broken, white dorsal line, a broad, white lateral stripe, and widely spaced short hairs. The head is dark green. There is also a pink-brown variant.

Actual size

FAMILY	Noctuidae
DISTRIBUTION	United States, from southern Maine, south to Florida, and west to Texas and Illinois
HABITAT	Woodlands and forest edges
HOST PLANTS	Greenbrier (*Smilax* spp.)
NOTE	Caterpillar that is a group-feeding leaf defoliator
CONSERVATION STATUS	Not evaluated, but secure and regularly occurring

ADULT WINGSPAN
1³⁄₁₆ in (30 mm)

CATERPILLAR LENGTH
Up to 1⅜ in (35 mm)

PHOSPHILA TURBULENTA

TURBULENT PHOSPHILA

HÜBNER, 1818

631

The Turbulent Phosphila caterpillar is a common species, typically encountered at early instars in tight clusters feeding gregariously on the underside of greenbrier leaves. Young larvae are pale green or tan in color, with a shiny black head and thoracic shield and indistinct lines traversing the length of their bodies. The more recognizable black-and-white, pinstriped "referee shirt" appearance manifests itself as the caterpillar matures. Later instar larvae may wander off to feed on their own but prefer to regroup into layered colonies, alternated back to front, when at rest.

On the posterior of the Turbulent Phosphila, a swollen fake head and "face" are designed to thwart predation. When threatened, the caterpillar curls its true head into the body and displays the false one as an alternative. It may also release its grip on the vine and drop to the ground to hide in leaf litter below. Two generations are typical throughout most of its range, with larvae present from May to November. The pupa overwinters, wrapped in silk within an enclosed leaf.

Actual size

The Turbulent Phosphila caterpillar has a head and posterior that are strikingly similar in appearance. Both are black with white eyespots and other markings. The upper torso is pinstriped in black and white. Yellow-orange coloration extends along the length of the lower body.

FAMILY	Noctuidae
DISTRIBUTION	Eastern North America, from Ontario to Florida, west to Montana and Texas
HABITAT	Woodlands and forest edges
HOST PLANTS	Greenbrier (*Smilax* spp.)
NOTE	Caterpillar that has been described as "fantastically bizarre"
CONSERVATION STATUS	Not evaluated, but sometimes common

ADULT WINGSPAN
1⅛–1⅜ in (28–35 mm)

CATERPILLAR LENGTH
1³⁄₁₆–1⅜ in (30–35 mm)

PHYPROSOPUS CALLITRICHOIDES
CURVE-LINED OWLET MOTH
GROTE, 1872

632

Curve-lined Owlet Moth caterpillars are defended by their bizarre appearance. With prolegs missing on the third and fourth abdominal segments and their typical, almost doubled-up pose on the host plant leaf or stem, they resemble withered leaves and, to further that mimicry, move from side to side as a leaf would in a slight breeze. They may be found from late June onward. When fully grown, the final instar larva pupates, manufacturing a cocoon from silked-together stems of the host plant.

The pupa overwinters. There are probably two generations in northerly areas of the species' range and possibly three in the south. Unlike the caterpillar, the adult Curve-lined Owlet moth is rather dull and unremarkable, although it, too, resembles a dead leaf. The adults fly year-round in the south of the range and from March to September in northern areas. There are two species of the genus in North America.

The Curve-lined Owlet Moth caterpillar is reddish brown with darker markings and a white area laterally where prolegs clasp the substrate. The anterior horn is black with a white area, and the legs may be red or black. The head is red with black markings.

Actual size

FAMILY	Noctuidae
DISTRIBUTION	From Spain and the British Isles across Europe, including the Mediterranean and southern half of Scandinavia; Asia Minor, the Caucasus, southern Siberia, and east to Russian Far East and Japan; across western and southern central Asia to eastern China
HABITAT	Calcareous grasslands, moorlands, open woodlands, low-lying damp grasslands, coastal sand dunes, and shingle; also urban habitats, including gardens
HOST PLANTS	Many, including rest-harrow (*Ononis* spp.), sainfoin (*Onobrychis* spp.), Henbane (*Hyoscyamus niger*), and Sea Sandwort (*Honkenya peploides*)
NOTE	Variably colored caterpillar that feeds on flowers and unripe seeds
CONSERVATION STATUS	Not evaluated, but widespread

ADULT WINGSPAN
1⁵⁄₁₆–1⁵⁄₈ in (33–41 mm)

CATERPILLAR LENGTH
1⁵⁄₁₆–1⁷⁄₁₆ in (33–37 mm)

PYRRHIA UMBRA

BORDERED SALLOW

(HUFNAGEL, 1766)

633

The Bordered Sallow caterpillar hatches from a ribbed, whitish egg laid on a leaf or flower. It feeds openly on the flowers and immature seeds of its food plant, thereby picking up different pigments from the wide variety of host plants favored. This probably explains the great variability of color in this species. When fully fed, the caterpillar forms a pupa in the ground, which overwinters. There is usually a single generation in the late summer and early fall; in warmer conditions, a partial second generation may occur.

The adult encloses between May and September and is similar in appearance to the closely related *Pyrrhia exprimens*, whose caterpillar is, however, more boldly marked than the Bordered Sallow. The range of *P. umbra*—the most extensive of its genus—was once thought to include North America. However, in 1996, the United States lookalike was described as a new species, *P. adela*, but is still commonly known as the Bordered Sallow or American Bordered Sallow.

The Bordered Sallow caterpillar can be light green, dark green, pinkish brown, gray, or blackish. Four white or yellow stripes occur along the back, with a darker stripe down the middle and a broader, white or yellow stripe on the sides, often dark-edged above. Paler forms are often peppered with white dots; darker forms frequently have conspicuous black warts. The head is plain brown or green.

Actual size

FAMILY	Noctuidae
DISTRIBUTION	Southern United States, in California, and from Kentucky and Maryland in the east, Oklahoma and Texas in the west, south to the Caribbean, Central America, and South America to Argentina
HABITAT	Diverse, including agricultural land, woodlands, grasslands, and wetlands
HOST PLANTS	Wide range of wild and cultivated plants, such as Sweet Potato (*Ipomoea batatas*), and even chemically defended plants such as rattlebox (*Crotalaria* spp.), greenbrier (*Smilax* spp.), and nightshade (*Solanum* spp.); also, in Florida, recorded on a semiaquatic plant, Dotted Knotweed (*Persicaria punctata*)
NOTE	Armyworm from a genus of highly destructive crop pests
CONSERVATION STATUS	Not evaluated, but common

ADULT WINGSPAN
1⅜₆ in (40 mm)

CATERPILLAR LENGTH
1⅜₆–2⅜ in (40–60 mm)

634

SPODOPTERA DOLICHOS
SWEET POTATO ARMYWORM
(FABRICIUS, 1794)

The Sweet Potato Armyworm caterpillar has a sausage-like appearance, smooth, thick, and segmented. It is brown, with black markings dorsally and a darker spiracular line, although the spiracles may have lighter markings. The markings on the first thoracic and last abdominal segments point anteriorly and posteriorly and are eyelike, an impression accentuated by additional posterior markings on every segment and white markings along the subdorsal line.

Sweet Potato Armyworm caterpillars hatch from as many as 4,000 eggs laid by a single female and feed on host plants from more than 40 families, illustrating their potential to be an explosive pest. However, their pest status is minor or negligible compared to some other species in the genus. The caterpillars are extensively parasitized by tachinid flies, and, while their eyespots may scare some predators, their only other defense is to fall to the ground when disturbed. The larvae are initially gregarious but disperse at later instars. They develop from hatching to pupation in around 23 days, with female larvae taking longer than males and undergoing six or seven instars. The larvae pupate in the soil within a silk-lined cell.

The genus *Spodoptera*, which evolved between 5 and 11 million years ago, includes more than 30 species. A number of species, such as the Fall Armyworm (*S. frugiperda*) and the Beet Armyworm (*S. exigua*), are highly destructive. Identification of both adults and larvae in the genus is difficult, and tiny characteristics may be used, such as the band on the thorax of the *S. dolichos* adult, which gave the species another common name, Banded Armyworm.

Actual size

FAMILY	Noctuidae
DISTRIBUTION	Throughout Africa, Macaronesia, southernmost Europe (rare migrant farther north to the British Isles), Madagascar, the Middle East, and western Asia
HABITAT	Rain forests, moist tropical forests, and many open, warm habitats, including farmland, rough ground, and gardens
HOST PLANTS	Many from more than 40 families, including Brassicaceae, such as cabbage (*Brassica* spp.); Asteraceae, such as Lettuce (*Lactuca sativa*); Fabaceae (legumes); Poaceae (grasses and cereals); and Euphorbiaceae (spurges)
NOTE	Significant, mainly African pest that is variable but quite distinctive
CONSERVATION STATUS	Not evaluated, but widespread and common

ADULT WINGSPAN
1⅜–1⁹⁄₁₆ in (35–40 mm)

CATERPILLAR LENGTH
1⁹⁄₁₆–1¾ in (40–45 mm)

SPODOPTERA LITTORALIS

AFRICAN COTTON LEAFWORM
(BOISDUVAL, 1833)

635

The eggs of the African Cotton Leafworm, also known as the Mediterranean Brocade, are laid in batches of several hundred, covered with brownish hairs from the tip of the female's abdomen. The caterpillar hides under a leaf when small, but when larger it may leave the plant to hide during the day. It feeds externally on the leaves, but on some hosts it also burrows into stems, such as Maize (*Zea mays*), or fruiting bodies, such as Tomato (*Solanum lycopersicum*). Severe attacks can destroy entire crops. The full-grown caterpillar pupates just beneath the soil surface.

Within two to four days of eclosing, the female will lay up to 2,000 eggs. The length of the life cycle can vary from 19 to 144 days, with up to seven generations a year recorded. This adaptable and prolific species is an important crop pest throughout its range, more so in the warmest areas, and is often accidentally transported on produce to other parts of the world. The Asian Cotton Leafworm (*Spodoptera litura*), with a more easterly range, is almost identical in all life stages.

The African Cotton Leafworm caterpillar is plump with a rather small head and somewhat hunched posture. It can be gray, brown, or gray green in color. The markings vary but often include a pair of bold, black spots on the fourth segment, sometimes forming a broken band, and often pairs of black or yellow spots, or both, along the back and sometimes yellow stripes.

Actual size

FAMILY	Noctuidae
DISTRIBUTION	Europe, including Portugal and northern Spain, southern Scandinavia, and most of the Mediterranean; Asia Minor, the Caucasus, and across central Asia to Korea and Japan
HABITAT	Open, rough vegetation in damp habitats, such as meadows and field edges, woodland edges, and banks
HOST PLANTS	Herbaceous plants, mainly orache (*Atriplex* spp.), goosefoot (*Chenopodium* spp.), knot-grass (*Polygonum* spp.), and dock (*Rumex* spp.)
NOTE	Striped caterpillar that is distinguished by two yellow spots
CONSERVATION STATUS	Not evaluated

ADULT WINGSPAN
1 9/16–2 in (40–50 mm)

CATERPILLAR LENGTH
1 11/16–1 7/8 in (43–48 mm)

TRACHEA ATRIPLICIS
ORACHE MOTH
(LINNAEUS, 1758)

636

The Orache Moth caterpillar hatches from a pale, domed, ribbed egg and feeds on the leaves, in one generation. It develops through all instars from July to September. When small, it rests under a leaf of the food plant; at later instars, it hides close to the ground during the day, emerging to feed at night. When fully fed, the caterpillar forms a pupa in a cocoon in the ground, and this stage overwinters.

Other noctuid caterpillars have markings that are broadly similar to those of the Orache Moth larvae. They include the Bright-line Brown-eye (*Lacanobia oleracea*) and Dog's Tooth (*L. suasa*), which can also be found on plants of the Chenopodiaceae family in similar situations. However, despite its wide range of coloration, the Orache Moth caterpillar is easily distinguished by the pair of yellow spots situated dorsolaterally near the tail end and the yellow spots at the sides near the head. These pairs of spots are absent in *Lacanobia* species.

The Orache Moth caterpillar is green, greenish gray, brown, or blackish, peppered with small, white dots, most noticeable in the darker forms. The head is usually brown. All color forms have two bright yellow spots on the back near the tail end. The caterpillar has a broad, yellow, whitish, or pinkish stripe along each side, replaced by yellow on the second and third segments.

Actual size

FAMILY	Noctuidae
DISTRIBUTION	North America, South America, Europe (except northern Scandinavia), much of Africa, and across Asia to China, Japan, Southeast Asia, and Malay Archipelago
HABITAT	Diverse, including tropical, subtropical, and temperate agricultural land
HOST PLANTS	Crucifers such as Cabbage (*Brassica oleracea*) and relatives; Beet (*Beta vulgaris*); cucurbits, such as watermelon (*Citrullus lanatus*); and solanaceous plants, such as Potato (*Solanum tuberosum*), Tomato (*Solanum lycopersicum*), and Tobacco (*Nicotiana tabacum*)
NOTE	Looper caterpillar with distinctive vestigal prolegs on two abdominal segments
CONSERVATION STATUS	Not evaluated, but common

ADULT WINGSPAN
1⅝₁₆–1½ in (33–38 mm)

CATERPILLAR LENGTH
1³₁₆–1⁹₁₆ in (30–40 mm)

TRICHOPLUSIA NI

CABBAGE LOOPER

(HÜBNER, [1803])

637

Cabbage Looper caterpillars hatch from yellowish-white or greenish eggs laid singly or in small groups on either side of a leaf. In early instars, they feed on the lower leaf surface but in the fourth and fifth instars make large holes in the center of host plant leaves. In cabbage, they can bore into the developing head, damage seedlings, or feed on wrapper leaves. They can consume three times their body weight daily, growing from egg to pupa in as little as 20 days; in some locations, there are seven generations per year. The fragile white cocoons are found under leaves, in leaf litter, or in soil.

Adults may migrate as far as 125 miles (200 km) from their breeding site; this, together with their polyphagous feeding, accounts for the species' extensive range. The caterpillars damage cruciferous crops and may occasionally attack many others. In addition to pesticides, their spread can be controlled by their natural enemies—wasp parasitoids, such as *Copidosoma truncatellus*, and tachinid flies, such as *Voria ruralis*. A nuclear polyhedrosis virus can also kill more than 40 percent of a caterpillar population.

Actual size

The Cabbage Looper caterpillar is mostly green but usually marked with a distinct white stripe on each side. It is thin and long throughout its development but becomes rather stout posteriorly approaching pupation, although remaining tapered toward the front end. It can be distinguished from other loopers by the vestigial prolegs located ventrally on abdominal segments three and four (the fully developed prolegs are on the fifth, sixth, and last segments).

FAMILY	Noctuidae
DISTRIBUTION	Pakistan, the Himalayas, India, Sri Lanka, Southeast Asia, China, Japan, Chinese Taipei, northern Australia, and Melanesia
HABITAT	Low- and medium-altitude mountains
HOST PLANTS	Mallows (Malvaceae), including *Hibiscus* spp., Okra (*Abelmoschus esculentus*), Caesarweed (*Urena lobata*), and Hollyhock (*Alcea rosea*)
NOTE	Semi-looper caterpillar of regional agricultural significance
CONSERVATION STATUS	Not evaluated, but widely distributed and common

ADULT WINGSPAN
1⅜–1⁹⁄₁₆ in (35–40 mm)

CATERPILLAR LENGTH
1⁹⁄₁₆–1¾ in (40–45 mm)

XANTHODES TRANSVERSA
TRANSVERSE MOTH
GUENÉE, 1852

Early instar Transverse Moth larvae are green and well camouflaged against host plant foliage and spend most of the time on the underside of leaves. Patterns and coloring change as they develop, with each molt becoming more aposematic (green, black, yellow, and red), and, by the final (sixth) instar, caterpillars spend all their time on the leaf surface. The final instar is also defined by a vibrant red spot on the rear abdominal segment, probably intended to distract potential predators from the head end.

Pupation occurs underground, and, in the northern range, caterpillars enter a prepupal diapause over the winter months. Otherwise, the Transverse Moth is multivoltine (having two or more broods annually) and completes larval development in four to six weeks. The species is a recognized commercial pest of Okra across its expansive range. The *Xanthodes transversa* moth is yellow with a geometric pattern of angular brown lines on the forewings.

Actual size

The Transverse Moth caterpillar is a semi-looper, with two pairs of prolegs plus anal claspers. It is green in color with dorsal and lateral yellow stripes, the latter incorporating the spiracles. All abdominal segments bear a trio of black dots either side of the midline, each the source of a lengthy black seta, between black dashes. Lateral setae are white. The head capsule is green with black spots and a yellow brow, and the dorsum of the rear segment is a vivid red.

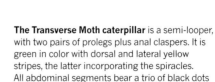

FAMILY	Noctuidae
DISTRIBUTION	Europe (including Iceland, Scotland, and southern Scandinavia), east across southern Russia to Russian Far East; Canary Islands, northwest Africa, Asia Minor, the Caucasus, the Middle East, and across central Asia to Korea and Japan
HABITAT	Northern moorlands, marshes, and upland pastures; also warm rocky slopes, open woodlands, lowland heathlands, and scrubby grasslands
HOST PLANTS	Many, including buttercup (*Ranunculus* spp.), hawk's-beard (*Crepis* spp.), dock (*Rumex* spp.), knapweed (*Centaurea* spp.), Bramble (*Rubus fruticosus*), and Bog Myrtle (*Myrica gale*)
NOTE	Beautifully marked caterpillar that can be found by day
CONSERVATION STATUS	Not evaluated

ADULT WINGSPAN
2¹⁄₁₆–2½ in (52–64 mm)

CATERPILLAR LENGTH
2⅜–2⁹⁄₁₆ in (60–65 mm)

XYLENA EXSOLETA
SWORD-GRASS
(LINNAEUS, 1758)

639

Sword-grass caterpillars hatch from brownish-yellow eggs laid in batches on low vegetation, in May, and are fully fed in July, when they form a pupa in a delicate cocoon in the ground. It does not always hide by day, even when large, and feeds openly by both day and night. There is a single generation, with the adults emerging in the fall, overwintering, then mating and living until May.

In the final instar, the strikingly marked Sword-grass caterpillar, with bold black, orange, yellow, and white markings, those on the back often like bicycle chain links, is usually unmistakable. However, the black-and-orange markings vary and do not appear until the final instar, so the caterpillar can resemble other noctuids, such as the closely related Red Sword-grass (*Xylena vetusta*), although that species' red or yellow spiracles differ from the whitish ones of *X. exsoleta*. The Sword-grass common name refers to an old term for sedge (Cyperaceae), thought to be the larval food plant when the species was first described.

The Sword-grass caterpillar is smooth and quite cylindrical, with a large head, and bright green or blue green in color. In the final instar, it has two dorsal lines of black marks, each containing two white spots and, below this, a yellow stripe. On the sides, there are groups of white spots and bright orange dashes, just above a white stripe.

Actual size

GLOSSARY

Aggregation Gathering of a large number of individuals, usually for protection.

Anal comb Sclerotized, ventral projection above the anus that larvae, mainly Hesperiidae, use to eject frass.

Anterior Toward the front.

Apex Highest point; the tip.

Aposematic, aposematically Brightly colored, usually as a warning defense by toxic organisms.

Bagworm Caterpillar of the Psychidae family that constructs a case around itself from materials within its habitat.

Barren Type of scrubby woodland usually dominated by one or two tree species.

Batesian mimicry Form of mimicry in which a harmless species has evolved to resemble a harmful species in order to avoid predation.

Biennial Having a two-year cycle.

Bifid Divided or branched into two parts.

Birdwing Butterflies in the Papilionidae family that belong to the genera *Trogonoptera*, *Troides*, and *Ornithoptera*.

Bluff Steep headland, promontory, riverbank, or cliff.

Boreal Of cool, northern climes.

Brood Group of young (larvae) hatched at one time from the same mother.

Brushfoot Butterflies in the family Nymphalidae.

Cauda (pl. **caudae**) Tail or any taillike process.

Caudal Of, or pertaining to, the tail or taillike process.

Chalaza Hardened lump on a caterpillar that bears more than one seta.

Chrysalis (pl. **chrysalids** or **chrysalides**) Pupal (resting) stage before adulthood.

CITES Convention on International Trade in Endangered Species of Wild Fauna and Flora, a multilateral agreement that came into force in 1975 and is designed to ensure that international trade in wildlife and plant specimens does not threaten any species' survival.

Claspers Posterior pair of prolegs, also known as anal prolegs.

Cloud forest Tropical or subtropical, evergreen, montane moist forest characterized by persistent cloud cover.

Clypeus Shield-like plate at the front of the head between the labrum and the frons.

Cocoon Silken sheath or envelope surrounding a pupa or chrysalis.

Cremaster Support hook or cluster of small hooks at the terminal end of a pupa.

Crochets Hooklike structures on abdominal prolegs.

Crypsis Ability of an organism to conceal or disguise itself to avoid detection.

Cryptic Concealed, camouflaged, or not easily recognized.

Cuticle Outer covering of an insect, containing chitin and wax.

Cuticular Of, or pertaining to, the cuticle.

Cyanogenic Capable of producing hydrogen cyanide; some caterpillars acquire toxic cyanogenic compounds from plants, making the larvae distasteful to predators, and some species are also able to create the chemicals themselves.

Diapause Period of physiologically controlled dormancy in which development or activity is suspended, and the organism is unresponsive to normally favorable stimuli, such as warmer temperatures.

Dimorphic Occurring in two different forms; a term used especially for sexual differences within a species.

Disruptive coloration Patterning and coloration serving to help an organism blend into the background.

Disturbed Describing a habitat or environment altered by man.

Diurnal Active during daylight hours.

Dorsal Pertaining to, or of, the dorsum (back).

Dorsolateral Located between the side and the back.

Dorsum On, or near, the back.

Ecdysial suture Thin areas in the cuticle that rupture during a molt and allow a new instar, or stage, to issue from the old integument.

Ecdysis Process of molting in which the old growth-limiting exoskeleton splits, which allows a larger stage to emerge and grow.

Eclosion Emergence of an adult insect from a pupa, or a larva from an egg.

Endophagous Feeding within a host.

Epicranium Upper part of the head capsule.

Epidermis Cellular layer of the integument that lies beneath and secretes the cuticle.

Estivation Period of summer dormancy or torpor to avoid stress in excessively hot or dry conditions.

Eversible Capable of being everted, turned outward or inside out; used in regard to a larval osmeterium.

Exudate Any substance discharged from a cell, organ, or organism.

Filament Threadlike, slender process of uniform diameter.

Flagellum (pl. **flagella**) Lash or filament-like appendage.

Flocculent Description of soft, woollike flakes covering a caterpillar body.

Frass Caterpillar fecal material.

Frass pier Narrow extension of a leaf vein or small stick, composed of frass and silk produced by a caterpillar, especially larvae of admiral butterflies of the Nymphalidae family, and on which the caterpillar may rest.

Frons Front part of the head above the clypeus and between the eyes.

Gall Abnormal growth or swelling of plant tissue caused by insects, mites, bacteria, fungi, or disease.

Glaucous Covered with a fine powdery or waxy bloom, causing the surface to appear whitish or bluish.

Gregarious Grouping together, usually for defense.

Hair pencil Cluster of hairlike scent scales, everted by some butterflies and moths during courtship, and from which pheromones are emitted.

Hammock Rich, tropical, hardwood forest on rocky upland sites in Florida.

Hilltopping Active aggregation of male butterflies at an elevated site (hilltop) seeking females for courtship.

Horn Enlarged spike or spine on the anterior or posterior end of some caterpillars.

Instar Larval growth stage between molts, or the caterpillar at that growth stage before molting.

Integument Outer covering or cuticle of the caterpillar's body, also known as the exoskeleton.

Iridoid glycoside Toxic chemical that caterpillars ingest from plants, making the larvae distasteful to predators.

Labium Lower hardened "lip" of an insect's mouth.

Labrum Upper hardened "lip" of an insect's mouth.

Leafminer Caterpillar that creates a tunnel between the upper and lower surfaces of a leaf.

Light gap Break in the forest canopy that allows young plants and shrubs to grow.

Littoral rain forest Coastal rain forest in eastern Australia adapted to salt-laden and drying winds.

Looper Common name for geometrid or other caterpillars that lack some or all of the middle abdominal legs, creating "looping" locomotion.

Mandibles Either of the two-toothed jaws of a caterpillar, used for biting and chewing food.

Melanic Darkened coloring, brown to black, often contrasting with less melanized tissue.

Melanization Increase in the amount of melanin (brown-black pigmentation), which results in darkening of the color of some individuals.

Meristem Plant tissue containing unspecialized cells that divide continually, enabling growth to take place.

Mesic Habitats with intermediate amounts of moisture; not decidedly wet or dry.

Mesophyll Internal part of a leaf, containing chlorophyll and located between the upper and lower epidermis.

Mesothorax Second (middle) segment of the thorax.

Metathorax Third (rear) segment of the thorax.

Mine Feeding tunnel of a caterpillar within a leaf or other part of a plant.

Molt Shedding of exoskeleton, or skin, to allow for growth.

Monocot Plant sprouting from a seed with a single leaf (grasses and sedges, for example).

Montane Of high altitude or mountainous habitats.

Morph Form.

Müllerian mimicry Form of mimicry in which two poisonous species that share a predator, or several predators, have evolved to resemble one another to avoid predation (*see also* Batesian mimicry).

Myrmecophilous Living symbiotically (in close association) with ants.

Ocherous Of the color of ocher, a natural earth pigment (light brownish yellow).

Oligophagous Using several closely related host plants.

Ontogeny Sequence of developmental events that collectively characterize an individual.

Osmeterium Y-shaped gland located behind the head of caterpillars in the family Papilionidae that can be everted to emit a chemical disagreeable to potential predators.

Oviposition Laying eggs, ovipositing.

Owlet Common name for moth species in the family Noctuidae, and used by some experts also for moth species of Nolidae, Erebidae, and Eutellidae.

Paddle Flattened or compressed, paddle-like hair or seta.

Paperbark woodland Woodland that is comprised of *Melaleuca* trees.

Parasitoid Organism that develops within a host and ultimately kills it.

Peduncle Stalk or stem supporting an organ or other kind of structure.

Peristaltic Having an action suggestive of peristalsis, the wavelike contraction that occurs along the gut and other body tubes as food is digested.

Petiole Stalk that attaches a leaf to the stem.

Pheromone Chemical produced by an insect that mediates behavior of other individuals of the same species.

Pier *see* Frass pier.

Pinna (pl. **pinnae**) Leaflet or primary division of a pinnate leaf on a plant, such as bracken, that has leaflets (pinnae) arranged opposite each other on a stem.

Plumose Feather-like or multiply branched, or divided.

Pocosin Inland wetland in the southern United States.

Polymorphic Occurrence of different forms or types in organisms of the same species, independent of sexual variations.

Polyphagous Feeding on a wide variety of host plants, including different plant families.

Proboscis Slender tubular feeding organ of adult butterflies and moths.

Process Projection from a surface, margin, or appendage.

Processional Describing the nose-to-tail processional movement of caterpillars of some moth species.

Prolegs Fleshy, unjointed structures found on the ventral surface of the abdomen of caterpillars; used for larval movement but generally lacking musculature.

Pronotal Of, or pertaining to, the dorsal plate of the prothorax, the first segment of the thorax.

Pronotum Dorsal plate of the prothorax.

Prothorax First (front), or prothoracic, segment of the thorax.

Pubescent Hairy or downy.

Puddling Feeding by butterflies and moths on wet mud or sand to imbibe salts and minerals.

Pupa (pl. **pupae**) Generally inactive developmental stage of butterflies and moths, intermediate between larva and adult, when the caterpillar pupates, or enters pupation, during which its cells are reorganized into the adult form.

Rangeland Grasslands, shrublands, wetlands, and deserts grazed by domestic livestock or wild animals.

Refugium (pl. **refugia**) Place providing protection or shelter.

Relict Organism or ecosystem that originally occupied a large expanse but is now narrowly confined.

Riparian Areas of land adjacent to streams or rivers.

Rugose Having wrinkles.

Satyrid, satyrine Denoting butterflies of the Satyrinae subfamily within the family Nymphalidae.

Scale Flat seta, many of which cover the wings of butterflies and moths and occur in various shapes and colors.

Scat Animal fecal dropping.

Sclerotized Part of the body that has been hardened by sclerotin.

643

Scolus (pl. **scoli**) Tubercles in the form of spinose projections of the body wall.

Senesce Decline with maturity, often hastened by environmental stress; to dry out seasonally (plants).

Sensilla Simple sense organ or sensory receptor that occurs on an insect body.

Sequester Ability of caterpillars to absorb plant toxins and then use them as defense in the adult stage.

Seta (pl. **setae**) Bristle or hair.

Skeletonize Pattern of leaf feeding by caterpillars leaving only a "skeleton" of veins.

Skipper Butterflies in the family Hesperiidae, named for their "skipping" flight.

Slug caterpillar Caterpillars of moths in the family Limacodidae that resemble slugs.

Speciose Rich in species.

Spicule Narrow, pointed, hardened, prickle-like projection.

Spiracle External tracheal opening on the side of an insect's thorax and abdomen through which air and water vapor pass in and out during respiration.

Spiracular Of, or pertaining to, the spiracle.

Stemmata Lateral eyes of caterpillars and arranged in a semicircle.

Stipule Small, stalk-like outgrowths at the base of a leaf petiole.

Swallowtails Butterflies in the family Papilionidae named for the short tail or tails on each hindwing.

Synchronous Occurring at the same time—for example, the hatching of a batch of eggs.

Thoracic Of, or pertaining to, the thorax.

Thorax Portion of the body between the head and the abdomen.

Tracheole Small, delicate tubules radiating out from the spiracles.

Treefall gap Distinguishable hole in a forest with vertical sides, usually caused by a falling tree or large limb.

Trichome Very fine bristle or hairlike structure.

True leg Jointed walking structure on the thorax of insect larvae and adults, also known as thoracic leg.

Tubercle Knobby protuberance on a caterpillar or pupa.

Urticating Stinging.

Vegetative succession Natural temporal sequence of flora in a habitat from grasses to bushes to trees.

Venter Midline of the underside or lower surface.

Ventral Of, or pertaining to, the lower side, or ventrum.

Ventrolateral Located between the side and the lower surface.

Ventrum Underside or lower surface.

Vermiform Wormlike.

Verrucae Elevated, chitinous plate on the integument, often bearing a group of radiating setae.

Webbing Silk produced by caterpillars for support or for defense, or for both.

Xeric Characterized by, or tolerant of, dry conditions.

CLASSIFICATION TERMS

Order Primary taxonomic rank or unit used for classifying organisms.

Suborder Division of an order higher than a family.

Superfamily Category below order and above family that includes a series of closely related family groups.

Family Level in the taxonomic hierarchy below the order and above the genus.

Genus Assemblage of species agreeing in one character or series of characters; the genus is indicated in the first part of the Latin species name—for example, by "*Danaus*" in the Monarch butterfly's scientific name, *Danaus plexippus*.

Species Primary biological unit; individuals similar in appearance and structure mating freely and producing young that themselves mate freely and bear fertile offspring resembling each other and their parents. This is indicated in the second part of the scientific name. For example, in the name *Danaus plexippus*, "*plexippus*" is the species within the genus *Danaus*.

Subspecies Geographical or host variation of a species, denoted by an additional defining scientific name, such as "*langei*" in *Apodemia mormo langei*.

Unassigned Designation when morphology and taxonomy of a species, family, or superfamily is too confused or insufficiently researched to be assigned full taxonomic status at all levels.

RESOURCES

The following is a selection of useful books, scientific journal articles, and websites currently available for those with an interest in caterpillars and Lepidoptera in general.

BOOKS

Allen, T. J., Brock, J. P., and J. Glassberg. *Caterpillars in the Field and Garden: A Field Guide to the Butterfly Caterpillars of North America* OXFORD UNIVERSITY PRESS, 2005

Carter, D. and B. Hargreaves. *Caterpillars of Britain and Europe* COLLINS FIELD GUIDE, 1986

Crafer T. *Foodplant List for the Caterpillars of Britain's Butterflies and Larger Moths* ATROPOS PUBLISHING, 2005

James, D. G. and D. Nunnallee. *Life Histories of Cascadia Butterflies* OREGON STATE UNIVERSITY PRESS, 2011

Miller, J. C. and P. C. Hammond. *Lepidoptera of the Pacific Northwest: Caterpillars and Adults* USDA, 2003

Miller, J. C., Janzen, D. H., and W. Hallwachs. *100 Caterpillars: Portraits from the Tropical Forests of Costa Rica* HARVARD UNIVERSITY PRESS, 2006

Minno, M. C., Butler, J. F., and D. W. Hall. *Florida Butterfly Caterpillars and their Host Plants* UNIVERSITY PRESS OF FLORIDA, 2005

Porter J. *The Colour Identification Guide to Caterpillars of the British Isles* BRILL, 2010

Scott, J. A. *The Butterflies of North America: A Natural History and Field Guide* STANFORD UNIVERSITY PRESS, 1986

Wagner, D. L. *Caterpillars of Eastern North America: A Guide to Identification and Natural History* PRINCETON UNIVERSITY PRESS, 2005

Wagner, D. L., Schweitzer, D. F., Bolling Sullivan, J., and R. C. Reardon. *Owlet Caterpillars of Eastern North America* PRINCETON UNIVERSITY PRESS, 2011

Waring, P. and M. Townsend. *Field Guide to the Moths of Great Britain and Ireland* (3rd edition) BLOOMSBURY, 2017

SCIENTIFIC JOURNAL ARTICLES

Greeney, H. F., Dyer, L. A., and A. M. Smilanich. Feeding by lepidopteran larvae is dangerous: A review of caterpillars' chemical, physiological, morphological, and behavioral defenses against natural enemies. *Invertebrate Survival Journal* 9: 7–34 (2012)

James D. G., Seymour, L., Lauby, G., and K. Buckley. Beauty with benefits: Butterfly conservation in Washington State, USA, wine grape vineyards. *Journal of Insect Conservation* 19: 341–348 (2015)

Stireman, J. O., Dyer, L. A., Janzen, D. H., Singer, M. S., Lill, J. T., Marquis, R. J., Ricklefs, R. E., Gentry, G. L., Hallwachs, W., Coley, P. D., Barone, J. A., Greeney, H. F., Connahs, H., Barbosa, P., Morais, H. C., and R. Diniz. Climatic unpredictability and parasitism of caterpillars: Implications of global warming. Proceedings of the National Academy of Sciences of the United States of America 102: 17384–17387 (2005)

Van Ash, M. and M. E. Visser. Phenology of forest caterpillars and their host trees: The importance of synchrony. Annual Review of Entomology 52: 37–55 (2007)

NATIONAL AND INTERNATIONAL ORGANIZATIONS DEDICATED TO THE STUDY AND CONSERVATION OF LEPIDOPTERA AND OTHER INSECTS

Amateur Entomologists' Society [UK]
https://www.amentsoc.org

Australian National Insect Collection
http://www.csiro.au/en/Research/Collections/ANIC

Buglife [UK]
https://www.buglife.org.uk

Butterfly Conservation Europe
http://www.bc-europe.eu

Butterfly Conservation [UK]
http://butterfly-conservation.org

Les Lépidoptéristes de France
https://www.lepido-france.fr

The Lepidopterists' Society of Africa
http://www.lepsoc.org.za

The Lepidopterists' Society [USA]
https://www.lepsoc.org

McGuire Center for Lepidoptera & Biodiversity [USA]
https://www.flmnh.ufl.edu/index.php/mcguire/home

Xerces Society [USA]
http://www.xerces.org

USEFUL WEBSITES

Australian Caterpillars and their Butterflies and Moths
http://lepidoptera.butterflyhouse.com.au

BugGuide
http://www.bugguide.net

Butterflies and Moths of North America
http://www.butterfliesandmoths.org

Butterflies of America
http://butterfliesofamerica.com

Eggs, Larvae, Pupae and Adult Butterflies and Moths [UK]
http://www.ukleps.org/index.html

HOSTS – a Database of the World's Lepidopteran Hostplants
http://www.nhm.ac.uk/our-science/data/hostplants

The Kirby Wolfe Saturniidae Collection
http://www.silkmoths.bizland.com/kirbywolfe.htm

Larvae of North-European Lepidoptera
http://www.kolumbus.fi/silvonen/lnel/species.htm

Learn about Butterflies
http://www.learnaboutbutterflies.com

Lepidoptera and their ecology
http://www.pyrgus.de/index.php?lang=en

The Monarch Larva Monitoring Project
http://monarchlab.org/mlmp

Monarch Watch
http://www.monarchwatch.org

Sphingidae of the Western Palearctic
http://tpittaway.tripod.com/sphinx/list.htm

645

CLASSIFICATION
of the LEPIDOPTERA

When the Swedish naturalist Carl Linnaeus first classified Lepidoptera in his *Systema Naturae* (1758), he divided them into just three genera. His genus *Papilio* included all known butterflies, while large hovering hawkmoths were placed in the genus *Sphinx*, and the rest of the moths in the genus *Phalaena*, which Linnaeus divided into seven subgroups.

By contrast, the classification table shown here (based on van Nieukerken *et al.*, 2011) lists 45 superfamilies and 137 families that today contain about 16,000 genera with an estimated 160,000 named species (the number of species in each family appears in square brackets). The 37 families marked with an asterisk (*) are represented in this book; most of the others are comprised of obscure and tiny moth species, whose larval life cycle is frequently unknown.

The system initiated by Linnaeus is being continuously refined as scientists seek to establish a classification that reflects the evolutionary history of Lepidoptera, which began approximately 200 million years ago and continues today. The few available Lepidoptera fossils help indicate how early certain defining characteristics were present. For instance, some primitive moths living now have chewing mandibles instead of a proboscis and feed on a solid diet such as fern spores, as did the small moth of the earliest known fossil, dating back some 190 million years. From fossils of host plant leaves, it is also hypothesized that many Lepidoptera were initially tiny "leafminers," whose larvae tunneled through the mesophyll layer of a leaf, as caterpillars of *Leucoptera erythrinella* and many other species still do now. The oldest known skipper (Hesperiidae) butterfly fossil is 56 million years old, and other 30 to 40 million-year-old fossilized butterflies, while representing extinct species, share many traits with their present-day relatives.

Taxonomists use various means for defining species, such as morphology (internal and external structural features), DNA sequences, and ecology (how a species interacts with its surroundings). Each new species receives a two-part name, for example, "*Danaus plexippus*"—the first part reflecting its genus and the second the species within that genus. The author and year of its original description may also appear, such as "(Linnaeus, 1758)"—in parentheses because Linnaeus originally placed this species, the Monarch butterfly, in the genus *Papilio*, and it was later assigned the genus name *Danaus*. Square brackets in an author citation indicate uncertainties about the author or date of description.

The classification of Lepidoptera remains very much a work in progress, with thousands of species still to be described. Despite the general advancement of science, according to Professor James Mallet, who was awarded the prestigious Darwin–Wallace Medal for major advances in evolutionary biology, there is still "no easy way to tell whether related geographic or temporal forms belong to the same or different species." His insight, which comes from studying evolution in *Heliconius* butterflies, echoes Darwin himself, who better than most understood the enormity of the taxonomist's task.

ORDER LEPIDOPTERA

Unassigned early lepidopterans
Family Archaeolepidae [1]
Family Mesokristenseniidae [3]
Family Eolepidopterigidae [1]
Family Undopterigidae [1]

Suborder Zeugloptera
SUPERFAMILY MICROPTERIGOIDEA
Family Micropterigidae [160]

Suborder Aglossata
SUPERFAMILY AGATHIPHAGOIDEA
Family Agathiphagidae [2]

Suborder Heterobathmiina
SUPERFAMILY HETEROBATHMIOIDEA
Family Heterobathmiidae [3]

Suborder Glossata
SUPERFAMILY ERIOCRANIOIDEA
Family Eriocraniidae [29]

SUPERFAMILY ACANTHOPTEROCTETOIDEA
Family Acanthopteroctetidae [5]

SUPERFAMILY LOPHOCORONOIDEA
Family Lophocoronidae [6]

SUPERFAMILY NEOPSEUSTOIDEA
Family Neopseustidae [14]

SUPERFAMILY MNESARCHAEOIDEA
Family Mnesarchaeidae [7]

SUPERFAMILY HEPIALOIDEA
Family Palaeosetidae [9]
Family Prototheoridae [12]
Family Neotheoridae [1]
Family Anomosetidae [1]
Family Hepialidae [606]

SUPERFAMILY NEPTICULOIDEA
Family Nepticulidae [819]
Family Opostegidae [192]

SUPERFAMILY ANDESIANOIDEA
Family Andesianidae [3]

SUPERFAMILY ADELOIDEA
Family Heliozelidae [123]
Family Adelidae [294]
Family Incurvariidae [51]
Family Cecidosidae [16]
Family Prodoxidae [98]

SUPERFAMILY PALAEPHATOIDEA
Family Palaephatidae [57]

SUPERFAMILY TISCHERIOIDEA
Family Tischeriidae [110]

SUPERFAMILY UNASSIGNED
Family Millieriidae [4]

SUPERFAMILY TINEOIDEA
Family Eriocottidae [80]
Family Psychidae* [1,350]
Family Tineidae* [2,393]

SUPERFAMILY GRACILLARIOIDEA
Family Roeslerstammiidae [53]
Family Bucculatricidae [297]
Family Gracillariidae [1,866]

SUPERFAMILY YPONOMEUTOIDEA
Family Yponomeutidae* [363]
Family Argyresthiidae [157]
Family Plutellidae [150]
Family Glyphipterigidae [535]
Family Ypsolophidae [163]
Family Attevidae [52]
Family Praydidae [47]
Family Heliodinidae [69]
Family Bedelliidae [16]
Family Lyonetiidae* [204]

SUPERFAMILY UNASSIGNED
Family Prodidactidae [1]
Family Douglasiidae [29]

SUPERFAMILY SIMAETHISTOIDEA
Family Simaethistidae [4]

SUPERFAMILY GELECHIOIDEA
Family Autostichidae [638]
Family Lecithoceridae [1,200]
Family Xyloryctidae [524]
Family Blastobasidae [377]
Family Oecophoridae [3,308]
Family Schistonoeidae [1]
Family Lypusidae [21]
Family Chimabachidae [6]
Family Peleopodidae [28]
Family Elachistidae* [201]
Family Syringopaidae [1]
Family Coelopoetidae [3]
Family Stathmopodidae [408]
Family Epimarptidae [4]
Family Batrachedridae [99]
Family Coleophoridae [1,386]
Family Momphidae [115]
Family Pterolonchidae [8]
Family Scythrididae [669]
Family Cosmopterigidae [1,792]
Family Gelechiidae [4,700]

SUPERFAMILY ALUCITOIDEA
Family Tineodidae [19]
Family Alucitidae [216]

SUPERFAMILY PTEROPHOROIDEA
Family Pterophoridae* [1,318]

SUPERFAMILY CARPOSINOIDEA
Family Copromorphidae [43]
Family Carposinidae [283]

SUPERFAMILY SCHRECKENSTEINIOIDEA
Family Schreckensteiniidae [8]

SUPERFAMILY EPERMENIOIDEA
Family Epermeniidae [126]

SUPERFAMILY URODOIDEA
Family Urodidae [66]

SUPERFAMILY IMMOIDEA
Family Immidae [245]

SUPERFAMILY CHOREUTOIDEA
Family Choreutidae [406]

SUPERFAMILY GALACTICOIDEA
Family Galacticidae [19]

SUPERFAMILY TORTRICOIDEA
Family Tortricidae* [10,387]

SUPERFAMILY COSSOIDEA
Family Brachodidae [137]
Family Cossidae* [971]
Family Dudgeoneidae [57]
Family Metarbelidae [196]
Family Ratardidae [10]
Family Castniidae* [113]
Family Sesiidae [1,397]

SUPERFAMILY ZYGAENOIDEA
Family Epipyropidae [32]
Family Cyclotornidae [5]
Family Heterogynidae [10]
Family Lacturidae [120]
Family Phaudidae [15]
Family Dalceridae [80]
Family Limacodidae* [1,672]
Family Megalopygidae [232]
Family Aididae [6]
Family Somabrachyidae [8]
Family Himantopteridae [80]
Family Zygaenidae* [1,036]

SUPERFAMILY WHALLEYANOIDEA
Family Whalleyanidae [2]

SUPERFAMILY THYRIDOIDEA
Family Thyrididae* [940]

SUPERFAMILY HYBLAEOIDEA
Family Hyblaeidae [18]

SUPERFAMILY CALLIDULOIDEA
Family Callidulidae [49]

SUPERFAMILY PAPILIONOIDEA
Family Papilionidae* [570]
Family Hedylidae [36]
Family Hesperiidae* [4,113]
Family Pieridae* [1,164]
Family Riodinidae [1,532]
Family Lycaenidae* [5,201]
Family Nymphalidae* [6,152]

SUPERFAMILY PYRALOIDEA
Family Pyralidae* [5,921]
Family Crambidae* [9,655]

SUPERFAMILY MIMALLONOIDEA
Family Mimallonidae [194]

SUPERFAMILY DREPANOIDEA
Family Cimeliidae* [6]
Family Doidae [6]
Family Drepanidae [660]

SUPERFAMILY LASIOCAMPOIDEA
Family Lasiocampidae* [1,952]

SUPERFAMILY BOMBYCOIDEA
Family Apatelodidae* [145]
Family Eupterotidae [339]
Family Brahmaeidae* [65]
Family Phiditiidae [23]
Family Anthelidae* [94]
Family Carthaeidae [1]
Family Endromidae* [59]
Family Bombycidae* [185]
Family Saturniidae* [2,349]
Family Sphingidae* [1,463]

SUPERFAMILY GEOMETROIDEA
Family Epicopeiidae* [20]
Family Sematuridae [40]
Family Uraniidae [686]
Family Geometridae* [23,002]

SUPERFAMILY NOCTUOIDEA
Family Oenosandridae [8]
Family Notodontidae* [3,800]
Family Erebidae* [24,569]
Family Euteliidae [520]
Family Nolidae* [1,738]
Family Noctuidae* [11,772]

647

INDEX *by* COMMON NAME

649

INDEX *by* SCIENTIFIC NAME

653

NOTES *on* CONTRIBUTORS

DAVID G. JAMES is an associate professor of entomology at Washington State University, United States, who developed a passion for the subject in England at the age of eight, rearing caterpillars in his bedroom. He studied zoology at the University of Salford, and then migrated to Australia to work for the New South Wales Department of Agriculture on ways of controlling agricultural pests without pesticides. After completing a PhD on the winter biology of Monarch butterflies in Sydney, he pursued a career as a biocontrol scientist in horticulture, developing a number of successful conservation biological control systems. David has published almost 180 peer-reviewed scientific papers on a wide range of entomological subjects, focusing on insect biology and management. He recently coauthored *Life Histories of Cascadia Butterflies*, a comprehensive and highly regarded study on the immature stages of Pacific Northwest butterflies.

Contributions: pages 46, 54, 56–7, 61, 64, 70, 78, 83–4, 87–91, 99, 102, 104–5, 113, 117–18, 122, 125, 129–30, 132–5, 137–8, 143, 145, 147–50, 153, 155, 164, 169, 170, 173–4, 179–81, 186, 189–93, 197, 199, 204, 210–11, 216, 218, 221, 225, 227–8, 241, 246, 259–60, 267, 269, 271, 280–2, 296, 326, 335, 350, 352, 370, 385, 396, 405, 413, 416–18, 435–7, 440–2, 444, 446–7, 449, 456, 459, 462–6, 468–9, 471, 475–6, 478, 539, 558, 592, 602, 621, 632

DAVID ALBAUGH is a part-time educator and a full-time hobbyist with a passion for conservation, raising, and releasing local species of butterflies and moths. He visits many schools and libraries to teach about the importance of insects and arachnids, using both preserved and live specimens. He has also raised many exotic species and currently manages the horticulture department at Roger Williams Park Zoo in Providence, Rhode Island, United States.

Contributions: pages 44, 369, 374, 383, 386, 402

BOB CAMMARATA is a freelance photographer and passionate naturalist, whose award-winning images and articles have been published in travel brochures, calendars, and business journals, and featured on myriad websites chronicling wildlife and nature photography. Bob also organizes macro-photography lectures and local "bug safari" field trips. His decades-long interest in lepidopteran larvae is fueled by the ostensibly endless flood of behavioral diversity and photographic potential that they have to offer.

Contributions: pages 60, 298, 301, 303, 305, 311, 504, 511–12, 515, 523–4, 527, 532, 543, 554, 613, 631

ROSS FIELD undertook his undergraduate studies in agricultural science at the University of Melbourne, Australia, and in 1981 completed his PhD in entomology at the University of California, Berkeley, focusing on pest management and biological control. He worked for 45 years in the Victorian Public Service in research and management positions, including the Director of Natural Sciences at Museum Victoria. Ross has spent more than 60 years collecting and studying the life histories of Australia's butterflies and in 2013 published an award-winning guide on the biology of Victoria's 130 butterfly species, *Butterflies: Identification and Life-history*. He has published widely on butterflies, pest management, and conservation.

Contributions: pages 40, 42, 45, 47, 50, 72, 79, 85, 92, 94–5, 97, 100, 109, 114, 119, 121, 123–4, 151–2, 157–9, 162, 165–7, 176, 205, 207, 220, 234, 238–9, 242, 256, 265–6, 279, 283, 293, 295

HAROLD GREENEY is a broadly trained natural historian with degrees in biology, entomology, and ornithology. He created the Yanayacu Biological Station & Center for Creative Studies in the Ecuadorian Andes, working there for more than 15 years; research into caterpillars and parasitoids of the Andes is one of the station's longest ongoing projects. Harold is the author of over 250 scientific papers and a recipient of the Alexander & Pamela Skutch Award and a Guggenheim Fellowship. He lives in Tucson, Arizona, United States.

Contributions: pages 66–7, 69, 73, 75–6, 80–2, 86, 98, 107, 110–11, 116, 128, 131, 139, 140–2, 184–5, 203, 212–13, 215, 219, 226, 231, 233, 235–7, 240, 244, 248, 252, 255, 262–4, 273, 275–7, 537, 551–3, 557, 559, 562, 565, 569, 571, 574, 576, 578, 580

JOHN HORSTMAN is an Australian living in Yunnan, China. A teenage passion for the biological sciences, entomology, and photography evolved into an all-consuming pastime, with a particular focus on China's rich diversity of Lepidoptera, specifically at the immature stages. From his relatively isolated but biodiverse location, John delivers unfamiliar and exotic visuals through social media platforms such as *itchydogimages* and *SINOBUG*. These include a large, diverse, and spectacular collection of mostly unidentified Limacodidae caterpillar images.

Contributions: pages 51, 58, 74, 178, 195–6, 209, 214, 222–3, 229, 243, 257, 272, 274, 278, 300, 306–7, 309, 310, 312, 314, 316–17, 319, 479, 505, 518, 521, 541, 550, 585, 587, 638

SALLY MORGAN is an author and photographer. A fascination with the natural world led her to study biologearcl sciences at Cambridge University, United Kingdom. She has written more than 250 books, covering a wide range of topics on natural history and the environment, and has traveled the world in her search of the exotic and unusual to photograph and feature. Sally owns an organic farm in the county of Somerset, England, which has an impressive diversity of lepidopterans, including one-third of all the species on the British butterfly list.

Contributions: pages 38, 43, 52–3, 55, 63, 65, 77, 96, 101, 106, 120, 126–7, 144, 146, 156, 160–1, 163, 171–2, 175, 177, 182, 187–8, 198, 200–2, 206, 208, 224, 245, 247, 250–1, 253, 258, 270, 288, 292, 294, 313, 315, 320–5, 329, 331, 339, 340–1, 345, 347–9, 351, 353–9, 362, 366, 373, 391, 398, 419, 428–30, 438, 443, 450, 453, 455, 457, 461, 467, 472, 480–7, 489–1, 493–4, 496–7, 500–3, 506–7, 509–10, 514, 516, 520, 522, 525–6, 531, 533–6, 538, 540, 542, 544, 555–6, 561, 564, 566–8, 572–3, 575, 577, 579, 582, 586, 588–91, 593, 600, 603–4, 607–8, 610–12, 614, 627–30

TONY (A. R.) PITTAWAY grew up in several European countries, enjoying their rich diversity of insects, and developed an early interest in butterflies and hawkmoths, especially at their immature stages. He gained an MSc and PhD in entomology at Imperial College, London, England. Tony has worked as an entomologist in the Middle East and written or coauthored three books and a number of scientific papers on butterflies, hawkmoths, and dragonflies, including *Insects of Eastern Arabia* and *The Hawkmoths of the Western Palaearctic*. Between traveling in search of specimens and images, and maintaining websites on the Saturniidae and Sphingidae of the Palaearctic region, he is Data Architect at the Centre for Agriculture and Biosciences International (CABI) in the United Kingdom.

Contributions: pages 433–4, 439, 445, 448, 452, 454, 460, 470, 473, 477

JAMES A. SCOTT received a PhD in entomology from the University of California, Berkeley, after researching butterfly behavior. He has since published several books on butterflies, including *Butterflies of North America: a Field Guide and Natural History*.

James has spent considerable time in the field, accumulating several thousand records for the host plants of mostly Rocky Mountain butterflies and more than 40,000 records of butterflies visiting flowers (recently published by the Gillette Museum at Colorado State University). His collection also includes thousands of photographs of eggs, larvae, pupae, and adults.

Contributions: pages 59, 68, 71, 103, 108, 112, 136, 194, 217, 230, 261, 268

ANDREI SOURAKOV developed a strong interest in Lepidoptera at the age of ten and later pursued undergraduate studies in Moscow, Russia. In 1997, he obtained his PhD from the University of Florida, United States. He currently works at the McGuire Center for Lepidoptera and Biodiversity in the Florida Museum of Natural History, home to one of the largest Lepidoptera collections in the world. Andrei is the author or coauthor of more than 100 scientific and popular articles on the taxonomy and biology of butterflies and moths. He also teaches a course entitled "Insects and Plants" and studies interactions between caterpillars and their host plants.

Contributions: pages 39, 41, 48–9, 62, 93, 115, 154, 168, 183, 232, 249, 254, 286, 289, 297, 302, 304, 308, 318, 327, 330, 360, 365, 367, 378, 395, 451, 458, 488, 492, 495, 498–9, 508, 513, 547, 560, 570, 581, 583–4, 618–19, 634, 637

MARTIN TOWNSEND is a freelance professional entomologist and invertebrate ecologist, who has been fascinated with moths and butterflies since the age of ten. He has carried out a wide variety of field surveys in the United Kingdom for conservation, government, research, and commercial bodies and agencies. He is coauthor of the *Field Guide to the Moths of Great Britain and Ireland*, the *Concise Guide to the Moths of Great Britain and Ireland*, and *British and Irish Moths*. He has also written more than 100 journal articles and reports.

Contributions: pages 287, 290–1, 299, 328, 332–8, 342–4, 346, 474, 517, 519, 528–30, 545–6, 548–9, 563, 594–9, 601, 605–6, 609, 615–17, 620, 622–6, 633, 635–6, 639

KIRBY WOLFE is a Research Associate of the Natural History Museum of Los Angeles County, California, and of the McGuire Center for Lepidoptera and Biodiversity in the Florida Museum of Natural History. He has spent more than 30 years photographing moths and their immature stages and is the author or coauthor of a number of books and research articles on insect behavior and development.

Contributions: pages 361, 363–4, 368, 371–2, 375–7, 379–82, 384, 387–90, 392–4, 397, 399–401, 403–4, 406–12, 414–15, 420–7, 431–2

655

ACKNOWLEDGMENTS

DAVID G. JAMES

I would like to express my gratitude to the editorial and design team at 3REDCARS—Rachel Warren Chadd, Jane McKenna, and John Andrews—for their guidance, vision, and dedication to this book. In particular, I must humbly thank Rachel for keeping me on task and providing numerous ideas and suggestions that substantially improved the book. I would also like to thank Kate Shanahan and her colleagues at the Ivy Press for developing the original idea and commissioning the book.

This international caterpillar book would not have been possible without the contributions from other caterpillar experts. I gratefully thank my many coauthors—Harold Greeney, Andrei Sourakov, Bob Cammarata, James Scott, and David Albaugh from the United States; Sally Morgan, Martin Townsend, and Tony Pittaway from the United Kingdom; Kirby Wolfe from Costa Rica; Ross Field from Australia; and John Horstman from China—for their excellent and substantial contributions. The involvement of these regional experts has allowed our coverage of caterpillars to be truly worldwide and comprehensive. Some of the caterpillars covered in our book have not featured in any other caterpillar book because of their rarity. Some have only recently been discovered.

I thank the numerous lepidopterists who contributed to this book as photographers and sources of information. In particular, I would like to thank Bob Pyle, Jon Pelham, David Nunnallee, and David Wagner for their insights on caterpillar biology and the systematics of Lepidoptera.

Finally, I thank all the people in my life who have encouraged and supported my studies and writings on caterpillars that began with woolly bears in my eighth year and continue with the Monarchs I rear today. The most important of these people are my mother, Doreen, and late father, Alan, along with my wife Tanya and daughters Jasmine, Rhiannon, and Annabella.

3REDCARS would also like to thank Paul Oakley and Charlotte Ward for their contribution to the production of this book and Jane Roe for her proofreading.

656

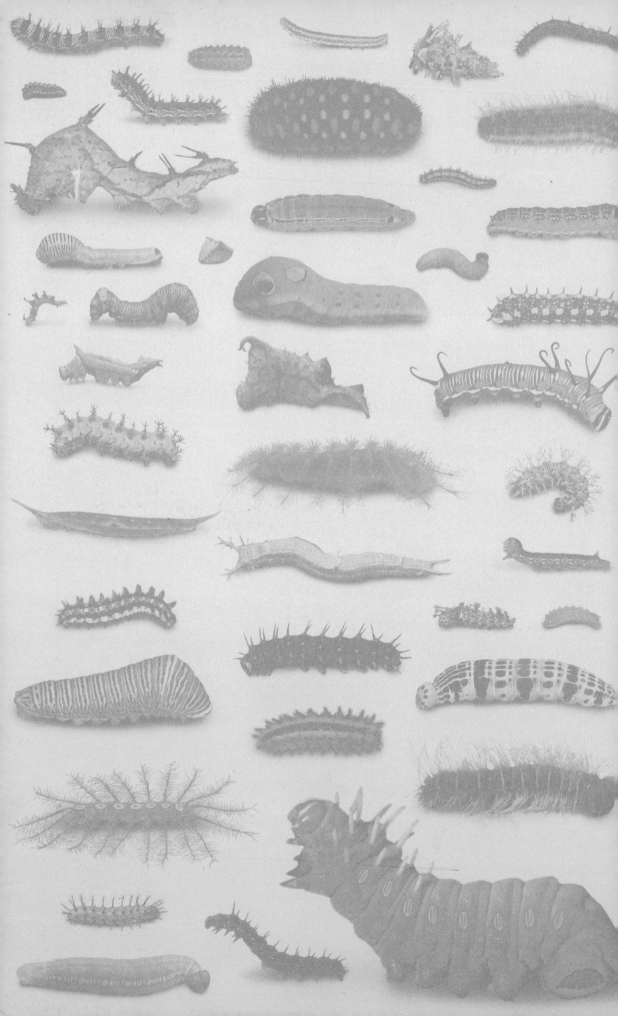